SCUTTLE FLIES: THE PHORIDAE

SCUTTLE FLIES: THE PHORIDAE

R.H.L. Disney

Field Studies Council Research Fellow, Cambridge University, UK

Phorids are the least studied of all the large families of Diptera
(Naumov, 1985)

SPRINGER-SCIENCE+BUSINESS MEDIA, B.V.

First edition 1994

Originally published by Chapman & Hall in 1994

Typeset in 10/11 Bembo by Type Study, Scarborough

ISBN 978-94-010-4562-9 ISBN 978-94-011-1288-8 (eBook)
DOI 10.1007/978-94-011-1288-8

A catalogue record for this book is available from the British Library

Library of Congress Catalog Card Number 94-70254

∞ Printed on acid-free text paper, manufactured in accordance with ANSI/NISO Z39 48-1992 (Permanence of Paper)

Contents

Advance in the knowledge of the natural history of a group of insects is dependent upon taxonomic knowledge. The following have served phoridology well with their taxonomic expertise, both by publication and by identifying specimens for others. The importance of the latter service is too often taken for granted.

Theodor Becker 1840–1928
John Henry Wood 1841–1914
William Lundbeck 1863–1941
Filippo Silvestri 1873–1949
John Russell Malloch 1875–1963
Hermann Schmitz 1878–1960
Charles Thomas Brues 1879–1955
Thomas Borgmeier 1892–1975
Charles Norman Colyer 1908–1970

Foreword

Today the topic of 'biodiversity' is a matter of general interest; national leaders refer to it in international meetings, such as the one at Rio de Janeiro. However, the assessment of biodiversity depends on detailed work of a careful and often unglamourous type. We are very uncertain as to the actual biodiversity of the Earth. All biologists will accept that many species are undescribed, a majority will agree that we know considerably less than half the living species. If we are really to understand our 'house', our '*oikos*' that gives us the name of the now popular science of ecology, we must be able to do more than put names on the organisms that share our Earth, we must know about their biology, their life style – their bionomics.

Insects constitute more than half the known species of living organisms and probably a greater proportion of the unknown ones. Entomologists therefore face a formidable task. Fortunately there are those rare persons who devote themselves to a particular group and over their lifetime gain a real understanding of its species and their biology. Henry Disney is one such person and we are extraordinarily fortunate that after 20 years of study he has provided us with this overview of his chosen group, it is an assemblage of the 'world knowledge' on the Phoridae. We will all hope that it will not be another couple of decades before he provides a second *oeuvre*.

The magnitude of the task he has achieved, as well as that ahead, can be appreciated when it is recognized that there may be five times as many species of scuttle fly as birds in the world. The family evolved a long time ago, at least 80 million years; scuttle flies coexisted with the dinosaurs and were undoubtedly a diverse group when the first hominoids appeared around 10 million years ago. This long evolutionary history expresses itself in the wide diversity of habit and form described in this volume. Although in the 230 years since a phorid was first described many entomologists have made contributions to our knowledge of the members of the family, it has been relatively neglected until Dr Disney turned his attention to it. This comprehensive account of existing knowledge, so much of which is due to the author, should encourage others to join in the study of this widespread and abundant, yet hitherto obscure, group.

Professor Sir Richard Southwood FRS
Oxford
February, 1994

Preface

I am well aware that I have probably overlooked important observations published in unexpected places. I should welcome references to any such omissions which readers may encounter. Some omissions have been deliberate, where I have doubts about their correctness or the identity of the species involved. In some of the latter cases, however, I have cited the observations but indicated the doubts. In many cases those who refer to a work cited will find the species referred to by a subsequently discarded synonym or an identification of the species which has been subsequently revised. These changes are covered by the catalogues cited in section 1.5 or the literature recommended for the genus (8.3). I have only rarely burdened the text by the citation of synonyms. I remain particularly aware that I have probably not done justice to the ecological literature. I would especially welcome notification of interesting quantified data I have overlooked.

A glance at the bibliography will quickly reveal that observations on Phoridae have been published in some decidedly obscure journals. Indeed some references cited in the literature defeated the library services of the United Kingdom. Some articles could not be traced, and in some cases the journal cited was not listed in any of the standard reference works.

A late aunt, Sheila Disney, encouraged my interest in natural history when I was a boy, and her legacy enabled me to buy my dream compound microscope. Without this good microscope I would never have been in a position to tackle the taxonomy of such small insects as the scuttle flies. With it I have been able to produce many detailed descriptions and the hundreds of drawings used to illustrate both these and the identification keys I have published. While a less sophisticated microscope is adequate for the identification of phorids, one needs a superior instrument to produce detailed drawings and descriptions.

I thank Professor G. Horn FRS and Dr K. A. Joysey for facilities in the Cambridge University Department of Zoology and the Zoology Museum.

The writing of this book was made possible by funding from the Isaac Newton Trust (Trinity College, Cambridge) and the Harold Hyam Wingate Foundation (London). I am most grateful for this support.

I am also very grateful to Dr Jeff Barnes (Biological Survey, Cultural Education Center, Albany, New York) for his constructive comments on early drafts of most chapters, and to Dr Sally Corbet (Department of Zoology, University of Cambridge) for casting her critical eye over a complete draft. Dr Zakaria Erzinçlioğlu (Department of Zoology, University of Cambridge) also commented on some sections, and has been a source of encouragement over many years. Ken G. V. Smith (formerly of the Natural History Museum, London) has also been a considerable encouragement, especially in the early years of my attempts to get to grips with phorid taxonomy. As a journal editor and referee he has shown himself not to be allergic to unorthodoxy, for which I am grateful.

I am grateful to numerous librarians, in particular to those of the Balfour Library, Department of Zoology, the Scientific Periodicals Library and the University Library at Cambridge University, the Royal Entomological Society of London and the libraries of the Natural History Museum, London.

I am grateful to Margaretha Harriott for the translation of some papers in Dutch.

Dr John Linley (University of Florida, Institute of Food and Agricultural Sciences) kindly allowed me to use his superb SEM photographs of the egg of *Megaselia scalaris* (Fig. 2.1). Dr Michael E. N. Majerus (Cambridge University, Department of Genetics) kindly allowed me to use his photograph of the living phorid in Plate 1. John W. Rodford (Cambridge University, Department of Zoology) drew Figs 9.2 and 9.3. Mr Peter V. York (Photographic Unit, The Natural History Museum, London) kindly took the photograph in Fig. 1.1(a). Mr Neal C. Maskell (Photographic Section, Cambridge University Department of Zoology) kindly took the photograph in Fig. 1.1(b).

I am grateful to Donald Azuma (Academy of Natural Sciences, Philadelphia) for the loan of a male *Neodohrniphora* and to Dr Brian Brown (Natural History Museum, Los Angeles) for information regarding a character of *Parasyneura*, and for allowing citation of some of his unpublished observations.

The following have kindly permitted me to reproduce the figures given in brackets:

American Naturalist (3.6)

E. Schweizerbart'sche Verlagsbuchhandlung – *Archiv für Hydrobiologie* (3.3, 4.4)

Archivos do Instituto de Biologia Vegetal, Jardin Botanico Rio de Janeiro (3.8, 4.6)

Royal Entomological Society of London – *Ecological Entomology* (6.13)

Dr K. Kaneko – *Japanese Journal of Sanitary Zoology* (2.2a)

Makunagi (6.11, 6.15)

Mushroom Growers Association – *MGA Bulletin* (7.1)

E. J. Brill, Leiden – *Nematologica* (6.16)

Treubia Deel (4.3a–d)

The Orion Publishing Group Ltd (incorporating Weidenfeld and Nicolson) (8.95)

Professor I. Eibl-Eibesfeldt – *Zeitschrift für Tierpsychologie* (3.9)

Gustav Fischer Verlag – *Zoologische Jahrbücher (Anatomie)* (5.1)

The rest of the figures are by myself. Most of these have appeared in previous publications. I acknowledge, with thanks, the following journals and organizations for allowing me to reproduce the figures given in brackets:

Annales Entomologici Fennici (2.3b,c, 8.5a)
Aquatic Insects (Swets Publishing Service) (2.2b, 4.3e,f, 8.9a)
Museum Alexander Koenig, Bonn – *Bonner zoologische Beiträge* (1.4, 6.4, 8.23b,c, 8.24, 8.63a, 8.84b, 8.85)
BENH Society – *British Journal of Entomolgy and Natural History* (2.6a, 8.21b,c, 8.40c, 8.54a, 8.70a, 8.75b, 8.78c–e, 8.91b)
International Institute of Entomology – *Bulletin of Entomological Research* (3.7, 8.2b,c)
Entomologica Fennica (8.13a, 8.81)
Entomologica Scandinavica (3.1–3.4, 3.5a, 4.1, 4.5, 6.1d, 6.7, 8.2a, 8.4a,b, 8.6a,b, 8.25b, 8.31a, 8.40a, 8.46, 8.56b, 8.57a, 8.60d, 8.61, 8.80b)
Royal Entomological Society of London – *The Entomologist* (1.2)
The Gem Publishing Company – *Entomologist's Monthly Magazine* (1.1b, 6.5a, 8.6c, 8.33a, 8.56a, 8.62, 8.64, 8.76, 8.77, 8.96)
The Entomologist's Record (4.2)
Giornale Italiano di Entomologia (8.10, 8.19, 8.20, 8.26, 8.29, 8.43b,c, 8.44, 8.49a, 8.60a–c)
Glasgow Naturalist (8.12a–b)
Journal of African Zoology (8.2d, 8.30, 8.56c–e, 8.57b, 8.58, 8.71, 8.72, 8.93c)
Taylor and Francis Ltd – *Journal of Natural History* (1.3a, 2.7, 3.5b, 5.2, 6.5c, 6.8–6.10, 8.47a,b, 8.49, 8.50, 8.59, 8.66–8.68, 8.92, 8.93a,b)
The Zoological Society of London – *Journal of Zoology, London* (8.7, 8.8, 8.15a)
Naturalist (8.9b, 8.14a,b)
Associated Publishers, Gainesville – *Oriental Insects* (3.10, 6.14, 8.78a,b)
Sociobiology (2.6b, 6.3, 6.5b, 6.6, 8.11a,b, 8.16, 8.17a, 8.34, 8.35, 8.38, 8.42a,c,d, 8.45b, 8.51b,c, 8.52–8.54, 8.74a, 8.79d, 8.80a)
Royal Entomological Society of London – *Systematic Entomology* (8.4e, 8.11c, 8.12c, 8.13b, 8.39f, 8.41b,c, 8.63b–d, 8.69a,b,d, 8.70c,d, 8.73b, 8.82a–d, 8.83, 8.84c,d, 8.86–8.90)
University of Newcastle upon Tyne, Agriculture Environment Research Group (9.2, 9.3)
Dunger and Humblot GHBH – *Zeitschrift für angewandt Zoologie* (1.1a, 2.5a,b, 8.18b, 8.27, 8.28a)
Academic Press Ltd – *Zoological Journal of the Linnean Society* (6.1a–c, 6.12, 8.37, 8.65)
Gustav Fischer Verlag – *Zoologischer Anzeiger* (8.31b,c, 8.32)

Note: Throughout, all scale bars on the figures are 0.1 mm.

R.H.L.D.
Cambridge, November, 1993

1
Introduction

. . . it is very doubtful whether the life histories of most . . . conspicuous Diptera can compare in interest with those of the tiny, dull-colored Phoridae (Wheeler, 1901)

1.1 INTRODUCTION

This book aims to introduce a family of flies whose diversity of larval lifestyles is apparently without rival among insect families. Because of taxonomic difficulties, and a widely scattered literature, the ecological significance of this fact is currently underrated. In order to encourage others to investigate these extraordinary flies, I have limited myself to a summary of what little we know of phorid natural history and to the provision of a new key to world genera along with a guide to the literature required for species-level identifications.

The scuttle flies have been known by various other common names, such as humpbacked flies and manure flies. But all of these names have also been applied to other families of flies. Oldroyd (1964) therefore considered it desirable to avoid confusion by discarding these names. He proposed the name 'coffin flies' instead. However, this name was already applied to the genus *Conicera* only, and to *C. tibialis* in particular. In view of this the name 'scuttle flies' was proposed, as a new term of more general application to the whole family (Disney, 1976b). The alternative general name for the family is plain 'phorids'. I use both of these general names in the following pages, as well as making occasional use of names such as 'coffin fly' or 'con fly' for particular species.

The scope of my review of the biology of phorids is largely restricted to natural history, here understood as the branch of biology concerned with the life histories and ways of life of organisms. A chapter on ecology (the study of the interrelationships between organisms and the environment) is included, as ecology is essentially an extension of certain aspects of natural history (Bates, 1950).

Whereas the objective of much biological research is to uncover the general principles and processes underlying the diversity of organisms, natural history is a branch of biology whose principal objective is the production of better and better monographs critically reviewing and illuminating the details of the diversity. That which orders, and gives significance to, these details is evolution, the fundamental explanatory principle and process of biology.

Biological colleagues concerned with general principles and processes have offered the opinion that the countless short publications reporting observations on small details of the natural histories of individual species do not constitute 'proper' science. By themselves, such notes do indeed amount to little. However, when critically evaluated, collated and synthesized, they become the individual bricks from which substantial scientific monographs are built.

A definitive monograph on Phoridae is not yet possible. The present work merely attempts to bring together what little is known to provide a point of departure for future studies. At this stage what is primarily needed is a marshalling of the significant observations that have been made by numerous workers, in order to stimulate others to investigate the many gaps in our knowledge. The family Phoridae is perhaps the most biologically diverse family of insects on the planet (Disney, 1990a). In evolutionary terms the family Phoridae is an outstanding success. In the long term a detailed knowledge of scuttle flies is likely to tell us much about evolution.

The main impediments to the study of Phoridae have been their small size and difficulties with the identification of species. Taxonomic advances are reducing these difficulties, so that studies of the natural history of scuttle flies have now become immense opportunities for original investigations. No natural history research is possible without a basis of acceptable taxonomy. Whereas some ecology can be accomplished without needing to identify every specimen to the species level, in natural history the focus of attention at all times is on the individual species. In practice the naturalist frequently investigates several species, partly because comparisons between species tend to highlight the significant.

A synopsis of a history of phoridology is, of necessity, initially dominated by the attempts to master the taxonomy of scuttle flies. It is the publication of identification keys that allows advance in the study of natural histories. This collation is essentially a crude sketch of what little is known of scuttle fly natural history. It resembles an incomplete tapestry, with most of the fabric still unadorned with coloured yarns. The emerging picture, however, is of an unusually rich and intricate design. Clearly many surprises await the investigator of phorid natural history. Hopefully some readers may feel encouraged to have a go at filling in some of the gaps which remain. No special qualifications, other than enthusiasm and a critical awareness, are required. The amateur still contributes, on equal terms with the professional, in the field of natural history. The natural history of most species of Phoridae remains unknown.

1.2 HISTORICAL PERSPECTIVE

In 1763 Scopoli described a small fly and named it *Musca festinans*. In fact this was a scuttle fly, and was the first to be described. The description, while allowing recognition that the fly was a phorid, does not allow recognition of the species today. In 1794 Fabricius described *Musca aterrima*, a species whose males are often to be observed in small swarms in Europe and North America. In 1796 Latreille concluded that it did not belong to *Musca*, and he erected the new genus *Phora* In 1802 he formally placed *Phora aterrima* in this genus Nearly two centuries later there was published the discovery that the name *aterrima* had not been available in the genus *Musca* in 1794, as (unknown to Fabricius) in 1789 Villers had named a tachinid *Musca aterrima* (in Crosskey, 1980). Unaware of all this, Meigen in 1804 had failed to recognize the species described by Fabricius and redescribed it under the name *Trineura atra*. The fly is, therefore, now known as *Phora atra* (Meigen), in strict accordance with the 'rules' of zoological nomenclature (enshrined in the International Code of Zoological Nomenclature) but contrary to commonsense and the provisions for conserving long-established names (Erzinçlioğlu, 1983a)!

In 1833 Curtis designated the family Phoridae for the genus *Phora*, along with its subsequently described relatives placed in other genera The name *Phora* is from the Greek word 'phora', meaning rapid motion, and refers to the fast scuttling gait of phorid flies (as correctly stated by authors such as Hardy and Beyer, 1964). Some nomenclaturists, however, have incorrectly derived the name from the Greek word 'phor', meaning a thief (e g. Jaeger, 1959).

This tale of confusions, with regard to both classification and nomenclature, has been sketched in order not only to indicate how far we have travelled since Scopoli but also to highlight an unfortunate feature of the literature on Phoridae The early workers had no idea that these small flies were a major family of insects Most species have still to be described and named Collections from many parts of the world are still dominated by undescribed species. Conservative estimates suggest that there may be around ten times as many species as there are species of mammal in the world.

We are clearly a long way from having a complete inventory of the species of animals and plants on the planet. The indications are that only about one tenth are known to science However, the estimates of the actual total remain the subject of conjecture and controversy (e g Stork, 1988; May, 1989; Hodkinson and Casson, 1991) Furthermore it seems probable that many species are becoming extinct, because of the accelerating rate of destruction of their dwindling habitat systems, before these species are even known to exist. To support the estimate that there are perhaps as many as five times as many species of Phoridae as of birds in the world, we will probably get nearer the truth if we restrict our attention to the British Isles. These islands have the most fully documented fauna in the world, including the best-documented phorid fauna (Disney, 1983a, 1989a, 1991j, 1993c). The Phoridae represent almost 1% of the known species of the British fauna. Furthermore I have recorded more than 60 species in my quite ordinary English suburban garden in Cambridge.

There are still many problems of both classification and nomenclature to be resolved. The problems of phorid taxonomy are due partly to the complexities of the giant genus *Megaselia*, partly to problems of small size (e g the genus *Metopina*) and partly to a marked sexual dimorphism in many genera. Still more than half of the 229 genera, reviewed in section 8.3, are known only in one sex. Undoubtedly many of these will be found to be the opposite sexes of markedly sexually dimorphic genera, whose 'missing' sex is currently treated as a different genus.

The family is not only large but also ancient. The earliest fossils known date from some 80 million years ago (Grimaldi, 1989). Associated with this antiquity is the extraordinary radiation of lifestyles exhibited by modern scuttle flies. The average textbook or reference work, however, dismisses them as mere muck breeders (saprophages). The persistence of this untenable view derives in part from the large and scattered literature on the family, and in part from the tendency of textbook writers to depend too much on previous textbooks rather than the primary literature.

1.3 WHAT ARE SCUTTLE FLIES?

The limits of the family Phoridae are, surprisingly, still open to dispute. Some authors (e.g. Grassé, 1986) have continued an earlier practice of treating the aberrant female phorids in the subfamilies Termitoxeniinae and Thaumatoxeninae as separate families. Indeed Rohdendorf (1974) placed the Termitoxeniinae not only in a separate family but in a separate infraorder as well. However, the discovery of the males of these two aberrant groups supports the case for treating them as true phorids (Disney, 1992a; Disney and Cumming, 1992). Some authors (e.g. Hennig, 1971; Brown and Pike, 1990) include some Cretaceous genera, originally assigned to the Sciadoceridae, in the family Phoridae. The Sciadoceridae evidently had already separated from the typical Phoridae in the Cretaceous and are today represented by a few relict species only. There are undoubted Phoridae and undoubted Sciadoceridae represented among the Cretaceous fossils. Some fossils, however, seem to bridge the gap between the two families, thus raising the possibility that the Sciadoceridae are really only a subsection of the Phoridae. The placement of these fossils depends upon one's interpretations of certain character states (section 8.5).

I am following most recent authors, such as Colless and McAlpine (1991), by treating the Phoridae as the largest family of the superfamily Phoroidea, which also includes three other families – the Ironomyiidae, Platypezidae and Sciadoceridae.

Winged specimens of adult Phoridae are easily recognized by their distinctive wing venation (Fig. 1.1a). This is characterized by shortened costal and radial veins (the so-called 'thick veins'), followed by very fine 'thin veins'. The first two, the medial veins, are unattached basally. The other posterior veins are linked only by cross-veins, if at all, at the extreme base. The result is that there are no cross-veins in the wing beyond the apex of vein R1. The only exceptions are throw-back mutations (Fig. 1.1b) (Disney, 1991c).

Numerous Phoridae have reduced wings or no wings, at least in the female sex. These may resemble similarly modified flies belonging to other families.

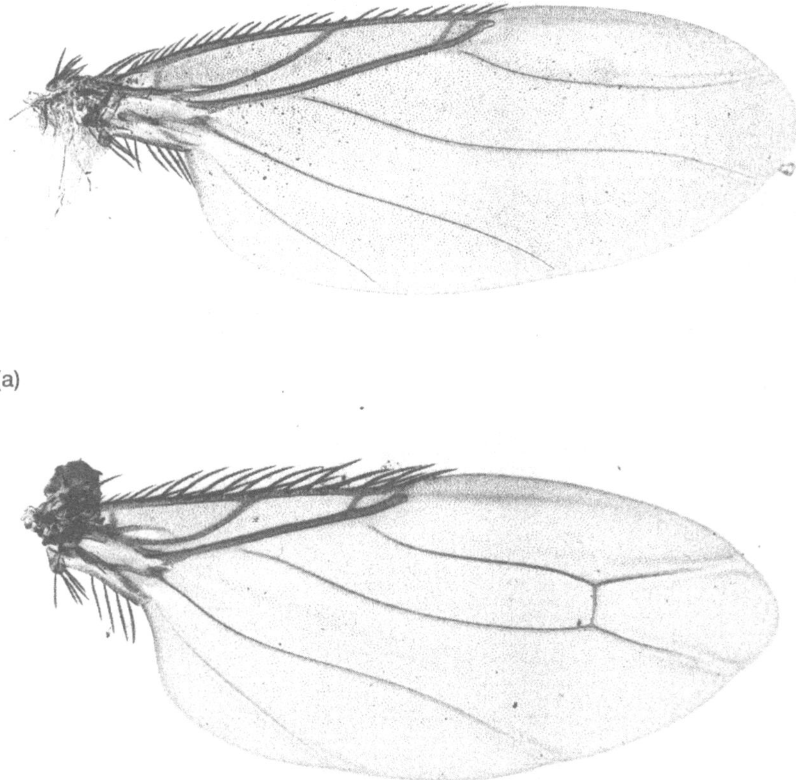

(a)

(b)

Figure 1.1 Right wings of scuttle flies (a) *Megaselia audreyae* male (b) *Megaselia pleuralis* male, a specimen with a throw-back mutation restoring the outer cross-vein of its ancestors (Photographs by Peter V York (a) and N C Maskell (b))

The most reliable recognition character for the Phoridae is the greatly enlarged apical swelling (conus) on the second antennal segment (pedicel) (Disney, 1988i) This conus is larger than the rest of the pedicel and is embraced by the third antennal segment (the first flagellar segment), to form a ball-and-socket articulation (Fig 1 2) The rest of the pedicel may be so reduced that the antenna appears to have only two basal segments (the scape and first flagellar segment) plus an arista The arista is normally three-segmented, but may be reduced or even absent Another characteristic of Phoridae, allowing recognition of flightless forms, is the hairing of the hind metatarsus There are 1–4 longitudinal hair palisades and usually, but not always, transverse combs of hairs as well (Fig 1 3) While both these features

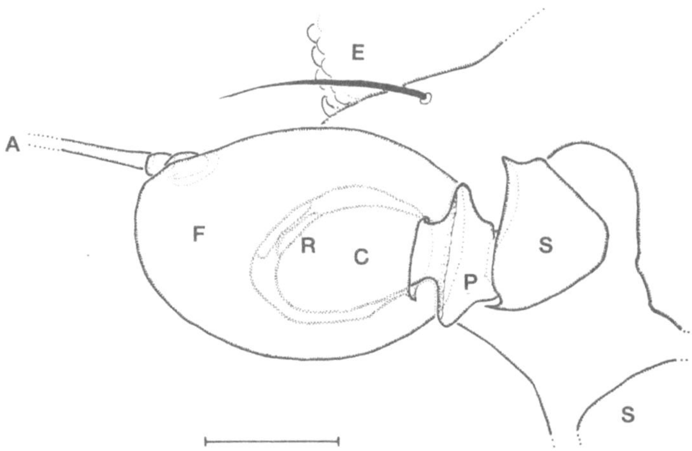

Figure 1.2 Right antenna of a male *Diplonevra abbreviata* A = base of arista, C = conus, E = eye, F = first flagellar segment, P = pedicel, R = apical ring of conus, S = scape

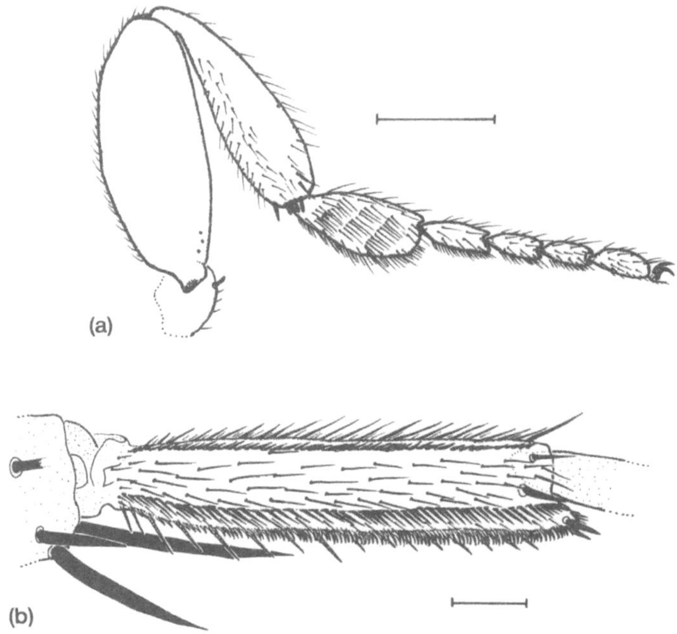

Figure 1.3 Hair palisades and transverse combs on hind metatarsi of Phoridae (a) Posterior face of whole leg of *Metopina ciceri* male (b) Anterior face of metatarsus of *Triphleba collini* male

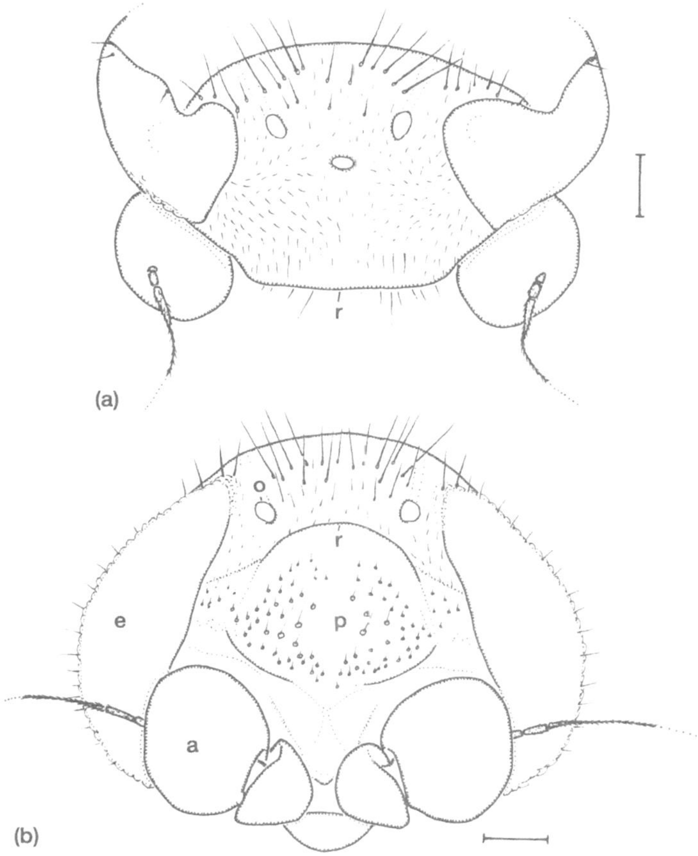

Figure 1.4 The head of *Aenigmatistes foveolatus* from above (a) and from front (b). a = third (first flagellar) segment of antenna; e = eye; o = posterior ocellus; p = ptilinum; r = post-ptilinal ridge.

are found in Sciadoceridae and some Platypezidae, the antennae will distinguish the latter. No flightless female Sciadoceridae are known.

By citing these recognition features one thereby excludes genera of flies which may be members of the same monophyletic group (i.e. a group of species representing *all* the descendants of a single ancestral species), but which retain ancestral (plesiomorphic) character states. The Phoridae as presently conceived may be polyphyletic (i.e. a group of species descended from two or more ancestral species which were not closely related).

'Phoridization', notably the reductions in the wing veins and the enlargement of the conus at the expense of the rest of the pedicel, might be a process that happened in several clades of the Aschiza flies independently. In the opinion of all recent authors, the balance of probability is that the Phoridae are a true monophyletic group.

The problem of how scuttle flies relate to other Aschiza, and to the Platypezidae and Sciadoceridae in particular, and to lower Diptera is not dealt with in this book, but is touched on in section 8.5. Recent evidence suggests the Aschiza are probably not a monophyletic group (Wada, 1991). The distinction between the Aschiza (= without a split) and Schizophora (= slit bearer) is the absence or presence of the ptilinal suture (a furrow above the base of the antennae, marking the position of the invaginated ptilinum – a balloon-like structure used when the adult emerges from its puparium). This distinction, however, is not absolute. At least one genus of phorids, *Aenigmatistes*, possesses a ptilinum, which, however, remains exposed to view (Fig. 1.4) and probably evolved independently of the evolution of the ptilinum of the Schizophora and some Syrphidae (hoverflies) (Disney, 1991k). The Syrphoidea (the hoverflies and related families, that is the Aschiza minus the Phoroidea and Lonchopteridae) plus the Schizophora probably constitute a monophyletic group distinct from the rest of the Aschiza (Wada, 1991).

For the purposes of reference I give the currently accepted classification of the Phoridae. The subfamilies and tribes are currently a matter of dispute (Chapter 8). A number of genera are likely to be reclassified in the future, as the subfamilies and tribes are better characterized in terms of the phylogenetic reconstruction of the family.

Order DIPTERA
Suborder Brachycera
Division Cyclorrhapha
Series Aschiza
Superfamily Phoroidea
  Family Platypezidae
  Family Ironomyiidae
  Family Sciadoceridae
  Family PHORIDAE
    Subfamily Phorinae
    Subfamily Aenigmatiinae
    Subfamily Thaumatoxeninae
    Subfamily Termitoxeniinae
    Subfamily Metopininae
      Tribe Megaseliini
      Tribe Metopinini
      Tribe Beckerinini

The tribe Beckerinini, however, is evidently polyphyletic as presently conceived in the literature. The genus *Beckerina* shows affinity with *Plectanocnema*, *Woodiphora* and some *Megaselia*, while being excluded from the Megaseliini. Other genera currently placed in the Beckerinini, such as *Aphiura* and *Rhopica*, clearly belong elsewhere. Likewise the Aenigmatiinae and the Metopinini are evidently polyphyletic, and in need of reappraisal.

Living Phoridae are characterized by their fast, scuttling, gait with its frenetic stop–go pattern of short bursts of activity interrupted by brief halts. Miller (1979b) has studied these activity patterns using high-speed filming.

He suggested they serve to enhance both visual and chemosensory mechanisms. Sciarid midges and some Platypezidae tend to exhibit similar patterns and, being of the same order of size, may be initially misidentified as Phoridae by the unsuspecting collector.

1.4 WHY STUDY PHORIDAE?

Phoridae are worthy of study in their own right as a biologically diverse family, of ecological significance, with a long evolutionary history. They are worthy of study because the larvae of large numbers of species are parasitoids or specialist predators (3.5), and therefore some may prove useful as agents of biological control for pests. Some species are themselves pests, such as the notorious spoiler of cultivated mushrooms *Megaselia halterata*. Some species are involved in cases of myiasis (invasion of the living vertebrate host by larvae) in both man and his livestock (3.5.7). Some species are useful in forensic entomology (3.2.3). Other species are important pollinators of plants (6.2). Because of their diversity in most terrestrial habitats, combined with their biological diversity, they are an excellent family to employ in conservation evaluation exercises (7.12). Some species have aquatic larvae, mainly in container habitats, and there is at least one intertidal species (7.2.1).

1.5 A BRIEF HISTORY OF PHORIDOLOGY

Workers in the nineteenth century described numerous species and began to highlight the biological diversity of the family. However, as with most other groups of small insects, the quality of the species descriptions was generally inadequate in the light of modern knowledge. In most cases species described in the nineteenth century have only been recognized subsequently when type specimens have survived and been re-examined in the twentieth century. In many cases the biological observations are difficult to evaluate because of uncertainties over species identifications.

Modern Phoridology begins with the monograph of Becker (1901), dealing primarily with the fauna of the Palaearctic Region. This was followed by the work of Brues (1903b) on the North American fauna and a treatment of world genera (Brues, 1906a). Wood (1906, 1907, 1908, 1909, 1910, 1911, 1912, 1914) built upon the work of Becker in his treatment of the British fauna, and Malloch (1910) also contributed to knowledge of the British fauna, before moving to America and building upon the earlier monograph of Brues (Malloch, 1912b). Meanwhile Brues, having taken a growing interest in the world fauna, provided the first world catalogue (Brues, 1915d). Thus by the time of the First World War the achievements of the earlier workers had at least been collated, and the monographs had been produced that would provide the springboard for the next stage.

Among the early contributors special mention must be made of Silvestri. He was not primarily interested in Phoridae, but he carried out a series of studies of insects found in termite nests. In a number of papers (1906, 1911,

1920, 1945, 1947a,b) he provided descriptions of bizarre new species and genera, setting a new standard in the quality of his illustrations. Regretfully his example was not followed by most of his contemporaries, whose collective fault was either a failure to provide figures of critical features or else a tendency to provide figures of inferior quality. In most cases only the wing was chosen for illustration, which was unfortunate, as the minor differences highlighted have tended to prove more variable than had been supposed.

After the First World War Lundbeck (1922) made an important contribution to the knowledge of the European fauna. Brues continued to describe exotic species as Schmitz was beginning his study of Phoridae. The latter's first major contribution was his *Revision der Phoriden* (1929g), with its key to world genera. Among other achievements, this work dismantled much of the confusing proliferation of genera proposed by Enderlein (1924), but retained (unfortunately) the untenable division of the giant genus *Megaselia* into the subgenera *Aphiochaeta* and *Megaselia*. This arbitrary division has tended to impede recognition of the true monophyletic subgenera within the genus. He continued publishing numerous papers until his death in 1960. His other major contribution was his, still incomplete, monograph on the Palaearctic fauna (Schmitz, 1938n, 1941c, 1943, 1949c, 1951c, 1953a, 1955b, 1956b, 1957a, 1958b; Schmitz and Beyer, 1965, 1974; Schmitz and Delage, 1974, 1981). He also performed an invaluable service by his critical re-evaluations of the collections of Phoridae made in the nineteenth and early twentieth centuries.

Borgmeier, a specialist on Neotropical ants, contributed a number of papers on Phoridae, mainly species associated with ants, from the 1920s onwards. In 1955 he turned his formidable energies to the Phoridae. He made massive contributions to our knowledge of Neotropical species, provided a new monograph on the Nearctic fauna and made a major contribution to knowledge of the Australasian and Oriental faunas in particular. All these works are cited in his revised world catalogue (1968b) and its supplement (1971b). Robinson (1971) collated much of the information on the natural histories of the species in the giant genus *Megaselia*. Baumann (1976a) provided an incomplete supplementary bibliography (which, however, is strong on published records for Phoridae in caves) and started collating and extending knowledge of scuttle fly natural histories, before he suddenly abandoned entomology. These collations have been recently updated (Brown, 1990c; Disney, 1991e).

Apart from his work on Termitoxeniinae (Borgmeier, 1964a) and a number of small papers, Borgmeier contributed relatively little to our knowledge of the Afrotropical fauna. Bridarolli (1951) made a start on the Afrotropical *Megaselia*, but this was published posthumously following his untimely death. Beyer, after a number of smaller papers, produced a large monograph on the Afrotropical fauna (Beyer, 1965b). However, Beyer's descriptions tend to be somewhat sketchy and his keys tend to be difficult to use with confidence.

The death of Borgmeier in 1975 marked the end of the Schmitz/Borgmeier epoch of Phoridology. They had made substantial contributions in terms of the volume of species descriptions. Schmitz had begun the process of synthesis for both the basic biology and the interpretation of morphology. Both were viewed in the perspective of evolution. Borgmeier, despite his view that evolutionary ideas have nothing to contribute to systematics (Borgmeier, 1957), performed an invaluable service in bringing some order to the widely scattered literature. Despite certain limitations in his choice of over-variable characters, he attempted to provide identification keys for many genera, and for the *Megaselia* of three biogeographic regions in particular.

The one remaining task, without which all but the most intrepid tend to be deterred from embarking on studies of exotic Phoridae, was the lack of an updated key to genera. By the time of Borgmeier's death the number of recognized genera had more than doubled since Schmitz's key of 1929 (Schmitz, 1929g). This lack was acutely felt by Colyer. While his published contributions amount to only a few relatively short papers, he circulated an unpublished first draft of a key to the genera of the world. This preliminary draft was put together from the literature as a working document, but was never revised and published before Colyer's death in 1970. It is certainly not publishable in its existing form. However, it has served to enable a number of phoridologists, in the last quarter of the twentieth century, to find their way around the family. It facilitated Peterson's (1987) revised key to Nearctic genera. My own copy is well worn and heavily annotated, and it has been my main means of running down exotic genera. Chapter 8 attempts a completely new key to genera, adopting an entirely fresh approach to the problem.

Among contributions in the last quarter of the twentieth century I single out a number that are significant. Delage rescued Schmitz's incomplete monograph on the Palaearctic fauna from the infelicities of Beyer's efforts (Schmitz and Delage, 1974, 1981), but other duties have obliged her to leave this project still uncompleted. W. H. Robinson gave greater emphasis to the taxonomic value of the male hypopygium, which he examined by means of temporary preparations for microscopic scrutiny (Robinson, 1978b, 1981). I went further in the break with the tradition of pin-mounted specimens by switching to slide-mounted specimens (Disney, 1979j). This allowed a radical revision of the British fauna (Disney, 1983a, 1989a) and initiated reappraisals of the suprageneric classification (e.g. Disney, 1987b, 1989f). The acceptance of the value of the male hypopygium allowed Gotô to carry out careful work on Japanese Phoridae, and on the genus *Phora* in particular (Gotô, 1986a). Brown has likewise made a significant contribution with his studies of the male genitalia, of the genus *Gymnophora* in particular, and his use of phylogenetic cladistic analysis (e.g. Brown, 1987a,b, 1989a). However, an agreed classification is still a long way off (8.5), not least because of the degree of convergent and parallel evolution that has evidently taken place in the family (e.g. Disney, 1991m; Brown, 1993a).

The physiology of Phoridae has been little studied and is not covered by this book, except by highlighting certain extraordinary larval environments (e.g. some records of *Megaselia scalaris* (3.2)) or properties (such as the use of a pheromone by *M. halterata* females (6.5) or the presence of the elastic protein resilin in the ovipositor of *Iridophora clarki* (2.5)). Readers primarily interested in physiological studies should refer to Fairbanks and Burch (1970).

My own interest in the Phoridae springs from a fascination with the astonishing biological diversity of these extraordinary flies. To some extent I have tended to see taxonomic work as a means of getting at the basic natural histories of the different species. The result is that the primary emphasis of my taxonomic studies has been to provide the means for others to identify species of scuttle fly with greater confidence. This book is an expression of my concern with the advance in knowledge of the diversity of the habits of these remarkable flies.

In view of the extent of our ignorance of scuttle fly natural history, I have concentrated on the presentation of established facts. Some preliminary generalizations and syntheses are included and some points requiring further investigation are indicated. The latter, along with the obvious gaps in knowledge, should serve to highlight fruitful directions for future research. It is my hope that readers who embark on the investigation of some aspect of phorid natural history for themselves will derive as much satisfaction from their explorations as I have from mine. In 1974, when I started trying to identify a few scuttle flies I had reared, I had no idea that I had embarked upon a commitment that has grown into a sustained enthusiasm for one of the most remarkable products of evolution – the Phoridae.

Advance in the understanding of both evolution and ecology depends as much on establishing the details of the existing diversity of individual natural histories as on the testing of particular theoretical ideas. The two approaches must be interdependent, if valid conclusions are to be drawn. In our present state of knowledge, all discussion of the evolution of the diversity of the habits of both larval and adult scuttle flies is largely a matter of speculation.

In order to reconstruct the true sequences of events in the evolution of phorid natural histories, it is first necessary to construct a classification that reflects the phylogeny of the family. This task is currently in its infancy because of our very imperfect knowledge of the taxonomy of the species at the most basic descriptive level. Most species are still undescribed. Many are poorly described. Most genera are poorly characterized and so far only a few are defined in terms of uniquely evolved character states (autapomorphies). Our knowledge of the natural history of most described species is rudimentary or nil. As knowledge of both taxonomy and natural histories advances, the future research programme will increasingly include the mapping of data on habits onto our emerging phylogenetic classification. Without the latter, speculations on the evolution of particular habits are likely to differ little from Kipling's *Just So Stories*.

Phylogenetic systematics is the only key to an understanding of the historical dimensions of the ecology and behaviour of present-day species (e.g. Brooks and McLennan, 1991). Without the unifying perspective of evolution our growing knowledge of natural history is in danger of becoming a mere catalogue of curious facts. However, a catalogue of the facts is a prerequisite for any more illuminating synthesis. The following preliminary survey certainly highlights some curious facts. Undoubtedly many more bizarre habits await discovery. The future of phoridology, however, will be best served by a growing synthesis grounded in the perspective of the evolutionary history of these remarkable flies.

2

Eggs

The fly is only an egg's way of making another egg (to paraphrase Samuel Butler's famous quip about hens).

2.1 INTRODUCTION

The egg is the least studied stage of scuttle flies. Egg production, however, is the principal function of the adult fly.

The morphology of scuttle fly eggs has not been studied systematically, but a preliminary review is given by Ferrar (1987). It is evident that phorid eggs are very varied in form. Detailed studies are likely to be rewarding in terms of adaptations of form to suit different ecological situations. The best-known egg is that of *Megaselia scalaris* (Fig. 2.1). The boat-shaped form, and gunwale-like palisade of flat platelets surrounding the respiratory plastron with its scattered tubercles, is associated with deposition of the egg into relatively liquid media (Ferrar, 1987). A similar egg is produced by the saprophagous *M. spiracularis* (Furukawa and Kaneko, 1981) and by the aquatic *M. deningi* (Fig. 2.2a) (Kaneko and Furukawa, 1983; Disney, 1991i). A contrasting egg is that of the aquatic *M. kovaci*. Its plastron is restricted to the anterior region, in a subterminal position (Fig. 2.2b). It is probably laid in the meniscus on the wall of the bamboo internode, with its anterior end upwards (Disney, 1993b). Even relatively simple eggs of closely related species may vary considerably in shape (length to breadth ratios) (Fig. 2.3).

The eggs of the Termitoxeniinae *Clitelloxenia hemicyclia*, *Javanoxenia punctiventris* and *Odontoxenia brevirostris* are known to be good mimics of the eggs of their termite host, *Odontotermes javanicus*. Indeed the worker termites collect the fly eggs and place them among their own. Furthermore they regularly groom the eggs along with their own, and thus keep them free of mould (Franssen, 1933b). The relatively enormous size of these eggs, in relation to the size of the fly (Kemner, 1926; Reichensperger, 1931), may be related to the curtailment of the larval stages (5.3). Furthermore Michl (1911) reported that the egg of *Termitoxenia havilandi* measures 1.15 × 0.40 mm compared with the egg of its host termite, *Odontotermes latericius*, which measures only 0.75 × 0.37 mm.

There is clearly much scope for critical studies of the varieties of form and

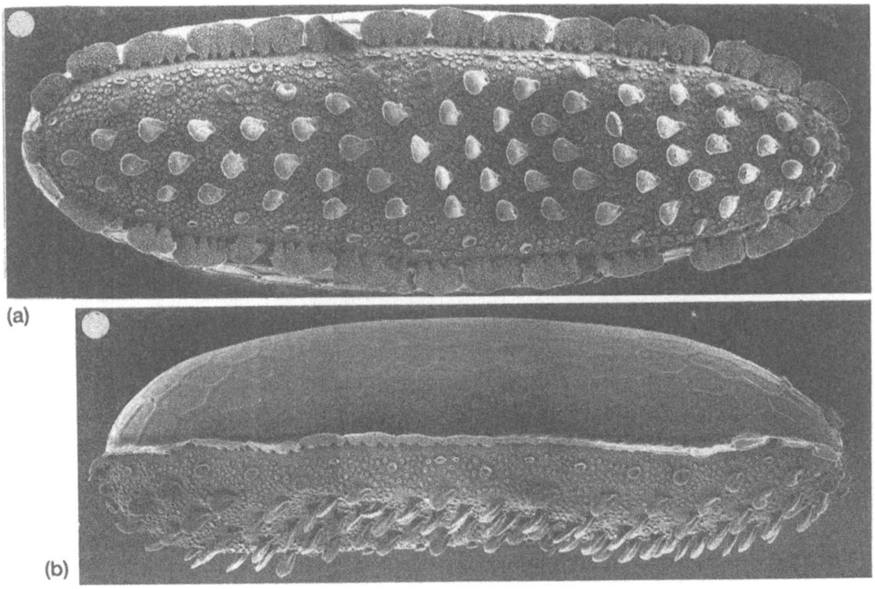

Figure 2.1 Eggs of *Megaselia scalaris*, with anterior ends facing top of page. (a) From above. (b) From side. (Photos by John Linley.)

relative sizes of phorid eggs in relation to oviposition site and the subsequent larval habits.

The study of phorid sperm is an almost totally neglected field. Curtis *et al.* (1989) have laid the foundations for future comparative studies, with their detailed account of the ultrastructure of the sperm of *Megaselia scalaris*. The spermatozeugmata (spermatophore-like aggregations of sperm) are discussed by Benner (1991). They lack the capsule, around the sperm mass, of a true spermatophore.

2.2 FECUNDITY, BATCH SIZE AND CLUTCH SIZE

Data on fecundity exist for only a few species. These are reviewed below. A female may lay a succession of egg batches, with a pause after each batch. The total number of eggs laid, in all the batches added together, represents the actual fecundity. The eggs of a single batch may be laid together in the same place, as a single clutch, or divided into a number of separate clutches. In the former case the batch size and clutch size are the same. A clutch may be as little as a single egg.

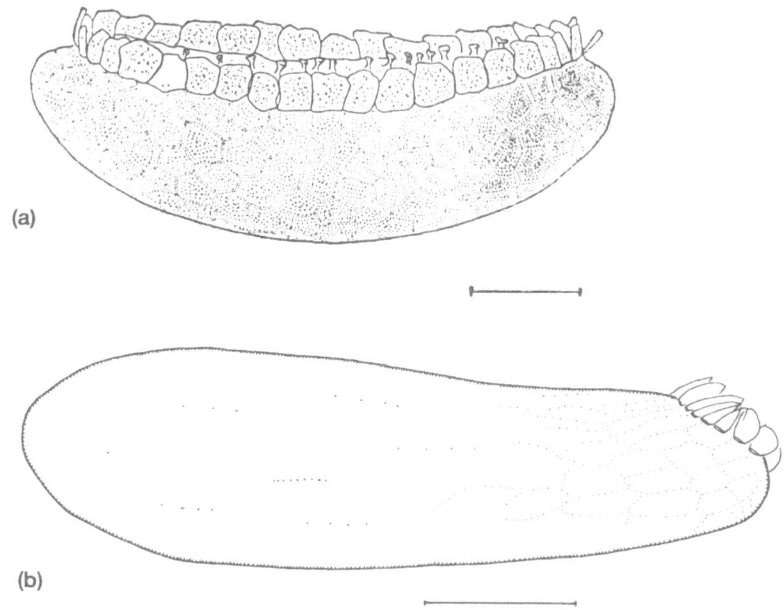

Figure 2.2 The contrasting eggs of two aquatic species of *Megaselia*. (a) *M. deningi*. (b) *M. kovaci*, with anterior end to right. ((a) after Kaneko and Furukawa, 1983.)

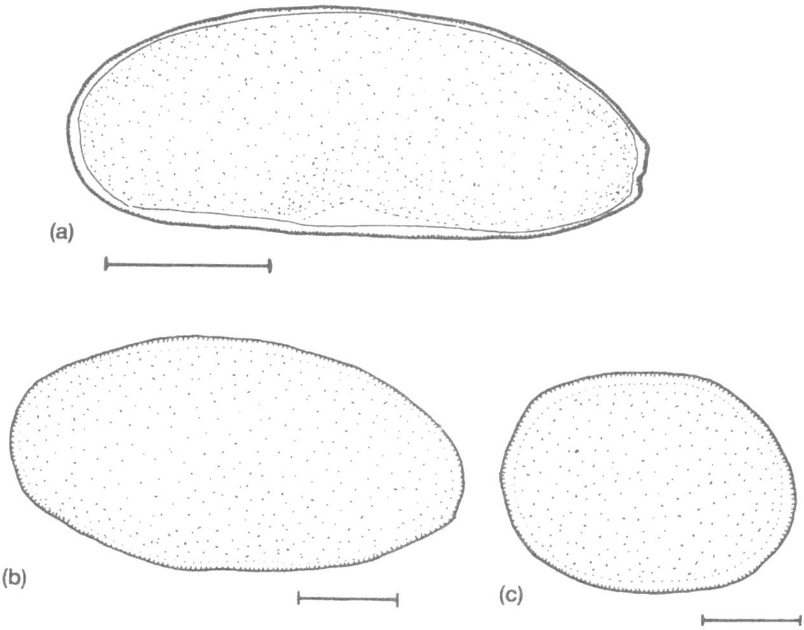

Figure 2.3 The eggs of *Megaselia* species. (a) *M. halterata*. (b) *M. stenoterga*. (c) *M. oxybeiorum*.

The number of eggs matured and laid at one time (that is the batch size) varies from 1 to 100, the number depending on the species and the condition of the female. Benner (1985) studied oocyte development in *Megaselia scalaris* and found that the number of mature oocytes ranged from 12 to 77, with oocyte development being synchronized in the different ovarioles and with only one mature oocyte per ovariole.

A low batch size would seem to be associated with two distinct phenomena. In some cases it is correlated with refined predatory or parasitic larvae, following accurate placement of each egg in relation to the target host or prey. The high chance of success compensates for risks associated with a concentrated investment. In the case of the con flies *Diplonevra mortimeri* and *D. watsoni*, parasitizing termites (3.5.6(f)), the one and only egg matured at one time is inserted into the comatose termite, which is then concealed with soil particles and guarded (Disney, 1986d). By contrast *Misotermes exenterans* scatters large numbers of eggs in its host-termite's nest, relying on chance encounters by the soldier termites, their target hosts (Kalshoven, 1938).

The other phenomenon associated with a reduced batch size is accelerated larval development and suppression of the free-living larval stage. For example, the larva of *Clitelloxenia hemicyclia* takes no food after emerging from the egg. Associated with this is an unusually long duration of the egg stage (Kemner, 1926; Franssen, 1933b) (5.3). In *Assmutherium rostratum* apparently only three eggs are matured at a time, and it seems the embryos start to develop before the eggs are laid (Fig. 2.4) (Disney, 1991m).

In some cases probably both phenomena are involved in batch-size reduction. This would seem to be the situation in the genus *Puliciphora*. Thus the polyphagous saprophage *P. togata* matures only eight eggs at a time, compared with about two dozen in the polyphagous saprophage *P. borinquenensis*. This reduced batch size probably relates to the relatively large egg size and a rapid larval development. In the specialized predator *P. boltoni*, whose adults live in the colonies of their ant hosts, and in *P. collinsi*, almost

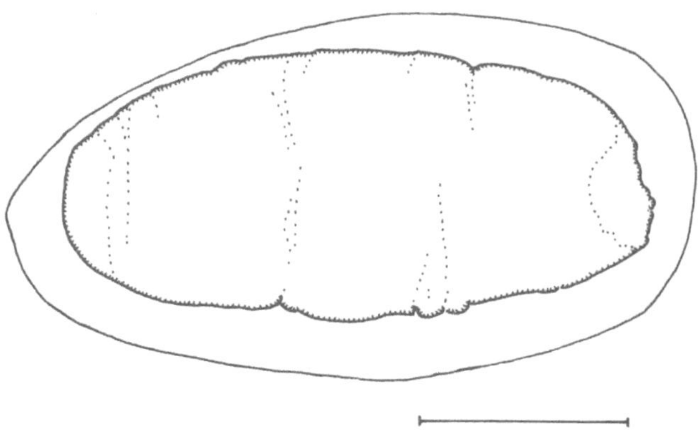

Figure 2.4 Embryo inside egg still retained by a gravid female of *Assmutherium rostratum*, with anterior end to right.

certainly a parasitoid of termites, only four eggs are matured at a time (Disney, 1988b). A reduced batch size in *Rhynchomicropteron nudiventer*, of 1–4 eggs (Table 2.1), probably relates to accelerated larval development, as the larva is known to be a saprophage within the colonies of its ant host (Brown, 1992a).

Some data on fecundity and reported batch or clutch sizes are reviewed below:

Apocephalus attophilus
The batch size is 10–16 eggs, but only one egg is laid in each ant host (Feener and Moss, 1990).

Dohrniphora cornuta
Weiss (1911) reported a batch size of 30–40 eggs in this species. At 23° ± 1° C, Barnes (1990a) reported that females laid 9.4–12.0 eggs per day. Fecundity ranged from 77–535 (mean 368 ± 143) eggs per female.

Megaselia aletiae
Balduf (1928) reported 50 eggs laid by a single female on the back of a caterpillar (3.5.6(g)).

Megaselia halterata
In captivity this species is reported to lay about 50 (40–60) eggs at a time (Moreton and John, 1955; Hussey, 1960a), and Hussey (1965) reported about 40 mature eggs in wild-caught females. Females parasitized by nematodes (6.8.2(b)) may lay as few as 2–3 eggs per batch (Hussey and Gurney, 1964a).

Megaselia rubricornis
1–5 (usually 2) eggs are laid on each beetle-host pupa (3.5.6(b)), and 30–40 eggs are laid within 10 days (Kanervo, 1946).

Megaselia scalaris
Semenza (1953) reported a mean batch size of 32.2 (643 progeny obtained from 20 pairs) at 23 °C. El-Miniawi and Moustafa (1965) reported a mean fecundity of 391 eggs (range 42–749) spread over a mean of 10 egg-laying periods (the first being 38 h after emergence) at 27 °C.

A later study only observed an output of two batches, each of 9–26 eggs, in the first seven days (Tumrasvin *et al.*, 1977). Prawirodisastro and Benjamin (1979) reported a mean of 664.8 eggs per female (range 192–1003) spread over 16 egg-laying days at 25 °C. The first batch was laid at an age of 45.6 h (1.9 days). The largest batches were laid after 17 days. Mean fecundity decreased to 591 (± 213.0) at 20 °C and 146.9 (± 51.3) at 15 °C.

Dias Macieira *et al.* (1983) reported that this species lays a first batch with a mean of 21.6 (± 11.8) 24–48 h after mating, at 25–30 °C. Given a constant supply of fresh medium, the second batch (laid within the next 24 h) averaged 21.5 (± 23.4) eggs; and a third batch (within the next 24 h) averaged 60.0

(± 23.2) eggs. The overall average was 103.2 eggs per female within 96 hours of mating.

Kneidel (1985) reported that this species laid 69.4 (± 7.6) eggs in 24 h at 27 °C, and also that fecundity is affected by body size. The latter was affected by the amount of competition for food as a larva.

McCrae's (1967) observations on this species breeding in a pot of paint (3.2) suggested that more than 350 eggs had been laid by a single female. This is of the same order as the fecundity estimates of El-Miniawi and Moustafa (see above).

Other Megaselia species

The predators of gall insects *M. chainensis* and *M. submarginalis* lay clutch sizes of 1–3 eggs and 1 egg, respectively, per gall (Robinson, 1970; Robinson and Brown, 1993). Their batch sizes are not known, but are presumably much larger.

Pseudohypocera kerteszi

Chaud-Netto (1980) reported fecundity figures of 31–102 eggs per female, with a mean of 66.4.

Some other records of batch sizes are given in Table 2.1.

2.3 DEVELOPMENT

The eggs of *Megaselia halterata* require a moisture film for successful development (Hussey and Gurney, 1964a).

2.4 EGG MATURATION

The experiments of Hussey (1960) suggest that *Megaselia halterata* will develop mature eggs prior to mating, but *M. nigra* females will only mature their eggs after mating. The eggs of *M. halterata* take 3–5 days to mature, from the time of eclosion of the adult (Hussey, 1965). *Phalacrotophora fasciata* has been reported to take 15 days to mature sexually in the laboratory (Hodek, 1973). It is possible that the reported laboratory diet (of sugar, agar, honey and water) was inadequate when compared with the normal diet of females. If they normally require a meal of insect haemolymph (6.3) to mature their eggs, then a lack of such a protein-rich meal might delay development.

Benner's (1985) studies of oocyte development in *Megaselia scalaris* showed that by the age of 34 h, at 25 °C, all females had mature eggs. At 24 °C this species normally starts laying on the second day after emergence. He also reported that virgin females will lay unfertilized eggs, but not until day 3. However, Ondraschek (1953) reported that mating did not occur until day 2, and that oviposition only starts 2–3 days after mating. Dias Macieira *et al.*

Table 2.1 Some recorded egg batch sizes for Phoridae

Species	Batch size	Reference
Assmutherium rostratum	3	Disney (1991m)
Ceylonoxenia bugnioni	6	Bugnion (1913)
Clitelloxenia peradeniyae	2	Bugnion (1913)
Dicranopteron oligoseta	1	Disney and Kistner (1990a)
D. philotermes	4	Disney and Kistner (1990a)
D. setipennis	1	Disney and Kistner (1990a)
Diplonevra florea	30–40	Schmitz (1949c)
D. mortimeri	1	Disney (1986d)
D. watsoni	1	Disney (1986d)
Maculiphora aenictophila	1	Disney and Kistner (1991)
Megaselia brevissima	20	Disney (1988h)
M. curtineura	16	Disney (1991a)
M. dentata	10	Compton and Disney (1991)
M. harteni	14	Disney (1991h)
M. longicostalis	4	Unpublished
M. malvinasensis	2	Disney (1989h)
M. microcurtineura	4	Disney (1991a)
M. miniseta	28	Disney (1991h)
M. orgaoa	24	Disney (1991h)
M. oxybelorum	2	Disney (1988h)
M. pagei	60	Disney (1987g)
M. parvorata	16	Compton and Disney (1991)
M. pseudobrevior	10	Disney (1988h)
M. pulicaria	20	Unpublished
M. sembeli	4	Disney (1986b)
M. sinuata	14	Unpublished
M. stenoterga	2	Disney (1988h)
M. verdensis	35	Disney (1991h)
Metopina galeata	6	Unpublished
Puliciphora boltoni	4	Disney (1988b)
P. borinquenensis	25	Paterson (1953)
P. collinsi	4	Disney (1988b)
P. jacobsonorum	8	Disney and Kistner (1989a)
P. malae	4	Disney (1990f)
P. rosei	2	Disney (1988b)
P. rufipes	14	Unpublished
P. tambopata	14	Disney and Kistner (1989a)
P. togata	8	Disney (1988b)
Rhynchomicropteron necaphidiforme	1	Disney (1990c)
R. nepalensis	1	Disney (1991g)
R. nudiventer	1–4	Steghaus-Kovac and Disney (1990), Disney (1990c, 1992b)

(1983) reported that this species lays its first batch of eggs about 24 h after mating. These differences probably reflect different temperature regimes or differences in the condition of the flies. At 22–24 °C, *Dohrniphora cornuta* starts laying eggs 1–3 days after eclosion, and mating frequently takes place on the day of eclosion (Barnes, 1990a).

2.5 OVIPOSITION

Most Phoridae oviposit directly onto or into, or at least close to, the larval pabulum. The well studied pests of cultivated mushrooms are examples. Having mated within 24 h of emergence, *Megaselia halterata* will start ovipositing within 2–3 days of emergence and will continue laying for about 10 days (Hussey and Gurney, 1964a). The eggs are laid about 1 mm from the tips of the growing hyphae (Hussey *et al.*, 1969). *M. nigra* will only oviposit in the light (Hussey and Wyatt, 1962). This species and *M. bovista* oviposit onto the freshly opened gills of mushroom sporophores (Hussey, 1961a). Preliminary investigations (Grove and Blight, 1983) suggest that *M. halterata* locates *Agaricus bisporus* by the C_8 aliphatic alcohols and ketones given off as volatiles by this mushroom.

Dohrniphora cornuta, breeding in trickling filter sewage systems, lays its eggs loose in protected crevices free from moisture and the sewage film (Kloter *et al.*, 1977). In laboratory cultures the eggs were randomly scattered in relatively dry areas of cotton wool (Barnes, 1990a), and are non-adherent (Jones, 1918). Likewise *D. diminuens*, whose larvae feed on the fungus combs of its termite hosts, has been observed ovipositing into the general nest debris (Disney and Kistner, 1989c).

The highly specialized physogastric females of the Termitoxeniinae (see 5.5), such as *Clitelloxenia hemicyclia*, *Javanoxenia punctiventris* and *Odontoxenia brevirostris*, are restricted to the fungus gardens surrounding the queen cell of their termite host, and they lay their eggs among the termite eggs.

In the case of *Puliciphora borinquenensis*, whose females are always flightless, the male has been reported locating a suitable pabulum for oviposition, memorizing the site, then searching for females. He then airlifts a female to the pabulum (Miller, 1984) (6.5).

Donisthorpe (1927) reported that *Pseudacteon formicarum* females will strike at one's hand if this is scented by the acid of its ant host. His observations suggested that the initial attraction to ants was by scent and subsequently it was by sight. Brown and Feener (1991b) provide experimental evidence that *Apocephalus paraponerae* is attracted to its host ants, *Paraponera clavata*, by olfactory cues. It then oviposits into dying or injured worker ants. This seems to be related to the frequency of aggressive encounters between *P. clavata* workers, and a consequent regular supply of maimed ants. Up to 25 eggs are laid into a single ant (Brown and Feener, 1991a). Whether these represent single clutches or layings by up to 25 separate females is not known. Other parasitoid species are reported to be attracted to injured individuals of their hosts (e.g. Berland, 1933; Disney and Schroth 1989).

Melaloncha females actually oviposit on bees while the latter are in flight, the larvae being parasitoids of the bees (3.5.6(e)) (Roubik, 1989). Most parasitoid species oviposit directly into or onto their hosts when the latter are

Plate 1 *Phalacrotophora fasciata* female ovipositing onto prepupa of the ladybird *Adalia 2-punctata* (Photo by Michael Majerus.)

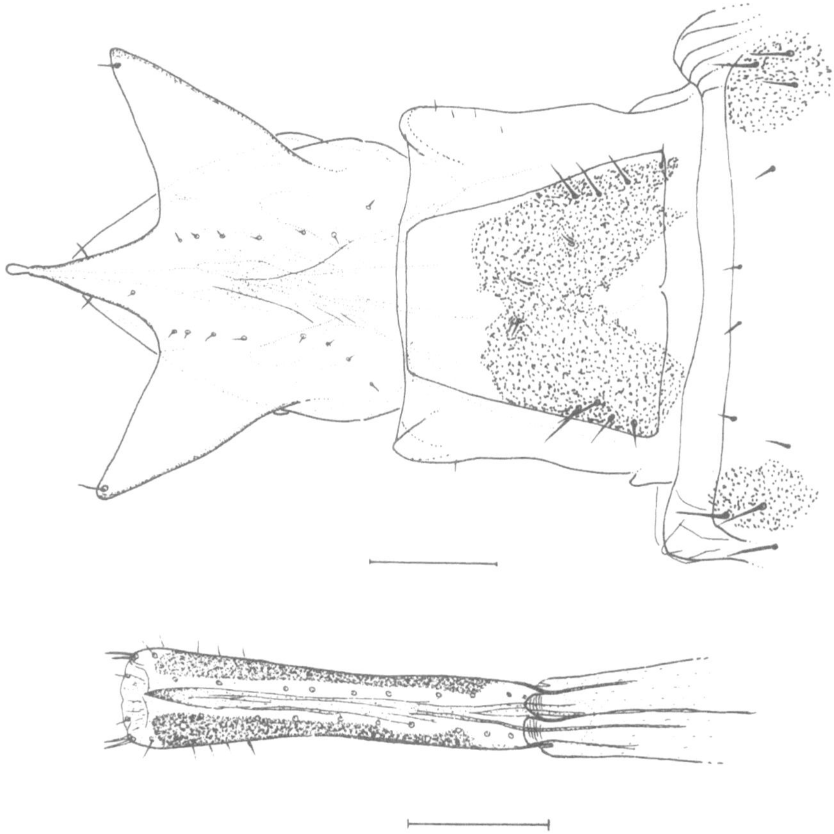

Figure 2.5 Dorsal views of ovipositors of *Auxanommatidia medleri* (above) and *Apocephalus wallerae* (below).

not in flight. In the case of *Iridophora clarki*, which very rapidly darts its eggs at its ant host as the latter are running up tree trunks or along the ground, the ovipositor is furnished with the elastic protein resilin (Disney, 1986b).

The con fly termite-parasitoids *Diplonevra mortimeri* and *D. watsoni* (3.5.6(f)) are attracted to exposures of termites, among which wounded individuals are present, but they only attack healthy workers. In an extraordinary pre-oviposition routine the female fly lands in the vicinity of termite workers and runs up to one of them. She prods the termite near its rear end. The termite normally responds by turning towards the direction from which the fly approached. The fly then races to the head of the termite and sets off running away from the termite colony. The termite, apparently stimulated by some odour emitted by the fly, follows her closely. Termites have been observed being led more than 2 m distance from the colony. The victim is then rendered comatose, an egg is laid in its abdomen, and it is

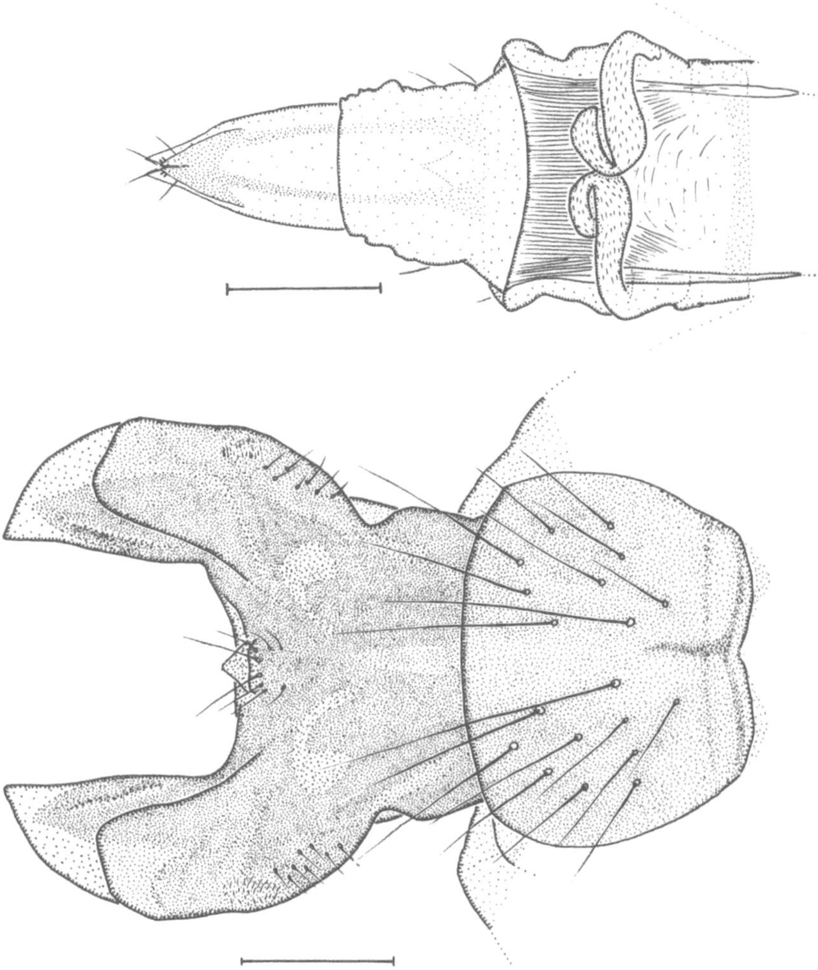

Figure 2.6 Ventral views of ovipositors of *Menozziola schmitzi* (above) and *Pseudacteon grandis* (below)

covered with soil particles and then guarded by the fly In Sulawesi mean attack rates at exposures of termites have been reported as being up to 10 1 per hour in the case of *D mortimeri* and up to 10 9 per hour for *D watsoni* (Disney, 1986d) This rapid recruitment to sudden spillages of termites onto the forest floor probably relates to the strength of the odours emanating from the wounded termites present on such occasions The possible evolutionary significance of attraction to wounded hosts is discussed below (3 1)

In some parasitoid phorids the ovipositor is remarkably modified (Figs 2 5, 2 6) While the varied forms of these ovipositors have proved to be of great

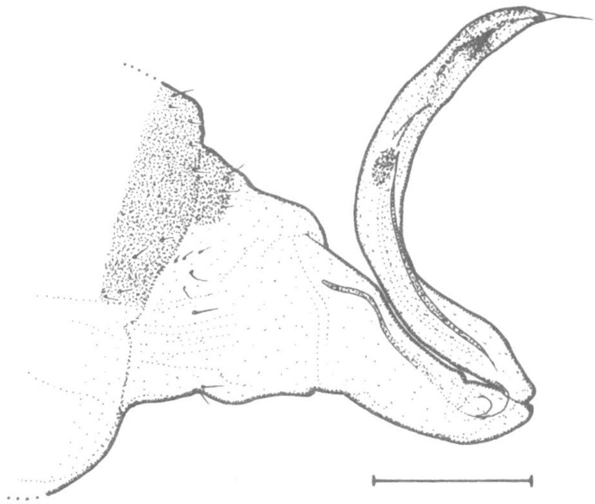

Figure 2.7 Side view of the 'spring-loaded' (resilin operated) ovipositor of *Iridophora clarki*.

taxonomic value, their adaptive significance is almost unexplored. However, the mode of use of the ovipositor of *Iridophora clarki* (Fig. 2.7) is described above. High-speed video might help in determining the mode of use of other types of ovipositor which are also employed with great speed. By contrast the ovipositor is only moderately modified in the *Phalacrotophora* species, which oviposit onto the static prepupae or pupae of their ladybird beetle hosts (Plate 1) in a more leisurely manner, taking up to 25 min per host (Disney *et al.*, 1994).

The access of *Conicera tibialis*, the coffin fly, to buried human corpses has been the subject of controversy (Colyer, 1954a). It is now clear that oviposition into the coffin before burial cannot explain all the data. It is significant that bodies buried in January, in England, are as heavily infested as those buried in July. In the British Isles adults of this species have only been reported in the months April to November (Disney, 1983a). Colyer (1954e,f) provided clear evidence that the adults emerging from the corpse of a dog buried about 1 m deep were able to make their way through the soil to the surface. The soil in question was moderatly loose, with abundant worm burrows. Lundt (1964) buried pieces of meat in soil (7.5.1) and recorded *C. tibialis* adults up to depths of 0.5 m after only 4 days from the time of burial. It seems probable, therefore, that gravid females can make their way down through the soil to a buried corpse. Adult *Triphleba nudipalpis*, emerging from corpses of white mice buried about 0.3 m deep in a heavy clay soil, have been found throughout the soil between the mice and the surface (Disney, 1991d). Miller (1979a) has seen the females of *Puliciphora borinquenensis* forcing their way through sand or among organic matter (such as decomposing insects) by the use of vigorous abdominal peristaltic movements. The abdomen became alternately long and thin with the ovipositor fully extended, and short and fat. Contraction rings were seen to pass anteriorly along the abdomen.

3

Larvae

> The more closely a species is studied, the more likely it is that some evidence will be found for ecological polymorphism or gradual ecological variation The variation is still largely ignored by ecologists, most of whom discuss the ecological requirements of species in a strictly typological manner (Mayr, 1970)

3.1 INTRODUCTION

The larvae of the Phoridae are very diverse A preliminary review of this diversity is provided by Ferrar (1987) Schmitz (1938n) discussed larval morphology The tracheal system was described for a species of *Megaselia* by Whitten (1955) Curtis *et al* (1987) have described the ultrastructure of the

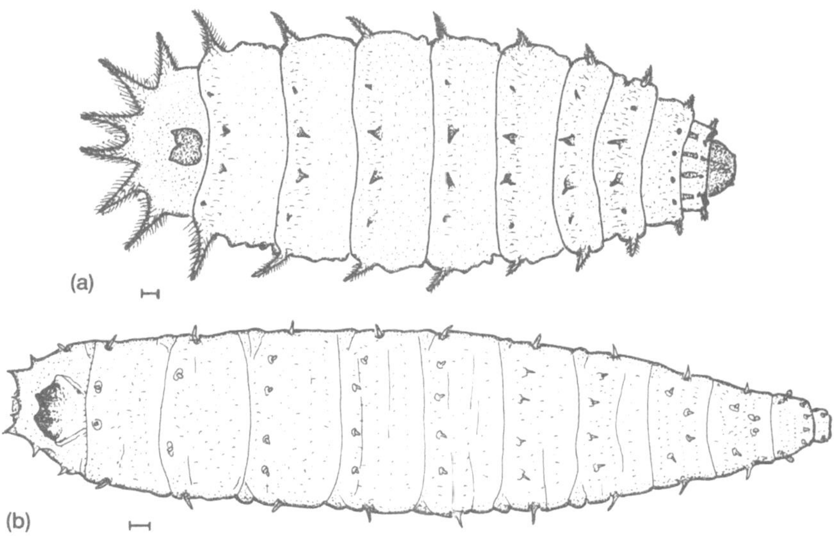

Figure 3.1 The larvae of two generalized saprophage species (a) *Dohrniphora cornuta* (b) *Megaselia rufipes*

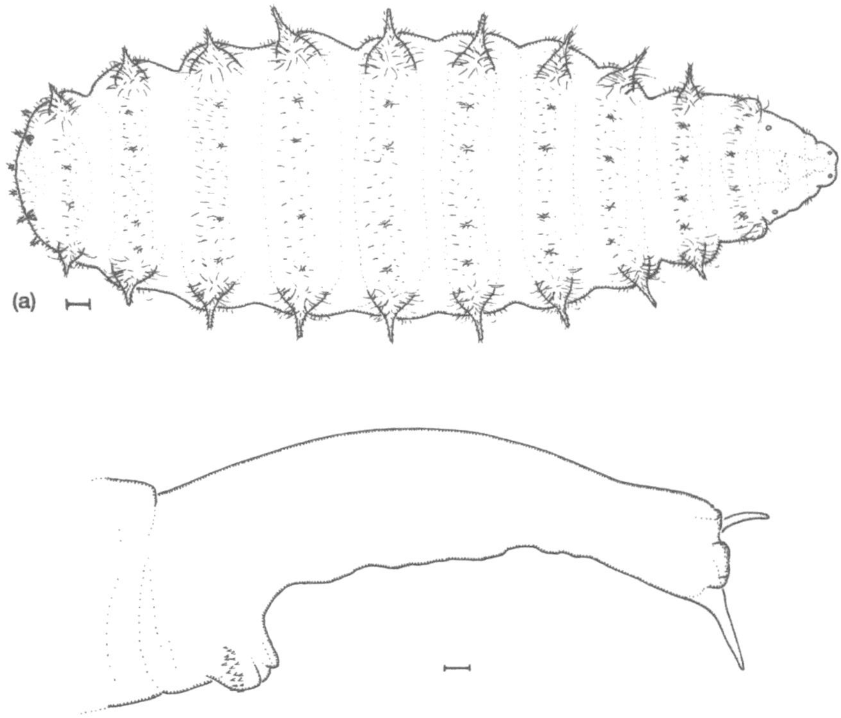

(a)

(b)

Figure 3.2 Aquatic *Megaselia* larvae. (a) *M. humida*, which lives in the meniscus. (b) The tail end of *M. schuitemakeri*.

larval salivary glands. Sinclair (1992) discusses the phylogeny of the larval mandibles (mouth hooks).

Figure 3.2 depicts the larvae of two generalized saprophage species. The range of form, however, is extensive. Even closely related species may differ in appearance. For example, the larvae of two species of *Chonocephalus* may inhabit the same fallen fruit (3.4), with the species inhabiting the more liquid parts having the end segments in the form of a long telescopic tail (as Fig. 3.8) and the species in the drier parts having normal end segments.

The aquatic larvae exhibit well the capacity for an adaptive radiation of forms among related species. Fully submerged larvae, such as *Megaselia bivesicata* and *M. deningi*, are very different from those living in the meniscus, such as *M. humida*, and at the water surface (e.g. Figs 3.2 and 3.3) (7.2.1).

The mouthparts of phorid larvae typically have paired mandibles (mouth hooks) and a median tooth supported by a cephalo-pharyngeal skeleton behind. They are directly comparable with those of other Cyclorrhapha, such as blowflies (e.g. Erzinçlioğlu, 1985a) and comparative studies suggest that the mouth hooks are derived from the mandibles of the larvae of the

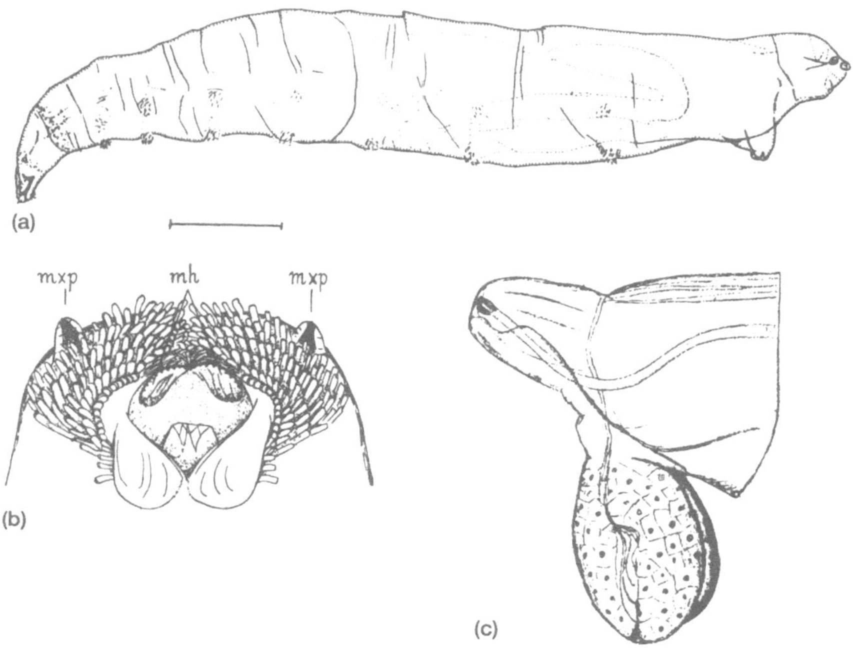

Figure 3.3 Further aquatic *Megaselia* larvae. (a) *M. deningi*, which lives fully submerged. (b) The underside of the head of *M. bivesicata* (mxp = maxillary palps, mh = 'mouth hooks' or mandibles). (c) The tail end of *M. bivesicata*. ((b) and (c) are after Schmitz, 1931d.)

Brachycera (Sinclair, 1992). In saprophage species (e.g. Fig. 3.4) the mandibles tend to have several teeth. In the predatory *Megaselia argiopephaga* the mandibles are heavily sclerotized and the terminal tooth of each is enlarged (Fig. 3.5a). In *Metopina ciceri*, whose larvae inhabit root nodules (3.4), the mandibles are fused to the sides of the median tooth (Fig. 3.5b). Ferrar (1987) illustrates several other modifications of the basic ground plan.

In the aquatic, fully-submerged, larvae of *Megaselia bivesicata* and *M. kovaci* pre-oral rows of flattened scales are developed (Fig. 3.3b) and evidently assist in food gathering.

Clearly a rich field for investigation is awaiting anyone who attempts to map details of larval habits onto a phylogenetic tree which incorporates data on larval morphology.

The larval habits of scuttle flies are astonishingly diverse. This probably reflects the antiquity of the family plus a tendency for evolution to favour smaller and smaller species, able to exploit ever smaller units of potential resources for the larvae. The impetus to evolve smaller adults may have initially come from competitive pressures for limited resources favouring those with a capacity for an accelerated development (7.9). This capacity may itself reside primarily in the habit, prevalent among female phorids, of

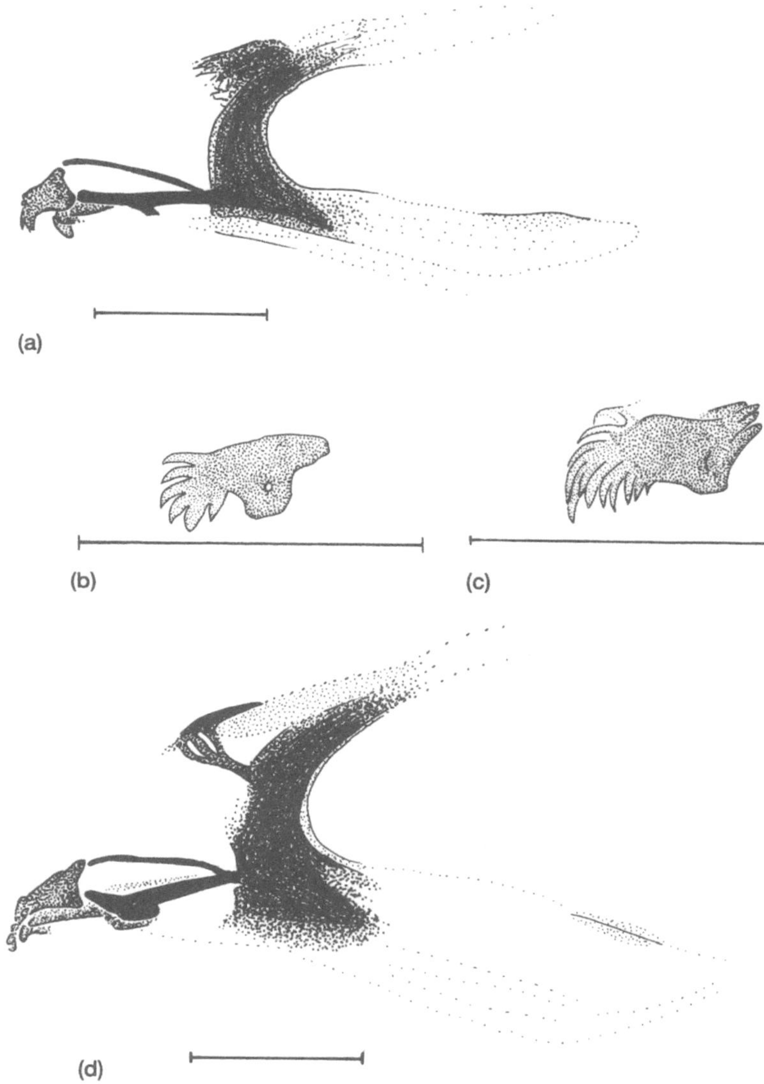

Figure 3.4 Cephalopharyngeal skeletons and mandibles of two saprophage *Megaselia* larvae (a), (b) *M rufipes* (c), (d) *M scalaris*

taking protein-rich meals (6 3) Such a habit reduces the need to accumulate resources in the pre-adult stages, in order to supply the nutrients to the eggs to be laid by the female This is analogous to the situation in predatory or blood-sucking Nematocera, such as mosquitoes, and this habit of adult females may indeed be an ancestral (plesiomorphic) feature retained from the

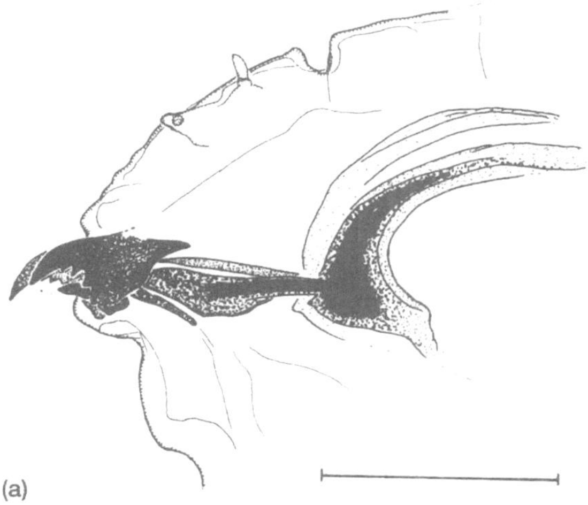

(a)

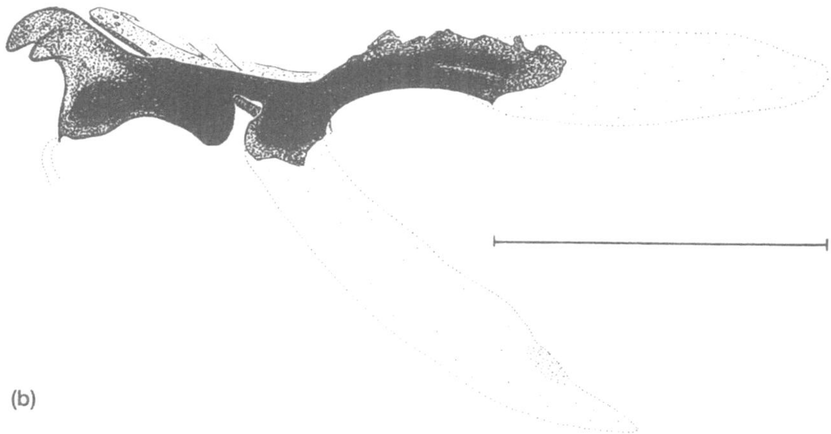

(b)

Figure 3.5 The cephalopharyngeal skeletons of two phorid larvae. (a) The predatory *Megaselia argiopephaga*; (b) *Metopina ciceri*, whose larvae live in root nodules.

Nematoceran roots of the Phoridae. However, unlike the biting Nematocera, the phorids have lost the piercing maxillae and mandibles. Where there is a need for piercing, the Phoridae utilize modifications of the labrum, the labium and/or the hypopharynx (6.2 and 6.3).

Perhaps the primitive larval habit in the family was either saprophage or fungivore, or perhaps a saprophage exploiting rotting fungi. This hypothesis cannot be evaluated until we have a far better understanding of the phylogeny of the family based on features other than larval natural history data.

Probably most larval Phoridae today are specialized predators or parasitoids. The evolution of such habits has probably occurred many times, independently, within the family. In each case the starting point of the evolutionary sequence is likely to have been a species with saprophage habits. Some sequences may have started from fungus feeders switching to feeding on other fungus feeders.

It may not be without significance that some notable polyphagous saprophage species (e.g. *Megaselia giraudii, M. rufipes* and *M. scalaris*) have been reported breeding in fungus sporophores (see below). Furthermore some species which primarily breed in fungus sporophores have also been reared from dead snails (e.g. *M. nigra*). Fungus sporophores clearly represent an enormous diversity of potential resources. However, they may be very seasonal in occurrence and experience marked fluctuations in abundance depending upon the weather conditions. In these circumstances phorid species able to exploit alternative patchily-distributed, ephemeral resources are likely to be favoured by natural selection. Periods of drought are likely to be associated with a dearth of fungus sporophores but an increase in dead or moribund snails. Probably the key to an ability to exploit contrasting larval resources lies in the digestive enzymes possessed by any particular fly larva or by the symbiotic micro-organisms in its gut. This, however, is an unexplored field.

An important consideration, in relation to whether the outcome of a switch from saprophagy or fungivory will end up in the larvae being predatory, parasitoid or parasitic, is the relative size of the phorid larva and its target organism. If the phorid larva is relatively small then it is likely to become a parasitoid. If the disparity is enormous then it is unlikely to kill its host, and so it will become a parasite. If, however, the phorid larva is approaching the same order of size as the target organism, or even exceeds it, then it is likely to become a predator.

The effect of relative size is exemplified by the following examples. The larvae of *Megaselia aequalis* feed on the eggs of the slug *Deroceras laeve*. The first-instar larva is essentially parasitoid, related to its relatively small size when compared with a single slug egg. The subsequent, relatively much larger, instars become predators on the rest of the eggs in the clutch (Robinson and Foote, 1968). The reverse occurs in *M. rufa*, whose larvae initially prey upon the eggs of its scale insect (Coccidae) host, the phorid larvae being relatively large in comparison with the eggs. Later the larvae become parasitoids of the female coccid, which is larger than they are (Schmutterer, 1952). The various saprophage species reported as facultative parasites of man (see below) clearly inhabit a host much larger than themselves. Some Phoridae whose larvae attack other Diptera larvae nicely illustrate this importance of the relative size of phorid larva and target organism. The parasitoid larva of *Megaselia flavicoxa* is significantly smaller than its sciarid-midge host larva *Bradysia bicolor* (Disney, 1976b). The predatory larvae of *Megaselia submarginalis* show a much smaller disparity in size in relation to their prey, the larvae of the gall midge *Contarinia negundifolia* (Robinson, 1970). The true-parasite larvae of *Megaselia paludosa* are very much smaller than their hosts, the larvae of the cranefly *Tipula paludosa* (Carter, 1977).

The switch from saprophagy or fungivory to predatory or parasitoid habits has probably occurred as a result of two principal mechanisms. One has probably been competition between the larvae and other insects. The other has probably been the attraction of gravid females to specific olfactory cues emanating from recently-dead invertebrates, these cues being the same as those emanating from the same organisms when wounded but still alive or when intact and alive. The potency of the emanation is likely to be strongest in the freshly-dead organism and weakest in the intact living organism.

The effect of interspecific competition causing facultative predatory behaviour is exemplified by the saprophage larvae of *Dohrniphora cornuta* when breeding in trickling-filter sewage beds. Normally the larvae feed on the film of micro-organisms coating the substratum. If, however, the populations of the larvae of the moth flies *Psychoda alternata*, doing the same, become too large the *D. cornuta* larvae will switch to facultative predation of the *P. alternata* eggs, larvae and pupae (Kloter *et al.*, 1977). Similarly intraspecific competition may cause a switch from normally saprophage habits. For example, *Megaselia scalaris* larvae scavenging in laboratory cultures of the cockroach *Periplaneta americana* may reach very high population levels. In these conditions some larvae will attack the living nymphs of *P. americana*. They initially feed on the latter's wing pads, and probably cause their death as the larvae start to invade deeper into their hosts (Robinson, 1975). In this case the competition has given rise to facultative parasitoidism.

The importance of olfactory cues emanating from recently-dead organisms is evidenced by the rapidity with which saprophage species such as *Megaselia giraudii, M. rufipes* and *M. scalaris* will locate insects that die in laboratory cultures or rearing containers. Indeed many claims in the literature that these species are facultative parasitoids are based on a failure to appreciate this (Disney, 1983a). The significance of wounds on potential living hosts was pointed out by Picard (1930), when he reported that the parasitoid *Megaselia elongata* lays its eggs into minute wounds in the integument of its host, the millipede *Schizophyllum sabulosum*. Berland (1933) then reported that experimentally-wounded ants and millipedes will readily attract ovipositing phorids. Subsequently it has been reported that the females of parasitoids of the larger species of ant are attracted to wounded individuals, which tend to be fairly frequent due to aggressive encounters between individuals from neighbouring colonies (Brown and Feener, 1991b).

In most cases a reliance on accidental wounds would seem a somewhat hazardous mechanism for ensuring host location. However, it is now known that many female phorids will attack other arthropods in order to feed on haemolymph (6.3). In the case of *Phalacrotophora berolinensis* and *P. fasciata*, which parasitize ladybird (Coccinellidae) pupae, the females feed on haemolymph by making small punctures in the integuments of the pupae. These punctures are thought to allow subsequent penetration by the first-instar larvae (Martelli, 1914; Lichtenstein, 1920; Richards, 1926; Colyer, 1952a; Delucchi, 1953; Disney *et al.*, 1994). It is also now known that various phorids are very quickly attracted to fresh exposures of termites. These exposures comprise intact living termites, wounded individuals and freshly dead termites. Many of these phorids feed on the haemolymph of the dead or wounded termites (6.3). There is an almost equally rapid recruitment of the con fly parasitoids of the termites *Diplonevra mortimeri* and *D. watsoni* for the purposes of oviposition into intact living termites (Disney, 1986d). This rapid recruitment of con flies may be related to the potent emanations from the associated wounded and freshly-dead termites. Wounding of termites is known to cause the release of alarm pheromones (e.g. Moore, 1969).

Clearly the importance of wounds in attracting parasitoid phorids to their hosts merits further investigation. What are the specific emanations that attract the phorids? Do scuttle flies respond to termite alarm pheromones? What are the frequencies and principal causes of small wounds in different host organisms? How do these relate to rates of recruitment?

It seems that another route from saprophagy to predatory habits, and perhaps even to parasitoidism, has been via larval kleptoparasitism. This term refers to the larvae stealing the provisions of another species, which is typically a social insect. In the case of Phoridae larval kleptoparasitism is especially associated with the stealing of the provisions of ants, bees, wasps and termites (3.2.5).

In the review of larval habits in the Phoridae, given below, the above discussion needs to be borne in mind. Under each category of host/prey species some reference is made to reports of saprophage species exploiting moribund specimens, but without listing all the detailed instances. Many reports present no hard evidence of facultative predation, parasitization or kleptoparasitism. These are omitted. In some cases the published data do not allow a decision as to whether the reported observations relate to saprophagy or parasitization. Such doubts are indicated. In some cases I have offered a different interpretation from the original authors, in the light of the rearing arrangements reported. It has to be constantly borne in mind that in laboratory cultures the conditions, and crowding in particular, may induce facultative predation or parasitoidism in a species that is normally saprophage.

Some larvae of the Termitoxeniinae, for example *Clitelloxenia hemicyclia*, do not feed at all during their brief free-living existence (Franssen, 1933b, 1936) (2.2). Delachambre (1965) describes some bizarre larvae which he attributed to *Termitostroma ivorensis*, on the basis of being found in the same fungus gardens in a nest of the termite *Protermes prorepens* as 30 females of this fly. He considered that the degenerate cephalo-pharyngeal skeleton (the sclerotized mouthpart apparatus) indicated that these larvae also do not feed. Further observations are needed. Delachambre also cited Silvestri's (1920) attribution of a larva to *Ptochomyia afra* on the basis of a similar association. However, this larva was almost certainly that of a *Dohrniphora* species (Borgmeier, 1964a).

3.2 FEEDERS ON DEAD OR DECAYING ORGANIC MATTER (SAPROPHAGE SPECIES)

Saprophage Cyclorrhaphan larvae essentially feed on particles of organic matter suspended in a fluid, the whole being rich in micro-organisms in most cases. Where the organic matter exists in solid form it will first have to be liquefied. This is normally accomplished by the secretion of digestive enzymes onto the potential food. Following ingestion of suspended food particles, a larva needs to concentrate them and to eliminate excess liquid in the process. In other Cyclorrhaphan larvae ridges in the floor of the pharynx facilitate this process. The dorso-lateral expansions (lamellae) of these ridges partly occlude the gutters between the ridges, functioning as a sieve. The concentrated particles pass backwards above the lamellae, while the excess fluid is carried forward in the gutters to be ejected (Dowding, 1967).

In Phoridae pharyngeal ridges are evident in the polyphagous saprophage species such as *Megaselia rufipes* and *M. scalaris*. Recently *Triphleba nudipalpis* larvae have been reported to possess these ridges also (Froese, 1992b), the larvae being known to breed in buried carrion (3.2.3). Both the *Megaselia* species are able to develop in some very liquid media. *Megaselia rufipes* has even been reported breeding in a sewage filter-treatment plant receiving a chemical (pharmaceutical) waste (Painter, 1980). The semi-aquatic larva of *M. humida*, living in the meniscus of waters held in the internodes of giant bamboos, also has pharyngeal ridges. Its gut content is detrital material rich in spores (Disney, 1991i).

Megaselia scalaris has been reared from an extraordinarily wide range of dead organisms and organic materials. Lever (1944) recorded larvae in a tin of boot polish, composed of wax, stearin (lard) and lamp black. Perhaps the most remarkable medium reported for this species was a pot of blue paint (McCrae, 1967). Three hundred and fifty larvae, pupae or drowned adults, probably representing the offspring of a single female, were present. Both the larvae and adults had the blue pigment present in their guts. The paint comprised 33% pigment (phthalocyanine blue), 43% solvent (water plus a little alcohol), 23.5% binder (vinyl acetate plus dibutyl phthalate, plus wetting agents) and 0.5% preservative (diphenyl mercury dodecenyl succinate). In the paint the dibutyl phthalate serves as a plasticizer, but elsewhere it has been used as an insect repellent! The phthalocyanine blue has a structure of nitrogenous and aromatic groups conjugated around a copper atom, and was therefore the most likely source of energy for the larvae. McCrae (1967) reared the same species from a snake preserved in alcohol. Likewise in the Hope Collections at Oxford University there are specimens of *M. rufipes* reared (by A. H. Hamm) from beet pickled in vinegar.

3.2.1 ORGANIC MATTER IN WATER

The aquatic phorids of container habitats whose larval habits are known feed on drowned insects, the fungi associated with these dead insects, spores and other organic detritus (Disney, 1991i, 1993b). The species recorded from these habitats are reviewed in section 7.2.1. Some of the species associated with the meniscus and water surface almost certainly ingest the surface film itself, which normally comprises a monomolecular layer of protein or lipoprotein plus micro-organisms (Guthrie, 1989).

3.2.2 DUNG

The indications are that Phoridae are more likely to exploit the faeces of invertebrates and small vertebrates than the larger accumulations of dung produced by larger vertebrates. The published records suggest that most cases are of general saprophage species extending the range of their acceptable pabula by opportunistically exploiting dung, when circumstances have given rise to reduced competition from the more characteristic dung-breeding insects. In the survey below, only the cave-dwelling *Triphleba* species and *Wandolleckia* are likely to be obligate specialist dung breeders There are probably many more species awaiting discovery of their larval habits. The systematic and critical investigation of the faunas of invertebrate dung and the droppings of small vertebrates remains an almost unexplored field.

(a) Invertebrate dung (frass)

Megaselia giraudii and *M. rufipes* have been reported breeding in larval Lepidoptera frass (Morley and Atmore, 1915; Aston, 1957). Schmitz (1957a) reports *M. limburgensis* from the frass of curculionid and scolytid beetles. Females of a *Chonocephalus* species have been caught in association with cerambycid beetle frass (Disney, 1986c).

Baer (1953) reared *Wandolleckia achatinae* from the dung of the snails *Achatina achatina* and *A. ventricosa*. The adult females are to be found running about on large snails (6.4.1).

(b) Bird dung

Puliciphora borinquenensis has been reared from chicken dung (Disney, 1988b) and so have *P. lucifera* and *Chonocephalus dorsalis* (Malloch, 1935b). *Megaselia rufipes* has been reared from swallow (*Hirundo rustica)* dung (Disney, 1979f) and *M. pleuralis* and *Conicera similis* from pigeon (*Columba livia*) dung (A. G. Irwin and Disney, unpublished data). The related *Conicera tibialis* has been reared from a laboratory culture medium of chicken and mink dung (K. Arevad and Disney, unpublished data).

(c) Mammalian dung

Triphleba antricola and *T. aptina* breed in bat guano (Leruth, 1939) and so does *Megaselia rufipes* (e.g. Disney, 1979f) and *M. meconicera* (A. M. Hutson and Disney, unpublished data). Hackman (1963) reared *M. giraudii*, *M. fusca* and *Triphleba bicornuta* from rodent droppings in rodent burrows.

Human faeces will support phorid larvae in the right circumstances, which require clarification. Lesne (1923) reported a *Megaselia* species (as *M. ruficornis*) breeding in human faeces and *M. rufipes* adults visiting it. Roberg (1915) reared *M. scalaris* from human faeces and it is reported breeding in dry-composting toilets in Africa below (7.12.2). Schmitz (1928a) reported *M. lanceolata* from human faeces.

Bohart (1947) reported *Megaselia suis* breeding in pig dung and *M. giraudii* has been reported from dog dung (Disney, 1979f).

Lundbeck (1922) and Schiner (1864) reported *Megaselia* species breeding in cow dung, but their specific identifications are open to doubt. Phoridae are certainly not a normal component of the cow-dung community in Britain (Skidmore, 1991). These early records need confirmation. There also needs to be clarification of the circumstances in which cow dung is an acceptable resource for phorids.

The dearth of larval phorids in exposed dung contrasts with their presence in dung in caves, rodent burrows, and enclosed toilets. Perhaps buried dung (e.g. of domestic cats) will be found to be favoured by phorid larvae, as is buried carrion in contrast to exposed carrion (3.2.3). This merits investigation. Kühlhorn (1987) reported *Conicera tibialis*, *Megaselia giraudii* and *Phora atra* visiting exposed cat dung.

3.2.3 VERTEBRATE CARRION

There are numerous records of Phoridae associated with vertebrate carrion, but many of these are almost certainly related to the adult females' need for a protein-rich meal (6.3). Hard data on actual breeding in the carrion are sparse. *Diplonevra peregrina* larvae feed upon freshly-dead bats in caves in Malaysia (McClure *et al.*, 1967). There are several records of phorids reared from dead nestling birds (7.5.3) and *Megaselia rufipes* has been reared from a dead rat (Disney and R. E. Evans, unpublished data). Payne *et al.* (1968) studied dead piglets which they had buried 50–100 cm deep in the soil. They observed phorid larvae on Day 7, but found them especially in evidence and active in Stage 4 (the disintegration stage) of the decay process. In reports of phorid larvae in vertebrate carrion it is often not clear whether they are actually feeding on the carrion or whether they are preying upon other invertebrates. However, Froese (1992b) reared *Conicera similis*, *Metopina oligoneura* and *Triphleba nudipalpis* from meat baits buried 15–20 cm in the soil of arable fields (7.2.3). *T. nudipalpis* has previously been reared from dead mice buried 30 cm deep in the soil of a vegetable garden (Disney, 1991d). The phorids known to breed in buried human corpses are reviewed below. It would appear that some Phoridae may be performing a role in buried carrion comparable to that of blowflies (Calliphoridae) in exposed carrion. There is a need for critical studies.

Because of the growing importance of forensic entomology (Smith, 1986; Erzinçlioğlu, 1989) I detail cases involving Phoridae, especially in corpses of known age. The faunal succession in human cadavers is briefly reviewed in section 7.5.1. In forensic entomology it is important to consider exposed and buried corpses as essentially different ecological resources. The duration of the development of the flies will also be influenced by temperature (5.3). For example, there is the report of phorid pupae in a corpse of an infant discovered behind a stove on a farm in Belgium. It was inferred that death had occurred only about three weeks previously and that the development of the insect fauna must have been accelerated by unusually high temperatures (Leclercq and Leclercq, 1948).

Conicera tibialis (the coffin fly)

The common name for this species derives from its frequent occurrence on buried corpses (Schmitz, 1928f, 1951c). It has often been found in the corpse about one year after burial (Smith, 1986). Blair (1922; see also Colyer, 1954a) reported the larvae feeding on a buried human corpse that had been buried for 10 months, especially on the backs of the hands and forearms. The pupae were most abundant in the hair, probably because the larvae quit the moist areas of the corpse in order to pupate. Schmitz (1928f) cites earlier reports of this species in buried corpses, often under incorrect names. In particular he summarizes Reinhard's (1882) classic studies, and his reports of literally millions of puparia and numerous adults of this species present in the coffins of some bodies buried up to five years previously. The evidence suggests that a sequence of generations are produced in these buried corpses. Schmitz's own records of this species relate to corpses buried for one year and for four years.

Colyer (1954e,f) reported adults of this species swarming over the place where a dead dog had been buried about a metre below the surface, 18 months previously. Exhumation of the dog revealed abundant larvae, along with young earthworms, in the lean blood-streaked parts of the corpse, but none in the adipose tissue. Pupae were found in both the lean parts and the hair. Furthermore there was a one-way traffic of adults from the buried dog to the surface. These observations confirm the opinion of Schmitz (1928f) that the adults of this species are able to move considerable distances through the soil (see also section 2.5).

Adults have been reported emerging from a four-month-old corpse wrapped in bedding and exposed beneath floorboards, in Yorkshire (Erzinçlioğlu, 1985b). This early occurrence in a corpse is probably related to the more rapid onset of mummification when it is exposed, and drying out is accelerated. By contrast Ardö (1953) reports exhuming a body, which had been buried for 3 years and 9 months, in March. He found many pupae, which were especially concentrated in the remains of the diaphragm. He reared adults from a sample of these pupae taken into the laboratory. The body had been buried in June in soil that was subjected to prolonged periods of cold (around freezing or below).

Webster (1890) reported numerous larvae, pupae and adults of *Conicera* (presumably *C. tibialis*) in the corpse of a 62-year-old man exhumed after two years of burial. He was found to have been poisoned with arsenic.

Megaselia abdita

In the USA this species has been reared from exposed corpses 90–98 days old (Disney, 1987k) and two months old, but in both cases the corpse was concealed in a building. In the second case it was wrapped in a quilt and plastic garbage bags and placed in a basement at about 10 °C (B. Greenberg and Disney, unpublished data).

Megaselia rufipes
Freshly-emerged adults have been obtained from a four-month-old corpse exposed beneath floorboards in Yorkshire (Erzinçlioğlu, 1985b). Pupae were obtained in a corpse exposed in a greenhouse in Belgium for 52 days (M. Leclercq and Disney, unpublished data).

Megaselia scalaris
In South Africa this species has been reared from exposed corpses that were 4–6 months old and had been treated with a bleaching agent. The corpses had begun to mummify (A. Prins and Disney, unpublished data). It was also in the same 90–98-day-old corpse that was infested with *M. abdita* (see above). Barnes (1990a) reports this species breeding in a black and tarry, partially skeletonized, corpse buried in Puerto Rico for about one year.

Triphleba hyalinata
In Europe this species has been reported breeding in buried corpses (Leclercq and Watrin, 1973). Schmitz (1943) reported it as frequent in coffins. Leclercq (1978) reported mature larvae and adults of this species on the body of a woman concealed in a thicket for 25 days.

Triphleba opaca
Adults have been reported on exhumed corpses (Brues, 1903b).

3.2.4 INVERTEBRATE CARRION AND GENERAL REFUSE

(a) Dead molluscs

Apart from the generalized saprophage species, such as *Megaselia scalaris, M. giraudii* and *M. rufipes*, a number of phorids are especially associated with breeding in dead molluscs. In particular I note the following records.

Chaetopleurophora, including *C. bohemanni, C. erythronota, C. pygidialis* and *C. spinosior* (Schmitz, 1917a, 1941c,d; Disney, 1980k).
Conicera formosensis (Beaver, 1987).
Dohrniphora cornuta (e.g. Bohart and Gressitt, 1951; Beaver, 1987).
Gymnophora luteiventris (Brown, 1987a).
Gymnoptera orientalis (Beaver, 1987).
Megaselia species, especially *M. abdita, M. aurea, M. brevicostalis, M. breviterga, M. compressa, M. fisheri* and *M. ruficornis* (Keilin, 1911b; Schmitz, 1917a; Lundbeck, 1922; Parmenter, 1965; Robinson, 1971; Beaver, 1972; Disney, 1979f and unpublished data; Kneidel, 1983, 1984a). In addition there are a few records of species which normally breed in fungi being occasionally reared from dead snails. For example, *M. nigra* has been reared from dead *Cochlodina laminata* (R. E. Evans and Disney, unpublished data).
Plethysmochaeta marmorata and *P. trinervis* (as *P. molluscivora*) (Schmitz, 1916b, 1929e).

Puliciphora, including *P. borinquenensis, P. exachatina, P. lucifera, P. obtecta, P. occidentalis, P. pallicauda, P. pulex* and *P. trisclerita* (Schmitz, 1916f, 1925a, 1929e; Brues, 1919c; Senior-White, 1924; Malloch, 1935b; Bohart and Gressitt, 1951; Borgmeier, 1963d; Beyer, 1967; Kneidel, 1983, 1984b; Disney, 1988b).

Spiniphora, including *S. atricostata, S. bergenstammi, S. conspicua, S. dorsalis, S. excisa, S. genitalis, S. maculata, S. punctipennis, S. signata* and *S. strobli* (e.g. Brues, 1903b; Schmitz, 1917a, 1938g, 1941c; Keilin, 1919; Lundbeck, 1922; Grensted, 1957; Colyer, 1955a; Disney, 1973; van Achterberg and Bin, 1981; Beaver, 1987; R. Hutterer and Disney, unpublished data).

(b) Dead arthropods

The corpses of arthropods, especially insects, are frequently exploited by the general saprophage species. The following reported rearings indicate the range of genera and species involved in the utilization of this resource.

Chonocephalus depressus (Meijere, 1912).

Conicera sp. (Richards, 1932).

Diplonevra armipes (e.g. Cuthbertson, 1936).

Dohrniphora cornuta (e.g. Bonnamour, 1921), *D. fisheri* (Disney, 1990g) and *D. paolii* (Schmitz, 1928b).

Gymnoptera neotropica (Borgmeier, 1960b).

Megaselia aletiae (e.g. Malloch, 1914c; Cross and McGovern, 1969), *M. aurea* (Rettenmeyer and Akre, 1968), *M. chapmani* (Borgmeier, 1967a), *M. curtineura* (e.g. Brunetti, 1912b; Malloch, 1935b; Schmitz and Delage, 1981; Disney and Nussbaum, 1990), *M. destructor* (Malloch, 1915a), *M. giraudii* (e.g. Lundbeck, 1922; Orosi-Pal, 1938; Schmitz and Delage, 1981); *M. goidanichi* (Goidanich, 1928a), *M. necrophaga* (Enderlein, 1912b), *M. nigriceps* (Kogan *et al.*, 1989), *M. pauculitincta* (Beyer, 1960e), *M. perdita* (e.g. Kneidel, 1984a), *M. ruficornis* (Kneidel, 1984a), *M. rufipes* (e.g. Lundbeck, 1922; Richards, 1932; Orosi-Pal, 1938; Britten, 1942; Hammond and Smith, 1955; Hackman, 1963; Eyles, 1965; Disney, 1979f), *M. scalaris* (e.g. Brunetti, 1912b; Enderlein, 1929; Creighton, 1936; Robinson, 1975b; Rocha *et al.*, 1984; Young, 1984) and *M. setacea* (Malloch, 1912b).

Plethysmochaeta nobilis, P. couvreuri (as *P. semilutea*) and *P. vectabilis* (Schmitz, 1933b, 1953d).

Pseudohypocera kerteszi (e.g. Salt, 1929).

Puliciphora beckeri, P. borinquenensis, P. lucifera, P. occidentalis, P. pulex, P. rufipes and *P. togata* (Meijere, 1912; Malloch, 1935b; Rettenmeyer and Akre, 1968; Kneidel, 1984a; Disney, 1988b).

Spiniphora maculata (Keilin, 1911a).

(c) The refuse in colonies and nests of Hymenoptera and Isoptera

The species associated with the nests of ants and termites are known as myrmecophiles and termitophiles respectively. In any survey of nests there will be occasional occurrences of general refuse species (3.2.4), which are only incidentally found in the nests of these social insects, alongside obligate myrmecophiles or termitophiles. The latter will include saprophages, specialized predators and parasitoids (6.4.3). The biologies of most species whose adults are found in the nests of other insects (e.g. 6.4.3(a) A and (b) A) remain to be explored. Those known to be saprophages (at least facultatively) are reviewed below.

The refuse of discarded dead insect prey, moribund brood, plant refuse, and other items associated with the nests of social insects, constitute a significant resource for saprophage phorids able to evade the attentions of their hosts. It is evident, however, that facultative predation of the host brood may be the norm for some of these phorid species. Thus *Triphleba lugubris* larvae are habitually scavengers on dead larvae and adults of the wasp *Vespula pensylvanica* as well as being predators on the living pupae (MacDonald *et al.*, 1975). Likewise consumers of the provisions of social insects (3.2.5) may be facultative predators of their host's brood also. There is thus no neat division into saprophage species breeding in refuse, species feeding on the host's provisions, and species preying on the host's pre-adult stages. The coverage of the species selected for review in this section and in sections 3.2.5, 3.5.6(e) and 3.5.6(f) owes as much to ignorance and arbitrary decision as to the application of clear-cut criteria. Ecological reality is more untidy than some ecologists suppose (hence the choice of quotation to head this chapter).

Hymenoptera

The following records refer to phorids whose presence in the nests of Hymenoptera seems to relate primarily to the scavenging habits of their larvae. Some, however, are also facultative predators on the brood of their hosts (3.5.6(e)) or feeders on their provisions (3.2.5). Species whose flightless females are essentially permanent residents in the colonies of social insects are reviewed in section 6.4.3. Only when their larval habits are known to be saprophagous are the same species also included here.

***Anthophoridae* (carpenter bees)**
Anthophora stanfordiana – *Megaselia* sp. (Linsley, 1945).

***Apidae* (bumblebees, stingless bees and honey bees)**

Apis cerana – *Puliciphora borinquenensis* larvae and pupae observed in a nest in Malaysia (N. Koeniger and Disney, unpublished data).

Apis mellifera (the honeybee) – The following are among the *Megaselia* species reported from bee hives: *M. giraudii* (Lundbeck, 1922; Orosi-Pal, 1938), *M. praeacuta* (Orosi-Pal, 1938), *M. rufipes* (e.g. Orosi-Pal, 1938; Taylor, 1944; Disney, 1979f), *M. scalaris* (Orosi-Pal, 1938; A. M. Zanon and Disney, unpublished data) and *M. ursina* (Robinson, 1971).

Bombus lapidarius – *Megaselia giraudii* (A. E. Stubbs and Disney, unpublished data).

Bombus pascuorum – *Megaselia giraudii* (R. Uffen and Disney, unpublished data).

Bombus ruderarius – *Megaselia giraudii* (Scott, 1920).

Bombus sonorus – *Megaselia* sp. and *Conicera* sp. (Ryckman, 1953).

Bombus terrestris (sensu Prŷs-Jones and Corbet, 1991) – *Gymnoptera longicostalis* (Collart, 1933). Whether the larvae of this species are facultative predators, like *G. vitripennis* in *Vespula* nests (Biegel, 1953) or prey on *Bombus* eggs (as suggested by Sladen, 1912) needs investigation. *Megaselia giraudii* (S. G. Goodwin and Disney, unpublished data).

Melipona favosa – *M. scalaris* (Salt, 1929).

Melipona interrupta – *Pseudohypocera kerteszi* larvae feeding on dead bees and dead pupae (Salt, 1929).

***Eumenidae* (burrowing wasps)**

These wasps have been treated as a subfamily of the Vespidae by some authors (e.g. Carpenter, 1982), but here I follow Gauld and Bolton (1988) in treating them as a family.

Monobia quadridens – unidentified phorids in nest (Krombein, 1967).

Pachodynerus erynnis – *M. aletiae* (Krombein, 1967).

Stenodynerus ammonia – *M. aletiae* (Krombein, 1967).

S. saecularis – unidentified phorids (Krombein, 1967).

***Formicidae* (ants)**

Dorylinae (driver ants)

Rhynchomicropteron nudiventer females have been recorded from a colony of *Aenictus* sp. (6.4.3(a) A). Brown (1992a) has provided evidence that the larvae of this fly breed in dead insects.

Ecitoninae **(army ants)**

These formidable ants have the habit of forming large colonies that alternate between staying in one place (statary phase) and moving from place to place, employing temporary bivouacs (the nomadic phase). They create refuse heaps of part-consumed prey, moribund brood, dead ants and much other debris. Many species locate these refuse heaps underground. *Eciton burchelli* and *E. hamatum*, however, locate theirs above ground (and have, therefore, been easier for entomologists to sample). These refuse heaps constitute a rich resource for insects able to co-exist with the ant hosts. A range of phorid genera have evolved to exploit this.

The phorid 'myrmecophiles' (or ectosymbionts) associated with some New World army ants have been studied by Rettenmeyer and Akre (1968). They reared a number of species in the laboratory and showed that their total development times were a little shorter than the average statary phase of their hosts. Thus the offspring of females ovipositing at the start of the statary phase not only develop in accumulating refuse heaps but are able to complete their development just before the ants embark on their next nomadic phase. Furthermore the flightless females, which characterize many of these genera, tend to emerge earlier than the fully-winged males. The faster development is presumably facilitated by dispensing with the need to produce functional wings and the associated musculature. The females run with the ants during their nomadic phase. Most follow the tail end of the emigration columns of the ants.

A detailed list of adult females recorded in or visiting the colonies of Ecitoninae is given in Chapter 6 (6.4.3(a)). The species shown to have saprophage larvae by Rettenmeyer and Akre are listed below.

Apterophora attophila, Dohrniphora cornuta, D. ecitophila, Ecitophora bruchi, E. collegiana, E. parva, Ecituncula halterata, E. tarsalis, Pulicimyia triangularis, Puliciphora borinquenensis and *Xanionotum hystrix.*

Formicinae
Brown (1992b) reared *Triphleba laticosta* from the nest of *Camponotus herculianus.*

Myrmicinae
Apterophora attophila females have been reported from a colony of *Atta sexdens* (6.4.3(a) A). The larvae have been shown to be saprophage (Rettenmeyer and Akre, 1968). *Chonocephalus heymonsi* adults have been reported with *Solenopsis saevissima* (6.4.3(a) A). The larvae of *C. heymonsi* are known to breed in rotting fruits (3.4).

Ponerinae (predatory ants)
The observations of Brown (1992a) indicate that *Rhynchomicropteron nudiventer* larvae breed in dead insects. The females have been recorded in the nests of *Leptogenys distinguenda, L. kittelli* and *L. mutabilis* (6.4.3(a) A).

Megachilidae (leaf-cutting bees)
Krombein (1967) recorded phorids in the nest of a *Megachile* sp.

Pompilidae (spider wasps)
Krombein (1967) recorded phorids in the nest of a *Dipogon* sp.

Proctotrupidae
Phaeoserphus vitiator – Megaselia giraudii (Schmitz and Delage, 1981).

Sphecidae (digger wasps)

Cerceris frontata – M. sphinx has been reared from the nest of this wasp (Robinson, 1971). What the larvae were feeding on is unknown.

Ectemnius cavifrons – M. rufipes (Warburton, 1919).

Ectemnius lituratus – Anevrina sp. (Lundbeck, 1922).

Rhopalum coarctatum and *Trypoxylon frigidum* nests were reported to be harbouring phorids by Krombein (1967).

Vespidae (social wasps)
The following review summarizes the records of phorids from the nests of these wasps.

Chartergus sp. – *Megaselia scalaris* (Borgmeier, 1932a); *Dolichovespula norwegica – M. giraudii* (Donisthorpe, 1927); *Leipomeles dorsata – M. scalaris*

(M. Cooper and Disney, unpublished data); *Polistes exclamans* – *M. aletiae* (Nelson, 1968);

Polybia sp.– *Puliciphora rufipes* (Borgmeier, 1960b);

Polybia occidentalis – *M. scalaris* (Bristowe, 1925); *Polybia simillima* – *M. scalaris* (Young, 1984); *Vespa crabro* – *Diplonevra funebris* (Collart, 1933), *Megaselia fuscinula* (Schmitz, 1926c) and *M. rufipes* (e.g. Collart, 1933); *Vespula* spp. – *Conicera similis* (Schmitz, 1952c) and *Diplonevra funebris* (Collart, 1933, Collin, 1939);

Vespula atropilosa – *Diplonevra funebris, Megaselia curticosta* and *M. rufipes* (MacDonald *et al.*, 1975);

Vespula germanica – *Conicera similis* (Richards, 1932), *Diplonevra* sp. (Tuck, 1896) and *Megaselia rufipes* (e.g. Lundbeck, 1922); *Vespula pensylvanica* – *M. curticosta* (Robinsòn, 1971); and *Vespula vulgaris* – *Diplonevra concinna* (Richards in Spradbery, 1973), *M. giraudii* (e.g. Lundbeck, 1922) and *M. rufipes* (Morley, 1934).

3.2.5 THE PROVISIONS OF SOCIAL HYMENOPTERA AND ISOPTERA (LARVAL KLEPTOPARASITISM)

Hymenoptera

The provisions, such as pollen and honey, found in the nests of social Hymenoptera are an open invitation to any fly that is able to avoid the attentions of the bees or wasps.

The stealing of provisions from another species is termed kleptoparasitism. However, among Phoridae, there is no clear-cut distinction between saprophage species, obligate larval kleptoparasites and facultative predators. For most species it is possible to say that the larvae are primarily saprophagous with a tendency to facultative kleptoparasitism or else they are primarily kleptoparasitic with a tendency to be facultative predators. The fact that the distinctions are not clearcut, however, suggests that larval kleptoparasitism is likely to have been a precursor in the evolution of obligate predatory behaviour.

A refined habit in the exploitation of the provisions of social Hymenoptera involves the larvae intercepting some of the food during its transmission from the workers to the brood. Wheeler (1901, 1907, 1910) described a striking example of a phorid doing this, and declared it to be a true case of commensalism, a situation in which two species live together and one derives benefit from the other, which is not adversely affected. However, it might be more accurate to describe this case as kleptoparasitic trophallaxis, trophallaxis being an exchange of food between different members of the host colony. In Wheeler's example the larva of *Cataclinusa pachycondylae* wraps itself around the neck of the larva of the ponerine ant *Pachycondyla harpax* (Fig. 3.6). It adheres to the ant larva with a sucker at the posterior end. The ant workers feed their larvae comminuted insects, rather than the regurgitated liquid fed by many ant species. The phorid larva intercepts some of the

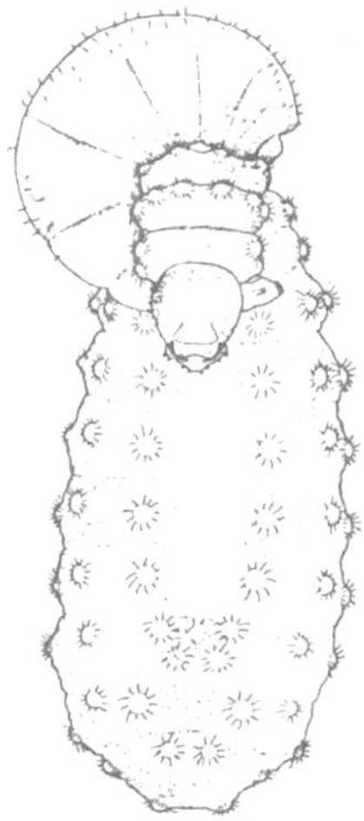

Figure 3.6 The larva of *Cataclinusa pachycondylae* wrapped around the neck of its ant larva host. (After Wheeler, 1901.)

food. If close enough it may steal from a neighbouring ant larva as well. The ants tend to groom their larvae after feeding them and will lick the fly larvae as well. Occasionally a fly larva transfers from one ant larva to another. When the ant larva spins its cocoon the fly larva positions itself at its rear, and the cocoon is slightly distorted as a consequence. The ant emerges first, from the front end of the cocoon. The 'vacated' cocoon is carried by a worker ant to a refuse heap, where the fly subsequently emerges. The fly puparium is flattened and adheres to the ant's cocoon by its ventral surface. The fly has a spring and autumn generation, just like its ant host. Another case concerns the larvae of *Tubicera lichtwardti* in the nests of the ant *Plagiolepis pygmaea*. Each larva may be surrounded by 2–6 workers, each of which may rest its feet on the larva for up to 15 min at a time. They groom the larva and feed it with regurgitated liquid nourishment (Le Manse, 1941). Some adult phorids also indulge in kleptoparasitic trophallaxis (6.3).

The species whose larvae exploit the provisions of social Hymenoptera are generally different from those acting as scavengers exploiting refuse (3.2.4(c)), and both may also be facultative predators of the host's brood (3.5.6(e)). Species whose feeding mode is in doubt are reviewed in section 3.2.4(c) above.

Apidae (bumble bees, stingless bees and honey bees)

Several workers have reported phorids in bee nests. For example, Roubik (1989) reports phorid larvae in various bee nests, the eggs having been laid in batches on the surface of cells and between brood combs, in pollen pots, or on the involucrum (protective envelope of the nest), in crevices and in other narrow spaces. Such reports, however, can only be evaluated, from a biological standpoint, when one knows what the larvae are feeding on in the nests of their hosts.

A range of apid bees have their provisions exploited by the larvae of *Pseudohypocera kerteszi* in the Neotropical Region. Pickel (1928) reported the larvae feeding on pollen and only found adult flies in cells with pollen. He suggested they were attracted to the hive by the scent of fermented pollen. G. E. Robinson (1981) reported infestation rates of 0.10 for honeybee colonies ($n = 900$) in Colombia; but rates of 0.75 for abandoned hives ($n = 180$), in which the fly larvae were feeding on both pollen and the abandoned bee brood. He concluded that *P. kerteszi* was a principal cause of bees abandoning hives. Reyes (1983) reported infestation rates of honeybee colonies in Mexico of 0.08–0.10. He confirmed that the larvae feed on the pollen stores, but that as the population levels of the phorid increase the larvae become facultative predators feeding on the prepupae and pupae of the bees. In addition to the records in the colonies of the honeybee (*Apis mellifera*) (Pickel, 1928; Borgmeier, 1930b; Prado, 1976; G. E. Robinson, 1981; Reyes, 1983) this phorid has been recorded in the nests of the following apid bees: *Melipona beecheii* (M. J. Sommeyer and Disney, unpublished data), *M. clavipes* (Borgmeier, 1930b), *M. interrupta* (Brues, 1928a), *M. pallida* (Pickel, 1928; Prado, 1976; Salt, 1929; Borgmeier and Schmitz, 1923a), *M. quadrifasciata* (Prado, 1976), *M. quadripunctata* (Borgmeier, 1930b), *M. ruficrus* (Pickel, 1928; Prado, 1976), *M. seminigra* (Portugal, 1977), *Tetragona clavipes* (Nogueira-Neto, 1970; Prado, 1976), *T. amalthea* (Salt, 1929; Prado, 1976), *T. capitata* (Salt, 1929; Nogueira-Neto, 1970; Prado, 1976), *T. ruficrus* (Salt, 1929) and *T. tataira* (Nogueira-Neto, 1970; Prado, 1976).

Salt (1929) reported pupae of *Melittophora salti* in the sealed cells in the nest of *Trigona amalthea*. Each cell contained one (or rarely two) fly pupa and one bee pupa. It was inferred that the larvae feed on the provisions in the cell.

Pompilidae (spider wasps)

A *Megaselia* species has been reared from the paralysed spider prey of the wasp *Anoplius viaticus* (3.5.3(2)).

Sphecidae (digger wasps)

Ectemnius paucimaculatus provisions its nest's cells with Diptera (Dolichopodidae, Acalyptrata and Muscidae). In Krombein's (1964) study 30 out of 93 cells examined from 18 nests had the prey consumed by *Megaselia aletiae* larvae. Indeed this fly, as a facultative larval kleptoparasite, was the commonest enemy of this wasp. Medler (1965) recorded the same fly in the nests of *Chlorion mexicanum*, there being up to 35 larvae per cell. In five nests the phorid larvae had destroyed the provisions in one-third of the cells, the latter being provisioned with bush crickets (Tettigoniidae) and tree crickets (Gryllidae). Krombein (1967) also recorded *M. aletiae* in the nests of *Podium rufipes, Trypoxylon collinum* and *T. striatum*.

Megaselia oxybelorum has been reared from the paralysed prey (the fly

Fannia scalaris) of the wasp *Oxybelus uniglumis* (Chevalier, 1925; Schmitz, 1928d; Disney, 1988h).

Isoptera (termites)

The adults of many scuttle flies in the colonies of termites are associated with their host's fungus gardens. In a few cases their larvae are known to feed on the fungi (*Termitomyces* species) (3.3 below). Probably many more of the species reviewed in section 6.4.3 have larvae which exploit the same resource.

3.3 FUNGUS-FEEDING/FUNGUS-BREEDING LARVAE

The fungi comprise an immense resource for insects, made up of an enormous diversity of species, each with its own peculiar chemistry. Furthermore each species is subdivided into a hidden mycelium and a sporophore (the familiar mushroom or toadstool), which is usually exposed to view. The mycelium is the vastly greater resource. The exploitation of the mycelium by Phoridae has been investigated only for the cultivated species of *Agaricus*. Fungivory by insects in general has been reviewed by Hanski (1989).

The overwhelming majority of the records of Phoridae reared from fungi are for those hatched from sporophores. The detailed studies of the phorid fauna of *Agaricus bisporus*, one of the principal cultivated mushrooms, have indicated that there is a clear niche separation into species whose larvae primarily exploit the sporophore (e.g. *Megaselia nigra*) and those that primarily exploit the mycelium (e.g. *M. halterata*) (Brauns, 1950; Hussey, 1961a; Robinson, 1978a). It is probable that many more species exploit the mycelium than are currently recognized. Indeed it is probable that some of the records which are based on one or two flies only, from a number of rearings from sporophores of the same fungus species which have yielded dozens of specimens of other phorid species, represent a few stray larvae of species that primarily exploit the mycelium rather than the sporophore.

It is apparent that some species breed in underground sporophores, such as the truffles (see under *Tuber* below). Underground fungi include the fungus combs in termite nests (see under *Termitomyces* below). The larvae of *Dohrniphora diminuens* have been observed feeding on termite fungus combs (Disney and Kistner, 1989c) and the larvae of *Javanoxenia punctiventris* and *Odontoxenia brevirostris* have been reported feeding on the microtubers (conidia) of *Termitomyces* (Franssen, 1933b). Presumably many more species of termitophilous phorids have fungivorous larvae. The mutualism between the Macrotermitinae termites and *Termitomyces* is reviewed by Wood and Thomas (1989).

In the leaf-cutter ants, *Acromyrmex* and *Atta*, fungi are cultured on leaf fragments in underground fungus gardens, comparable to the fungus gardens of termites. A greater range of fungus genera are employed by the ants, but *Attomyces* is one of the more important. Cherrett *et al.* (1989) review the mutualism between leaf-cutter ants and their fungi. It seems probable that some myrmecophile phorids associated with these ants will prove to be fungivorous, as either larvae or adults.

The following lists attempt to provide a comprehensive reference of reliable records of named phorids reared from named fungi. A major problem has been the unstable state of fungal classification and nomenclature. At the specific level my primary reference has been Phillips (1981), but with modifications derived from Pacioni (1985), Bon (1987) and Buczacki (1989). The generic names, however, follow (with few exceptions) the work by Hawksworth *et al.* (1983). The supra-generic classification follows Buczacki (1989).

BASIDIOMYCOTINA

Agaricales

Agaricaceae

Agaricus sp – *Dohrniphora perplexa* (Borgmeier, 1963d).

A. augustus – *Megaselia bovista* (Disney and Evans, 1990; Notton, 1991a), *M. hirtiventris* (Disney and Evans, 1988, 1990, unpublished data), *M. nigra* (Dely-Draskovits and Disney, unpublished data).

A. bernardii – *M. nigra* (Dely-Draskovits and Disney, unpublished data).

A. bisporus (= *A. brunnescens*) – *M. agarici* (Robinson, 1977; Disney, 1993d), *M. bovista* (Colyer, 1954b), *M. halterata* (Colyer, 1954b; Hussey, 1961a), *M. iroquoiana, M. longipennis* (Robinson, 1977), *M. nigra* (Colyer, 1954b; Hussey, 1961a).

A. bitorquis – *M. berndseni* (Dely-Draskovits and Disney, unpublished data), *M. halterata* (Clift and Larsson, 1987), *M. nigra* (Dely-Draskovits and Disney, unpublished data).

A. bresadolianus – *M. nigra* (Dely-Draskovits and Disney, unpublished data).

A. campestris – *M. agarici* (Malloch, 1912b), *M. berndseni* (Dely-Draskovits and Disney, unpublished data), *M. bovista* (Colyer, 1954b), *M. halterata* (Colyer, 1954b; Brown and Marshall, 1984), *M. hirtiventris* (Eisfelder, 1956), *M. nigra* (Colyer, 1954b; Disney and Evans, 1982; Dely-Draskovits and Disney, unpublished data), *M. rubella* (Yakovlev, 1986), *M. rufipes* (Engel, 1916; Lundbeck, 1922; Brown and Marshall, 1984), *M. scutellaris* (Dely-Draskovits and Disney, unpublished data).

A. elvensis – *M. nigra* (Dely-Draskovits and Disney, unpublished data).

A. flavescens – *Megaselia* sp. (Canzanelli, 1941).

A. langei – *M. hirtiventris* (Disney and Evans, 1990).

A. macrosporus – *M. hirtiventris* (Disney and Evans, 1994).

A. maskae – *M. berndseni, M. lata, M. nigra*, and *Megaselia* sp. (Dely-Draskovits and Disney, unpublished data).

A. nivescens – *M. nigra* (Louise Vet and Disney, unpublished data).

A. praeclaresquamosus (= *A. placomyces*) – *M. hirtiventris* (Disney and Evans, 1994).

A. romagnesii – *M. nigra* (Dely-Draskovits and Disney, unpublished data).

A. silvaticus – *M. hirtiventris* (Disney and Evans, 1988), *M. longipennis* (Robinson, 1971).

A. silvicola – *M. hirtiventris*, *M. lata* and *M. nigra* (Disney and Evans, 1994).

A. vaporarius – *M. hirtiventris* and *M. nigra* (Disney and Evans, 1994).

A. xanthodermus – *M. nigra* (Dely-Draskovits and Disney, unpublished data).

Amanitaceae

Amanita aspera – *M. lata* (Dely-Draskovits and Disney, unpublished data).

A. citrina – *M. lata* (Eisfelder, 1956), *M. scutellaris* (Disney and Evans, 1988; Dely-Draskovits and Disney, unpublished data).

A. excelsa – *M. flavicans* (Disney and Evans, 1982), *M. lata*, *M. lutea* and *M. scutellaris* (Eisfelder, 1956).

A. farinosa – *M. gotoi* (Disney, 1989i).

A. fulva – *M. berndseni* (Disney and Evans, 1994).

A. gemmata – *M. flavicans, M. lata, M. lutea, M. scutellaris* and *Megaselia* sp. (Eisfelder, 1956).

A. muscaria – *M. lata* (Eisfelder, 1956), *M. rubella* (Yakovlev, 1986).

A. pantherina – *M. lata* (Eisfelder, 1956; Dely-Draskovits and Disney, unpublished data), *M. lutea* (Eisfelder, 1956), *M. rubella* (Yakovlev, 1986).

A. phalloides – *M. berndseni* and *M. scutellaris* (Eisfelder, 1956; Dely-Draskovits and Disney, unpublished data), *M. lata* (Eisfelder, 1956).

A. rubescens – *M. lata* (Eisfelder, 1956; Disney and Evans, 1979, 1990; Dely-Draskovits, Louise Vet and Disney, unpublished data), *M. lutea* and *M. scutellaris* (Eisfelder, 1956).

A. spissacea – *M. gotoi* and *M. kanekoi* (Disney, 1989i).

A. vaginata – *M. berndseni* (Dely-Draskovits and Disney, unpublished data), *M. lata* (Eisfelder, 1956).

Termitomyces spp.– larvae feeding on fungus combs and microtubers (conidia): *Dohrniphora diminuens* (Disney and Kistner, 1989c), *D. vorax* (Schmitz, 1915e), *Javanoxenia punctiventris* and *Odontoxenia brevirostris* (Franssen, 1933a,b, 1936). Larvae feeding on fresh sporophores: *Megaselia scalaris, M. termitomyca* and *M. zariaensis* (Disney, 1989b). Larvae feeding on rotting sporophores: *Chonocephalus heymonsi* and *Puliciphora borinquenensis* (J. Deeming and Disney, unpublished data).

Bolbitiaceae

Agrocybe praecox – *M. berndseni* (Disney and Evans, 1990).

A. semiorbicularis – *M. berndseni* (Schmitz, 1948b).

Bolbitius vitellinus – *M. rufipes* (Buxton, 1961).

Boletaceae

Boletellus russellii – *M. berndseni* (Bruns, 1984).

Boletus appendiculatus – *M. flavicans* (Dely-Draskovits and Disney, unpublished data).

B. edulis – *Conicera* sp. (Eisfelder, 1956), *Megaselia berndseni* (Bruns, 1984), *M. lata* (Eisfelder, 1956; Dely-Draskovits and Disney, unpublished data), *M. mixta, M. rubella, M. rubescens* and *M. virilis* (Yakovlev, 1986), *M. rufipes* (Canzanelli, 1941).

B. luridus – *M. berndseni* (Bruns, 1984), *Megaselia* sp. (Eisfelder, 1956).

B. pallidus – *M. berndseni* (Bruns, 1984).

B. piperatus – *M. berndseni* (Bruns, 1984).

B. regius – *M. berndseni* and *M. flavicans* (Dely-Draskovits and Disney, unpublished data).

B. reticulatus – phorid (Burla and Bächli, 1968).

B. rhodoxanthus – *Conicera similis* (Dely-Draskovits and Disney, unpublished data).

Gyroporus castaneus – M. lutea (Bruns, 1984), *M. scutellaris* (Disney and Evans, 1990).

G. cyanescens – *M. lutea* (Bruns, 1984).

G. purpurinus – *M. lutea* (Bruns, 1984).

Leccinum (= *Krombholziella*) sp.– *M. berndseni* and *M. longipennis* (Bruns, 1984).

L. carpini – *M. berndseni* (Dely-Draskovits and Disney, unpublished data).

L. chromapes – *M. berndseni* (Bruns, 1984).

L. holopus – *M. berndseni* (Bruns, 1984).

L. scabrum – *M. berndseni* (Eisfelder, 1956), *M. flava, M. mixta, M. rubella, M. rubescens* (Yakovlev, 1986) and *M. flavicans* (Disney and Dely-Draskovits, unpublished data).

L. subglabripes – *M. berndseni* (Bruns, 1984).

L. versipellis – *M. rubella, M. rubescens* and *M. rufipes* (Yakovlev, 1986).

Strobilomyces floccopus – *M. rufipes* (Bruns, 1984).

Suillus americanus – *M. longipennis* and *M. lutea* (Bruns, 1984).

S. aeruginascens – *M. lutea* (Bruns, 1984).

S. bovinus – *M. lutea* (Schmitz, 1948b) and *M. rubescens* (Yakovlev, 1986).

S. granulatus – *M. berndseni* (Dely-Draskovits and Disney, unpublished data), *M. lata* (Eisfelder, 1956), *M. lutea* (Eisfelder, 1956; Bruns, 1984) and *M. scutellaris* (Eisfelder, 1956).

S. grevillei – *M. lutea* (Bruns, 1984; W. Hackman and Disney, unpublished data) and *M. rubescens* (Yakovlev, 1986).

S. luteus – *M. mixta, M. rubella, M. rubescens* and *M. virilis* (Yakovlev, 1986).

S. pictus – *M. berndseni* and *M. lutea* (Bruns, 1984).

S. subluteus – *M. lutea* (Bruns, 1984).

S. variegatus – *M. lutea* (Schmitz, 1948b), *M. discreta, M. rubella*, and *M. rubescens* (Yakovlev, 1986).

Tylopilus felleus – *M. berndseni* (Bruns, 1984).

Xerocomus badius – *M. scutellaris* (Eisfelder, 1956).

X. chrysenteron – *M. berndseni, M. flavicans* and *M. lutea* (Eisfelder, 1956).

X. rubellus – *M. berndseni* (Disney and Evans, 1988) and *M. flavicans* (Dely-Draskovits and Disney, unpublished data).

X. spadiceus – M. berndseni (Bruns, 1984).

X. subtomentosus – M. berndseni (Bruns, 1984), *M. pulicaria* (Disney, 1980d) and *M. rubella* (Yakovlev, 1986).

X. truncatus – M. berndseni (Bruns, 1984).

Coprinaceae

Coprinus comatus – Megaselia sp. (Malloch, 1912b).

C. micaceus – M. berndseni and *M. giraudii* (Buxton, 1961), *M. latior* (Disney and Evans, 1979), *M. spinicincta* (Eisfelder, 1956).

C. radians – M. latior (Buxton, 1961).

Panaeolus sp.– *M. lutescens* (Meijere, 1947; Buhr, 1965). The fungi were galled.

P. campanulatus – M. berndseni and *M. lutescens*, in the cases with the latter species the fungi were galled (Disney and Evans, 1990, 1994).

P. papilionaceus – M. berndseni (Disney and Evans, 1979).

P. subbalteatus – M. lutescens, the fungus being galled (Disney and Evans, 1988).

Psathyrella candolleana – M. berndseni (Disney and Evans, 1979, 1990), *M. okazakii* (Disney, 1989i), *M. rufipes* and *M. spinicincta* (Eisfelder, 1956).

P. gossypina – Megaselia sp. (Eisfelder, 1956).

Cortinariaceae

Cortinarius sp.– *M. rubescens* (Yakovlev, 1986).

C. delibutus – M. scutellaris (Eisfelder, 1956).

C. fulmineus – M. lata (Schmitz, 1948b).

C. glaucopus – M. flavicans (Dely-Draskovits and Disney, unpublished data).

C. mucosus – M. lata and *M. scutellaris* (Eisfelder, 1956).

C. pseudosalor – Megaselia sp. (Dely-Draskovits and Disney, unpublished data).

Gymnopilus hybridus – M. flava (Disney and Evans, 1988).

G. junonius – Triphleba minuta (Colyer, 1952b, 1954b; Parmenter, 1965). Colyer (1954b) also recorded *M. spinicincta* visiting this species.

G. sapineus – M. maura (Eisfelder, 1956).

Hebeloma crustuliniforme – M. flavicans (Eisfelder, 1956).

Ionocybe aeruginascens – M. berndseni (Dely-Draskovits and Disney, unpublished data).

I. agardhi – M. berndseni and *M. scutellaris* (Dely-Draskovits and Disney, unpublished data).

I. caesariata – M. flavicans (Schmitz, 1948b).

I. fastigiata – phorids (Burla and Bächli, 1968).

I. lacera – M. rubella (Yakovlev, 1986).

I. patouillardi – M. berndseni (Eisfelder, 1956).

I. serotina – M. berndseni (Disney and Dely-Draskovits, unpublished data).

Kuehneromyces (= *Galerina*) *mutabilis – M. rubella* and *M. virilis* (Yakovlev, 1986).

K. vernalis – *M. virilis* (Yakovlev, 1986).
Pholiota sp.– *M. rubella* (Yakovlev, 1986).
Rozites sp.– *M. rubescens* (Yakovlev, 1988b).
R. caperatus – *M. lata, M. pygmaea* and *M. scutellaris* (Eisfelder, 1956).

Entolomataceae

Entoloma sericeum – *M. flavicans* (Disney and Evans, 1978, 1990).
Nolanea farinolens – *M. flavicans* (Disney and Evans, 1990).

Gomphidiaceae

Chroogomphus rutilus – *M. lutea* (Eisfelder, 1956).

Hygrophoraceae

Hygrophorus eburneus – *M. scutellaris* (Dely-Draskovits and Disney, unpublished data).
H. penarius – *Megaselia* sp. (Dely-Drakovits and Disney, unpublished data).

Lepiotaceae

Lepiota cristata – *Megaselia* sp. (Dely-Draskovits and Disney, unpublished data).
L. procera – *M. flavicans* and *M. lata* (Dely-Draskovits and Disney, unpublished data), *M. longipennis* (Malloch, 1912b).
L. rhacodes – *M. hirtiventris* and *M. nigra* (Dely-Draskovits, W. Hackman and Disney, unpublished data).
Leucoagaricus leucothites – *Megaselia* sp. (Dely-Draskovits and Disney, unpublished data).

Paxillaceae

Omphalotus olearius – *Megaselia* sp. (Dely-Draskovits and Disney, unpublished data).
Paxillus involutus – *Megaselia* sp. (Canzanelli, 1941).
Phylloporus bellus – *M. murakamii* (Disney, 1989i).

Pleurotaceae

Lentinus lepideus – *M. rubella* (Yakovlev, 1986).
Pleurotus sp.– *M. chaetoneura* (Brown and Marshall, 1984), *M. sylvatica* (Disney and Evans, 1979).
P. cornucopiae – *M. frameata* (Buxton, 1961; Chandler, 1973; Disney and Evans, unpublished data), *M. giraudii* (Buxton, 1961), *M. plurispinulosa* (Chandler, 1973).
P. ostreatus – *M. rubescens* (Yakovlev, 1986).

P. sajor-caju – *Megaselia* spp. (Somchoudhury *et al.*, 1988), *Megaselia* n. sp. and *M. scalaris* (Johal and Disney, in press).

Plutaceae

Pluteus sp.– *M. plutei* (Borgmeier, 1971a).
P. cervinus – *Conicera similis* (Eisfelder, 1956), *M. flava* (Disney and Evans, 1982), *M. longipennis* (Robinson, 1971), *M. sylvatica* (Disney and Evans, 1988).
P. minutissimus – *M. sylvatica* (Disney and Evans, 1982).
P. petasatus – *M. berndseni* (Dely-Draskovits and Disney, unpublished data).
P. umbrosus – *M. sylvatica* (Buxton, 1961).
Volvariella sp.– I have examined *Chonocephalus depressus* and *C. heymonsi* caught visiting this fungus genus in India.

Russulaceae

Lactarius azonites – *M. lutea* (Dely-Draskovits and Disney, unpublished data).
L. deliciosus – *M. rufipes* (Engel, 1916).
L. flava – *M. rubella* (Yakovlev, 1986).
L. glyciosmus – *M. lutea* (Schmitz, 1948b).
L. helvus – *M. lata* and *Megaselia* sp. (Eisfelder, 1956).
L. mitissimus – *M. lutea* (Schmitz, 1948b).
L. necator – *M. rubescens* (Yakovlev, 1986).
L. pallidus – phorids (Burla and Bächli, 1968).
L. piperatus – *M. berndseni* (Eisfelder, 1956), *M. lutea* (Schmitz, 1948b).
L. quietus – *M. lutea* (Schmitz, 1948b).
L. resimus – *M. lutea* (Disney and Evans, 1978).
L. rufus – *M. berndseni*, and *M. eisfelderae* (Eisfelder, 1956), *M. lutea* (Schmitz, 1948b; W. Hackman and Disney, unpublished data), *M. rubescens* (Yakovlev, 1986).
L. serifluus – *M. berndseni* (Dely-Draskovits and Disney, unpublished data).
L. torminosus – *M. rufipes* (Canzanelli, 1941).
L. vellereus – *M. latior* (Disney and Evans, 1979).
Russula sp.– *M. rubescens* (Yakovlev, 1986).
R. adusta – *Megaselia* sp. (Eisfelder, 1956).
R. aeruginea – *M. flava* and *M. flavicans* (Eisfelder, 1956), *M. rubella* (Yakovlev, 1986).
R. alutacea – *M. lutea* (Disney and Evans, 1990).
R. amethystina – *M. lutea* (Eisfelder, 1956).
R. atropurpurea – *M. lutea* (Schmitz, 1948b).
R. castanopsidis – *M. furukawae* (Disney, 1989i).
R. consobrina – *M. berndseni* (Eisfelder, 1956).
R. cyanoxantha – *M. flavicans* (Disney and Evans, 1982), *M. lutea* (Eisfelder, 1956).
R. decolorans – *M. lata, M. lutea* and *M. scutellaris* (Eisfelder, 1956).

R. delica – *M. flavicans* (Disney and Evans, 1994).
R. depallens – *M. lutea* (Canzanelli, 1941).
R. elegans – *M. lutea* (Schmitz, 1948b).
R. emetica – *M. lutea* (Eisfelder, 1956).
R. fellea – *M. lutea* (Louise Vet and Disney, unpublished data).
R. foetens – *M. berndseni* (Dely-Draskovits and Disney, unpublished data), *M. lata* and *M. lutescens* (Eisfelder, 1956) and *M. lutea* (Disney and Evans, 1990).
R. fragilis – *Megaselia* sp. (Eisfelder, 1956).
R. furcata – *M. berndseni* (Dely-Draskovits and Disney, unpublished data).
R. grisea – *M. flavicans* (Eisfelder, 1956).
R. heterophylla – *M. berndseni, M. flava, M. flavicans, M. lata* and *M. lutea* (Schmitz, 1948b).
R. ionochlora – *M. berndseni, M. flavicans* and *M. lutea* (Disney and Evans, 1990).
R. lepida – *M. berndseni* (Eisfelder, 1956) and *M. lutea* (Eisfelder, 1956; Disney and Evans, 1982).
R. lilacea – *M. scutellaris* (Dely-Draskovits and Disney, unpublished data).
R. linnaei – phorids (Burla and Bächli, 1968).
R. lutea – *M. scutellaris* (Dely-Draskovits and Disney, unpublished data).
R. maculata – *M. lutea* (Disney and Evans, 1990).
R. nauseosa – *M. scutellaris* (Eisfelder, 1956).
R. nigricans – *M. berndseni* (Eisfelder, 1956; Disney and Evans, 1982).
R. vinosa – *M. scutellaris* (Buxton, 1961).
R. ochroleuca – *M. berndseni, M. scutellaris* (Buxton, 1961) and *M. lutea* (Buxton, 1961; Disney and Evans, 1990).
R. olivacea – phorids (Burla and Bächli, 1968).
R. paludosa – *M. eisfelderae, M. flavicans, M. lata, M. lutea* and *M. scutellaris* (Eisfelder, 1956).
R. pectinata – *M. berndseni* (Dely-Draskovits and Disney, unpublished data) and *M. lutea* (Eisfelder, 1956).
R. pectinatoides – *M. lutea* (Dely-Draskovits and Disney, unpublished data).
R. polychroma – *M. lutea* and *M. scutellaris* (Dely-Draskovits and Disney, unpublished data).
R. puellaris – *M. berndseni, M. lutea* and *M. scutellaris* (Eisfelder, 1956).
R. romellii – *M. flavicans* (Dely-Draskovits and Disney, unpublished data).
R. sardonia – *M. lutea* (Eisfelder, 1956).
R. vesca – *M. berndseni, M. flavicans, M. lata, M. lutea* and *M. scutellaris* (Eisfelder, 1956; and *M. lutea* confirmed Disney and Evans, 1979, 1982, 1990).
R. virescens – *M. berndseni* and *M. lata* (Eisfelder, 1956).
R. xerampelina – *M. eisfelderae, M. flavicans* and *M. lutea* (Eisfelder, 1956).

Strophariaceae

Hypholoma fasciculare – *M. maura* (Dely-Draskovits and Disney, unpublished data).
H. rugocephalum – *M. longipennis* (Robinson, 1971).

Stropharia coronilla – *M. berndseni* (Dely-Draskovits and Disney, unpublished data).

S. semiglobata – *M. berndseni* (Eisfelder, 1956).

Tricholomataceae

Armillaria matsudake – *M. flava* (Sasaki, 1935).

Clitopilus prunulus – *Megaselia* sp. (Brues, 1903b).

Collybia butyracea – *M. scutellaris* (Dely-Draskovits and Disney, unpublished data).

C. dryophila – *M. berndseni* (Dely-Draskovits and Disney, unpublished data).

Dermoloma atrocinereum – *M. scutellaris* (Buxton, 1961).

Hohenbuehelia petaloides – *M. rubella* (Yakovlev, 1986).

Lepista nuda – *M. scutellaris* and *M. uliginosa* (Dely-Draskovits and Disney, unpublished data).

L. sordida – *M. uliginosa* (Disney and Evans, 1990; Dely-Draskovits and Disney, unpublished data).

Leucopaxillus giganteus – *Megaselia* sp. (Canzanelli, 1941).

L. lepistoides – *M. berndseni* (Dely-Draskovits and Disney, unpublished data).

Lyophyllum fumatofoetens – *M. flavicans* (Disney and Evans, 1988).

Mycena galericulata – *M. scutellaris* (Disney and Evans, 1988).

M. pelianthina – *M. scutellaris* (Dely-Draskovits and Disney, unpublished data).

Oudemansiella sp.– *M. rubella* and *M. rubescens* (Yakolev, 1988b).

O. radicata – *M. flavicans* (Eisfelder, 1956), *M. kanoi* (Disney, 1989i) and *M. longipennis* (Robinson, 1971).

Tricholoma argyraceum – *M. scutellaris* (Dely-Draskovits and Disney, unpublished data).

T. edodes – *M. flava* (Kiyoku, 1958).

T. gambosum – *M. berndseni* (Buxton, 1961).

T. luridum – *Megaselia* sp. (Schmitz, 1948b).

T. terreum – *M. flavicans* and *M. scutellaris* (Eisfelder, 1956).

Tricholomopsis platyphylla – *M. eisfelderae* (W. Dreyer and Disney, unpublished data) and *M. scutellaris* (Disney and Evans, 1988).

Aphyllophorales

Cantharellaceae

Cantharellus cibarius – *M. berndseni* (Dely-Draskovits and Disney, unpublished data).

Coniophoraceae

Coniophora puteana – *M. frameata* (Disney and Evans, 1988).

Corticiaceae

Merulioporia (= *Meruliopsis*) *corium* – *M. cinereifrons* (Buxton, 1961).
Plicaturopsis crispa – *M. frameata* (Eisfelder, 1956).

Ganodermataceae

Ganoderma sp.– *M. armata* (Colyer, unpublished notebooks).

Polyporaceae

Bjerkandera fumosa – *M. frameata* (Buxton, 1961).
Fomes fometarius – *Megaselia* sp. (Robinson, 1971).
Meripilus giganteus – *M. frameata* (Colyer, 1954b), *M. hyalipennis* (Chandler, 1966).
Ovinus (= *Caloporus*) *cristatus* – *M. cinereifrons* (Dely-Draskovits and Disney, unpublished data).
Piptoporus betulinus – *M. nigra* (Disney and Evans, 1994).
Polyporus squamosus – *M. frameata* (Buxton, 1961).
P. volvatus – *M. polyporicola* (Borgmeier, 1966a).
Poria medulla-panis – phorid (Burla and Bächli, 1968).
Scutiger (= *Albatrellus*) *ovinus* – *M. cinereifrons* (W. Hackman and Disney, unpublished data).
Trametes pecki – *M. fungicola* (Coquillett, 1895).

Thelephoraceae

Thelephora terrestris – *M. frameata* (Disney and Evans, 1982).

Lycoperdales

Lycoperdaceae

Calvatia excipuliformis – *M. bovista* (Disney and Evans, 1990).
Langermannia gigantea – *M. bovista* (Gimmerthal, 1848; Disney and Evans, 1982).
Lycoperdon granulosum – phorid (Burla and Bächli, 1968).
L. perlatum – phorids (Burla and Bächli, 1968).
Vascellum pratense – *Megaselia nigra* (Disney and Evans, 1994).

Phallales

Phallaceae

Phallus impudicus – *M. rufipes* (Smith, 1956). *Conicera dauci* has been recorded visiting this fungus (Smith, 1956; Nielsen, 1968).

Sclerodermateles

Sclerodermataceae

Scleroderma citrinum – *M. berndseni* (Dely-Draskovits and Disney, unpublished data), *M. hirtiventris* (Disney and Evans, 1988) and *M. lutea* (Eisfelder, 1956).

Tremellales

Tremellaceae

Exidia truncata (= E. glandulosa) – *Megaselia nigra* (Disney and Evans, 1994).

ASCOMYCOTINA

Clavicapitales

Clavicapitaceae

Paraisaria (= conidial *Cordyceps*) *dubia* growing on dead larva of swift moth (*Hepialus lupulinus*, Hepialidae) – *M. hirtiventris* (Disney and Evans, 1990).

Pezizales

Humariaceae

Aleuria sp. – *M. pulicaria* (Yakovlev, 1980).

Morchellaceae

Gyromitra esculenta – *M. albicans, M. discreta, M. mixta, M. pulicaria, M. rubella, M. rubescens* and *M. rufipes* (Yakovlev, 1986).
G. gigas – *M. rubescens* (Yakovlev, 1986).
Morchella esculenta – *M. flavicans* (Eisfelder, 1956; Dely-Draskovits and Disney, unpublished data).
M. rotunda – *M. flavicans* (Dely-Draskovits and Disney, unpublished data).
M. conica – *M. flavicans* (Dely-Draskovits and Disney, unpublished data), *M. mixta* and *M. rubella* (Yakovlev, 1986).
Verpa bohemica – *M. mixta, M. rubella* and *M. rubescens* (Yakovlev, 1986).

Pezizaceae

Peziza sp.– *M. rubescens* (Yakovlev, 1986).
P. micropus – *M. flava* (Disney and Evans, 1982).

Table 3.1 Rearings of *Megaselia* species from fungi in Europe. F = fungus families; G = fungus genera; S = fungus species

Megaselia species	Numbers of host fungi				
	F	G	S	S/F	S/G
M. albicans	1	1	1	–	–
M. armata	1	1	1	–	–
M. berndseni	12	20	46	3.8	2.3
M. bovista	2	3	6	3.0	2.0
M. cinereifrons	2	3	3	–	–
M. discreta	2	2	2	–	–
M. eisfelderae	3	3	4	–	–
M. flava	5	5	9	1.8	1.8
M. flavicans	8	14	29	3.6	2.1
M. frameata	6	8	8	1.3	1.0
M. giraudii	2	2	2	–	–
M. halterata	1	1	3	–	–
M. hirtiventris	3	3	10	3.3	3.3
M. hyalipennis	1	1	1	–	–
M. lata	6	9	24	4.0	2.7
M. latior	2	2	3	–	–
M. lutea	5	7	41	8.2	5.9
M. lutescens	2	2	3	–	–
M. maura	2	2	2	–	–
M. mixta	2	6	6	3.0	1.0
M. nigra	5	5	17	3.4	3.4
M. plurispinulosa	1	1	1	–	–
M. pulicaria	3	4	4	–	–
M. pygmaea	1	1	1	–	–
M. rubella	8	16	20	2.5	1.3
M. rubescens	6	10	16	2.7	1.6
M. rufipes	7	8	9	1.3	1.1
M. scutellaris	7	16	32	4.6	2.0
M. spinicincta	2	3	3	–	–
M. sylvatica	2	2	4	–	–
M. uliginosa	1	1	2	–	–
M. virilis	2	3	4	–	–

P. repanda – *M. flava* (Buxton, 1961; Disney and Evans, unpublished data).
P. varia – *M. flava* (Disney and Evans, 1982).
P. vesiculosa – *M. flava* (Disney and Evans, 1994).

Tuberaceae

Tuber aestivum – *Conicera similis* (Disney and Evans, 1988).

Sphaerales

Xylariaceae

Xylaria (= *Hypoxylon*) *multiforme* – *Megaselia frameata* (Buxton, 1961).

It is immediately apparent from this review that the bulk of the records are from Europe and most concern the genus *Megaselia*. For the rest of the world the reporting of named Phoridae from named fungi can only be described as sporadic, apart from the pioneering systematic rearings carried out in North America by Bruns (1984). The available data do not yet allow broad generalizations. However, despite the bias in the selection of fungus species used in rearing experiments (with a marked preference for Agaricales) and despite the bias in the naming of the phorid species (leading to a dearth of records for species whose recognition has been, or continues to be, beset with taxonomic difficulties), the data for the rearings of *Megaselia* in Europe merit a summary tabulation. The data in Table 3.1 indicate that some species are catholic in their choice of hosts. However, the high ratio of species to the number of genera and also to the number of families for *M. lutea* indicates this species is less catholic than is *M. scutellaris*. Such differences may reflect differences in the feeding habits of the larvae. Some, such as *M. nigra*, are known to feed on fungus tissue. Some of the more catholic species, at the generic and family levels, may be predatory or parasitoid upon other fungivorous insects.

3.4 PLANT-FEEDING/PLANT-BREEDING LARVAE

Numerous reported rearings refer to saprophage species exploiting plants which are decaying or locally damaged. These records include *Chonocephalus depressus*, *C. globipygus*, *C. hibisci* and *C. wirthorum* (Disney, 1980h, 1991h; Paulian, 1958); *Conicera formosensis* (unpublished data); *Megaselia rufipes* (e.g. Parmenter, 1965; Szwejda, 1980), *M. seticauda* (Borgmeier, 1962c) and *M. scalaris* (e.g. Borgmeier, 1969a; Lever, 1944; Malloch, 1912b); *Metopina oligoneura* bred from residues of the leaves and roots of *Beta vulgaris* (Froese, 1992c); *Plethysmochaeta trinervis* (Le Pelley, 1959); and *Puliciphora borinquenensis* (unpublished data).

Several records refer to phorids in leaf mines. Most of these are for familiar saprophage species and therefore almost certainly represent secondary invasions of pre-existing leaf mines made by other organisms. The phorid species recorded include *Megaselia giraudii* (Wood, 1911), *M. limburgensis* in the leaf mine of the agromyzid *Phytomyza soenderupi* on *Caltha palustris* (Ranunculaceae) (Hering, 1951) and *M. rufipes* (e.g. Malloch, 1909b; Parmenter, 1965). In the case of *M. arietina* more than 60% of the available leaves of its host plant, the chickpea (*Cicer arietinum*, Leguminosae), have been reported to be mined in Macedonia (Fig. 3.7). However, the plants were already subjected to infestations of the leaf-spot fungus *Ascochyta rabiei* and

Figure 3.7 A chickpea leaf mined by *Megaselia arietina*. (Photo by Lj. Čakar.)

the rust fungus *Uromyces ciceri arietini* (Čakar and Disney, 1991). Furthermore the adult females ingest the spores of the *Uromyces* (6.3) and probably transmit this fungus from plant to plant (7.11.1). The extent to which *M. arietina* will attack plants not debilitated by fungus attack requires investigation.

Familiar saprophage species have also been reared from a variety of seeds. The records include *Dohrniphora cornuta* (Malloch, 1912b), *Megaselia rufipes* (e.g. Lundbeck, 1922; Öberstein, 1916), and *M. scalaris* from groundnuts (*Arachis hypogaea*), from a seed of *Koompassia malaccensis* (Leguminosae) and from stored seeds of *Nephelium lappaceum* (Sapindaceae) (unpublished data). Of greater significance are the observations of Robinson (1970) on the larvae of *M. winnemanna* feeding on the seeds of the previously healthy capsules of *Viola pensylvanica*, *V. pubescens* and *V. sordida* (Violaceae). The larvae quit the capsules to pupariate in the soil. Hering (1951) reported larvae of a *Megaselia* species feeding on the embryos within the seeds of a *Pinus* species. I have examined a *Chonocephalus* male (probably an undescribed species) reared (by Karen Kool) from a seed of a *Pelaquium* species (Sapotaceae) in Kalimantan. Larvae of a *Megaselia* species have been recorded feeding on the developing and mature seeds, together with the placenta of the pods, of *Capsicum annuum* (Solanaceae) in Nigeria (Iheagwam and Nwankiti, 1980).

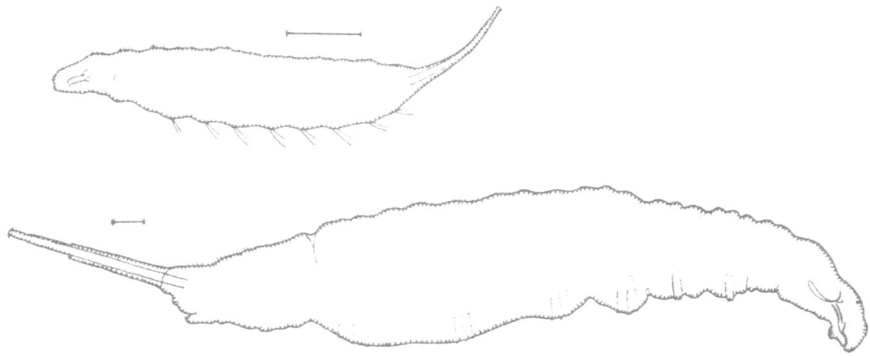

Figure 3.8 The first (above) and last (below) larval instars of *Chonocephalus heymonsi*. (After Borgmeier, 1935g.)

Robinson (1967) reported the females of a *Megaselia* species laying 1–5 eggs into the unopened flowers of *Smilacina racemosa* (Liliaceae). The first-instar larvae feed on the enclosed leaves and still-soft flowers, for 1–2 days. The second and third instars feed on any available tissue (lasting 4–5 days and 10–15 days respectively) before leaving the plant to pupariate on the ground. The larvae of *M. aristolochiae* develop in the flowers of *Aristolochia labiata* (Aristolochiaceae) (Hime and Costa, 1985). Both these species are truly phytophagous. In the case of the two species invading the leaves or roots of chickpeas (*Cicer arietinum*) the position is not so clear. In India the invasion of the root nodules by the larvae of *Metopina ciceri* can exceed 70% of the available nodules (Disney, 1988d), but the larvae could be feeding mainly on the nitrogen-fixing bacteria rather than the nodule tissues. The leaf-mining species is discussed above.

Ears of green corn (maize – *Zea mays*, Gramineae) have been reported suffering economic damage by larvae of *Megaselia scalaris* in Texas, the eggs being deposited on the silks just beneath the edge of the tip of the husk (Walter and Wene, 1951). In Mexico *Megaselia seticauda* has been reared from an ear of green corn (Borgmeier, 1962c).

Fruits are exploited by phorid larvae, especially by the genus *Chonocephalus*. The larvae of some species (e.g. *C. heymonsi*) have a telescopic tail, bearing the posterior spiracles at the tip (Fig. 3.8), whereas other species have short tails. Fragmentary evidence suggests the long-tailed larvae invade the more liquid parts of over-ripe fruits, whereas short-tailed species inhabit the drier parts, especially when a short-tailed and a long-tailed species are co-existing in the same fruit. Phorids breeding in fruits are reviewed below.

CACTACEAE

Opuntia dilleni – *Puliciphora borinquenensis* (Paulian, 1949).

CARICACEAE

Carica papaya (paw-paw) – *Chonocephalus blackithorum, Chonocephalus* species B and D (Disney, 1986c).

CUCURBITACEAE

Cucumis melo (melon) – *Megaselia* sp. (Nishida, 1955).

LEGUMINOSAE

Gleditsia sp. – *Chonocephalus* sp. (Paulian, 1958).

MELIACEAE

Dysoxylum gaudichaudianum – *Chonocephalus* species D (of Disney, 1986c). The females arrive at the fruits on the third or fourth day from the initiation of decay, the fruit often being still on the tree (Bryan Turner and Disney, unpublished data).

MORACEAE

Artocarpus altilis (breadfruit) – *Chonocephalus heymonsi, C. pallidulus, C. simiolus* and *C. wirthorum* (Hardy and Beyer, 1964, Disney, unpublished data).

Artocarpus heterophyllus (jackfruit) – *Chonocephalus* sp., *Puliciphora* sp. (Winder, 1977).

Ficus species (figs) – *Chonocephalus* species B and C (Disney, 1986c). *Megaselia dentata, M. ficaria, M. parvorata, M. surophila* and *M. vorata* have been reared from nine species of fig in South Africa (Compton and Disney, 1991). There is no evidence of any host specificity and it is thought that the larvae may be preying upon the developing fig wasps (Agaonidae) (3.5.6(e)).

MUSACEAE

Musa sp. (banana) – *Chonocephalus subglaber* (Bohart, 1947).

PALMAE

Borassus sambiranensis – *Chonocephalus dimakae* (Paulian, 1958).

SAPINDACEAE

Alectryon macrococcus – unidentified phorids (Swezey, 1934).

SOLANACEAE

Solanum sodomeum – *Plethysmochaeta vectabilis* (Brues, 1924b).

3.5 PREDATOR, PARASITOID AND PARASITIC LARVAE

Parasitic or parasitoid larvae are clearly common among the Phoridae. While the details of life histories are poorly known for most cases, the evidence taken together indicates that considerable sophistication has been achieved in terms of synchronization with the host, location of the right host, and other aspects.

The term 'parasitoid' has been used in a variety of senses in the literature. In this book the term is used in the sense given by Lincoln *et al.* (1982), who define a parasitoid as 'an organism with a mode of life intermediate between parasitism and predation; usually a species of hymenopteran in which the larva feeds within the living body of another organism eventually causing the death of the host'. Recently Eggleton and Gaston (1990) have critically reviewed the concept and defined it a little more broadly as 'an organism which develops on or in another single ("host") organism, extracts nourishment from it, and kills it as a direct or indirect result of that development'.

How parasitoid larval Phoridae suppress or evade the defence reactions of their hosts is an unexplored field.

The following review is primarily concerned with the recorded hosts or prey of the different species. Where fuller details are available these are indicated.

3.5.1 ANNELIDA

Lumbicridae (earthworms)

While *Triphleba nudipalpis* has been reared from a dead earthworm (Colyer, in Disney, 1979f), it is evident that some Phoridae parasitize living worms. Just as earthworms parasitized by blowfly (Calliphoridae) larvae are known to autotomize (deliberately shed) tail-end segments that are infested (Thomson and Davies, 1973), so a similar case has been observed with a phorid (Disney, 1991d). This habit complicates attempts to assess infestation rates in the field.

The established cases of parasitization involve an unknown species of earthworm parasitized by *Diplonevra pilosella* (Colyer, 1950) and *Lumbricus terrestris* parasitized by *D. nitidula* (Disney, 1991d).

3.5.2 MOLLUSCA

Whether any of the species that breed in dead snails (section 3.2.4(a)) will attack living molluscs is not known. Species known to attack living molluscs are detailed below.

Gastropoda (slugs and snails)

(a) Egg predators and parasitoids

The detailed study by Robinson and Foote (1968) established that the first-instar phorid larvae are essentially parasitoids, feeding on the perivitelline fluid of the slug egg. The second-instar attacks the slug embryo and then leaves the remains of the egg to moult to the third instar. This initially feeds on gelatinous material surrounding the egg capsule of another egg. After consuming the egg contents it then preys on at least four more eggs before it is ready for pupariation. When two or more second instar larvae enter the same egg only one larva survives. The following host eggs have been recorded.

Arion sp. attacked by *Megaselia ciliata* (Disney, 1979f), *Deroceras* sp. by *M. ciliata* (Disney, 1977b, 1979f), and *D. laeve* by *M. aequalis* (Robinson and Foote, 1968). Larvae of the closely related *M. nasoni* prey on spider eggs (3.5.3(2)).

(b) Snail predators and parasites

Juvenile *Achatina fulica* were reported to be parasitized by the type series of *Megaselia javicola* (Beyer, 1959h). The citrus tree snail, *Drymaeus dormani*, is preyed upon by larvae of a *Megaselia* sp. and up to five flies have been reared from a snail, having been 'suspected of being parasitic' (Muma, 1954, 1955). *Vitraea contracta* and *V. crystallina* are definitely parasitized by *M. fuscinervis* (Disney, 1982h and unpublished data). The larvae pupariate in the shell of the host (Fig. 4.2).

3.5.3 ARACHNIDA

(1) Acari (mites and ticks)

Megaselia scalaris has been reared from dead ticks (*Boophilus microplus*, Ixodidae) (Rocha *et al.*, 1984).

(2) Araneida (spiders)

Megaselia giraudii has been reared from a moribund spider (*Theridion redimitum*, Theridiidae) (Bristowe, 1941). A *Megaselia* sp. has been reared from a spider

(*Trochosa terricola*, Lycosidae) paralysed by a pompilid wasp (*Anoplius viaticus*) (Bristowe, 1928).

The rearing of *Apocephalus borealis* from *Latrodectus mactans* (Theridiidae) seems to represent a true case of parasitization (Borgmeier, 1963d). Further investigation is needed.

Three females of *Megaselia arachnophila* were reported from a nest-tunnel of *Trechona venosa* (Dipluridae) in Brazil (Bristowe, 1925; Borgmeier, 1931b). What these flies were doing is not clear. It seems most probable that they were seeking out the eggs of the spider (see below).

The larvae of a number of Phoridae are now known to be predators on spider eggs. The known cases are detailed below.

Araneidae

Argiope sp. (*A. picta* or *A. reinwardti*) – *Megaselia argiopephaga* (Disney, 1982m).

A. aemura – *M. araneivora* (Gotô, 1985a).

A. aurantia – *Megaselia* sp. – Kaston and Jenks, 1937.

Gasteracantha cancriformis - *Phalacrotophora epeirae*. 43.7% of 216 egg sacs were infested, the mean number of larvae per sac ranging from 7 to 42. The maximum number of adults successfully reared from one sac was only 17, suggesting there is some intraspecific competition between the larvae. The female flies laid 1–61 eggs, but the maximum on one sac was 34 (Muma and Stone, 1971).

Nuctenea sclopetaria – *P. epeirae* (Auten, 1925).

N. sericata – *P. epeirae* (Brues, 1903b).

Linyphiidae

Pityohyphantes costatus – *P. epeirae* (Manuel, 1984).

Metidae

Meta menardi – *Megaselia melanocephala*, 3–8 pupae per sac (Decou-Burghele, 1961).

Salticidae

Phidippus audax – *Phalacrotophora epeirae* (Jones, 1940, Manuel, 1984).

Theridiidae

Probably *Enoplognathus* or *Robertus* – *Megaselia nasoni*. There were 1–5 (mean 2.4) flies reared per sac (Disney and Evans, 1980). The closely related *M. aequalis* and *M. ciliata* are predators of slug eggs (see above).

Probably *Enoplognathus* or *Robertus* plus egg sacs of unknown spiders – *M. pulicaria*. 1–20 (mean 5.0) adult flies reared per sac. Recorded infestation rates

11.6–20.0%. In some cases both *M. nasoni* and *M. pulicaria* were reared from the same egg sac (Disney and Evans, 1980, Disney, 1989a and unpublished data).

Unknown spiders

- *Megaselia diptychogastra* (Schmitz, 1940a).
- phorid larvae of Sulawesi species (unpublished data).

(3) Opiliones (harvestmen)

Borgmeier (1931b) reported nine females of *Megaselia opilionidis* pursuing the harvestman *Opiliones laniatores*. This behaviour requires further investigation. He also (1969a) recorded a female *M. luteicauda* on a dead opilionid.

3.5.4 CHILOPODA (CENTIPEDES)

Borgmeier (1961a) recorded *Megaselia juli, M. luteizona* and *M. usticolor* on a dead *Scolopendra* sp. It seems most likely that these flies were feeding on fluids from the centipede's corpse (6.3).

3.5.5 DIPLOPODA (MILLIPEDES)

Megaselia juli was the first phorid recorded ovipositing into millipedes (Brues, 1908c; Knab, 1913; Brues, 1932a; Meyers, 1934). The observations of Picard (1930) and Berland (1933) suggest that pre-existing minute punctures in the integument of the host are necessary if the phorid larvae are to obtain entry. Another possibility is that the millipedes may be vulnerable just after moulting, when their integuments are still soft. Further observations are required. Some phorid adults are found on the backs of giant millipedes (6.4.2). The larval biology of these species remains unknown. The following list details the known host and phorid parasitoid records.

Ommatoiulus diplurus – *Megaselia elongata* (P. Bailey, A. Kirk-Spriggs and Disney, unpublished data).

O. moreletii – *M. elongata* (Baker, 1985, and P. Bailey and Disney, unpublished data).

Paraiulus sp. – *M. juli* females darting at millipede 'wriggling and twisting on dead leaves' (Banks, 1911).

Schizophyllum sabulosum – *M. elongata* (Picard, 1930).

Spirobolus sp. – *M. juli* (Borgmeier, 1961a, 1963d).

S. marginatus – *M. juli* (Schmitz, 1939c).

S. spinigerus – *M. juli* (Remy, 1950).

3.5.6 INSECTA

3.5.6 (a) Blattodea (cockroaches and allies)

In America Robinson (1975b) reared *Megaselia scalaris* from cultures of a variety of cockroach species. In crowded conditions some larvae become facultative parasites (3.1). The same species and *Puliciphora borinquenensis* breed in cultures of cockroaches in England (e.g. Miller, 1979a). Edmunds (1952) reared a *Megaselia* species from the egg pods of *Parcoblatta pennsylvanica*.

3.5.6 (b) Coleoptera (beetles)

One of the earliest records of parasitoid habits by a phorid is the report of *Phalacrotophora fasciata* parasitizing pupae of ladybird beetles (Coccinellidae) (Curtis, 1833). The wide range of coccinellid species (see below) parasitized by this species, and by *P. berolinensis*, taken with the small number of records for hosts from other families of beetles, is striking. Likewise the range of Lampyridae parasitized by *Apocephalus antennatus* (see below), with no records from other beetles, also calls for comment. It seems likely that olfactory cues peculiar to the beetle families in question are responsible for this family specificity, despite the host-species catholicity within the family of the fly's choice.

There are numerous records in the literature of saprophage phorid species reported as 'parasites' of beetle larvae, pupae or adults (e.g. Eisenschmidt, 1958). Most of these can be discounted as cases of facultative parasitoidism (cf. 3.5.6(g) below). However, in a few cases the reported observations justify the claim of parasitoidism, as in the case of *Megaselia aletiae* attacking *Scolytus ventralis* (see below). The following review collates the known records considered to be cases of obligate parasitoidism or predation.

Cantharidae (soldier beetles)

Chauliognathus fallax – *Apocephalus mortifer* (Borgmeier, 1958).

Cerambycidae (long-horned beetles, timber beetles)

Brues (1907c) reported a mating pair of *Phalacrotophora braunsi* in an old willow stump frequented by these beetles and a large carpenter bee (3.5.6(e)).

Chrysomelidae (leaf beetles)

Chrysomela aenea – *Megaselia rubricornis* larvae parasitize the pupae, 1–5 eggs (usually 2) being laid per pupa. The first-instar larvae enter the host pupa, usually at the boundary of two segments or at the base of a leg. Infestation rates of between 0.30 and 0.80 of available pupae have been recorded (Kanervo, 1946).

Diabrotica balteata – *Phalacrotophora* sp. larvae preying on eggs and parasitizing pupae, in a laboratory culture (Saba, 1970). The significance of this report requires further investigation.

Coccinellidae (ladybirds)

With regard to the species parasitizing ladybird-beetle pupae most authors (see below) record little more than the rearing of the flies from the named hosts. Few give any quantitative data. Menozzi (1927) recorded rates of parasitization of pupae of *Adalia 2-punctata* by *Phalacrotophora fasciata* of 0.48; and of 0.30 for *Chilocorus 2-pustulatus* parasitized by the same species. Ford (1979) recorded 10 out of 12 (0.83) pupae of *Coccinella 7-punctata* to be parasitized by *P. fasciata*. Near St Petersburg (Leningrad) Lipa and Sem'yanov (1967) recorded rates of no more than 0.05 for *P. fasciata* parasitizing pupae of this species and likewise for pupae of *Calvia 14-guttata*; but a rate of 0.25 for pupae of *Myzia oblongoguttata*. About half the pupae of *Aphidecta obliterata* collected by Wylie (1958) were parasitized by *P. berolinensis*. Limited data on *Phalacrotophora* obtained in Wiltshire in August 1977 (Disney, 1979f) indicated rates of parasitization of *Adalia 2-punctata* by *P. berolinensis* of about 0.30 and by *P. fasciata* of 0.13. The rate for *Coccinella 7-punctata* parasitized by *P. fasciata* was 0.20.

In New Caledonia the recorded rates of parasitization of the pupae of the introduced *Olla v-nigrum* by *P. quadrimaculata* ranged from 0.15 at one locality to 0.79 at another, with a mean of 0.39. The mean number of fly pupae per host pupa was 4.3 (range 2–7). In laboratory rearings a mean of 0.79 flies emerged per ladybird pupa (Disney and Chazeau, 1990). Only 2 out of 64 (0.03) pupae of *Cleobora mellyi* collected by New (1975) were infested by *Phalacrotophora* sp.

Lichtenstein (1920) gives details of the natural history of *Phalacrotophora fasciata*. He noted that the eggs were laid between the legs of the host pupae. The larvae fed for two days in the pupa, emerging to pupariate. Wylie (1958) gives details for *P. berolinensis*. The eggs were laid externally on the ventral surface of the host, beneath its wing pads. The larvae left the host to pupate in the soil, within 2 cm of the surface. Richards (1926) noted that parasitized pupae may be known by a change in the normally white parts to a yellowish white.

Martelli (1914) observed the adults of *P. fasciata* sucking fluid from pupae of *Psyllobora 22-punctata*. Lichtenstein (1920) confirmed these observations. He observed *P. fasciata* sucking fluid from both larvae and pupae of *P. 22-punctata* and *Vibidia 12-guttata*. Richards (1926) observed the same species sucking from a pupa of *Adalia 2-punctata*. Colyer (1952a, 1954g) observed *P. berolinensis* sucking at the ventral parts of the pupae of *Aphidecta obliterata*. Delucchi (1953) also observed females of this species imbibing haemolymph from injured pupae of the same ladybird species. These authors have suggested that this feeding behaviour serves either as a source of a protein-rich meal or as a means of creating vulnerable points on the integument of the host, allowing penetration by the first-instar larva of the fly. It seems most probable that both functions are served by this behaviour. What is clear is that these small punctures, and the consequential loss of small amounts of haemolymph, do not prevent either the development of unparasitized pupae or the development of flies from parasitized pupae

More detailed data on infestation rates for *P berolinensis* and *P fasciata* in England have been provided recently (Disney *et al* , 1994), along with other observations These confirm and extend the information reviewed above Thus it was confirmed that the eggs are normally laid on the host pupa, usually just after the splitting of the pre-pupal larval skin, but sometimes before The eggs are laid between the legs, either on the surface of the pupa or internally Occasionally they are inserted through the dorsal surface One to five eggs are laid at a time, but usually one or two The largest number of phorid larvae recorded was 22 in a pupa of *Coccinella 7-punctata* *Calvia 14-guttata* and *C 7-punctata* were preferred to *Adalia* species However, no evidence of niche separation emerged from these data Indeed three cases of a pupa (of *A 2-punctata*) being parasitized by both phorid species were recorded, thus indicating that the single case recorded previously (Disney, 1979f) was not a freak occurrence Females of *P fasciata* were observed feeding on pupae of *C 7-punctata* and those of *P berolinensis* on pupae of *A 2-punctata* The incision was made in the dorsal surface of abdominal segments 1 or 2 Feeding on the haemolymph took up to 27 mins Phorids subsequently emerged from these pupae

It is evident that the larvae and pupae of these beetles may be parasitized by a number of phorid species, some of which are clearly catholic in their choice of host species, within the restriction of being confined to coccinellids The known host records are summarized in Tables 3 2 and 3 3

Table 3.2 Records of Coccinellidae parasitized by *Phalacrotophora berolinensis* and *P fasciata* The numbers refer to the published sources of the records, listed at the foot of Table 3 3

Coccinellid host	*P berolinensis*	*P fasciata*
Adalia 2-punctata	8, 18, 37	5, 8, 22, 29, 30, 37
A 10-punctata	37	37
A variegata	–	18, 20
Anatis ocellata	18	18, 25, 32
Aphidecta obliterata	4, 5, 7, 18, 33	–
Calvia 14-guttata	37	18, 32, 35, 37
Chilocorus renipustulatus	14	18
C 2-pustulatus	–	8, 22
Coccinella 7-punctata	26, 37	8, 10, 17, 22, 27, 37
Exochomus 4-pustulatus	14	5
Harmonia 4-punctata	4, 17, 32	–
H 14-punctata	–	13
Myzia oblongoguttata	–	4, 18, 29, 35
Psyllobora 22-punctata	–	16, 21
Vibidia 12-guttata	–	16
Total number of host species	9	13
Total number of host genera	8	10

Table 3.3 Records of other Phoridae that attack Coccinellidae

Coccinellid host	Phorid species	Source (see foot of table)
Chilocorus distigma	*Megaselia* sp	15
The eggs of this fly were laid on the dorsal faces of the host larvae The larvae of the fly penetrated the integument of the host larva The host usually died in the pupal stage There was a mean of 4 7 (range 1–9) fly pupae produced from each infested host pupa The fly puparia are arranged in a regular fan around the anterior end of the host's dead pupa		
C 4-maculatus	*Megaselia* sp	31
Cleobora mellyi	*Phalacrotophora* sp	24
Two out of 64 larvae were infested, giving rise to 5 adult flies		
Coccinella 11-punctata	*Phalacrotophora* sp	23
Epilachna sp (cf *humeralis*)	*Megaselia* spp	1
Larvae of three fly species were reported preying on the beetle's eggs, destroying a high percentage of them		
E varivestis	*Megaselia*	11
The fly larvae, several per host, parasitize the host larvae and pupae		
Harmonia axyridis	*Phalacrotophora* sp	36, 38
In an initial study (38) in Japan, infestation rates ranging from 0 02 ($n = 68$) to 0 18 ($n = 269$) were recorded and means of 4 9 ($n = 34$) and 4 1 ($n = 192$) larvae per host pupa In a subsequent study an infestation rate of 0 18 ($n = 249$ pupae) was reported, but some of these were destroyed by cannibalism by the ladybirds (36)		
H eucharis	*P indiana*	6
H brieti	*Phalacrotophora* sp	28
Hippodamia 11-notata	*Phalacrotophora* sp	22
Neda marginalis	*P nedae*	3, 19
Neocalvia anastomozans	*P nedae*	2, 3
Olla v-nigrum	*P quadrimaculata*	9

1 – Berry and Parker, 1949, 2 – Borgmeier, 1961, 3 – Borgmeier, 1963, 4 – Colyer, 1952, 5 – Colyer, 1954, 6 – Colyer, 1961a, 7 – Delucchi, 1953, 8 – Disney, 1979, 9 – Disney and Chazeau, 1990, 10 – Ford, 1979, 11 – Howard and Landis, 1936, 12 – Iperti, 1964, 13 – Iperti, 1965, 14 – P Kearns and Disney, unpublished, 15 – Le Pelley, 1959, 16 – Lichtenstein, 1920, 17 – Lundbeck, 1922, 18 – Majerus and Kearns, 1989, 19 – Malloch, 1912b, 20 – Martelli, 1913, 21 – Martelli, 1914, 22– Menozzi, 1927, 23 – Merrit-Hawkes, 1927, 24 – New, 1975, – 25 Parmenter, 1965, 26 – Prescher and Disney, unpublished, 27 – Pugh, 1941, 28 – Rao and Ghani, 1972, 29 – Richards, 1926, 30 – Richerson, 1970, 31 – Schmutterer, 1972, 32 – Van Emden, 1950, 33 – Wylie, 1958, 34 – Data on museum label (Oxford), 35 – Lipa and Sem'Yanov, 1967, 36 – Osawa, 1992, 37 – Disney *et al* , 1994, 38 – Maeta, 1969

Lampyridae (fireflies, glow worms)

In North America Lloyd (1973) reports the rearing of *Apocephalus antennatus* from the adult beetles of the eight species detailed below. From 1 to 6 larvae emerged from each beetle, either from the tip of its abdomen or from between the pronotum and the head. The recorded hosts were *Photinus consanguineus*, *P. curtatus* x *marginellus*, *P. ignitus*, *P. macdermotti*, *P. obscurellus*, *Photuris* sp., *Photuris congener* and *Pyractomena borealis*.

In South America Lampyridae are parasitized by *Apocephalus longistylus* (Brian Brown, *in litt.*).

Scarabaeidae (chafers and dung beetles)

Brischke (1868) reared the normally saprophage *Megaselia rufipes* from a living adult *Osmoderma eremita*. Brown (1988) reported a *Menozziola* species in a burrow of a *Peltotrupes* species and Robinson (1978b) reported *Megaselia marginalis* in the cell of *P. youngi*, the cell being almost 2 m deep. Robinson (1971) reported *M. cavernicola* to be a parasite of *Phyllophaga implicata*. The latter requires further investigation, along with critical study of these other reported associations with Scarabaeidae.

Scolytidae (bark beetles)

In North America Ashraf and Berryman (1969) reported that 1.2% of adults in a population of *Scolytus ventralis* had been parasitized by *Megaselia aletiae*, there being 1–4 larvae in the host's head and abdomen. Parasitized females failed to lay eggs. *M. aletiae* is normally a saprophage (3.2.4(b)).

3.5.6(c) Diptera (flies, midges and gnats)

Diptera, other than scuttle flies, are proving to be the hosts or prey of Phoridae. The number of cases now authenticated suggests there are probably many more to be discovered. Other reported cases need further investigation. For example, MacDougall (1920) reported phorid larvae, mainly under cover of the leaf sheaths, on oat plants (*Avena sativa*) infested with frit fly larvae (*Oscinella frit* – Chloropidae). The identity and biology of these larvae, and the relation (if any) between them and the chloropid larvae, require investigation.

Bibionidae (St Mark's, fever, heather and blossom flies)

Bibio marci larvae are parasitized by larvae of *Borophaga incrassata* (Morris, 1922). *Penthretia holosericea* larvae are parasitized by *B. germanica* (Gemesi and Disney, 1991).

Cecidomyiidae (gall midges)

The females of *Megaselia submarginalis* oviposit into the galls of *Contarinia negundifolia* on the leaves of the box elder (*Acer negundo*). The larvae prey upon the cecid larvae and vacate the gall to pupariate in the soil (Robinson, 1970; Robinson and Brown, 1993).

Schmitz (1957a) reported *Megaselia pleuralis* reared from a gall of *Dasyneura urticae* infested by the sciarid *Lycoriella mali*.

Psychodidae (moth flies)

The eggs, larvae and pupae of *Psychoda alternata* are preyed upon by the larvae of *Dohrniphora cornuta* if the population levels of the psychodid, in trickling-filter sewage beds, exceed critical levels (Kloter *et al.*, 1977).

Sciaridae (lesser fungus gnats or sciarid midges)

Megaselia sciaricida was the first reported case of a phorid parasitizing larval Sciaridae (Leefmans, 1927; Schmitz, 1932d). Circumstantial evidence suggests many sciarid larvae or pupae may be attacked by a range of phorid species. Established cases are listed below.

Bradysia bicolor – Megaselia flavicoxa
In his survey of Diptera in vole burrows, Hackman (1963) found *M. flavicoxa* only in burrows also occupied by *B. bicolor*. It has subsequently been discovered that the larvae of this phorid are true parasitoids of this *Bradysia* species. There is one phorid larva per host larva. The latter spins its cocoon, in preparation for its pupation, but is then killed by the phorid larva, which proceeds to purloin the host's cocoon for its own pupariation (Disney, 1976b).

Bradysia confinis – Megaselia mallochi
The pupae of this midge are preyed on by larvae of the phorid. A single larva will consume a series of pupae. 50% of the prey pupae may be attacked. There may be 2–3, or even 4, larvae attacking one pupa. In one estimate there were 16.3 prey pupae to every phorid larva and 8.3 adult midges emerging for every adult phorid that emerged (Krespi *et al.*, 1985; Deleporte, 1986).

Trichosia caudata
Tuomikoski (1957) first reported this species to be parasitized by a phorid. Subsequently it was reported that larvae of *Megaselia luminifrons* and *M. obscuripennis* were 'predators' of this species (Schmitz and Beyer, 1974). However, it is now established that *M. obscuripennis* larvae are true parasitoids. There is one larva per host larva. The mature phorid larva vacates the remains of the host to pupariate (Disney, 1980g).

Trichosia trochanterata larvae are also parasitized by *M. obscuripennis* larvae (Disney, 1980g).

Tipulidae (craneflies)

Edwards (1929) reported parasitization of a *Tipula* larva by larvae of a *Megaselia* species (which F. W. Edwards determined as '*M. brevicostalis*'; the specimens having been lost, I have been unable to check the identification). Grandi (1951) reported *M. paludosa* parasitizing a tipulid larva, which is confirmed below.

Tipula luna larva – phorid (possibly *Megaselia paludosa*) (Carter, 1977). A single phorid larva was found in the fat body of the host larva.

Tipula paludosa larvae – *Megaselia paludosa* (Coggins, 1970; Carter, 1977). Carter found 3 out of 12 larvae to be parasitized. There were 3 phorid larvae in one of these which he dissected. The other 2 larvae were kept alive. Two phorid larvae emerged from one and 8 from the other. The pupal period was about 9 days. The host larvae were still live and healthy after 14 days, when they were sacrificed for other purposes. It is evident, however, that the *M. paludosa* larvae were truly parasitic rather than parasitoid.

3.5.6(d) Homoptera (bugs)

Circumstantial evidence suggests Homoptera constitute the hosts or prey of scuttle flies more frequently than is currently realized. The observations of Yarkulov (1972) on phorid predation on root aphids highlight that attacks on Homoptera may be concealed from view. The specific records are detailed below.

Aphididae (aphids, blackflies, greenflies)

Phora holosericea larvae prey on *Brachycaudus helichrysi* (Yarkulov, 1972).

Coccidae (scale insects)

Megaselia rufa lays its eggs among the eggs under a female scale insect. Its larvae initially prey upon the eggs and subsequently parasitize the coccid itself (Schmutterer, 1952). Recorded hosts are *Eulecanium mali, E. tiliae* and *Parthenolecanium corni* (Schmutterer, 1952; Goantsa, 1966).

Eriococcidae (scale insects)

Kermes sp. is parasitized by *Megaselia brokawi* (= *Kerophora brunnea*) (Disney, 1992b).

Margarodidae (scale insects)

Syneura cocciphila parasitizes *Icerya purchasi* (and probably *I. brasiliensis* also) and *Neocoelostoma* sp. (Coquillett, 1895; Brues, 1903b; Malloch, 1912b; Leonardi, 1927; Autuori, 1928; Schmitz, 1928a; Borgmeier, 1963d).

Pemphigidae (aphids, woolly aphids)

Megaselia chainensis larvae prey upon *Colopha ulmicola* in its cockscomb galls on American elm (*Ulmus americana*) and pupariate in the galls (Robinson, 1970; Robinson and Brown, 1993).

Phora holosericea larvae prey upon the following root pemphigids: *Forda formicaria, Kaltenbachiella pallida, Pemphigus fuscicornis* and *P. lichtensteini* (Yarkulov, 1972).

3.5.6(e) Hymenoptera (ants, bees, wasps and sawflies)

Among the earliest authentic observations of parasitoidism by phorids are reports in the 1860s of scuttle fly larvae parasitizing honey bees. Assmuss (1860, 1865) reported attacks on bee larvae and Amerling (1862) reported parasitization of adults. In both cases the identification of the phorid species involved was erroneous (Orosi-Pal, 1938). Amerling's observations almost certainly refer to *Melaloncha ronnai* (Borgmeier, 1968b, 1971b), but the identity of the phorid observed by Assmuss remains uncertain (see below). The parasitization of honey bees (*Apis mellifera*) by *M. ronnai* in Brazil is certainly recent, as honey bees were introduced by man. The native bee hosts of *M. ronnai* remain to be discovered. Other species of *Melaloncha* are known to parasitize *Trigona* and *Bombus*.

The larvae of *Melaloncha* species enter the thorax of their bee hosts, after the egg has been laid on the flying bee (2.5).

Gymnoptera larvae were considered by Sladen (1912) to be predators of the eggs of bumblebees (*Bombus*). Alford (1975), however, considers this to be unlikely. The question invites critical investigation. Biegel (1953) provides evidence for believing that the larvae of *Gymnoptera vitripennis*, in the nests of *Vespula* wasps, are essentially scavengers during the summer months but become predators upon the wasp brood in the autumn. He attributes the same habits to *Triphleba lugubris*.

The following review collates the reported cases of parasitoidism or predation by phorids.

Aculeata

Agaonidae (fig wasps)
Megaselia dentata, M. ficaria, M. parvorata, M. surophila and *M. vorata* breed in fig syconia (*Ficus* species), but it is thought that the larvae are probably feeding on the developing fig wasps rather than the figs (3.4).

Anthophoridae (carpenter bees)
Three dead females of *Melissodes floris* have been reported with a single phorid puparium in the thorax of each (Batra and Schuster, 1977). Further investigation is required.

Apidae (bumblebees, stingless bees and honey bees)
Enderlein (1924) reported a possible case of parasitization of an apid bee by *Stichillus insperatus*.

Apis cerana – a female of a *Megaselia* species has been reared (along with larvae and pupae not reared through) from an adult (of the subspecies *indica*) from India (unpublished data).

Apis mellifera (the honey bee)
Early records of phorids parasitizing honey bees are bedevilled by misidentifications, ever since Assmuss (1860, 1865) misidentified the phorid he observed. All records of bee parasites prior to Orosi-Pal's (1938) critical

review need re-evaluation. The identity of the phorid observed by Assmuss remains a mystery. He figured a drosophilid larva in error. His figure of the puparium depicts a phorid, but of a species too small for the genus to which he assigned his species. He misidentified the phorid adults from Meigen's figures and then proceeded to illustrate his own work with copies of these figures, while utilizing Schiner's description of the species (Orosi-Pal, 1938). Despite these errors his observations are worth citing. The females were seen ovipositing through the integuments of living bee larvae. He observed three larval instars developing within the host larvae, with the third instars leaving to pupariate either in the cell or in the soil outside the hive. It is possible that the phorid species involved will not utilize bees in modern hives subject to modern systems of management. It may survive in those parts of Europe where pre-modern types of hive are still in use.

In Brazil *Melaloncha ronnai* parasitizes adult bees, the larvae feeding mainly in the host's thorax and pupariating there also (Ronna, 1936), or rarely leaving the bee's corpse to pupariate (Ronna, 1937b). The eggs are laid between abdominal tergites 1 and 2, and the young larvae then move into the thorax (Ronna, 1937a). Up to 50% of the bees in a hive may be parasitized (Ronna, 1937a).

Pseudohypocera kerteszi larvae are facultative predators of prepupae and pupae of the bees (Chaud-Netto, 1980), probably when the density of fly larvae exceeds a critical level (Reyes, 1983).

Bombus mexicanus – parasitized by *Melaloncha* sp. (Ramirez, 1982).

Bombus vosnesenskii – parasitized by larvae of *Apocephalus borealis*, the larvae feeding mainly on the muscles of the thorax. A few are found in the base of the abdomen. When the host bee dies the larvae emerge, at the junction of the head and thorax, and feed for a further 4–5 days on the remains of the bee. They then move away from the corpse to pupariate. A single bee will support 10–15 fly larvae. An infestation rate of about 0.10 has been recorded (W. O. De Prins and Disney, unpublished data). Brown (1993b) has reared the same phorid from a *Bombus* sp.

An adult *Bombus terrestris* (sensu Prŷs-Jones and Corbet, 1991) – parasitized by a *Megaselia* species (Alford, 1975).

Tetragona and *Trigona* – *Pseudohypocera kerteszi* larvae have been reported to kill the brood of these bees, and even to destroy entire colonies (Nogueira-Neto, 1970). These are clearly cases of facultative predation, as reported under honeybees above.

Trigona angustula – as host of *Melaloncha flava* (Simões *et al.*, 1980).

Trigona postica – *Melaloncha sinistra* (Borgmeier, 1971a; Borgmeier and Prado, 1975). Simões *et al.* (1980) report detailed observations. The larvae are in the abdomens of forager bees. Oviposition takes place outside the nest entrance, a female phorid curving her ovipositor forwards beneath the bee's body. Infestation rates for 16 colonies ranged from 0 to 0.37 of forager bees. In another two large samples of forager bees infestation rates were 0.31 (n = 833) and 0.34 (n = 412). Typically there is a single larva per bee, but rarely there are two. The larva consumes the host's abdominal tissues first, including the gut contents of pollen, before invading the thorax. Pupariation is usually in the thorax, but occasionally in the abdomen.

Trigona ruficrus – Melaloncha clavata. An adult male and two pupae were reared from the subspecies *corvina* (Borgmeier, 1938b).

Trigona spinipes – Melaloncha clavata (Simões *et al.*, 1980).

Xylocopa caffra – Brues (1907c) reported a mating pair of *Phalacrotophora braunsi* in an old willow stump frequented by cerambycid beetles and this large carpenter bee. The significance of this observation requires further investigation. In view of the reports cited below it seems possible that the fly species may be a parasitoid of the bee.

The above records suggest that the *Melaloncha* species tend to exhibit considerable host specificity, with only *M. clavata* being recorded from two host species. In addition *M. ronnai* must have two host species, but its native host is unknown. By contrast the primarily kleptoparasitic *Pseudohypocera kerteszi* is recorded from the nests of 14 species belonging to three genera (3.2.5).

Brian Brown (*in litt.*) reports an *Apocephalus* species parasitizing males of a stingless bee in Costa Rica.

Eumenidae **(burrowing wasps)**

Borgmeier (1934a) reported *Phalacrotophora pruinosa* parasitizing an eumenid.

Formicidae **(ants)**

The supra-generic classification of ants is currently undergoing reassessments (e.g. Bolton, 1990; Urbani *et al.*, 1992). In the following review of the records of phorids which attack ants the generic names and classification of the ants follows Hölldobler and Wilson (1990), except for the Dolichoderinae, which follows Shattuck (1992a,b).

Ant larvae as prey

Williams and Whitcomb (1974) reported that the larvae of an *Apodicrania* species parasitize the larvae of the fire ant *Solenopsis saevissima*. Parasitized larvae are recognizable by their more yellow appearance. The fly pupae are tended by the worker ants, who will even carry them to safety if the nest is disturbed.

In the laboratory *Xanionotum hystrix* has been reared when offered live ant larvae (Rettenmeyer and Akre, 1968). They readily preyed on larvae of *Iridomyrmex analis*, *Crematogaster* sp. and *Pheidole* sp. They also fed on dead *Neivamyrmex* larvae.

Ant pupae as hosts

Formica exsecta – Aenigmatias lubbocki (Schmitz, 1914d).

F. fusca – Aenigmatias lubbocki adults were reared from ant pupae by Donisthorpe (1913a, 1927).

F. glebaria – Aenigmatias dorni (Schmitz, 1941f; Delage and Lauraire, 1971c).

F. rufibarbis – Aenigmatias dorni (Schmitz, 1941f) and *A. lubbocki* (Donisthorpe, 1927) have been reared from ant pupae.

F. sanguinea – Aenigmatias lubbocki (Schmitz, 1941f).

F. transkaucasica – Aenigmatias lubbocki adults have been reared from fly pupae found in nest (Donisthorpe, 1927).

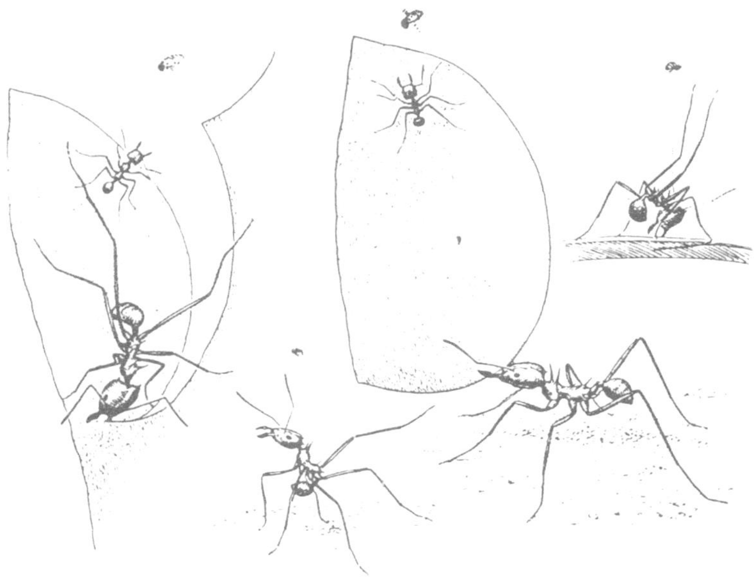

Figure 3.9 Workers of *Atta cephalotes* (leaf-cutter ants) defending themselves from attack by *Neodohrniphora* sp. (After Eibl-Eibesfeldt and Eibl-Eibesfeldt, 1967.)

Lasius niger – *Aenigmatias franzi* has been caught in the vicinity of this species of ant and the next (Smith, 1977).

Myrmica ruginodis – *Aenigmatias franzi* has been caught in the vicinity of this species, and the above species (Smith, 1977).

Pheidole dentata – *Apocephalus* sp. females have been observed landing on pupae being carried by the workers (LaBerge, 1953).

Ant adults as hosts (worker ants unless indicated otherwise)

Large numbers of Phoridae are reported as parasitoids of ants. While the detailed life histories have been worked out for only a few species, oviposition, attempted oviposition or behaviour associated with the preliminaries to an oviposition attack have been observed for many species. Likewise ants have frequently been observed trying to parry the attacks by phorids, taking evasive action or engaging in frantic grooming behaviour following a successful attack.

Ants which are prone to attack by phorids and which habitually use the same trails tend to convert these to covered ways. Some ants utilize covered ways purloined from termites, such as those ascending tree trunks. It has been suggested that the use of covered ways has evolved as a direct consequence of the vulnerability of these species to attack by Phoridae (Disney, 1986b).

The leaf-cutting ants (*Atta*) have workers of more than one size. The smaller, 'minim', workers are often to be observed hitch-hiking on the leaf fragments being carried by larger workers. Observations in Trinidad gave rise to the hypothesis that this was as a defence against attacks by parasitic Phoridae (Eibl-Eibesfeldt and Eibl-Eibesfeldt, 1967) (Fig. 3.9). An alternative hypothesis proposed that this hitch-hiking behaviour reduced transport costs for minim workers, which tend to collect plant fluids rather than leaf fragments (Stradling, 1978). Feener and Moss (1990) investigated these two hypotheses for *Atta colombica*, subject to attacks by the phorid *Apocephalus attophilus*. Their data indicate energy savings by hitch-hiking minims to be less than 10%. By contrast the presence of hitch-hiking minims significantly reduces the time the phorids spend on leaf fragments and also the probability that they will land in the first place. The hitch-hiking minims snap at approaching flies with their mandibles and chase them off the leaf fragments. Occasionally they will catch a phorid, and dismember it. Orr (1992) has found that attacks by phorids reduce the amount of daytime foraging and also influence the size of the ants foraging by day.

The following review lists the known hosts of Phoridae oberved 'attacking' ants. Unless indicated otherwise the host records relate to female flies observed 'attacking' ant workers outside the nest.

Unidentified ants

Pseudacteon curriei (Borgmeier, 1963d), *P. javensis* (Brues, 1915b).

Dolichoderinae

Azteca sp. – *Apocephalus aztecae* (Borgmeier, 1961a).

Dolichoderus lugens – possibly the workers of this ant are parasitized by *Rhyncophoromyia maculineura* (Brown and Feener, 1993).

Dorymyrmex pyramicus – *Pseudacteon dorymyrmecis* females observed at nest entrance (Borgmeier, 1925b).

Linepithema (= *Iridomyrex*) *humile* – *Apocephalus silvestrii* and *Pseudacteon pusillus* (Borgmeier, 1938c).

Liometopum occidentale – *Pseudacteon californiensis* (Disney, 1982k), the females ovipositing into the gaster.

Philidris (= *Iridomyrmex*) *cordatus* – *Iridophora clarki*, the female darting eggs into worker ants (seemingly into their heads) while in flight, and *Megaselia sembeli*, the female landing and racing up to a worker ant to oviposit into its gaster (Disney, 1986b).

Tapinoma erraticum – *Pseudacteon formicarum* (Donisthorpe, 1927).

Ecitoninae (army ants)

Eciton sp. – *Borgmeieriphora greigae* (Brown, 1993a). It is not certain that this fly parasitizes the ants.

E. burchelli – *Acanthophorides clavicercus, A. condei* and *A. divergens* (Borgmeier, 1928c, 1967h), *Apocephalus cultellatus* (Borgmeier, 1961a), *Diocophora appretiata* (Schmitz, 1923b; Borgmeier, 1959; Disney and Kistner, 1989d) and *D. disparifrons* (Borgmeier, 1959). Brown *et al.* (1991) collected *Styletta crocea* in the vicinity of this species. Brown (1993a) reported

Borgmeieriphora kempfi with a raiding column of this ant and numbers of *B multisetosa* over this ant, but it is not known whether the ants are parasitized by this fly

E coecum – *Cremersia spinicosta* (Schmitz, 1924b, Greene, 1938) and *Myrmosicarius grandicornis* (Borgmeier, 1931a)

E hamatum – *Diocophora* sp (Rettenmeyer and Akre, 1968)

E quadriglume – *Apocephalus quadriglumis* (Borgmeier, 1961a) and *Diocophora appretiata* (Borgmeier, 1959)

E schlechtendali – *Cremersia brasiliensis, C pilosa* and *C salesiana* (Borgmeier, 1928c)

Labidus coecus – *Apocephalus analis, A comatus* (Borgmeier, 1958), *A inimicus* (Borgmeier, 1961a), *Cremersia longipes* (Borgmeier, 1971a) and *Dacnophora setithorax* (Borgmeier, 1961a)

L praedator – *Acanthophorides labidophilus* (Borgmeier, 1967h), *A longicornis* (Borgmeier, 1928c), *A pilosicauda* (Borgmeier, 1967h), *Apocephalus ecitonis* (Borgmeier, 1928c), *A platypalpis* (Borgmeier, 1963d), *A praedator* (Borgmeier, 1971a), *Cremersia setitarsus* (Borgmeier, 1971a), *C spinosissima* (Borgmeier, 1928c), *Myrmosicarius biarticulatus* (Borgmeier, 1931a) and *Pradea iniqua* (Borgmeier, 1927b)

Neivamyrmex diversinodis – *Cremersia australis* (Borgmeier, 1961a)

N legionis – *Cremersia costalis* (Borgmeier, 1925c) and *Dacnophora legionis* (Borgmeier, 1961a)

N pilosus – *Pseudacteon simplex* (Borgmeier, 1969b)

N pseudops – *Cremersis costalis* (Borgmeier, 1961a)

N raptans – *Dacnophora legionis* (Borgmeier, 1961a)

Nomamyrmex esenbecki – *Apocephalus infradentatus, A persecutor, Cremersia brasiliensis, C crassispina, C pilosa* (Borgmeier, 1961a), *C spinicosta* (Rettenmeyer and Akre, 1968), *Diocophora disparifrons* (Borgmeier, 1959) and *Myrmosicarius grandicornis* (Borgmeier, 1931a)

N hartigi – *Cremersia brasiliensis, C crassispina, C pilosa* and *C salesiana* (Borgmeier, 1961a)

Table 3.4 Phorid parasitoids of army ants (Ecitoninae)

Phorid genus	No phorid species (P)	No of ant species (A)	Mean no ant spp per fly sp	Mean no fly spp per ant sp (P/A)
Acanthophorides	5	2	1 0	2 5
Apocephalus	10	5	0 5	2 0
Borgmeieriphora	3	2	0 7	1 3
Cremersia	10	8	1 7	1 3
Dacnophora	2	3	1 0	0 7
Diocophora	2	4	1 5	0 5
Myrmosicarius	3	3	1 0	1 0
Pradea	1	1	1 0	1 0
Pseudacteon	1	1	1 0	1 0

The penultimate column is a preliminary index of catholicity of host choice The last column is a preliminary index of niche overlap

These data indicate a high degree of host specificity, with only species of *Cremersia, Dacnophora* and *Diocophora* being reported attacking two species, and only *C. brasiliensis* reported attacking three species. Table 3.4 summarizes the data.

Formicinae (wood ants and relatives)

Acropyga acutiventris – *Puliciphora boltoni* (Disney, 1988b), the larvae being predatory upon the workers.

Camponotus (wood ants)

Camponotus sp. – *Apocephalus coquilletti* (Borgmeier, 1963d), *A. crucicauda* (Borgmeier, 1928c), *A. similis* (Schmitz, 1914d), *Diocophora appretiata* and *D. disparifrons* (Borgmeier, 1963d). In addition Fage (1933) reported a phorid attacking an ant of this genus and Colless and McAlpine (1991) report a phorid larva as an 'ectoparasite attached to ant's neck'.

C. abdominalis – *Rhyncophoromyia* sp. (Brown, 1992b).

C. acvapimensis – *Microselia deemingi* (Disney, 1983g); a female was attempting to oviposit on an ant attending the coccids *Udinia catori* on the tree *Cussonia barteri* (Araliaceae).

C. blandus – *Apocephalus mucronatus* (Borgmeier, 1958).

C. cingulatus – *Apocephalus bispinosus* (Borgmeier, 1928c), *Diocophora disparifrons* (Borgmeier, 1959), *Macrocerides neivai* (Borgmeier, 1928c) and *Pseudacteon borgmeieri* (Borgmeier, 1922a).

C. crassus – *Apocephalus setiventris* (Borgmeier, 1971a) and *Diocophora disparifrons* (Borgmeier, 1959).

C. cruentatus – *Microselia rivierae* (Della Santa, 1993).

C. femoratus – *Apocephalus laceyi* (Disney, 1981d) and *Rhyncophoromyia* sp. attracted to workers (Brown and Feener, 1993).

C. floridanus – *Diocophora trichogaster*, observed riding on the head of an ant (Brown, 1992b).

C. gigas – *Megaselia persecutrix* (Schmitz, 1932d; Disney and Schroth, 1989).

C. herculeanus – *Menozziola schmitzi* (Nielsen, 1934), the larvae were in the host's abdomen, and *Trucidophora camponoti* (Brown *et al.*, 1991), the queens being the preferred host.

C. maculatus – *Apocephalus coquilleti* (Schmitz, 1914d).

C. novaeboracensis – *Trucidophora camponoti* (Brown *et al.*, 1991); the queens were the preferred host.

C. pennsylvanicus – *Apocephalus coquilletti* (Schmitz, 1914d). *A. pergandei* was the first phorid to be established as a parasitoid in the heads of ants. Fox (1887) reported decapitated ants and found the detached heads on the ground, each with one fly larva inside. Pergande (1901) observed ants moving about aimlessly in a listless manner. Subsequently in captivity their heads fell off and revealed a single fly larva in each. He reared adults from some of these larvae. He later observed the same fly species darting at ants in the field. *Rhyncophoromyia conica* has been bred from the abdomen of this ant (Malloch, 1912b). A further parasitoid is *Trucidophora camponoti*, which attacks the queens. Six out of 40 queens have been reported infested, with 26–120 larvae per ant. Most larvae were in the host's gaster, but some were in the thorax and head. The larvae leave the remains of the host to pupariate, moving up to 1 m away (Brown *et al.*, 1991).

C. rufipes – *Apocephalus angustinervis* (Borgmeier, 1961a), *A. camponoti* and *A. lanceatus* (Borgmeier, 1925c), *A. setiventris* (Borgmeier, 1971a), *Auxanommatidia myrmecophila* (Borgmeier, 1925c), *Diocophora disparifrons, D. modesta, Pseudacteon borgmeieri,* and *Rhyncophoromyia gymnopleura* (Borgmeier, 1922a, 1959).

C. sansabeanus – *Apocephalus coquilletti* (Borgmeier, 1963d).

C. sericeiventris – injured workers are parasitized by *Rhyncophoromyia maculineura* and larvae of *Diocophora* sp. were found in one ant (Brown and Feener, 1993).

C. vagus – *Microselia southwoodi* (Disney and Shaw, in press).

C. vicinus – *Apocephalus similis* (Brian Brown, *in litt.*).

Schmitz (1924h) cautioned that some of Donisthorpe's host records for *Pseudacteon formicarum* (see below) might be for *P. lundbecki*. Donisthorpe (1945) accepted this possibility. Unless the latter species were to be recorded from Britain, this must remain mere speculation.

Formica (large wood ants)

F. fusca – Borgmeier (1963d) considered *Apocephalus wheeleri* to be 'probably parasitic' upon this ant, but Brown (1993b) doubts this.

F. sanguinea – *Pseudacteon formicarum* (Donisthorpe, 1927).

Lasius alienus – *Pseudacteon formicarum* (Donisthorpe, 1927).

L. flavus – *Pseudacteon formicarum* (Wasmann, 1918b; Donisthorpe, 1927; Disney, unpublished data).

L. fuliginosus – *Pseudacteon formicarum* (Donisthorpe, 1927).

L. niger – *Pseudacteon formicarum* (Wasmann, 1918b; Donisthorpe, 1927), the eggs being laid between free segments of gaster, and the larva developing within. *P. lusitanus* has also been recorded with this ant (Schmitz, 1938m). The '*Lasius niger*' of most authors is now known to comprise two sibling species, *L. niger* and *L. platythorax*. The latter is mainly found in damp woodlands, bogs and fens. Seifert's (1992) key allows identification of the Palaearctic species of the subgenus *Lasius*.

L. umbratus – *Pseudacteon formicarum* (Donisthorpe, 1927).

Paratrechina fulva – *Pseudacteon convexicauda* (Borgmeier, 1962d).

P. melanderi – *Microselia texana* (Disney, 1982k), the females landing on the heads of the workers.

These data indicate considerable host specificity, but with some notable exceptions. *Pseudacteon formicarum* is recorded from six host species, belonging to two genera. It is also recorded from a dolichoderine (see above) and a myrmicine (see below). *Diocophora appretiata* is recorded from only one formicine species, but also from two Ecitoninae (see above). *D. disparifrons* is recorded from four species of Formicinae and one ecitonine. *Apocephalus coquilletti* is recorded from four species of Formicinae and from one

Table 3.5 Phorid parasitoids of formicine ants

Phorid genus	No. phorid species (P)	No. of ant species (A)	Mean no. ant spp. per fly sp.	Mean no. fly spp. per ant sp. (P/A)
Apocephalus	12	10	1.3	1.2
Auxanommatidia	1	1	1.0	1.0
Diocophora	5	6	1.6	0.8
Macrocerides	1	1	1.0	1.0
Megaselia	1	1	1.0	1.0
Menozziola	1	1	1.0	1.0
Microselia	4	4	1.0	1.0
Pseudacteon	4	9	2.5	0.5
Puliciphora	1	1	1.0	1.0
Rhyncophoromyia	5	5	1.0	1.0
Trucidophora	1	3	3.0	0.3

The penultimate column is a preliminary index of catholicity of host choice. The last column is a preliminary index of niche overlap.

myrmicine, and *A. similis* and *Trucidophora camponoti* from two species. The data overall are summarized in Table 3.5.

Myrmicinae

Acromyrmex (leaf cutter ants).

Acromyrmex sp. – *Apocephalus luteihalteratus* and *A. neivai* (Borgmeier, 1937) and *Neodohrniphora acromyrmecis* (Borgmeier, 1925c; Prado, 1976).

A. aspersus mesonotalis – *Myrmosicarius infestans* (Borgmeier, 1931a).

A. disciger – *Myrmosicarius infestans* (Borgmeier, 1931a) and *Neodohrniphora acromyrmecis* (Borgmeier, 1929).

A. lundi – *Apocephalus barbicauda, A. luteihalteratus, Myrmosicarius catharinensis* and *M. persecutor* (Borgmeier, 1931a).

A. muticinodus – *Apocephalus lamellatus* (Borgmeier, 1926c), *A. luteihalteratus* (Borgmeier, 1931a), *Myrmosicarius cuspidatus, M. tarsipennis, Procliniella hostilis* and *Stenoneurellys laticeps* (Borgmeier, 1929).

A. niger – *Apocephalus lamellatus, A. luteihalteratus, Myrmosicarius cuspidatus, M. tarsipennis, Procliniella hostilis* and *Stenoneurellys laticeps* (Borgmeier, 1931a).

A. subterraneus – *Apocephalus dubitatus, A. rionegrensis, Myrmosicarius cuspidatus, M. gracilipes* and *M. infestans* (Borgmeier, 1929, 1931a).

Atta (leaf cutter ants)

Atta sp. – *Neodohrniphora declinata* (Prado, 1976).

A. cephalotes – *Apocephalus attophilus* (Feener and Moss, 1990), *Neodohrniphora* sp.1 (as '*Apocephalus*') (Eibl-Eibesfeldt and Eibl-Ebesfeldt, 1967), *Neodohrniphora* sp. 2 (Feener and Moss, 1990), *N. curvinervis* attacking foraging workers by day (Orr, 1992) laying one egg per ant into the backs of the heads of the larger ants which are not carrying pieces of leaves (Feener and Brown, 1993).

A. colombica – *Apocephalus attophilus* (Feener and Moss, 1990), the females ovipositing into the heads of the ants at the base of the mandibles. There is only one egg laid per ant but 10–16 are matured at a time.

A. laevigata – *Apocephalus attophilus* (Borgmeier, 1929). Piffer *et al.* (1972) noted that larvae are mainly found in the heads, occasionally in the thorax and only rarely in the abdomen. Workers carrying pieces of leaf are more frequently parasitized than soldiers, who are more frequently attacked than workers carrying soil out of the nest. Seemingly the degree of exposure outside the nest is the critical determinant of infestation rates. *Neodohrniphora declinata* also parasitizes this species (Borgmeier, 1928b).

A. mexicana – *Neodohrniphora* sp. (Feener and Moss, 1990).

A. sexdens – *Apocephalus attophilus* (Borgmeier, 1928b, 1958). Foresti and Pereira da Silva (1970) reported the larvae in the heads of the ants and reared some through to the adult flies. Other parasitoids include *Myrmosicarius crudelis*, *M. grandicornis*, *Neodohrniphora acromyrmecis*, *N. curvinervis* and *N. declinata* (Borgmeier, 1929; Prado, 1976).

A. texana – *Apocephalus wallerae* (Disney, 1980i), the females ovipositing between the host's mandibles. Waller and Moser (1990) reported that the female rarely hovers, preferring to run up to forager ants and tending to select the larger workers. They found a larva in one side of the head. *Myrmosicarius texanus* also attacks this ant, ovipositing into the back of the head (Greene, 1938; Disney, 1980i).

Crematogaster (tree ants)

Crematogaster sp. – *Pseudacteon onyx* (Disney, 1982k).

C. lineolata – *Pseudacteon onyx* (Borgmeier, 1963d).

C. parabiotica – *Pseudacteon* sp. (Diane Davidson and Disney, unpublished data).

C. scutellaris – *Menozziola schmitzi* (Menozzi, 1921).

Myrmica lobicornis – *Pseudacteon formicarum* (Donisthorpe, 1927).

M. ruginodis – *Pseudacteon brevicauda* (Donisthorpe, 1927).

M. scabrinodis – *Pseudacteon brevicauda* (Disney, 1979f).

Pheidole (harvester ants)

Pheidole constipata – *Microselia texana* (Disney, 1982k), the females attacking the major workers.

P. dentata – *Apocephalus feeneri* (Feener, 1981; Disney, 1982k), the females attacking the soldiers ('major workers') by darting an egg into the gaster. There is only one larva per host. *A. tenuipes* hovers over this ant and lands at rear of thorax or on gaster (Burges, 1979).

P. rotundata – *Plastophorides bequaerti* female observed on head of this ant (Schmitz, 1916c; Beyer, 1959f).

P. subdentata – a phorid whose females dart at soldiers (Feener, 1987).

P. titanis – *Apocephalus* sp. whose hovering females dart at soldiers and workers (Feener, 1988).

Pheidologeton (marauder ants)

Pheidologeton sp. – *Megaselia kodongi* (Disney, 1986b), the females landing and running up to workers and attacking their gasters.

Solenopsis (fire ants)
Trager (1991) has revised part of this genus of ants.

Solenopsis geminata – Pseudacteon antiguensis (Malloch, 1912b), *P. arcuatus* (Feener and Brown, 1992), *P. browni* (Disney, 1991f; Feener and Brown, 1992), *P. crawfordi* (Coquillett, 1907; Smith, 1928; Feener, 1987), *P. grandis* (Disney, 1991f), *P. longicauda* (Borgmeier and Prado, 1975), *P. solenopsidis* (Schmitz, 1914d; Borgmeier, 1925c) and *P. wasmanni* (Schmitz, 1914d).

S. invicta – P.obtusus (Williams and Banks, 1987), the egg being placed at the rear of the head of a worker. Winged females are also parasitized by a phorid (Wojcik *et al.*, 1987), whose pupa has been reported in the ant but not reared through to the adult fly.

S. richteri – Apocephalus normenti (Prado, 1976), *Pseudacteon curvatus, P. litoralis, P. obtusus* and *P. tricuspis* (Borgmeier, 1925c).

S. saevissima – Myrmosicarius crudelis (Borgmeier, 1928b) and *M. diabolicus* (Borgmeier, 1931a). *Pseudacteon affinis, P. borgmeieri, P. comatus, P. convexicauda, P. cultellatus, P. curvatus*★ , *P. dentiger, P. lenkoi, P. litoralis*★, *P. nocens, P. nudicornis, P. obtusus*★, *P. pradei*★, *P. solenopsidis, P. tricuspis*★ and *P. wasmanni* (Borgmeier, 1922a, 1923c, 1925c, 1926a, 1962c,d; Borgmeier and Prado, 1975; Williams *et al.*, 1973; Williams and Whitcomb, 1974). It seems extraordinary that this ant is attacked by so many species. Williams *et al.* (1973) observed that after a successful oviposition attack by a female *Pseudacteon* the ant appears to be visibly stunned, and it may remain motionless for one or two seconds. Pesquero *et al.* (1993) confirmed the records marked with an asterisk (★), but recorded the flies in mating swarms of the ants.

S. texana – Microselia texana has been observed in the vicinity of this ant species, but no evidence of attack was seen (Disney, 1982k).

S. xyloni – Apocephalus coquilletti (Greene, 1938), *Pseudacteon crawfordi* (Feener, 1987), the females hovering over larger workers, rapidly landing on the thorax and ovipositing into propleuron; and *P. spatulatus* (Feener, 1987; Disney, 1991f).

These data indicate that some species of ant are attacked by a large number of phorid species. *Apocephalus coquilletti* and *Pseudacteon formicarum* are both recorded only from a single species of Myrmicinae, but they are both recorded from several species of Formicinae (see above). The specificity with respect to the different phorid genera attacking Myrmicinae is summarized in Table 3.6. The penultimate column is a preliminary index of catholicity of host choice. The last column is a preliminary index of niche overlap in terms of host species, so that *Pseudacteon* species would seem to be more likely to share the same host species than are the *Apocephalus* species.

Ponerinae (predatory ants)

Paraponera clavata – Apocephalus paraponeraė (Borgmeier, 1958). Females are attracted by olfactory cues to injured workers and they oviposit into the

Table 3.6 Phorid parasitoids of myrmicine ants

Phorid genus	No. phorid species (P)	No. of ant species (A)	Mean no. ant spp. per fly sp.	Mean no. fly spp. per ant sp. (P/A)
Apocephalus	12	14	1.6	0.9
Megaselia	1	1	1.0	1.0
Menozziola	1	1	1.0	1.0
Microselia	1	1	1.0	1.0
Myrmosicarius	10	9	1.6	1.1
Neodohrniphora	6	8	1.8	0.8
Plastophorides	1	1	1.0	1.0
Procliniella	1	2	2.0	0.5
Pseudacteon	24	11	1.3	2.2
Stenoneurellys	1	1	1.0	1.0

head, thorax or gaster of their victims (Brown and Feener, 1991a,b). Brian Brown (*in litt.*) reports undescribed *Apocephalus* spp. attacking various ponerine ants in Costa Rica.

Ichneumonidae (ichneumon flies)

The larvae of *Megaselia mallochi* are predators of the pupae of the sciarid *Bradysia confinis* (3.5.6(c)). The same pupae are also parasitized by larvae of the ichneumonid *Stenomacrus laricis*. Parasitized pupae, along with any ichneumonid larva, are preyed upon by *M. mallochi* as readily as are unparasitized pupae (Krespi *et al.*, 1985; Deleporte, 1986).

Sphecidae (digger wasps)

These wasps provision the brood cells in their nests with prey insects for their larvae. These paralysed prey insects, plus any discarded prey accumulating in the nest away from the brood cells, are exploited by phorids (3.2). The observations of Krombein (1964), on the invasion of the brood cells of *Ectemnius paucimaculatus* by *Megaselia aletiae* larvae, suggest that the fly larvae also destroy the wasp's eggs and that the wasp larvae (deprived of their prey) die of starvation. These dead larvae are probably then consumed by the phorid larvae also. The same situation seems to prevail with *M. aletiae* in the nests of *Trypoxylon striatum* (Krombein, 1967).

Vespidae (social wasps)

The nests of these familiar wasps are invaded by a number of scuttle flies whose larvae are primarily saprophage, or thought to be so. However, Biegel's (1953) observations (see above) suggest that for some fly species there may be a seasonal shift to facultative predation in the autumn. Likewise MacDonald *et al.* (1975) reported that the larvae of *Triphleba lugubris* 'significantly lowered colony productivity by destroying developing queen pupae'. It is possible that *Diplonevra funebris* is also a facultative predator of wasp brood, but critical observations are lacking and so records of this species are reviewed in section 3.2.4(c). Only records of phorids reported attacking living wasp brood are reviewed below.

Apocephalus borealis larvae parasitize the adults of *Vespula arenaria, V. pensylvanica* and *V. vulgaris*. Up to eight larvae per wasp are to be found mainly in the thoracic haemocoel, but some may occur in the abdomen. The mature larvae leave the wasp, mainly from the cervical region, in order to pupariate. About 5% of the wasps in four nests of *V. pensylvanica* at a site in California were found to be parasitized (Ennik, 1973).

Gymnoptera vitripennis has been reported in the nests of the following species of wasp or hornet: *Vespa crabro* (Collart, 1933), *Vespula germanica* (Dunlop, 1930; Richards, 1932; Leclercq, 1945; Biegel, 1953; Disney, D. Henshaw and I. F. G. McLean, unpublished data) and *V. vulgaris* (Lundbeck, 1922; Dunlop, 1930; Richards, 1932; Biegel, 1953; Kidd and Brindle, 1959; Disney, D. Henshaw and D. A. Smith, unpublished data).

Triphleba lugubris has been reported in the the nests of the following species: *Dolichovespula norwegica* (unpublished data), *Vespa crabro* (Collart, 1933), *Vespula atropilosa* (MacDonald *et al.*, 1975), *V. austriaca* (Brown, 1945), *V. germanica* (Leclercq, 1945; Biegel, 1953; Borgmeier, 1963d; Disney and D. Henshaw, unpublished data), *V. media* (Lundbeck, 1922), *V. pensylvanica* (MacDonald *et al.*, 1975), *V. rufa* (Brown, 1945) and *V. vulgaris* (Richards, 1932; Biegel, 1953; Disney, D. Henshaw and D. A. Smith, unpublished data).

Symphyta (sawflies)

There are a number of records of Phoridae reared from the pupal cocoons of sawflies. However, these are mainly for well-known saprophage species and so they probably represent infestations of already moribund sawfly pupae. The species reported include *Diplonevra funebris, Megaselia agarici, M. giraudii, M. minuta, M. ruficornis* and *M. rufipes* (Aldrich, 1892; Azevedo e Silva and Serrão Nogueira, 1965; Brues, 1903b; Hsin, 1935; Liston, 1979, 1982; Lundbeck, 1922). A curious rearing is that of *Megaselia arietina* from moribund pre-pupae of the pamphiliid species *Cephalcia abietis* and *C. arvensis* (Dr N. J. Mills and Disney, unpublished data). This scuttle fly has been reported as a major leaf miner of chickpeas infested with a rust fungus (3.4).

Brues (1906b) reported *Phalacrotophora longifrons* hovering around the burrows of the siricid *Tremex columba* in an old stump, along with the siricid's ichneumonid parasitoids *Thalessa lunator* and *T. atrata*. The significance of this behaviour needs further investigation.

3.5.6(f) Isoptera (termites)

A number of authors have reported Dipteran larvae in the heads of termites. For example, Emerson (1928), Noirot (1955) and Ernst (1964) all reported such larvae in the heads of the soldiers of *Acanthotermes acanthothorax*. These larvae have subsequently been interpreted as being phorid larvae (Borgmeier, 1967b; Ferrar, 1987). The review below mainly covers the few cases where the identity of the phorids parasitizing termites has been established.

Undoubtedly many more of the adults listed in section 6.4.3 will turn out to have predatory or parasitoid larvae.

Misotermes exenterans parasitize the soldiers of *Macrotermes gilvus*. The female fly scatters large numbers of eggs at a time in the host termite's nest (Kalshoven, 1938). Silvestri (1926) originally reported a larva in this termite species, which he thought might be a conopid, but which is now considered to have been a *Misotermes* species (Ferrar, 1987). The females of *Palpiclavina kistneri* have been observed laying eggs on the nymphs of this same termite species (Disney and Kistner, 1989b, 1992).

Phorid larvae have been reported in the heads of the soldiers of *Macrotermes subhyalinus* (Geigy *et al.*, 1964). Adult phorids known to be associated with this species are listed in section 6.4.3.

Diplonevra mortimeri parasitizes the workers of *Bulbitermes sarawakensis*. The females are attracted to exposures of termites and lay a single egg into the antero-dorsal region of the abdomen. Pupariation takes place in the remains of the abdomen, with the anterior end of the puparium projecting from its rear end (Disney, 1986d). The pre-oviposition routine is extraordinary (2.5). Likewise from the behaviour of the females of *D. mortimeri* and *D. watsoni* it is inferred that both flies parasitize workers of *Nasutitermes culasiensis*. Furthermore a female of *D. mortimeri* has been observed covering a comatose termite, presumably containing her egg, with soil particles and then standing guard over it. Both *D. mortimeri* and *D. watsoni* have also been observed leading workers of *Nasutitermes latus* and *N. makasserensis* in their pre-oviposition behavioural routine. Likewise *D. mortimeri* has been observed behaving the same way with *N. mollis* and development of the fly larva in the termite was subsequently confirmed (Disney, 1986d).

Kemner (1922, 1926) suggested that the larvae of *Javanoxenia punctiventris* prey upon the eggs of *Odontotermes javanicus*. As reported by Ferrar (1987), this hypothesis has been generally discounted. This was because Franssen (1933a,b, 1936) observed the same species feeding on fungus conidia (of *Termitomyces* – see 3.3). However, it remains a possibility that the larvae may feed on both termite eggs and fungus conidia.

In a colony of *Odontotermes sundaicus* inhabited by the females of *Puliciphora collinsi* the head of a soldier termite was found to contain two phorid larvae (Fig. 3.10), which were almost certainly the parasitoid larvae of *P. collinsi* (Disney, 1988b).

Dicranopteron setipennis females have been observed ovipositing into the abdomens of the young nymphs of *Odontotermes takensis* (Disney and Kistner, 1990a).

3.5.6(g) Lepidoptera (butterflies and moths)

There are many records of Phoridae reared from the larvae and pupae of Lepidoptera, especially from Noctuidae. There are also a few records of rearings from adults. Most of these observations refer to phorid species well known to be primarily saprophage, such as *Megaselia giraudii, M. rufipes, M. scalaris* and *Puliciphora pulex*. Indeed many authors have noted that the Lepidoptera were already dead (e.g. Disney and Evans, 1990; Enderlein,

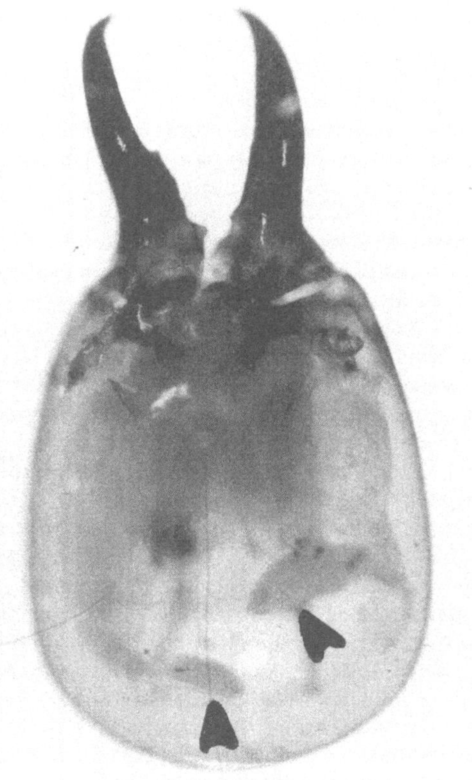

Figure 3.10 Larval phorids (arrowed), probably *Puliciphora collinsi*, in head of termite soldier, *Odontotermes sundaicus*.

1929; Eyles, 1965; Lundbeck, 1922; Meijere, 1912; Morley, 1934; Robinson, 1971; Schmitz, 1953d). Furthermore Richards (1947) specifically noted that *M. rufipes* adults are attracted to dead larvae of *Ephestia* sp. (Pyralidae). Likewise Eyles (1965) emphasized that the larvae of this species were scavengers on the pupae of *Wiseana cervinata* (Hepialidae) only after the latter had died of other causes. However, a number of authors have assumed the 'host' Lepidoptera had been alive at the time of oviposition, and have claimed facultative parasitism at least.

In some cases the Lepidoptera larvae reported as hosts were definitely alive when collected in the field and when placed in rearing containers in the laboratory, where phorids subsequently emerged from them. However, some authors (e.g. Balduf, 1928) have noted that in such cases the living caterpillars tend to be already diseased.

Where nothing is stated regarding the state of health of the Lepidoptera hosts, the published information is rarely sufficient to permit any firm conclusions regarding the claim of parasitoidism by the Phoridae. An exception is the detailed paper by Hesjedal (1984). He reported the rearing of large numbers of *M. rufipes* from large samples of pupae of three species of moth, which were collected in the field as healthy larvae. He concludes that *M. rufipes* 'seems to be a facultative parasite as it is able to parasitize apparently healthy noctuid and geometrid larvae'. At first sight this would seem to be clear evidence of the facultative parasitism of healthy larvae. However, his rearing containers (which were placed at outdoor temperatures, in an insectary) were closed with *perforated* plastic sheeting. In addition he had up to 250 moth larvae per container. These conditions tend to favour the spread of the virus and bacterial infections to which caterpillars are prone. As Balduf (1928) wrote, 'inasmuch as diseases of Lepidopterous larvae and pupae are common, and because their presence is often not indicated externally until they have reached an advanced stage of development, it is plausible that most, if not all instances of attack by Phoridae may be upon organisms more or less debilitated by internal disorders'.

It seems more likely, therefore, that Hesjedal's results were due to an epidemic disease; perhaps due to the crowded conditions favouring a high incidence of diseased and moribund moth larvae and pupae. This attracted *M. rufipes* females, who oviposited through the perforations in the plastic sheeting closures of his rearing containers. The larvae then developed as saprophages in the already dead larvae and as facultative parasites in the already diseased larvae.

I conclude that facultative parasitism of diseased larvae and pupae of Lepidoptera, by normally saprophage species of scuttle flies, has been demonstrated. Oviposition into healthy larvae or pupae by these species has yet to be established, but it remains a possibility. An intriguing report by Carpenter (1937) is certainly arresting. He collected an adult Burnet Moth (*Zygaena trifolii*) on 12 July. He killed it with ammonia. Three larvae of *Megaselia rufipes* emerged from the dead moth on 15 July. Less than three days for the egg and larval development would seem too short to allow the obvious explanation that a female laid its eggs on the dead moth.

Balduf's (1928) observation of facultative parasitization, of an already diseased caterpillar of *Achatodes zeae* (Noctuidae) by *Megaselia aletiae*, is of particular interest. The eggs were scattered on the back of the host's abdomen. The larvae invaded the hind gut of the caterpillar, and were observed both coming out and re-entering through the latter's anus. The posterior end of the host was distended by the presence of the fly larvae within. These larvae not only damaged the hind gut but reduced the adjacent fat bodies. Seemingly they were feeding on the fatty tissues, and not on the faeces in the hind gut. By contrast *M. giraudii* and *M. rufipes* have been reared exclusively from the frass (faeces) of the moth *Tyria jacobaeae* (Arctiidae) (Morley and Atmore, 1915).

Scuttle flies may also develop in caterpillars debilitated or killed by parasitization by tachinid flies. For example, a *Bombyx mori* (Bombycidae) larva (silkworm) parasitized by *Exorista sorbillans* has been reported giving rise to *Dohrniphora cornuta* adults as well (Coles, 1890, Bigot, 1890). Likewise a phorid has been reared from *Chilo partellus* (Pyralidae) parasitized by a tachinid (Butani, 1957).

Despite its name, the reports of *Megaselia parasitica* reared from larval noctuids (*Heliothis* sp. and *Spodoptera mauritia*) (Malloch, 1915a; Borgmeier, 1967e) may be only further cases of parasitisation of diseased hosts. Other cases, such as *M. densior* reared from the geometrid *Bupalis piniaria* (Rohlfien and Ewald, 1975), cannot be evaluated until further observations are available.

A case of facultative parasitization appears to be Wildermuth's (1914) report of pupae of the butterfly *Colias eurytheme* (Pieridae) attacked by *Megaselia perdita*. The pupae were clearly alive initially, before the emergence of the flies in the laboratory. However, this fly is also known to breed in dead arthropods (Kneidel, 1984a).

The one undoubted case of the parasitization of healthy Lepidoptera larvae by a phorid is that of *Megaselia chlumetiae* larvae attacking the larvae of the mango shoot-borer, *Chlumetia brevisigna* (commonly misidentified as *C. transversa* – see Holloway, 1985) (Noctuidae) (Disney *et al.*, 1992). Another reported case is that of a *Megaselia* species ovipositing into fourth- and fifth-instar larvae of the owl butterfly (*Caligo eurilochus*, Nymphalidae) and the completion of the larval development in the host pupae, producing up to 158 adults from a single butterfly pupa (Malo and Willis, 1961). The caterpillars of this butterfly are a pest of banana. Further investigation is clearly required, and the species of *Megaselia* needs determination.

3.5.6(h) Orthoptera (grasshoppers, locusts and relatives)

There are no unequivocal records of obligate parasitisation of Orthoptera by Phoridae. However, there are some cases of facultative parasitism by normally saprophage species such as *Megaselia giraudii*, *M. rufipes*, and *M. scalaris*. But most records for these species, and for others, are for rearings from dead Orthoptera (see Robinson, 1971). A clear case of facultative parasitism was described for *Tettigonia viridissima* (Tettigoniidae) parasitized by *M. giraudii* (Timon-David, 1938). A dozen fly larvae emerged from the mesothoracic spiracle of the living grasshopper, whose spiracle was enlarged and deformed. A dissection of the grasshopper showed that the phorid larvae had consumed the thoracic muscles.

Megaselia scalaris has been reared from adults of *Zonocerus variegatus* (Acrididae) debilitated by being first parasitized by the sarcophagid fly *Blaesoxipha filipjevi* (Gregorio and Leonide, 1980).

There are several reported cases of supposed parasitism, or facultative parasitism, of the egg pods of Acrididae (e.g. Schmitz, 1930b; Jack, 1935; Cuthbertson, 1936; Blanchard, 1937; Disney, 1988h). Greathead (1963) noted that Phoridae do not constitute a significant source of locust egg mortality, and regarded the reported cases as being likely to be secondary attacks by essentially saprophage species. He based this opinion upon the observations of Popov (1959) that a *Megaselia* sp. only attacked egg pods already damaged by carabid beetle larvae. Popov's actual data were 1.1% of the egg pods of *Locusta migratoria* examined ($n = 553$) were infested with phorid larvae. However, the infestation rate was 1.8% for egg pods already damaged by carabid larvae ($n = 282$). In the case of *M. oxybelorum* reported from egg pods by Schmitz (1930b), the same species has been reared from an adult fly, *Fannia scalaris* (Fanniidae), paralysed by the sphecid wasp *Oxybelus uniglumis* (Chevalier, 1925; Schmitz, 1928d). The common feature would seem to be a living, immobilized, slightly-wounded, insect resource buried in the soil.

3.5.6(i) Trichoptera (caddis flies)

The larvae of *Megaselia alsea* have been recorded as predators in the egg mass of the caddis fly *Hydatophylax hesperus*; the egg mass being deposited on vegetation above the water (Robinson and Wisseman, 1983).

3.5.7 VERTEBRATES

(1) Amphibia

Villa (1980) reported *Megaselia scalaris* larvae in decaying egg clutches of the frogs *Agalychnis callidryas* (Hylidae) and *Centrolenella fleischmanni* (Centrolenidae), and cited an unpublished-thesis report (by Snyder) of phorids ovipositing into the egg clutches of the salamander *Aneides aeneus* (Plethodontidae). Subsequently *M. scalaris* larvae were reported preying on viable eggs of *Agalychnis annae* and *Eleutherodactylus cogui* (Villa and Townsend, 1983).

Recently the larvae of an undescribed *Megaselia* species have been found preying on the viable eggs of the frog *Leptodactylus fuscus* (Leptodactylidae), which lays its eggs in nests out of water (R. Downie, Disney, *et al.*, in preparation).

(2) Mammals

The infestation of living vertebrates by fly larvae is called *myiasis*. The identity of the species of Phoridae involved in cases of myiasis in people and domesticated animals has been bedevilled by taxonomic confusions. Consequently the uncritical citation of the primary literature by textbook writers has led to the propagation of a considerable amount of misinformation. The following review of cases is based on the primary literature, including subsequent revisions of opinion as to the correct identity of the phorid species involved.

People (*Homo sapiens*)

Cases of myiasis due to Phoridae can be confidently attributed to particular species of fly only if larvae are reared to the adult stage. Many reported cases of myiasis are only identified as being due to Cyclorrhaphan larvae. How many of these are due to Phoridae can only be guessed at. The following review, therefore, is probably a very poor indication of the medical significance of Phoridae as a cause of myiasis. The observations of Ogawa *et al.* (1959) in Japan are of particular interest, as they were concerned with a mass treatment programme against hookworm and only incidentally discovered that 1.3% (n = 224) of the women they treated suffered from intestinal myiasis due to *Megaselia spiracularis* larvae.

In South America *Megaselia scalaris* has been reported in a case of cutaneous myiasis (Brues, 1915c). Larvae of *Megaselia* sp. (probably *M. scalaris*) have been reported in crusting lesions on the back of the neck and upper trunk of a 6-year-old child in Texas (Biery *et al.*, 1978).

Wright (1927) reported a case of larvae of *M. scalaris* in a crusted keratitis of the cornea of the eye, as a secondary complication.

Austen (1910) reported 63 larvae of *M. scalaris* in a wound on a man's foot in Belize. In Hawaii Hardy (1951) reported larvae of the same species invading the site of a year-old operation on the chest of a patient.

Larvae of a *Megaselia* (probably *M. scalaris*) have been reported in the nasal aspirate of a child in Texas (Carpenter and Chastain, 1992). About 30 larvae of a *Megaselia* (possibly *M. spiracularis*) have been extracted from the lungs of a 19-year-old man in Japan (Komori *et al.*, 1978).

Megaselia scalaris has been recorded in cases of intestinal myiasis in Australia (Lee, 1968), in Burma (Patton and Evans, 1929), in India (Stricklan and Roy, 1941), in East Africa (seven cases) (Rhodes-Jones, 1957), in Ghana (Ingram, 1922) and in Trinidad (Laurence, 1910). In Japan *M. spiracularis* has been reported in six cases of intestinal myiasis and *Dohrniphora* sp. (probably *D. cornuta*) in one case (Ogawa *et al.*, 1959; Kano *et al.*, 1962).

In these cases of intestinal myiasis it is surmised that infestation normally occurs as a consequence of ingesting eggs or early instar larvae in uncooked or cold food. The fate of eggs and larvae in the hostile conditions of the stomach has not been investigated.

Patton's Burmese case (in Patton and Evans, 1929) described the extraordinary situation of a man thought to have acquired an intestinal infestation, of *M. scalaris*, from eating candied bael fruit (the Bengal quince) contaminated with eggs or larvae. For about a year the patient passed newly-hatched larvae and puparia at about two-monthly intervals, and on one occasion passed about ten adult flies. On the basis of 'great care being taken to prevent the further ingestion of eggs and larvae', following the initial discovery of the infection, Patton proposed that the flies were mating and ovipositing while still in the intestine. In support of this hypothesis he reported having managed to breed more than one generation of *M. scalaris* in small corked vials (only opened to remove dead flies and to add more food for the larvae). This hypothesis, however, rests on the assumption that the precautions taken to prevent re-infestation were completely effective. Zumpt (1965) did not accept this. What is not in doubt is that *M. scalaris* is able to complete its larval development, produce viable pupae, and occasionally give rise to freshly-emerging adults, all inside the human intestine. This further enhances the reputation of this species as being one of the most extraordinary saprophages known to science (see above). Clearly further investigation is required into the apparent ability of this species to survive conditions verging on the anaerobic.

In the light of the above discussion along with our present knowledge of development times, some published, and frequently cited, records of intestinal myiasis by this species must be called into question. For example, Van Slype (1932) reported three cases of stool samples infested with eggs of this species, out of 341 cases examined. However, from the developmental times given (5.3), it seems more likely that these samples were contaminated after expulsion by the patients. The alternative explanation, that the patients swallowed the eggs and that these not only survived passage through the gut but did not hatch until after expulsion with the faeces, seems highly improbable. This is especially unlikely in view of the shortened development time to be expected at the temperature regime that would be experienced by the eggs while in the human gut. However, it is possible that the low oxygen regime may counteract the acceleration of development due to the relatively high temperature.

The passing of live larvae in the urine indicates infestation of the urogenital system. It seems that such infestations are normally a secondary consequence of some obstruction to the normal flow of the urine, such as adjacent contusions or stone formation in the urinary system. Urogenital myiasis due to *M. scalaris* has been reported in S.E. Asia (Priyanond *et al.*, 1973), in India (Singh and Rana, 1989) and Malawi (Meinhardt and Disney, 1989). A case of vaginal myiasis in Texas was caused by a *Megaselia* (probably *M. scalaris*) (Biery *et al.*, 1978). A case of urogenital myiasis in Japan was caused by *M. deningi* (Disney and Kurahashi, 1978; Disney, 1991i), the species being more characteristic of pitcher plants and the water retained in the internodes of giant bamboos (7.2.1).

Other Mammals

Larvae of *Dohrniphora cornuta* have been recorded from a wound on the tail of a bullock and in wounds on cattle (Patton, 1922; Zumpt, 1965, with amendments to the fly's identity by Borgmeier, 1968b). *Megaselia scalaris* larvae have been recorded in a gangrenous tail and in a wound on the neck of two different bullocks, and in a wound on the horn of a bull calf (Patton, 1922).

3.6 MISCELLANEOUS REARING RECORDS

Yakovlev (1986) reared a *Megaselia* (identified as *M. pulicaria*) from the slime mould (Myxomycota) *Fuligo* sp. The normal larval habit of *M. pulicaria* is as a predator on spider eggs (section 3.5.3). This scuttle fly, however, belongs to a sibling-species complex (Disney, 1989a) and so a closely related species may be the slime-mould breeder.

A number of Phoridae have been reared from bird nests. These are discussed in Chapter 7.

Malloch (1908b) found about 40 pupae in a mole's nest in Scotland. The adults he reared were about 30 *Anevrina urbana* and the rest *A. thoracica*. Lundbeck (1922) also procured the pupae of these two species in mole nests. Schmitz (1953a) reported the rearing of *Conicera floricola* from a mole's nest.

These species are among those reported as adults in mole's nests (see Chapter 7).

3.7 ENEMIES

3.7.1 PREDATORS

The larvae of the beetle *Rhizophagus parallelocollis* are thought to be predators of the larvae of *Conicera tibialis* feeding on buried human corpses (Blair, 1922; Colyer, 1954a). The earwig *Chelesoches morio* has been reported preying on the larvae of a *Megaselia* species breeding in decaying melon fruits (*Cucumis melo*) in Hawaii (Nishida, 1955). Larvae of the cecidomyiid midge *Lestodiplosis syringopais* prey upon the young aquatic larvae of *M. schuitemakeri* (Beaver, 1979). Predatory mosquito larvae of the genera *Armigeres* and *Toxorhynchites* have been observed preying on *M. deningi* larvae in bamboo internodes (Kovac and Streit, in press). The larvae of the muscid fly *Muscina assimilis* have been observed preying on *M. rufipes* larvae (Keilin, 1917). The nymphs and adults of the mirid bug *Deraeocoris ruber* prey upon pupae of the coccinellid beetles *Coccinella 7-punctata* and *Adalia 2-punctata* parasitized by the larvae of *Phalacrotophora berolinensis* and *P. fasciata*, killing both the host pupae and the phorid larvae (Disney *et al.*, 1994). These few records are probably a very poor indication of the mortality of phorid larvae due to predation.

3.7.2 PARASITES AND PARASITOIDS

The species known to attack Phoridae are listed below, along with the species of scuttle fly victims.

(1) Hymenoptera

Braconidae (braconid wasps)

Synaldis concolor parasitizes *Megaselia nigra*. While this wasp is reported as a pupal parasite (4.4.3), the evidence indicates that oviposition is into the young larvae (Hussey, 1960). Probably this applies to at least the majority of the records of Braconidae reported in section 4.4.3. The observations of Notton (1991a) support this conclusion, but in the case of *M. bovista* parasitized by Diapriidae he reported that the parasitized puparia had fully extruded respiratory horns. It is concluded that the larval wasps are, in such cases, endoparasites of the pupae. However, for convenience all observations on parasitoid Hymenoptera reared from puparia are treated in section 4.4.3.

(2) Nematoda (round worms)

Howardula husseyi – *Megaselia halterata*. The pre-adult females invade the larvae of the host. The full life cycle and effects of this parasite are outlined in section 6.8.2(b).

Nematodes have also been reported competing with the larvae of *Megaselia fuscinervis* in the same snail host (Disney, 1982h).

(3) Protozoa

Trypanosomatidae

Herpetomonas megaseliae – *Megaselia scalaris* (Dagget *et al.*, 1972). This parasite is found in the intestine of the host larva. Its ultrastructure has been studied by Janovy *et al.* (1974a, b).

4

Pupae

The pupa 'is a unique invention of insects, and its intercalation in the life history has made possible the extreme divergence between the young and the adult' (Bates, 1950)

4.1 INTRODUCTION

The pupa is enclosed in the sclerotized integument of the third-instar larva, the puparium. The puparia are typically as in Fig. 4.1, with paired eclosion plates penetrated by the pupal respiratory horns. An example of a modified puparium is that of *Megaselia fuscinervis*, which is always located in the same position in the shell of its snail host (3.5.2) (Fig. 4.2). The paired eclosion plates are evidently fused together and also fused to the anterior segments, the whole being heavily sclerotized. The rest of the puparium, protected by the host's shell, is only weakly sclerotized (Fig. 4.2). These modifications of the typical *Megaselia* state would seem to be a clear case of adaptations relating to the habit of pupariating in the host's shell, reminiscent of the distorted puparia of the snail-parasitizing Sciomyzidae (e.g. Knutson, 1970; Barnes, 1990c). Other variations are considered below.

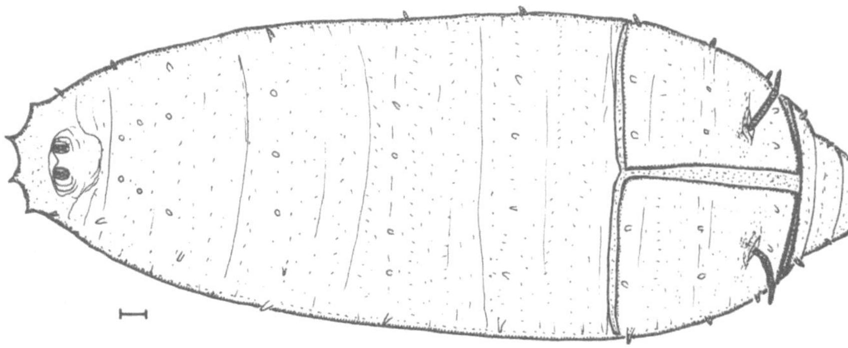

Figure 4.1 Pupa of *Megaselia rufipes*.

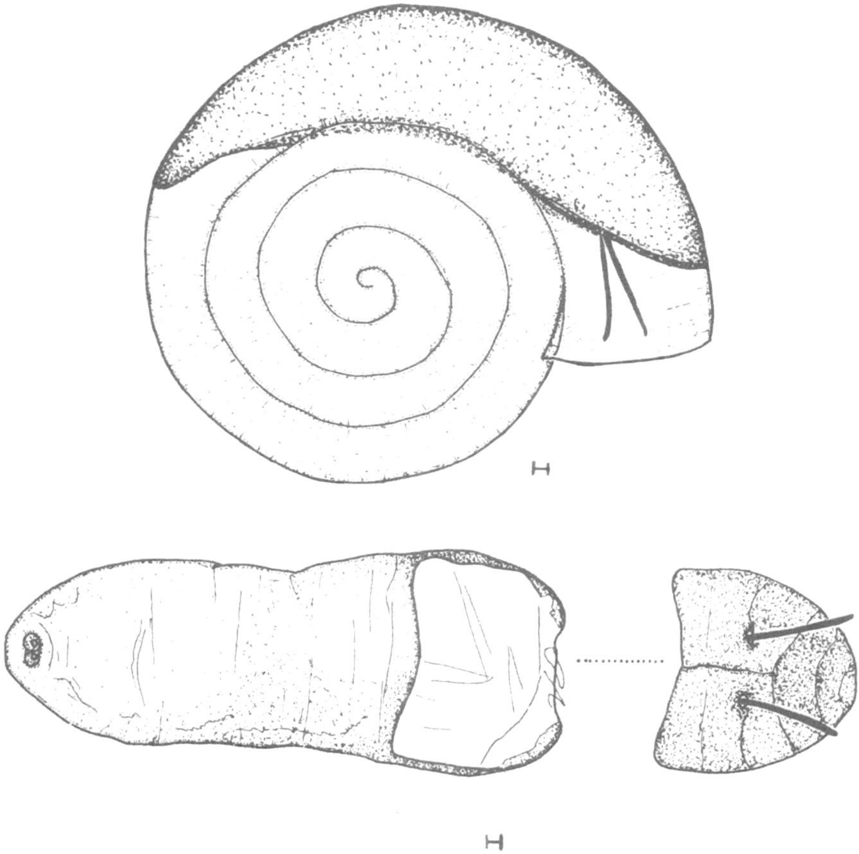

Figure 4.2 Pupa and puparium of *Megaselia fuscinervis* Above Pupa in shell of its larval host Below Puparium, with its detached eclosion plate, from above

In some phorids, e g *Misotermes* and *Megaselia anomaloterga* (Fig 4 3), there is a single dorsal plate and a separate anterior cap Schmitz (1938n, 1941c) considered this to be the ancestral (plesiomorphic) state and the double dorsal plates of typical *Megaselia*, and related genera, to be the derived (apomorphic) state Other modifications are the fusion of a single dorsal plate to the anterior cap, e g in *Pseudohypocera kerteszi* (Salt, 1929) In the semi-aquatic *Megaselia anomaloterga* the single dorsal plate is almost certainly derived by fusion of two plates, as in *M fuscinervis* In the fully aquatic *M campylonympha* there is a single dorsal plate and the anterior cap is no longer detachable It seems likely that in this case also the single plate is derived by a secondary fusion The puparium of *M campylonympha* is much modified in relation to its aquatic habit (see below) In *Spiniphora* the cap is also no longer detachable, except dorsally its rear portion has become fused to the single,

Figure 4.3 Puparia with a single dorsal plate and detachable anterior cap. (a) *Misotermes exenterans*, with detached anterior cap and eclosion plate above, and with posterior end and whole puparium below. (b) *Megaselia anomaloterga*, with detached anterior cap and single eclosion plate to right. ((a) after Schmitz, 1938i.)

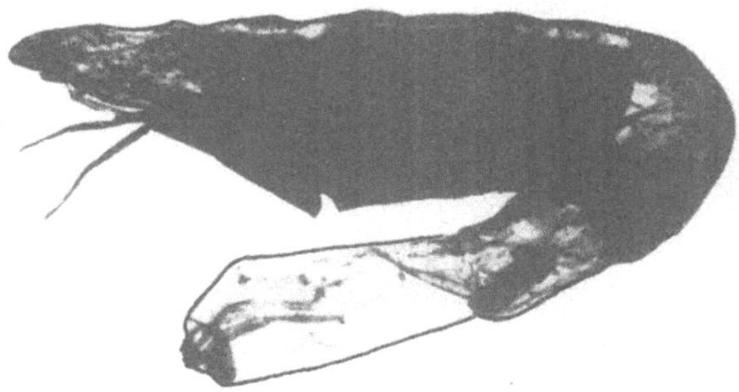

Figure 4.4 Puparium of *Megaselia campylonympha* in side view. (After Schmitz, 1931d.)

detachable dorsal plate. These different means of producing the exit hole from the puparium, for the emerging adult, merit further study. It is while attempting to escape from the puparium that Cyclorrhapha experience a moment of high risk. In the Schizophora the escape is facilitated by the pumping of haemolymph back and forth into the ptilinum (1.3) (Hinton, 1946). This adaptation no doubt decreases the risk of death through failure to escape from the puparium. An alternative, or supplementary, adaptation, serving the same end, is the evolution of more efficient escape hatches from the puparium, such as we see in those reviewed above. The independent evolution of a ptilinum in *Aenigmatistes* (1.3) suggests that the unknown puparium of this phorid genus may have a less sophisticated escape hatch.

Most of the aquatic species breeding in container habitats pupariate well clear of the water on the walls of the container. Their puparia are typical of *Megaselia*. However, some species, e.g. *M. campylonympha* and *M. schuitemakeri*, have the tail end of the puparium bent around to form a U (Fig. 4.4). It is inferred that this is an adaptation to facilitate respiration in the event of flooding, but critical observations are lacking. Another possible adaptation in relation to being in a liquid medium is the development of curious, thin-walled, flanges on two sides of each respiratory horn in *Dohrniphora cornuta* (Fig. 4.5). The detailed structure is described and discussed by Meijere

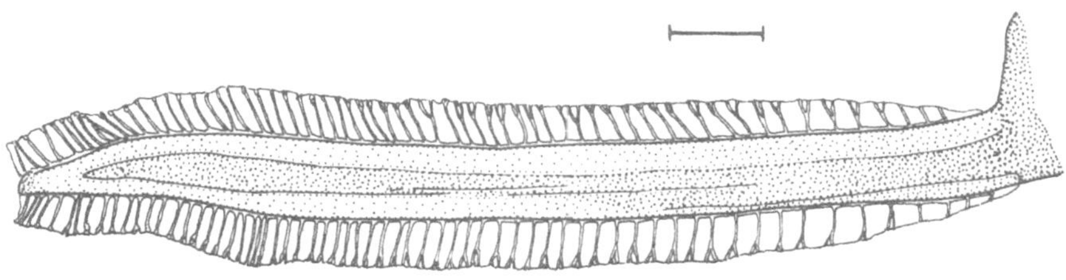

Figure 4.5 Pupal respiratory horn of *Dohrniphora cornuta*.

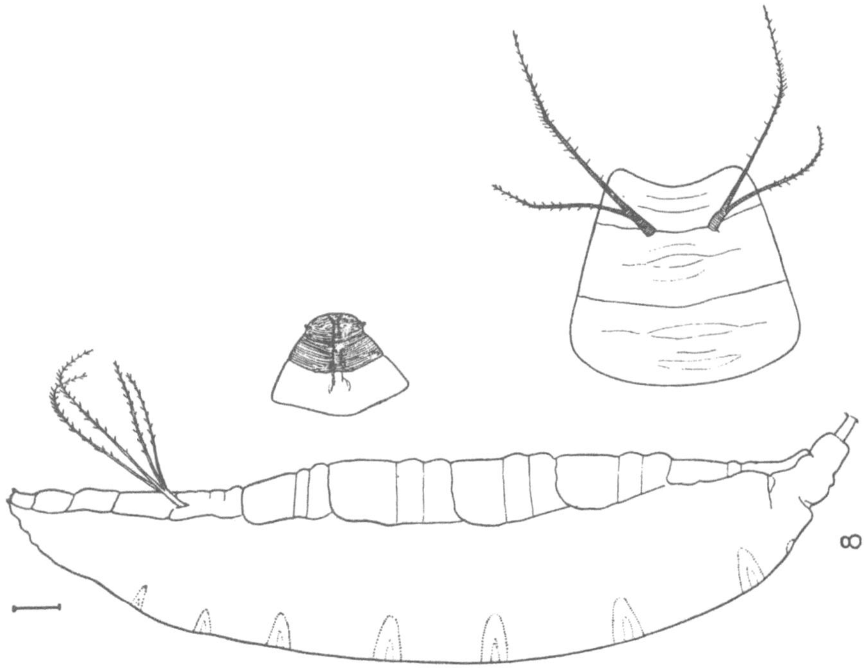

Figure 4.6 Puparium of *Chonocephalus heymonsi* in side view and dorsal plate and anterior cap from above. (After Borgmeier, 1935g.)

(1938). In *Chonocephalus heymonsi*, whose larvae are known to invade the more liquid parts of rotting fruits (3.4), the puparium has two-branched respiratory horns, a single dorsal eclosion plate and a detachable anterior cap, and the whole puparium is somewhat boat-shaped (Borgmeier, 1935g) (Fig. 4.6). Presumably these features are adaptations related to the liquid state of the pabulum. In *Spiniphora* the puparia are also boat-shaped and, in addition, they are cemented to the substratum. This is typically on the inside of the shell after consumption of the dead snail by the larvae (e.g. Lundbeck, 1922). Such shells are liable to flooding with rain water. The larvae of some species will develop in milk dregs left in discarded bottles, and the puparia are subsequently cemented to the walls of the bottle (e.g. Oldroyd, 1964).

The extraordinary *Metopina ciceri*, whose larvae infest the root nodules of chickpeas (see Chapter 3), is unusual. The puparium possesses the respiratory-horn tubercles on the eclosion plates, complete with the central weak spot for the penetration of the respiratory horns. However, the latter are short and weakly sclerotized and fail to puncture the central weak spots (Disney, 1988d). The figure of the puparium of *Metopina oligoneura* given by Froese (1992c) suggests this may be a characteristic of the genus. Normally a puparium with the pupal respiratory horns not protruding is one where the larva has been parasitized by a Hymenopteran or other parasitoid (4.4.3).

However, it seems that the respiratory horns of healthy pupae of Termitoxeniinae also fail to penetrate the puparium (Bridarolli, 1937b, 1940a), thus suggesting there is no need for the horns to penetrate the puparium when the pupa is small and the puparium is weakly sclerotized.

The sexual dimorphism in the size of adult *Megaselia scalaris* is reflected in the sizes of the puparia, but sizes and length:breadth ratios are variable and the ranges for the two sexes extensively overlap (Carareto and Mourao, 1988).

4.2 PUPATION

Pupation takes place after pupariation, and is normally signalled by the extrusion of the respiratory horns through the specially weakened spots on the eclosion plates of the puparium. It takes place about one day after pupariation in *Megaselia aequalis* (Robinson and Foote, 1968), one to two days in *Dohrniphora cornuta* (Barnes, 1990a) and three days in *M. chainensis* and *M. submarginalis* (Robinson, 1970; Robinson and Brown, 1993).

4.3 OVERWINTERING

About one month after pupation the adults of *Megaselia chainensis* and *M. submarginalis* are fully formed, but they then overwinter within the puparia until the following spring (Robinson, 1970; Robinson and Brown, 1993). This probably applies to many phorids which are reported to overwinter 'as pupae'. It would be more accurate to say that these phorids overwinter within the puparium. Examples of such reports are the following:

Anevrina thoracica in Finland (Hackman, 1963).

Megaselia rubricornis in Germany, the pupal period being almost a year (Kanervo, 1946).

Megaselia rufipes in Europe (e.g. Eisenschmidt, 1958).

Megaselia winnemanna in America (Robinson, 1970).

Menozziola schmitzi in Italy (Goidanich, 1928b).

Phalacrotophora berolinensis and *P. fasciata* in England (Disney, 1979f).

Spiniphora maculata in England (Disney, 1973).

Triphleba hyalinata in Belgium. This species diapauses as a larva and then as a pupa (Leclercq and Watrin, 1973).

Triphleba lugubris larvae collected in Belgium in October pupariated in the same month and gave rise to adults from 6 February to 10 June. Peak emergences were in April for pupae kept at 100% relative humidity, but in May for pupae at lower relative humidities (5.4) (Leclercq, 1946b).

4.4 ENEMIES

4.4.1 FUNGI

Cephalosporium sp. This fungus has been reported attacking pupae of *Megaselia halterata*. It was suggested, however, that it may be saprophytic on already moribund pupae (Hussey and Gurney, 1963). The matter requires investigation.

4.4.2 DIPTERA

Müller (1919) reared the scatopsid fly *Swammerdamella brevicornis* from a phorid pupa. This merits further investigation.

4.4.3 HYMENOPTERA

Parasitic Hymenoptera undoubtedly exert a significant control on the populations of many Phoridae. For example, Bin (1972) reported that 45% ($n = 600$) of the pupae of *Spiniphora dorsalis* he studied had been parasitized by three species of Parasitica.

Brauns (1950) reported *Synaldis concolor* parasitizing the pupae of *Megaselia nigra*, the parasitized pupae being recognizable by the failure of the respiratory horns to protrude. However, the evidence indicates that oviposition is into young larvae of the host (Hussey, 1960). Indeed, a pupa whose respiratory horns fail to penetrate the integument of the puparium is normally not viable. Frequently this failure is a sign of parasitized larvae which have pupariated but not pupated. The parasitisation of the larvae almost certainly renders them incapable of pupation. However, the adults of these parasitoid Hymenoptera emerge from the puparia and so, as a matter of convenience, they are treated here rather than in section 3.7.2(1). The observations of Notton (1991a) indicate that the Diapriidae are indeed endoparasites of the pupae but the Braconidae are endoparasites of the larvae (3.7.2).

Braconidae

The Alysiinae are clearly an important group parasitizing Phoridae. The taxonomy of these small wasps, however, is still poorly known. Fischer's (1974) paper needs evaluating in the light of that by van Achterberg (1988). The following review is clearly a very incomplete picture and there is much scope for further investigations.

Aspilota spp. – *Megaselia flava, M. giraudii, M. hirtiventris, M. rufipes, Spiniphora bergenstammi* and *S. dorsalis* have all been reported as hosts of unidentified species (Bin, 1972; Disney, 1979f; Disney and Evans, unpublished data; Lundbeck, 1922; Postner, 1952; Schmitz, 1938c).

A. ahrburgensis – a phorid (Fischer, 1974).

A. fuscicornis – *Spiniphora* sp. (Schmitz, 1935j) and *Conicera dauci* (Fischer, 1974).

A. intermedia – *Megaselia bovista* (Notton, 1991a,b).

A. latifemur, A. nervulata, A. perlustranda and *A. reductidens* – unidentified phorids (Fischer, 1974).

Dinotrema spp. – *M. eisfelderae, M. flava, M. lutescens* and *M. scutellaris* (Disney, W. Dreyer and R. E. Evans, unpublished data).

D. amplisignata – *Spiniphora dorsalis* (van Achterberg and Bin, 1981). Bin (1972) reported an infestation rate of 0.26 ($n = 600$ pupae).

D. mesocaudatum – *M. hirtiventris* (van Achterberg, 1988).

D. phoridarum – *Spiniphora signata* (Bin, 1972; Fischer, 1974; van Achterberg and Bin, 1981).

Opius leptostigma – ex phorid (Lundbeck, 1922).

Orthostigma sp. (cf *lucidum*) – reared from same fungus as *Megaselia flava* (Disney and Evans, unpublished data).

O. pumilum – *M. rufipes* (Scott, 1920; Lundbeck, 1922; van Achterberg, 1988).

Panerema (= *Aspilota*) *inops* – *M. fuscinervis* (Disney, 1982h; van Achterberg, 1988).

Synaldis concolor – *M. nigra* (Brauns, 1950). The same wasp has been recorded in nests of the ant *Solenopsis fugax* inhabited by *Metopina formicomendicula* (Hölldobler, 1928).

Diapriidae

The observations of Notton (1991a) indicate that these wasps are solitary endoparasites of the phorid pupae, the pupal respiratory horns of the latter being fully extruded through the puparium. Amerling (1862) reared '*Diapria phorae*' from phorid pupae. This diapriid was most probably a species of *Trichopria*. The following are the available records of named diapriids reared from phorid puparia.

Aneurhynchus lativentris – *Spiniphora maculata* (Lundbeck, 1922).

A. phorivora – *S. maculata* (Lundbeck, 1922).

Basalys sp. – *Megaselia bovista* (Notton, 1991a).

B. helicola – *Spiniphora dorsalis* (Bin, 1972), with an infestation rate of 0.14 ($n = 600$).

B. longipennis – *S. bergenstammi* (Disney and R. E. Evans, unpublished data).

B. parva – *Megaselia bovista* (Notton, 1991a) and *M. abdita* in laboratory experiments (Notton, *in litt.*).

B. pedisequa – reared from phorid puparia in spider egg sacs (see 3.5.3 (2)) (Fitton *et al.*, 1987; Disney and R. E. Evans, unpublished data).

Idiotypa nigriceps – two *Megaselia* species (D. G. Notton, *in litt.*).

Spilomicrus sp. – *Spiniphora bergenstammi* (Lundbeck, 1922).

S. abnormis – *Megaselia* sp. (D. G. Notton, *in litt.*).

S. crassicornis – *Diplonevra oldenbergi* (Lundbeck, 1922).

Synacra brachialis – *Megaselia* sp. (Canzanelli, 1941).

Termitopria sheasbyi – *Syntermitoxenia pseudonanna* (Muesebeck, 1965).

Trichopria sp. – *Megaselia* sp. (Beaver, 1979).

Encyrtidae

Beaver (1979) reared an encyrtid from pupae of *Megaselia schuitemakeri* and Thienemann (1932) reported an encyrtid from *M. tobaica*. *Exoristobia philippinensis* has been reared from *M. curtineura* and a *Megaselia* sp., there being one larva per third-instar fly larva, which pupariates before the adult wasp emerges (Beaver, 1987). *Homalotylus flaminius* has been recorded as a parasitoid of a larval *Phalacrotophora* species, the adult wasp emerging from the fly puparium (du Buysson, 1917).

Eucoilidae

Didictyum zigzag has been reared from *Megaselia aletiae* (Lundbeck, 1922) and *Trybliographa* sp. from *M. deningi* (Disney, 1982n).

Eulophidae

Entedon costalis has been reared from *Spiniphora* sp. (Lundbeck, 1922) and *Tetrastichus* sp. from *Phalacrotophora epeirae* pupae, the infestation rate being 0.70 ($n = 53$) (Muma and Stone, 1971).

Ichneumonidae

A *Gelis* and a *Hemiteles* species have been reared from *Spiniphora* sp. and *Phygadeuon flavicans* from *Anevrina* (probably *A. thoracica*) (Lundbeck, 1922); and *Pimpla annulipes* from egg sacs of the spider *Nuctenia* sp. parasitized by *Phalacrotophora epeirae* (Brues, 1902a).

Platygastridae

Platygaster aegeus has been reared from *Megaselia rufipes* (Morley, 1934).

Proctotrupidae

Undetermined species have been reported from pupae of *Chonocephalus depressus* and *Megaselia gregalis* (Rao, 1961; Meijere, 1910).

Pteromalidae

Eupteromalus sp. parasitizes pupae of *Spiniphora dorsalis*, with a reported infestation rate of 0.05 ($n = 600$) (Bin, 1972; van Achterberg and Bin, 1981). *Spalangia nigroaenea* was reported from a culture with larvae of *Megaselia* sp. and the blowfly *Lucilia sericata* (Roberts, 1935), it not being clear which fly was the host of the wasp.

4.4.4 NEMATODA

The adult females and eggs of *Howardula husseyi* occur in the haemocoel of the pupae of *Megaselia halterata* (see section 6.8.2(b) for full details of the life cycle and effects on the host).

5

Development

As the animals came to live in new (drastically different) environments . . . some of them managed to adapt themselves by changing their developmental timing, by either acceleration or retardation in the onset of development of structures (heterochrony) (Matsuda, 1987)

5.1 INTRODUCTION

The information on the development of scuttle flies is scattered in the literature and is patchy in coverage. There is a need for more detailed comparative studies. Development within the egg is a largely unexplored field, but sometimes starts before the egg is laid (2.2, 2.4). Wolf and Blackman (1992) describe the ultrastructure of the developing gonads in the pupa of *Megaselia scalaris*.

5.2 NUMBER OF LARVAL INSTARS

There are typically three, free-living, larval instars. This generalization is derived from the rearing of species in the laboratory. For example, this has been established for *Spiniphora bergenstammi* and *Megaselia rufipes* (Keilin, 1911b), *Wandolleckia achatinae* (Baer, 1953), *Phalacrotophora berolinensis* (Wylie, 1958) and *Dohrniphora cornuta* (Barnes, 1990a). The third instar tends to last longer than the other two. Thus in *D. cornuta*, at 22–24 °C, the first instar lasts 1–2 days, the second 1–2 days, and the third 3–4 days before pupariation and a further 1–2 days before pupation (Barnes, 1990a).

In the Termitoxeniinae, whose females spend most of their lives in termite nests, there is a curtailment of the free-living larval stages associated with the evolution of enormous long-lived eggs. The free-living larvae last only a few minutes in *Clitelloxenia hemicyclia*, but several hours in *Javanoxenia puntiventris* and *Odontoxenia brevirostris* (Kemner, 1926; Franssen, 1933b; Bridarolli, 1937b, 1940a). In the case of *C. hemicyclia* the free-living larvae take no food, but the other two species feed on the termite fungus-garden microtubers (conidia) (Franssen, 1933a-b).

Kemner (1926) noted that the mouth hooks of *J. punctiventris* are most strongly developed in the hatchling larva, presumably to facilitate escape from the egg. It moults soon after hatching. Furthermore this moult seems to coincide with pupariation, so that the last instar experiences no free-living existence. It is the instar emerging from the egg, of this species and of *O. brevirostris*, which lasts several hours (Bridarolli, 1937b). This instar is naked, whereas the next (whose integument forms the puparium) is hairy (Franssen, 1936). Furthermore it can be inferred that the larvae are fully formed in the egg some time before hatching, as Franssen (1933b) found that if batches of fly eggs are immersed in water an unexpectedly large proportion of small larvae will emerge within a few minutes.

In the case of *C. hemicyclia* the hatchling larva emerges already part moulted to the last instar, so that it has two skins (Bridarolli, 1937b; summary reproduced in Schmitz, 1938b). The outer skin is initially distended on parts of the ventral face. During its brief existence the larva rapidly loses these swellings and then moults by a longitudinal split from the front to rear. The inner skin then forms the puparium. Bridarolli considered the two larval stages outside the egg to be the first and second, the third having been suppressed. Some (e.g. Oldroyd, 1964) have accepted this interpretation, but others (e.g. Ferrar, 1987) consider the pupariating instar to be the third. This implies that the first instar is either suppressed or confined to the egg. The latter interpretation is the more likely, but only further research can settle the matter.

Prior to Kemner's (1922, 1926) discovery of larval and pupal Termitoxeniinae, Wasmann (e.g. 1902a, 1913) advanced his Ametabolism Theory to account for their apparent lack. He suggested the young adult females (stenogastrics – see 5.5) emerged from the egg. This theory was rendered redundant by Kemner's discovery, but some textbooks have continued to repeat it along with other discarded theories (see 5.6).

5.3 DEVELOPMENT TIMES

The relative durations of the egg, larval and pupal stages in any species would appear to be the result of evolution, and therefore primarily under genetic control. Within this constraint, however, in practice the actual duration of development is largely a function of temperature, at least up to the onset of any diapause. Data on development times are sparse in the literature. In cases of the infestation of food or human corpses by phorid eggs, larvae or pupae the interpretation of forensic evidence will frequently demand knowledge of the temperature regime experienced by the food sample or corpse. The relatively constant temperature regimes of laboratory experiments are unnatural.

What little we know of development times, and their modulation by temperature, is reviewed below. Most larval and pupal periods recorded in the literature give no information on the temperature regime. Only where both periods are recorded are these times noted below. The 'pupal periods' are from pupariation to eclosion of the adult. Pupation strictly begins after pupariation and is usually signalled by the extrusion of the respiratory horns

(4.2); furthermore, in 'overwintering pupae' the adult may be fully formed within the puparium months before eclosion (4.3).

In this review the sex is noted only when data for males and females were recorded separately; the temperatures (*t* – °C) are expressed as means, ranges or, where unknown, as 'lab'; the times are in days for E = the egg, L = the larva, P = the pupa and T= the total.

Apocephalus paraponerae: t – *c*.26, E – 0.27–0.29, L – 2.3–4.9, P – 17–26, T – 19–23, the larvae only feeding for about two days (Brown and Feener, 1991a).

Borophage germanica: t – 15, P – 34–37 (Gemesi and Disney, 1991).

Chonocephalus depressus:

female, *t* – 23–26, E – 2, L – 8–14, P – 7–8, T – 17–24

male, *t* – 23–26, E – 2, L – 8–14, P – 9–10, T – 19–26 (Rao, 1961).

Diplonevra armipes: lab, E – 0.75–1.0, L – 6–7, P – 12–14, T – 19–22 (Cuthbertson, 1936).

Diplonevra mortimeri: lab, E+L – 8, P – 10, T – 18 (Disney, 1986d).

Dohrniphora cornuta:

lab, L – 15–21, P – 25–40, T – 40–60 (Weiss, 1911).

t – 22–24, E – 1, L – 5–8, P – 13–15, T – 21–28 (Barnes, 1990a).

Gymnophora luteiventris: lab, E – 0.7–1, L – 15, P – 15, T – 31 (Brown and Feener, 1991a).

Gymnoptera molluscivora: t – 29, L – 12, P – 4 (Bohart and Gressitt, 1951).

Megaselia aequalis: t – 20–25, E – 2–4, L – 6–9, P – 10–12, T – 23–31 (Robinson and Foote, 1968).

M. agarici:

t – 7, P – 85, T – 197 (Snetsinger, 1972).

t – 13, T – 58 (Snetsinger, 1972).

t – *c*.18, E – 3, L – 14, P – 21, T – 38–39 (Snetsinger, 1972).

t – *c*.24, E – 2, L – 7, P – 14, T – 20–23 (Snetsinger, 1972).

M. aletiae: lab, E+L – 7–11, P – 7–11, T – 21 (Balduf, 1928).

M. chainensis: lab, E – 0.5–1, L – 11–17, P – 30+ (overwinters) (Robinson and Brown, 1993).

M. curtineura: t – 25, E – 3, L – 4–6, P – 6–12, T – 13–21 (Yagi *et al.*, 1977).

M. halterata:

t – winter, T – 42–56 (Moreton and John, 1955).

t – 10, E – 7.5, L – 33.0, P – 65.5, T – 106 (Hussey, 1961b).

t – 16, T – *c*.45 (van Zaayen, 1983).

t – 18.3, E – 3.4, L – 12.3, P – 21.0, T – 37 (Hussey, 1961b).

t – summer, T – 21–28 (Moreton and John, 1955).

t – 19–21, T – 24–28 (Richardson and Hesling, 1977a).

t – 20–24, E – *c*.2, L – 10–14, P – 12–21, T – 24–37 (Hussey and Gurney, 1964a).

t – 21.1, E – 3.0, L – 8.0, P – 12.0, T – 23 (Hussey, 1961b).

t – 23.9, E – 2.0, L – 4.5, P – 7.5, T – 14 (Hussey, 1961b).

t – 24, E – <2 (Hussey and Gurney, 1964a).

t – 24, T – 15 (White, 1983).

t – 24, T – 13 (van Zaayen, 1983).

Note: I have omitted the data reported for a temperature of 28 °C by Somchoudhury *et al.* (1988), because of doubts about the correct identity of the flies (Johal and Disney, in press).

M. longicostalis: lab, E+L – 7–14, P – 30–43 (Schmitz, 1938c).

M. nigra:

t – 18–19, E – 3, L – 4, P – 5, T – 12 (Hussey, 1960).

in field, P – 26 (Hussey, 1960).

Note: I have omitted the data of Somchoudhury *et al.* (1988) for the reasons given above.

M. rubricornis: t – 17.7–23.2, E – 1.5–2, L – 5.5–7, P – 'almost a year' (Kanervo, 1946).

M. scalaris:

t – 15, E – 2, L – 19.7 (Prawirodisastro and Benjamin, 1979).

female, *t* – 20, E – 1.1, L – 10.6, P – 17.5, T – 29.2 (Prawirodisastro and Benjamin, 1979).

male *t* – 20, E – 1.1, L – 10.6, P – 17.7, T – 29.4 (Prawirodisastro and Benjamin, 1979).

t – 21, E+L – 16.7, P – 20.1, T – 37.0 (Trumble and Pienskowski, 1979).

t – 18–20, T – 21–27 (Patton, 1922).

t – 20–25, E – 1, L – 6–9, P – *c.*14, T – 20–25 (Robinson, 1975).

t – 22, E – 1.3+, L – 5.5+, P – 15.1+, T – 22+ (Greenberg, 1991).

t – 23, E+L – 5–6, P – 11–12, T – 17–18 (Semenza, 1953).

t – *c.*25, E+L – 9, P – 14, T – 23 (Van Slype, 1932).

female, *t* – 25, P – 18.2, T – 23+ (Benner and Ostermeyer, 1980).

male, *t* – 25, P – 12.1, T – 21+ (Benner and Ostermeyer, 1980).

t – 25, E – 0.7, L – 7.3, P – 9.8, T – 18 (Prawirodisastro and Benjamin, 1979).

female, *t* – 27, E – 0.7, L – 5–7, P – 9, T – 15.5 (El-Miniawi and Moustafa, 1965).

male, *t* – 27, E – 0.7, L – 4–6, P – 9, T – 14.5 (El-Miniawi and Moustafa, 1965).

t – 27, E+L – 6.8–6.9, P – 10.5, T – 17.4 (Trumble and Pienskowski, 1979).

female, *t* – 26–28, E – <1, L – 5–9, P – 6–9, T – 16–22 (Tumrasvin *et al.*, 1977).

male, *t* – 26–28, E – <1, L – 5–9, P – 6–9, T – 15–20 (Tumrasvin *et al.*, 1977).

t – 29, E – 0.7+, L – 3.5+, P – 6.9+, T – 11.1+ (Greenberg, 1991).

t – 32, E+L – 5.4, P – 7.3, T – 13.5 (Trumble and Pienskowski, 1979).

Note: I have examined specimens of this species reared from contaminated food kept at −2 °C to −3 °C in the freezer compartment of a fridge in Hong Kong. Evidently this primarily warm-climate species can, at least, complete its development at these low temperatures.

M. submarginalis: lab, L – 10–17, P – 30+ (overwinters) (Robinson and Brown, 1993).

M. winnemanna: lab, E – 0.5–1, L – 12–18, P – 30+ (overwinters) (Robinson, 1970).

Melaloncha ronnai: lab, P – 21–22, T – 20–35 (Ronna, 1936, 1937a).

M. sinistra: lab, E+L – 13, P – 21 (Simões *et al.*, 1980).

Phalacrotophora berolinensis: lab, E – 1, L – 5 (Wylie, 1958).

P. epeirae: lab, E – 1–2, L – 4–8, T – 14–28 (Muma and Stone, 1971).

P. quadrimaculata: t – 21.5–25.5, E – 1, L – 4, P – 16–19, T – 20–24 (Disney and Chazeau, 1990).

Phora holosericea: t – 22–26, L – 11–14, P – 18–22 (Yarkulov, 1972).

Pseudohypocera kerteszi: lab, P – 4–6 (Salt, 1929).

Wandolleckia achatinae: t – 26, E – 2, L – 7, P – 9, T – 18–20 (Baer, 1953).

These data indicate that the ratio of the larval period to the pupal period (LP/PP) is typically between 0.5 and 1.0. The ratio in *Chonocephalus depressus*, however, is 1.18. It would seem that this is due to the larval stage being longer than expected, rather than an unusually short pupal period. By contrast the ratio is only 0.23 in *Phalacrotophora quadrimaculata*, indicating a relatively long pupal period. Comparison with other species in the above list indicates that it is the duration of the earlier stages that is curtailed relative to the pupal period. Females of the genus *Phalacrotophora* are known to feed on insect haemolymph (6.3). Perhaps the females of *Chonocephalus* do not take in protein-rich meals and consequently need to build up food reserves for their eggs during the larval stage.

The larval period can be subdivided into a feeding stage and a post-feeding stage (in the last phase of the third instar). The ratio of the durations of the feeding and post-feeding stages in *Megaselia scalaris* is 1.2:1 at 22 °C and 1.5:1 at 29 °C (Greenberg, 1991). Discrimination between feeding and post-feeding third-instar larvae may be useful in forensic cases.

Kemner (1926) established that the duration of the egg stage of *Javanoxenia punctiventris* is at least three weeks. Franssen (1933b, 1936) confirmed this and reported the same for *Clitelloxenia hemicyclia* and *Odontoxenia brevirostris*. This stands in marked contrast to the brief free-living larval stage (see above). Related to this exceptionally long incubation period is the relatively enormous size of the egg of *J. punctiventris*. It measures 0.64 × 0.32 mm, compared with the female's body length of 1.2–1.6 mm. Similarly the egg of *C. hemicyclia* measures 0.81 × 0.40 mm, compared with a female body length of just over 2 mm. By contrast in *Megaselia scalaris* the egg measures about 0.6 mm in length, compared with a female body length typically about 3.5 mm.

In his rearings of phorid flies from dead snails (*Achatina fulica*) Beaver (1987) recorded the total development times under comparable, but unspecified, conditions. These data allow the species to be ranked in terms of relative duration of development (Table 5.1).

Table 5.1 Total development of times of phorids reared under the same conditions

Species	Mean duration (+ range) in days
Dohrniphora cornuta	11.7 (11–13)
Megaselia curtineura	16.9 (14–24)
Spiniphora genitalis	19.3 (15–40)
Conicera formosensis	27.9 (26–32)
Gymnoptera orientalis	28.9 (21–30)

Evidently *Dohrniphora cornuta* develops more than twice as rapidly as *Gymnoptera genitalis*, under the same conditions. This difference presumably largely represents a difference in the genetic component of the determination of the duration of development.

5.4 PUPARIATION AND PUPATION

Experiments with *Megaselia halterata* showed that pupariation (indicated by darkening of the integument) was successful only if the relative humidity exceeded 90%. The pupae remained healthy at relative humidities above 86% (Hussey and Gurney, 1963). Likewise Leclercq (1946b) showed that pupae of *Triphleba lugubris* failed to complete development at relative humidities below 75%. At values of 75%, 86% and 95% between 50 and 75% of the pupae gave rise to adults. At 100% relative humidity almost all the pupae successfully produced adults.

With *Megaselia scalaris* the male larvae pupariated two days earlier than the female larvae, at 25 °C (Benner and Ostermeyer, 1980), and the pupal periods of the males were shorter also. Males emerging before females may be able to feed and mature their sperm before the females emerge. However, others failed to find a difference in the pupal periods of males and females (5.3).

5.5 ADULT DEVELOPMENT

A number of phorid species, especially in the Termitoxeniinae and other genera inhabiting termite nests, emerge from the puparium as stenogastric forms, and subsequently develop into physogastric forms (Fig. 5.1). While the bulk of the abdomen in the physogastric stage is occupied by the greatly enlarged ovaries, the evidence indicates that this development includes some enlargement of other organs as well (e.g. Mergelsberg, 1935). This phenomeon has been reviewed by Kistner (1982). In several Termitoxeniinae there is a post-emergence, post-mating, elongation of the rear part of the head. Likewise full pigmentation of the legs and other parts may not be completed until after mating and the shedding of the wing membranes. The flying-stage, early stenogastric, female is, in effect, teneral. It is possible that some early stenogastric termitoxeniine females are too teneral to be capable of flight, and so are completely dependent on males for aerial transport. In most early stenogastric females the sclerotization is more advanced in the thorax than in other parts, such as the legs, but the thorax is initially less heavily sclerotized than that of males or physogastric females. A comparison of the rates of post-emergence sclerotization of different parts of female Termitoxeniinae, with the process in species such as *Megaselia scalaris*, might reveal a considerable measure of heterochrony in some species.

Enlargement of the abdomen can be induced in a flying-stage stenogastric female by drowning it in water. The uptake of water leads to inflation of the abdomen and ventral deflection of its terminal segments, mimicking the changes observed during development (Disney and Cumming, 1992).

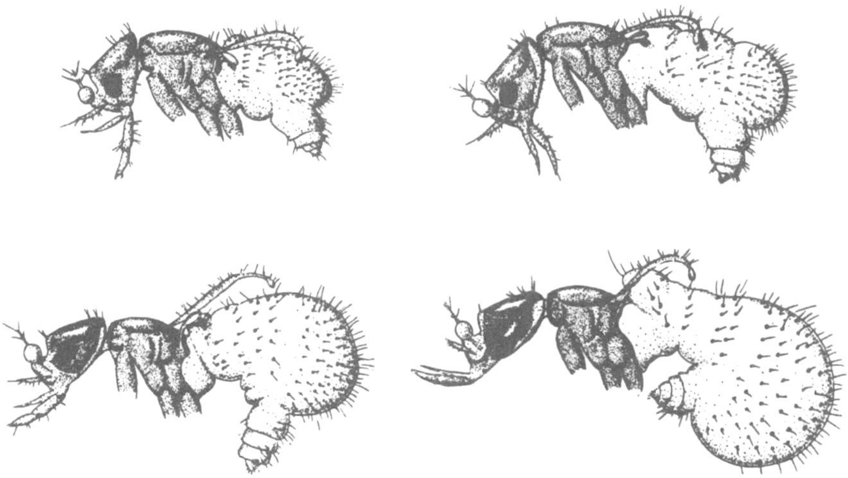

Figure 5.1 *Termitophilomyia gracilis* females (with legs omitted) showing development from a stenogastric to the mature physogastric stage. (From Mergelsberg, 1935.)

During normal development there is no evidence of any increase in glandular tissue (Malin and Kistner, 1970). Wilson (1971) has suggested that the distension serves to project the fatty tissues in all directions, and that the fatty exudates include substances which appease the host termites. He concluded that the abdomen has been 'reshaped into a kind of conciliatory device'. Whether the fatty tissues do indeed perform this function, or are solely concerned with the supply of nourishment to the enormous eggs, remains to be properly investigated. There is little doubt that the prime function of the physogastric inflation and distortion of the abdomen is to accommodate the relatively enormous eggs. The enormous size of the eggs is related to their long incubation period and to the mimicry of their host termite's eggs (2.1). The long incubation period allows curtailment of the free-living larval stage (5.2).

5.6 ALLEGED PROTANDROUS HERMAPHRODITISM AND PARTHOGENESIS

The story of the fates of Wasmann's various theories relating to the Termitoxeniinae is fully documented elsewhere (Disney, 1992c, 1994a; Dessart, 1993). Here I refer to his theory of protandrous hermaphroditism, according to which the stenogastric stages were functional males which then became functional females during the physogastric stage. This theory is still to be found in many textbooks. It was based, however, on inferences drawn mainly from preserved material and limited observations in the field. In particular it placed too much emphasis on what had **not** been observed in the field, giving undue weight to the failure to find any males in the termite mounds. The material had been collected exclusively by the excavation of termite nests. The evidence now available, however, indicates that the initial stage of the termitoxeniine female is not the flightless stenogastric stage collected in termite mounds, but the fully winged stenogastric stage with its anus directed rearwards. It appears that these females normally shed their wing membranes at the time they take up residence in the termite mound. Before that they can be found outside the mounds, along with the fully-winged males (Disney and Cumming, 1992). There is no evidence of hermaphroditism in any Phoridae.

A number of authors (e.g. Rohdendorf, 1974; Ferrar, 1987) suggested an alternative theory to explain the reported lack of males in Termitoxeniinae. They proposed that some might be parthenogenetic. There is no evidence to support this theory. The males of several Termitoxeniinae are now known and in several species where the males are still unknown, sperm has been observed in the spermatheca (sperm storage sac) of the female. Indeed this sperm-filled sac was central to Wasmann's hermaphroditism theory, according to which it was to be interpreted as a single testis. The rival school of thought (e.g. Bugnion, 1913; Silvestri, 1920) interpreted this sac as a spermatheca. Some authors (e.g. Grassé, 1986) accepted both theories, concluding that some genera had a testis and some a spermatheca. With the demise of the hermaphroditism theory we can conclude that the sperm-filled sac of stenogastric females is always a spermatheca. When sperm is observed in this spermatheca, even in species whose males are still unknown, we can safely conclude that they are not normally parthenogenetic. A more likely role for parthenogenesis is in the production of the blind morph of *Metopina ciceri*, and probably *M. psociformis* as well. These blind morphs may be produced parthenogenetically, while the sighted morphs are produced normally. However, this is mere speculation and the mechanisms underlying the production of these contrasting morphs require investigation (5.11).

5.7 SEX DETERMINATION

Three pairs of chromosomes have been described and characterized for *Megaselia scalaris* (Traut *et al.* 1990). In each pair the two chromosomes look the same (i.e. they are homomorphic). Thus the X and Y chromosomes, the sex-determining pair, are homomorphic like the other two pairs (the autosomes). However, molecular studies indicate that the Y chromosome contains a conserved segment in which crossover is suppressed, and restriction site differences have accumulated relative to the X chromosome. These chromosomes, therefore, conform to a theoretically expected early stage of sex chromosome evolution (Willhoeft and Traut, 1990). There is an epistatic male determining factor (M), which is normally located on the Y chromosome, but which occasionally appears to undergo transposition to what is usually an autosome (Traut and Willhoeft, 1990). These studies confirm earlier reports (e.g. Mainx, 1964) that a sex-determining linkage group is not completely fixed. Indeed Springer (1967) demonstrated that individuals with M located on an autosome are favoured under certain environmental conditions.

This observation would seem to have wider theoretical implications. It suggests the possibility of a mechanism for sympatric speciation (the origin of new species without geographical isolation of populations of the ancestral species). The proliferation of species in the giant genus *Megaselia*, estimated to include between 5000 and 20 000 present-day species (Disney, 1983b), is difficult to envisage as being entirely due to allopatric speciation. Parapatric speciation (contiguous but non-overlapping speciation) has evidently been important in groups, such as snails, with limited powers of dispersal (e.g. Cowie, 1992). It would seem unlikely to apply to phorids, with their very considerable powers of dispersal (7.10). However, the species with flightless females, being less capable of easy dispersal, may be more liable to parapatric speciation. Probably of more significance, when considering the evolution of phorid diversity in general, is the evidence of considerable ecological polymorphism in many species, such as *Megaselia nigra* breeding in both fungi and dead snails (see Chapter 3, including the quotation at its head!).

It has long been recognized (e.g. Mayr, 1942) that whenever similar ecological conditions recur in widely separated places, similar ecotypes (locally adapted races) are likely to appear. Furthermore many such habitat races are primarily examples of phenotypic adaptation. In terms of host choice, a specialized case of microhabitat choice, Phoridae evidently exhibit such ecological variation not only in different regions but also in the same region. If the females emerging from host species A are more likely to oviposit into other individuals belonging to species A than into host species B, then such ecological races may become established. However, unless males are preferentially inclined to mate with females derived from the same host species as themselves, these ecological races are likely to remain as phenotypic races only. There is evidence from other insects, however, suggesting that a 'chemical legacy' from the larval host may indeed influence adult behaviour (Corbet, 1985). While this evidence primarily relates to females tending to oviposit into the same host or pabulum as experienced by themselves as larvae, there is also some evidence that males tend to preferentially choose females carrying the same chemical legacy as themselves or seek mates in the vicinity of the same pabulum or habitat as that experienced by their own larvae. Without such a tendency the random mating by males, with respect to different sympatric ecological races, will prevent the evolution of genetic reinforcement of this primarily phenoypic variation. The relevance of Springer's (1967) observations is that they indicate that, even in a broad-spectrum saprophage species, different conditions of the larval pabulum may give rise to males with different chromosomal signatures. It seems highly likely, therefore, that less obvious genetic reinforcements of primarily ecological polymorphisms are like to occur. The observations of Springer, in the context of Corbet's chemical legacy hypothesis, merit further investigation. Sympatric speciation and host-mediated conditioning have been more recently discussed by Mackenzie (1992). Also Hoffman and O'Donnell (1992) have reported variation in resource fidelity in *Drosophila melanogaster*, in relation to different types of fruit in the field.

5.8 SEX RATIOS

The sex ratio of *Megaselia halterata* in mushroom cropping houses has been reported as 1.06 males:1 female (n = 42 000) (Hussey, 1965). This contrasts strikingly with collections obtained in traps outdoors (7.10). Some other reported sex ratios (number of males:1 female) obtained in laboratory rearings are as follows. *Megaselia ciliata* 1.0 (n = 16) (Disney, 1977b, 1979f), *M. obscuripennis* 0.87 (n = 71) (Disney, 1980g), *Melaloncha sinistra c.*1.0 (Simôes *et al.*, 1980), *Phalacrotophora berolinensis* and *P. fasciata* no significant departure from 1.0 (n = 91 and 868 respectively) (Disney *et al.*, 1994), and *Pseudohypocera kerteszi* 0.98 (n = 855) (Chaud-Netto, 1980). The data for *Megaselia scalaris*, however, indicate that the ratio is influenced by temperature. At 23 °C the ratio was reported as 1.28 (n = 643) (Semenza, 1953), at 25 °C as 1.18 (n = 1463) (Benner and Ostermeyer, 1980), at 25–30° C as 1.0 (Dias Macieira *et al.*, 1983) and at 27 °C as 0.86 (El-Miniawi and Moustafa, 1965). Ratios procured in field collections, as opposed to those obtained in rearings, are frequently dominated by one sex (7.10), presumably reflecting differences in behaviour between the sexes.

5.9 DIAPAUSE

In Europe the last-instar larva of *Triphleba hyalinata* is reported to have a diapause lasting several weeks (Leclercq and Watrin, 1973). In America the pupae of *Trucidophora camponoti* required a minimum period of nine weeks at 8 °C by day and 1 °C by night for successful completion of development in the laboratory (Brown *et al.*, 1991). This requirement would explain the occurrence, in the field, of only a single generation each year. Probably many other species have a winter diapause in the egg, larval or pupal stage. However, a number of overwintering 'pupae' are now known to be cases of adults overwintering within the puparium (4.3). The mechanism of delayed emergence (this diapause in the adult stage) needs elucidating.

5.10 EFFECTS OF CROWDING

Overcrowding of larvae affects survival to pupation and also the adult size in *Megaselia scalaris* (Kneidel, 1985). Likewise the adult size of *M. aurea* is affected by the amount of food available to the larvae (Kneidel, 1984b).

5.11 POLYMORPHISM

Sexual dimorphism is characteristic of Phoridae, but is especially marked in those genera whose females are flightless for all or part of their lives.

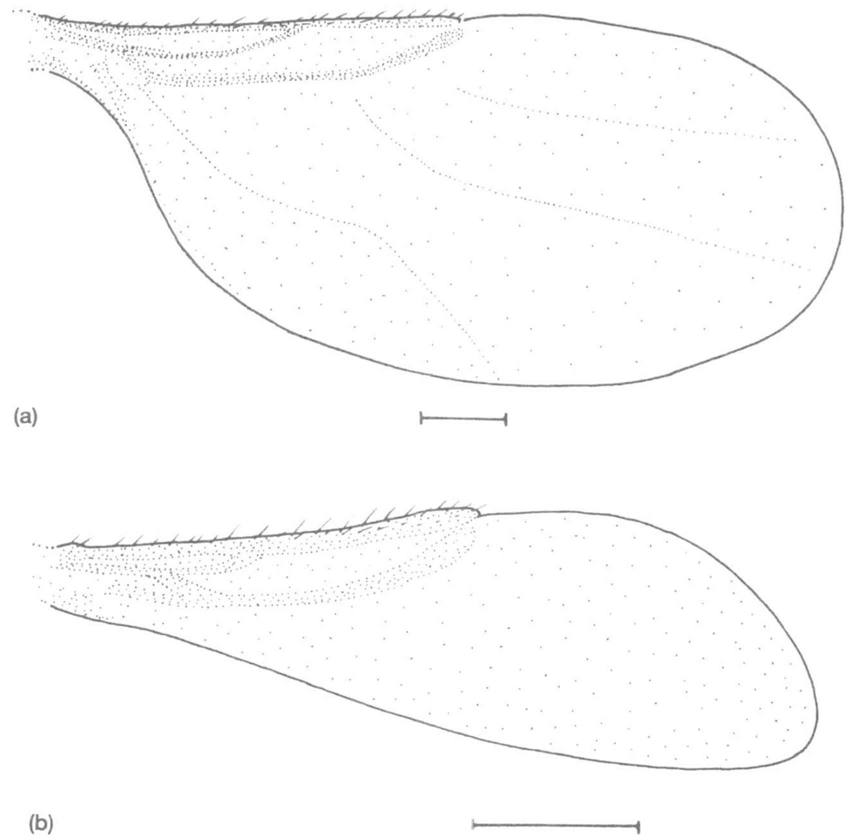

Figure 5.2 *Metopina ciceri* female, right wings of sighted morph (a) and blind morph (b).

A true polymorphism occurs in *Metopina ciceri*, whose larvae develop in chickpea root nodules (3.4). There are two distinct female morphs. Some females have fully developed wings (Fig. 5.2a), ocelli and eyes. Others have reduced wings (Fig. 5.2b), no ocelli and eyes, or the latter are represented by one or two vestigial ommatidia only. It has been suggested that the type of morph produced may depend upon the activation or inactivation of genes controlling the reduction and cessation of the secretion of juvenile hormone during development (Disney, 1988d). This hypothesis requires testing. A futher case of polymorphic females possibly occurs in the genus *Adelopteromyia*. It has been suggested that two 'species', one of which is a short-winged version of the other, seem to have identical ecologies and may therefore represent a single polymorphic species (Disney and Kistner, 1990b). Springer's (1967) observations (see 5.7) indicated that two different chromosome morphs of *Megaselia scalaris* are favoured under different rearing

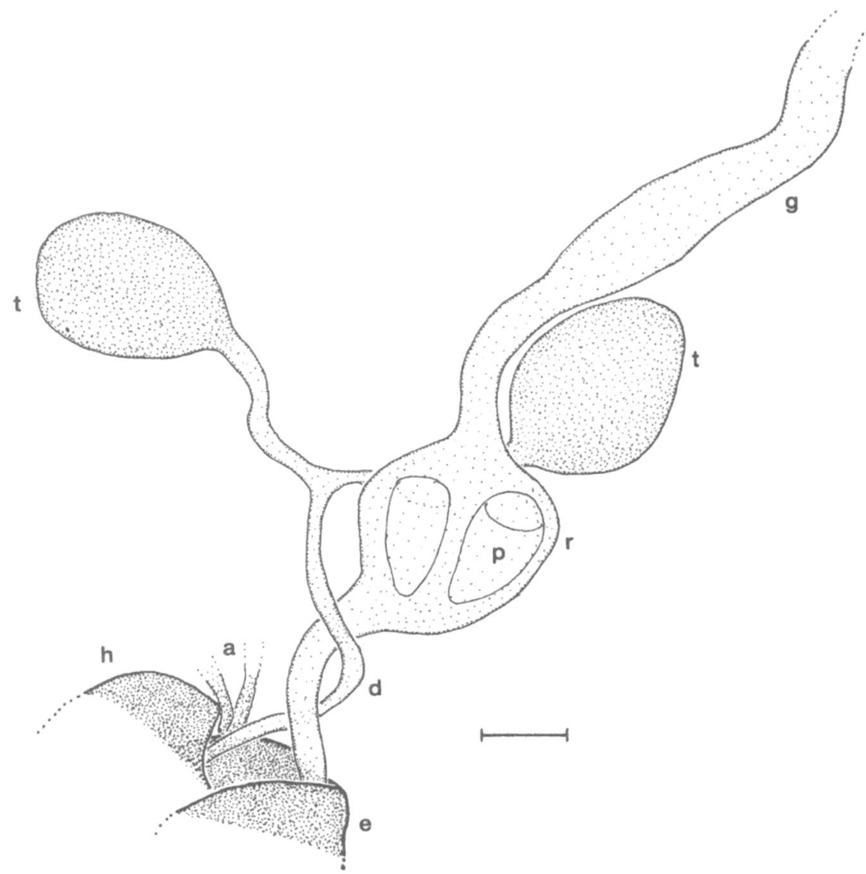

Figure 5.3 Dissection of end of abdomen of *Megaselia paludosa* male. a = ducts to accessory glands; d = ejaculatory duct; e = epandrium; g = hind gut; h = hypandrium; p = rectal papilla; r = rectum; t = testis.

conditions, one being favoured under poorly-ventilated over-crowded conditions and the other under well-ventilated less-crowded conditions.

On Ascension Island only a single phorid species, the widely distributed tropical saprophage *Megaselia curtineura*, has been reported in more than 400 scuttle flies collected. However, the species exists in two forms on this remote island (Disney, 1991b). One form has four rectal papillae and the other has only two papillae (cf. Fig. 5.3). Whether this represents discontinuous phenotypic variation, a true genetic polymorphism, or sibling species is not known. Normally the number of rectal papillae is constant for a species, apart from any difference in number between the sexes. In many species the males have fewer papillae than the females. In a few species the number of rectal papillae is known to be variable, for example, in *Megaselia pulicaria, M.*

sinuata and *M. spinolabella* (Disney, 1989a). In some cases such variation may represent sibling species. In one reared series of *M. pulicaria* from a single spider egg sac, however, two flies had four and one had three papillae. This is, therefore, most likely a case of polymorphism.

Modern techniques might be usefully applied to some sibling species/ polymorphism problems in phorid taxonomy. The first requirement is to define the problems in terms of morphological or ecological segregates. These segregates can then be examined with techniques such as cytotaxonomy, analysis of cuticular hydrocarbons or structural enzymes, or molecular techniques. The *Megaselia pulicaria* complex is a prime candidate for such investigations. Classic comparative morphology and simple biometrics have been unable to resolve all the problems with this complex (Disney, 1989a). The considerable variation in size, with its consequential allometric effects, has tended to swamp the differences between segregates in terms of simple measures such as wing length or costal ratios.

For most Phoridae there has been a dearth of critical comparative morphological investigations. Such studies remain the foundation of taxonomic studies, especially at the species level. These studies are still capable of unmasking sibling species, as in the case of the *Pseudacteon spatulatus* complex in which detailed morphological studies revealed there to be four not two species (Disney, 1991f). Likewise simple field observations, such as observing mating pairs, may resolve taxonomic problems. Thus the procurement of mating pairs of *Termitophilomyia braunsi* allowed the abolition of a subfamily (Disney and Cumming, 1992). Reared series are also invaluable. The detection of polymorphism in *Metopina ciceri* was revealed by reared series (i.e. samples of adults reared from single individual hosts or discrete pabulum units). If I had been presented with the two female morphs and the males as three separate samples they would have been assigned to three separate genera, in terms of the taxonomic literature at the time (Disney, 1988d). These examples suggest that caution is called for when considering whether polymorphism is rare or widespread in the Phoridae.

6

Adult natural history

The insect interests me much more when engaged in its work than when stuck on a pin in a cabinet (Fabre, 1918)

6.1 INTRODUCTION

The internal anatomy of adult Phoridae is less well known than the external morphology, which is covered in numerous taxonomic works on the family (Chapter 8). Detailed anatomical studies were carried out by Schmitz (1929g), who also reviewed the external morphology (1938n). The anatomy of female Termitoxeniinae has been intensively studied, with particular reference to the reproductive system. However, it was partly misinterpreted in terms of the, now discarded, hermaphroditism theory of Wasmann (5.6). The full literature on the anatomy of Termitoxeniinae is cited by Borgmeier (1964a). Benner and Curtis (1988) provide a detailed treatment of the internal reproductive organs of *Megaselia scalaris*.

The evolution of Phoridae has probably been characterized by a reduction in the size of the adults associated with an acceleration of larval development (2.2). One strategy evolved, independently, many times has been the production of flightless females. This is especially characteristic of species whose females inhabit the nests of ants and termites. The flightless condition has been attained in two distinct ways. One way is to shed part of the wing soon after emergence; the wings are in part deciduous (or caducous). In genera such as *Eutermiphora*, *Dicranopteron* and all Termitoxeniinae, the outer part of the wing is shed, leaving behind bases as wing stumps (e.g. Fig. 6.1).

The second method by which the flightless condition has been attained is the reduction of the wing, including its total failure to develop at all. The four basic conditions encountered in other flightless insects (see Hackman, 1964) are all found in the Phoridae, each in a variety of forms. The brachypterous condition, in which the wing is reduced in area by shortening, is found in most females of *Megaselia longicostalis*. The male has normal wings (Fig. 6.2a). In females the wing is normal or, more usually, variably shortened (Fig. 6.2b,c). In genera such as *Hypogeophora* (Fig. 6.2d) the female wing is always shortened. The stenopterous condition, where the wing is reduced by

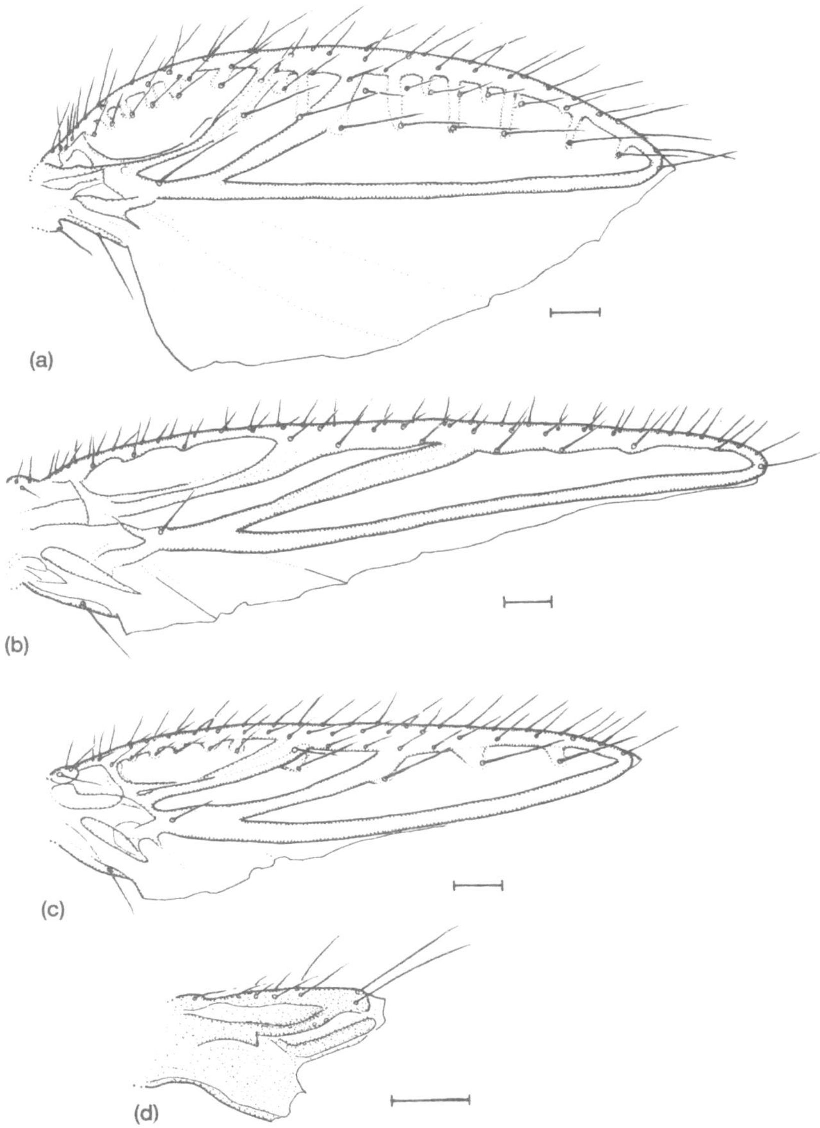

Figure 6.1 Wing stumps remaining after most of wing membrane has been shed. (a) *Dicranopteron setipennis*. (b) *D. oligoseta*. (c) *D. emersoni*. (d) *Eutermiphora abdominalis*.

narrowing, is found in genera such as *Franssenia* (Fig. 6.3). The micropterous condition, in which the wing is reduced to a rudiment, is found in genera such as *Aenigmatistes, Maculiphora*, *Rhynchmicropteron* and *Thaumatoxena* (Figs 6.4,

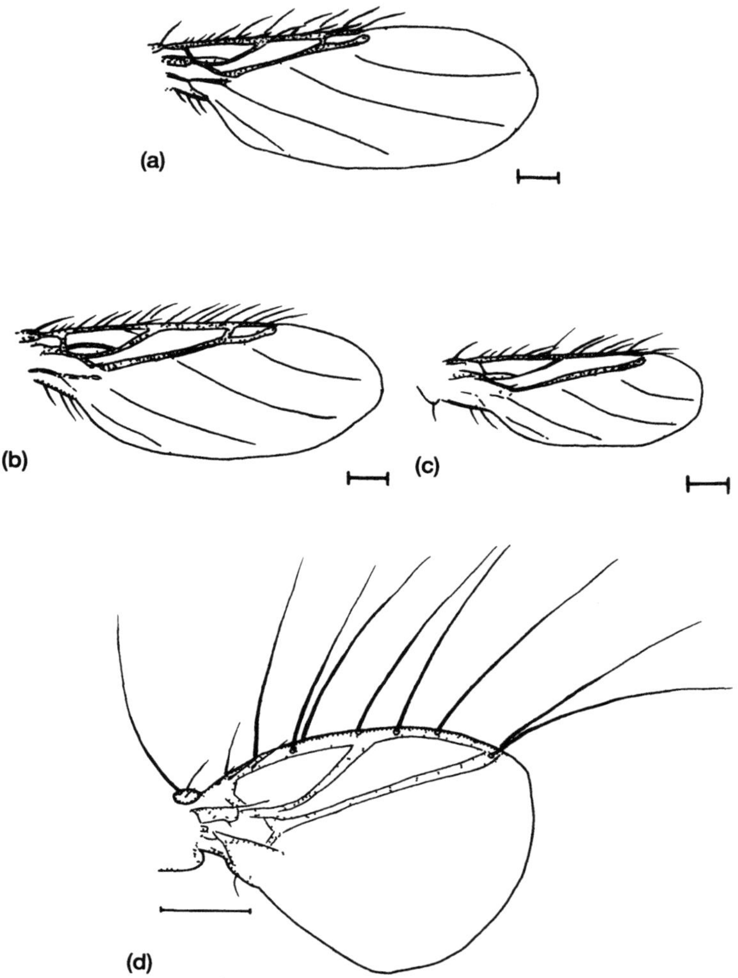

Figure 6.2 Right wings to show differing degrees of the brachypterous condition. (a) The normal wing of a *Megaselia longicostalis* male. (b) and (c) two different females of *M. longicostalis*. (d) *Hypogeophora macrothrix* female.

6.5 and 6.6). The apterous condition, where the wing is represented by a bristle at the most, is found in genera such as *Chonocephalus* and *Puliciphora* (e.g. Fig. 6.7).

The evolution of wing reduction possibly started with a variation such as in *Megaselia longicostalis* or as a polymorphism of fully-winged and short-winged forms produced by a single species, as in the case of *Metopina ciceri* (5.11) (Fig. 5.2). Indeed occasional mutations occur in populations of *M. scalaris*, producing forms with short wings or with an embayment of the wing margin (Johnson *et al.*, 1988).

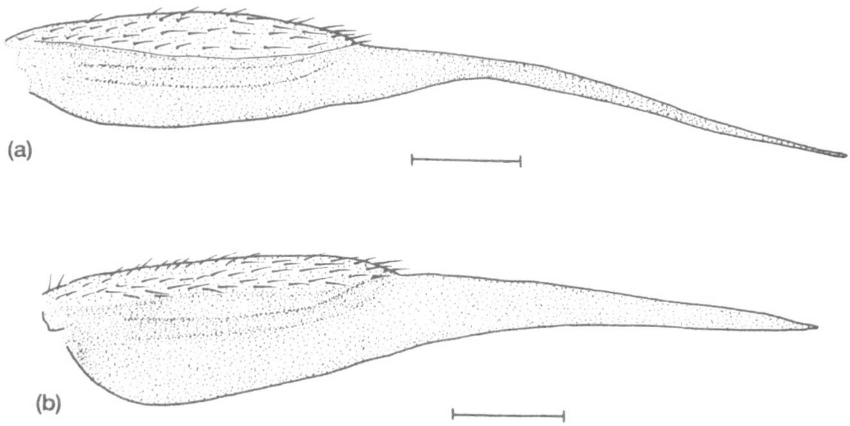

Figure 6.3 The stenopterous wings of *Franssenia* females (a) *F hirundella* (b) *F rocheorum*

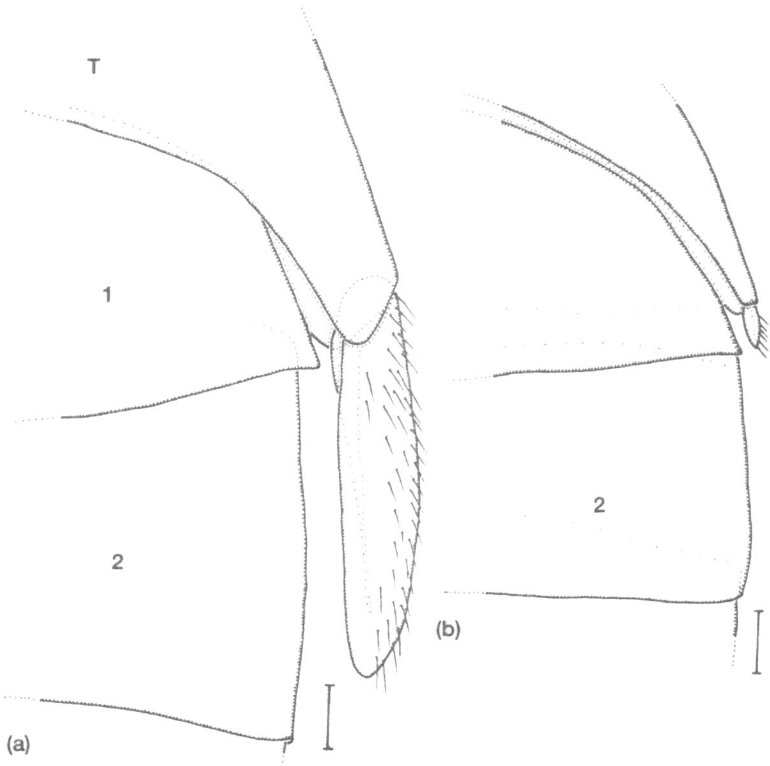

Figure 6.4 Micropterous wings in females of *Aenigmatistes* (a) *A herero* (b) *A latifrons* T = thorax, 1 and 2 = 1st and 2nd abdominal tergites

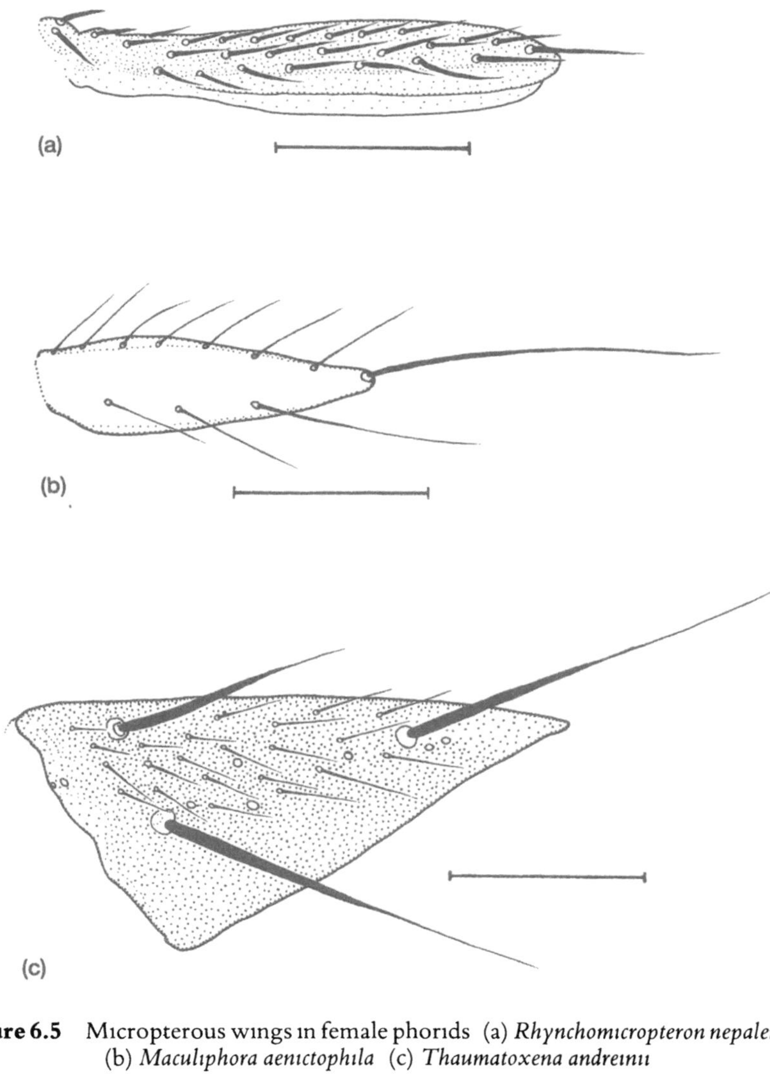

Figure 6.5 Micropterous wings in female phorids (a) *Rhynchomicropteron nepalensis* (b) *Maculiphora aenictophila* (c) *Thaumatoxena andreinii*

The variety of means by which the flightless condition has been attained indicates that the loss of flight has indeed evolved independently many times in the Phoridae In nearly all cases the males retain fully developed wings, but two Afrotropical genera have flightless males (8 2 (B)) In the case of *Aptinandria effeminata* this would seem to be a case of the flightless condition being associated with high altitudes (Mani, 1968), as the species lives on Mount Kenya between 3300 and 3700 m (Schmitz, 1921f) Normally genera with flightless females have flying males, which in many cases are known to transport the females while mating (see below) Hackman (1964) has

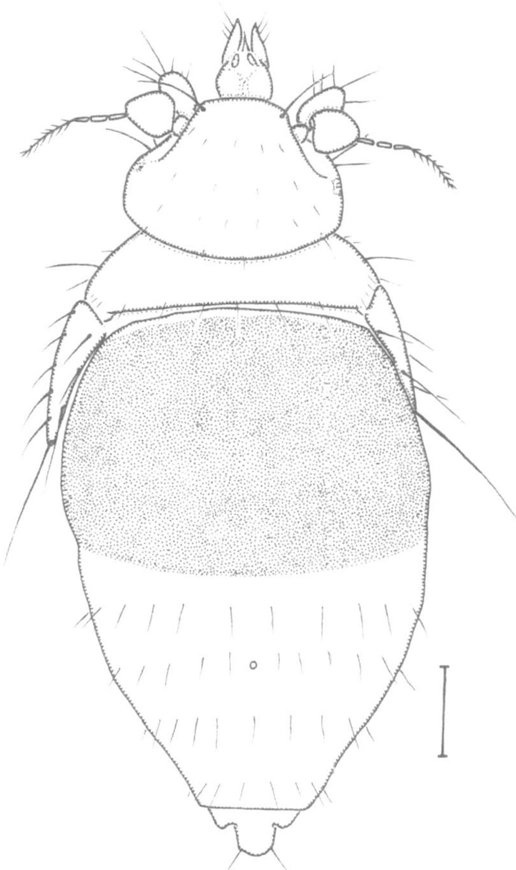

Figure 6.6 Flightless female of *Maculiphora aenictophila*, with micropterous wings.

emphasized that flying males make gene flow possible over a wider range than when both sexes are flightless. The reduction and loss of wings in females, and any associated reduction in the flight muscles, must save much energy and nutrients, and thus facilitate more rapid development.

6.2 FLOWER VISITING

Taxonomic complexities have tended to deter most workers trying to identify the phorid species visiting particular flower species. Furthermore scuttle flies depart very rapidly from flowers when disturbed. McAlpine (1965) has emphasized that most flower-visiting flies are easily disturbed by the presence of a human observer. This is especially true of Phoridae. By the time the more conspicuous insects have been picked off by a collector the

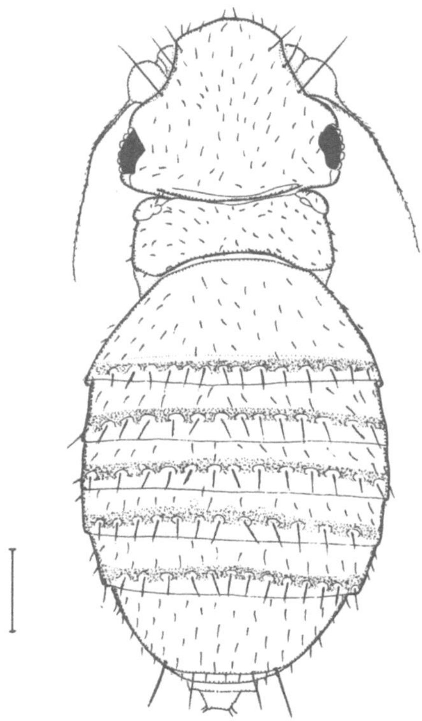

Figure 6.7 Flightless female of *Chonocephalus wirthorum* without wings (apterous).

Phoridae will have departed. The consequence is that scuttle flies are usually under-sampled.

The earlier literature with records of flower-visiting Phoridae has been reviewed by Baumann (1978a), who made original observations on the subfamily Phorinae visiting Umbelliferae (Baumann, 1978b). Subsequently the records of flower visiting in Britain have been reviewed and extended (Disney, 1980c).

The importance of Phoridae as pollinators has been indicated by the studies of Young (1984) on the insect visitors to the flowers of *Herrania* species. Furthermore his study of the times insects were attracted to the flowers of the related cacao-tree (*Theobroma cacao*) found about half the Phoridae were attracted at night (Young, 1986). However, while these studies clearly established that phorids are visiting the flowers of *Herrania* species in significant numbers, they were not observed actually visiting the flowers of *Theobroma* species, despite their clear attraction to the immediate vicinity of the flowers. It seems that a number of phorids, including species of the genera *Megaselia*, *Dohrniphora*, *Chonocephalus* and *Puliciphora*, are attracted to *Theobroma* floral oils (essences) (Young *et al.*, 1987a). Indeed 16% of the Diptera attracted to steam-distilled floral oils of *Theobroma* were Phoridae.

Furthermore some of the 'oily' hydrocarbons of cacao are also characteristic of *Herrania* species. It seems likely that it is these that account for the attraction of the scuttle flies that normally visit *Herrania* to the vicinity of *Theobroma* flowers (Young *et al.*, 1989). Hall and Brown (1993) provide evidence of pollination of *Aristolochia littoralis* by species of *Megaselia*, the latter seemingly being attracted to the flowers by an olfactory cue.

Some authors have reported the sexes of the Phoridae caught at flowers. For example, Borgmeier (1969b) noted that the 37 *Megaselia graminicola* and 136 *M. parianae* caught at *Pariana* flowers were all females, whereas the 37 records for *Diplonevra funebris* at various flowers are all of males (Baumann, 1978b; Disney, 1980c and unpublished data). There is a marked sexual dimorphism in the proboscis of this species, with that of the male being much more slender. Even when both sexes are reported from flowers the numbers are frequently significantly biased towards one sex. For example, Hall and Brown (1993) caught 164 males and 9 females of *Megaselia scalaris*, 54 males and 1 female of *M. aurea* and 21 males and 1 female of *M. perdita* in the flowers of *Aristolochia littoralis* in Florida. The many records for *Phora* and *Conicera*, apart from the distinctive *C. dauci*, have to be ignored, however, as the difficulties of identifying the females in these genera have prevented accurate recording of the sex ratios for individual species. Table 6.1 presents data for a few European species (based on records by Baumann and myself).

Table 6.1 Sex ratios of some Phoridae observed visiting flowers

Species	Males	Females	% Male	*p*-value
Conicera dauci	118	61	65.9	<0.1
Diplonevra nitidula	19	13	59.4	n.s.
Megaselia brevicostalis	79	71	52.7	n.s.
Metopina oligoneura	80	44	64.5	<0.05
Triphleba citreiformis	44	1	97.8	<0.001
Triphleba nudipalpis	16	15	51.6	n.s.

In the list above I attempt to present all critically evaluated records of visits to named flowers by named Phoridae. Numbers of early records have had to be discarded, including some noted by Baumann (1978a). For example, he cited Lundbeck (1922) for *Chaetopleurophora bohemanni* on *Aegopodium podagraria*, but reference to Lundbeck indicates this record was from the foliage of the plant, not the flower. The purposes of flower visiting by phorids need critical evaluation. Presumably many are seeking nectar. Pollen has not been recorded in the crop or stomach of any phorid, in contrast to the regular ingestion of whole pollen grains by many Syrphidae and Calyptrata (e.g. see Kearns, 1992). There remains the possibility that some phorids may pierce pollen grains and ingest the contents or discharge saliva onto the grains and then ingest the saliva plus any leachate from the pollen. Some visits to flowers are evidently not for anything the flower itself has to offer. For example, Stuckenberg (1953) observed phorids on a flower of *Foeniculum vulgare* feeding on the fluids oozing from the honey-bee prey of a mantid (*Polyspilota* sp.).

In the following list the generic nomenclature of the flowers and their family assignments follow Mabberley (1989). For the specific names of the British flora I have followed Clapham *et al.* (1987).

ACANTHACEAE

Acanthus mollis – *Conicera dauci* (Baumann, 1978a).

ACERACEAE

Acer spp. – *Phora* spp. (Gotô, 1984b).
Acer platanoides – *Diplonevra florea* (A. Nilsson and Disney, unpublished data).
A. pseudoplatanus – *Megaselia conformis* (Disney, 1988f).

ARACEAE

Amorphophallus sp. – *Chonocephalus secundus, Puliciphora lucifera* (Schmitz, 1928a).
Anthurium denudatum – *Chonocephalus globipygus* (Borgmeier, 1967f).
Arum sp. – *Puliciphora lucifera* (Brues, 1903b).
A. maculatum – *Megaselia pulicaria, M. sinuata* and *M. variana* (I. Beck and Disney, unpublished data).
Zantedeschia aethiopica – *Melaloncha genitalis* males 'waiting for the females' (Borgmeier, 1971a).

ARALIACEAE

Hedera helix – *Megaselia pleuralis* and *M. pulicaria* (Parmenter, 1965; Disney, 1980c) and *Triphleba nudipalpis* (unpublished data).

ARISTOLOCHIACEAE

Aristolochia altissima – *Megaselia pulicaria* (Baumann, 1978a).
A. labiata – *M. aristolochiae* (Hime and Costa, 1985).
A. littoralis (= *A. elegans*) – *Dohrniphora cornuta* (Brues, 1928c), *Megaselia aurea, M. perdita*, four species of *Megaselia* sp. (Hall and Brown, 1993) and *M. scalaris* (Borgmeier, 1925c; Brues, 1928c; Hall and Brown, 1993).
A. macrophylla (= A. sipho) – *Megaselia* sp. (Carr, 1924), *M. nigriceps* (Schmitz, 1957a).

A. pallida – *Megaselia sordida* (Baumann, 1978a).
A. tormentosa – *M. fungicola* (Robinson, 1971).

ASCLEPIADACEAE

Ceropegia racemosa – *Megaselia* spp. (A. Nilsson and Disney, unpublished data).

BOMBACACEAE

Ochroma lagopus – *Megaselia* spp. visiting fallen flowers (Rettenmeyer and Akre, 1968).

CANNACEAE

Canna speciosa – *Megaselia sibylla* (Borgmeier, 1967e).

CAPRIFOLIACEAE

Sambucus nigra – *Gymnoptera longicostalis* (D. Henshaw and Disney, unpublished data).

CARYOPHYLLACEAE

Cerastium fontanum – phorids (Baumann, 1978a).
Gypsophila hispanica – *Megaselia pseudobrevior, M. subnudipennis* and *M. xanthozona* (J. Blasco-Zumeta and Disney, unpublished data).
Silene acaulis – *Megaselia coei* (Schmitz, 1938a).
Stellaria holostea – *Megaselia brevicostalis* (Disney, 1980c).
S. media – *Megaselia pulicaria* (A. Nilsson and Disney, unpublished data).

COMPOSITAE

Anthemis cotula – *Metopina oligoneura* (Disney, 1980c).
Aster tripolium – *Phora atra* (Parmenter, 1965; L. Clemons and Disney, unpublished data).
Baccharis pilularis – *Megaselia peterseni* (Borgmeier, 1966a).
Carduus bourgeanus – *Megaselia xanthozona* (J. Blasco-Zumeta and Disney, unpublished data).
Chrysanthemum maximum – *Megaselia brevicostalis* (Disney, 1980c).
Cirsium arvense – *Megaselia pulicaria* (unpublished data).

Eupatorium cannabinum – *Borophaga incrassata* (unpublished data).

Hypochoeris radicata – *Diplonevra funebris, Megaselia brevicostalis* and *Metopina oligoneura* (Disney, 1980c).

Leontodon autumnalis – *Diplonevra funebris, Megaselia brevicostalis, M. pleuralis* and *Metopina oligoneura* (Disney, 1980c).

L. hispidus – *Megaselia brevicostalis* (Disney, 1980c).

Senecio jacobaea – *Borophaga carinifrons* (Aston, 1957).

Sonchus oleraceus – *Diplonevra funebris* (unpublished data).

Taraxacum officinale agg.– *Diplonevra funebris, Megaselia brevicostalis, M. pulicaria* and *Metopina oligoneura* (Disney, 1980c) and *M. heselhausi* (Schmitz, 1938c).

Tussilago farfara – *Megaselia brevicostalis* (Disney, 1980c) and *M. superciliata* (A. Nilsson and Disney, unpublished data).

CONVOLVULACEAE

Convolvulus arvensis – *Metopina oligoneura* (P. J. Chandler and Disney, unpublished data).

CORNACEAE

Cornus obliqua – *Diplonevra funebris* (Judd, 1975).

C. sanguinea – *Diplonevra nitidula* (A. Nilsson and Disney, unpublished data).

CRUCIFERAE

Crambe maritima – *Megaselia* sp. (Baumann, 1978a).

EUPHORBIACEAE

Euphorbia cyparissias – *Megaselia pulicaria* (A. Nilsson and Disney, unpublished data).

E. helioscopia – *Metopina oligoneura* (A. Nilsson and Disney, unpublished data).

GRAMINEAE

Pariana vulgaris – *Megaselia graminicola* and *M. parianae* (Borgmeier, 1969b; Soderstrom and Calderón, 1971).

P. radiciflora – phorids (Baumann, 1978a).

LABIATAE

Marrubium vulgare – *Metopina oligoneura* (J. Blasco-Zumeta and Disney, unpublished data).

Mentha aquatica – phorids (Baumann, 1978a).

LILIACEAE

Gagea lutea – *Diplonevra florea* (Baumann, 1978a), *Megaselia pulicaria* and *M. pumila* (A. Nilsson and Disney, unpublished data).

Narcissus pseudonarcissus – *Megaselia brevicostalis, M. ciliata, M. pulicaria, M. rufipes, Triphleba nudipalpis* and *T. opaca* (Disney, 1980c) and *Megaselia conformis* (Disney, 1988f).

LIMANTHACEAE

Limanthes douglasii – *Megaselia pulicaria* (Disney, 1980c).

MAGNOLIACEAE

Magnolia sp. – *Megaselia brevicostalis* and *M. ciliata* (Disney, 1980c).

ORCHIDACEAE

Listera ovata – *Megaselia pulicaria* (A. Nilssen and Disney, unpublished data).

OXALIDACEAE

Oxalis acetosella – *Megaselia mallochi* (Baumann, 1978a) and *M. rufipes* (Parmenter, 1965).

PALMAE

Cocos nucifera – *Megaselia setaria* (Sholdt, 1966).

PARNASSIACEAE

Parnassia palustris – *Megaselia brevicostalis, M. pulicaria, M. parnassia* and *Triphleba citreiformis* (Disney, 1980c, 1986e).

PLUMBAGINACEAE

Plumbago caerulea – *Metopina oligoneura* (Baumann, 1978a).

POLEMONIACEAE

Phlox decussata – *Metopina oligoneura* (Baumann, 1978a).

RANUNCULACEAE

Caltha palustris – *Megaselia brevicostalis* (Disney, 1980c) and *M. pulicaria* (A. Nilsson and Disney, unpublished data).

Myosurus minimus – phorids (Baumann, 1978a).

Ranunculus acris – *Diplonevra funebris, D. nitidula* and *Metopina oligoneura* (Disney, 1980c) and *Megaselia brevicostalis* (A. Nilsson and Disney, unpublished data).

R. bulbosus – *Megaselia brevicostalis* (Disney, 1980c).

R. ficaria – *Megaselia brevicostalis* and *M. pulicaria* (Disney, 1980c).

R. trichophyllus – *Megaselia manicata* (A. Nilsson and Disney, unpublished data).

RESEDACEAE

Reseda lutea – *Megaselia pulicaria* (A. Nilsson and Disney, unpublished data); *M. albocingulata, M. pseudobrevior* and *Metopina heselhausi* (J. Blasco-Zumeta and Disney, unpublished data).

ROSACEAE

Crataegus sp. – *Conicera schnittmanni* and *Diplonevra florea* (Baumann, 1978a).

C. monogyna – *Beckerina umbrimargo, Conicera dauci, Megaselia aequalis, M. brevicostalis, M. ciliata, M. clemonsi, M. collini, M. giraudii, M. involuta, M. latior, M. propinqua, M. pulicaria, M. sinuata, Phora atra, Triphleba distinguenda* and *T. intermedia* (L. Clemons and Disney, unpublished data).

Potentilla anserina – *Diplonevra funebris* (Schmitz, 1949c), *Megaselia brevicostalis* and *Metopina heselhausi* (Schmitz, 1938c).

P. argentea – *Megaselia subnudipennis* (A. Nilsson and Disney, unpublished data).

P. erecta – *Megaselia brevicostalis* (Disney, 1980c).

P. sterilis – *Diplonevra glabra* and *Megaselia brevicostalis* (Disney, 1980c).

P. tabernaemontani – *Megaselia brevicostalis* (A. Nilsson and Disney, unpublished data).

Prunus communis – phorids (Baumann, 1978a).

P. laurocerasus – *Phora atra* (Disney, 1980c).
Rubus fruticosus agg. – *Diplonevra funebris* (Disney, 1980c).

SALICACEAE

Salix sp. – *Phora stictica* (Baumann, 1978a).
S. caprea – *Megaselia ciliata*, *M. pectoralis*, *M. pulicaria*, *Triphleba nudipalpis* and *T. opaca* (Disney, 1980c).
S. cinerea – *Diplonevra nitidula* (Parmenter, 1965), *Megaselia mallochi* and *M. pulicaria* (Disney, 1980c).
S. discolor – *Borophaga femorata* (Baumann, 1978a).
S. purpurea – *Triphleba lugubris* and *T. papillata* (Baumann, 1978a).

SAPINDACEAE

Cupania vernalis – *Melaloncha ronnai* and *M. zikani* (Borgmeier, 1934b, 1935g; Ronna, 1936).

SAXIFRAGACEAE

Chrysosplenium alternifolium – phorid at floral nectary (Schlee, 1977).
Saxiphraga diapensoides – *Megaselia berndseni* and *M. pulicaria* (Disney, 1980c).
S. granulata – *Phora* sp. (A. Nilsson and Disney, unpublished).
S. hypnoides – phorid (Proctor and Yeo, 1973).
S. stellaris – *Megaselia valvata* (A. Nilsson and Disney, unpublished data).

SCROPHULARIACEAE

Euphrasia nemorosa – *Megaselia brevicostalis* and *M. longicostalis* (Disney, 1980c).
Veronica filiformis – *Megaselia brevicostalis* (Disney, 1980c).
V. persica – phorid (Baumann, 1978a).

STERCULIACEAE

Herrania albiflora, *H. nitida* and *H. purpurea* – *Dohrniphora* sp. and *Megaselia* sp. (Young, 1984).

TAMARICACEAE

Tamarix canariensis – *Megaselia albicaudata*, *M. berndseni*, *M. brevior*, *M. manicata*, *M. pseudobrevior*, *M. xanthozona* and *Metopina oligoneura* (J. Blasco-Zumeta and Disney, unpublished data).

UMBELLIFERAE

Aegopodium podagraria – *Conicera dauci* and *Phora stictica* (Baumann, 1978b).

Angelica archangelica – *Megaselia pleuralis* (Baumann, 1978a).

A. sylvestris – *Anevrina unispinosa* (Aston, 1957), *Conicera dauci* (Parmenter, 1965; Disney, 1977a, Baumann, 1978b), *C. floricola* (Baumann, 1978b), *C. schnittmanni* (I. F. G. McLean and Disney, unpublished data), *C. similis* (Baumann, 1978b), *Diplonevra florea* (unpublished data), *Gymnoptera longicostalis* and *G. vitripennis* (Baumann, 1978b), *Megaselia angelicae* (Wood, 1910; Disney, 1980c), *M. badia* (Disney, 1980c), *M. brevicostalis* (Wood, 1910; Disney, 1980c), *M. coei* (unpublished data), *M. discreta* (Wood, 1909), *M. giraudii* (Parmenter, 1965), *M. manicata* (Wood, 1910; Disney, 1980c), *M. pleuralis* (Disney, 1980c), *M. posticata* (Wood, 1908), *M. pulicaria* (A. Nilsson and Disney, unpublished), *Phora atra* (Disney, 1980c), *P. stictica* (Baumann, 1978b; Disney, 1980c), *Triphleba citreiformis* (Wood, 1906; Lundbeck, 1922; Baumann, 1978b; Disney, 1980c) and *T. disparinervis* (Baumann, 1978b).

Anthriscus sylvestris – *Conicera dauci* (Baumann, 1978b), *Diplonevra nitidula* (Parmenter, 1965), *Megaselia plurispinulosa* (A. Nilsson and Disney, unpublished data) and *Phora atra* (Parmenter, 1965).

Chaerophyllum bulbosum – *Conicera floricola, C. similis, C. tibialis, Diplonevra funebris, D. nitidula* and *D. pilosella* (Baumann, 1978b).

Conium maculatum – *Megaselia berndseni* (Parmenter, 1965).

Daucus carota – *Conicera dauci* (Meigen, 1830; Macquart, 1835; Baumann, 1978b), *C. floricola, C. similis, C. tibialis, Diplonevra funebris, D. pilosella* and *Gymnoptera longicostalis* (Baumann, 1978b).

Foeniculum vulgare – *Conicera dauci* (Parmenter, 1965).

Heracleum sphondylium – *Conicera dauci* (Grensted, 1945; Parmenter, 1965; Baumann, 1978b; Disney, 1980c, 1985c), *C. schnittmanni* (unpublished data), *C. floricola* and *C. schnittmanni* (Baumann, 1978b), *C. similis* (Baumann, 1978b; Disney, 1980c), *C. tibialis* (Baumann, 1978b), *Diplonevra funebris* and *D. nitidula* (Baumann, 1978b; Disney, 1980c), *Gymnophora nigripennis* (Baumann, 1979), *Gymnoptera longicostalis* (unpublished data), *Megaselia angelicae* (Wood, 1910), *M. brevicostalis, M. ciliata* and *M. giraudii* (Disney, 1980c), *M. hirsuta* (Wood, 1910), *M. involuta* and *M. latifrons* (unpublished data), *M. manicata* (Wood, 1910), *M. nasoni* and *M. pulicaria* (Disney, 1980c), *M. rufipes* (Grensted, 1945), *Metopina pileata, M. ulrichi* and *Phora edentata* (Disney, 1980c), *P. stictica* (Drabble and Drabble, 1927; Baumann, 1978b), *P. tincta* (unpublished data), *Spiniphora jugorum* and *Triphleba bicornuta* (Baumann, 1978b), *T. citreiformis* (Wood, 1906; Malloch, 1908a; Lundbeck, 1922; Baumann, 1978b; Disney, unpublished data), *T. disparinervis* and *T. distinguenda* (Baumann, 1978b).

Laserpitium latifolium – *Conicera dauci, Phora stictica* and *Triphleba citreiformis* (Baumann, 1978b).

Myrrhis odorata – *Conicera floricola* and *Diplonevra glabra* (Disney, 1980c), *Gymnoptera longicostalis* (Bauman, 1978b), *Megaselia ciliata*, and *M. pulicaria* (Disney, 1980c), *Phora stictica* and *Spiniphora strobli* (Baumann, 1978b).

Pastinaca sativa – *Diplonevra glabra* (Parmenter, 1965), *Megaselia ciliata* and *M. minor* (A. Nilsson and Disney, unpublished data).

Table 6.2 Summary of records of Phoridae reported visiting at least five species of flower

Fly species	Flowers: No. families (F)	Flowers: No. species (S)	Provisional catholicity index (F/S)
Conicera dauci	3	11	0.27
C. floricola	1	7	0.14
C. similis	1	5	0.20
Diplonevra funebris	4	11	0.36
D. nitidula	4	8	0.50
Gymnoptera longicostalis	2	6	0.33
Megaselia brevicostalis	9	24	0.38
M. ciliata	5	7	0.71
M. pulicaria	16	23	0.70
Metopina oligoneura	7	10	0.70
Phora atra	3	6	0.50
P. stictica	1	7	0.14
Triphleba citreiformis	2	5	0.40

Petroselinum crispum – *Conicera dauci* (Disney, 1980c).

Peucedanum ostruthium – *Anevrina thoracica, Conicera dauci, C. floricola, Diplonevra nitidula, Gymnoptera longicostalis, Phora dubia, P. holosericea, P. penicillata, P. pubipes, P. stictica, P. tincta, Spiniphora jugorum, S. strobli, Triphleba alpestris, T. bicornuta, T. citreiformis, T. distinguenda* and *T. nudipalpis* (Baumann, 1978b). Schmitz (1957a) recorded *Megaselia palmeni* swarming around this species.

Pimpinella major – *Conicera floricola, C. similis* and *Diplonevra nitidula* (Baumann, 1978b).

P. saxifraga – *Megaselia pulicaria* (A. Nilsson and Disney, unpublished data) and *Phora stictica* (Baumann, 1978b).

Seseli libanotis – phorid (Baumann, 1978b).

Torilis japonica – *Phora atra* (Parmenter, 1965).

Table 6.2 summarizes the records for fly species recorded visiting at least five species of flower. A more sophisticated analysis is not possible because of the unequal effort given to the collecting of phorids from different families and species of flowers. The Umbelliferae have been particularly favoured by those sampling phorids at flowers. However, even this simple tabulation suggests that species such as *Megaselia ciliata, M. pulicaria* and *Metopina oligoneura* are far more catholic in their choice of flower species and families visited than are species such as *Conicera dauci* or *Diplonevra funebris*. More rigorous studies are likely to prove rewarding.

An observation that is especially intriguing is the recording of 238 phorids, belonging to seven species, visiting daffodil flowers (*Narcissus*) at 400 m altitude in Yorkshire, but only a single record of a phorid visiting a daffodil at lower altitudes in England (Disney, 1980c). Perhaps daffodils are only utilized when other preferred sources of nectar are not available. Further investigation is clearly required. A possible explanation, however, is offered below (6.3).

6.3 FEEDING

Patient observation of almost any patch of vegetation is likely to reveal phorids engaged in leaf-searching behaviour. Typically they race rapidly around a series of leaves in quick succession. They often pause momentarily, especially at the leaf margins. Sometimes the proboscis is lowered to a leaf surface. It may be obvious that it is lowered into a spot of honeydew or is probing a dead insect lying on the leaf. Most often, however, it is not obvious what is being encountered by the tip of the proboscis. Sometimes a fly remains rooted to one spot, apparently 'attached' to the leaf by its proboscis. At other times a fly appears to be brushing the leaf surface with the proboscis. Indeed it may move methodically along the 'gutter' formed by a leaf vein, seemingly 'grooming' the gutter with its proboscis as it progresses.

During a couple of days at Malham Tarn (North Yorkshire) in late May/early June, I attempted to secure every phorid observed extending its proboscis while halted on a leaf of dog's mercury (*Mercurialis perennis*). In a few hours of observation 36 phorids were caught with a pooter and transferred to tubes of 70% alcohol. A male and a female *Megaselia pleuralis* were collected while they were probing evident spots of honeydew. A female *M. pleuralis* was procured while probing the abdomen of a dead cranefly (*Tipula meigeni*). The following were caught while apparently probing a bare leaf surface with the proboscis: *M. flava* male, *M. frameata* female, *M. hirticrus* male, *M. lucifrons* female, *M. pectoralis* male, *M. pleuralis* 9 males and 16 females, *M. pulicaria* male, *M. sinuata* female and an indeterminate female *Megaselia* (belonging to Schmitz and Beyer's (1965) 'der Abteilung IV, Zweite Reihe', a group in which only males can be named with confidence in most cases, in our present state of knowledge). Examination of the guts of these flies revealed no pollen grains or spores, in contrast to the numerous pollen grains of dog's mercury and occasional fungus spores evident in the guts of the abundant small yellow springtails (Sminthuridae) caught hunting the same leaf surfaces on the same occasion. Indeed I have never found pollen grains in the gut of a phorid, although they are often found adhering to their bodies. By contrast many Syrphidae and Calyptrata ingest pollen grains and are evidently important flower visitors, becoming especially significant for pollination at higher altitudes (e.g. Kearns, 1992). It seems, therefore, that leaf-searching phorids not observed at honeydew spots or dead insects are probably ingesting moisture, traces of honeydew or possibly minute seepages of sap oozing from abrasions or punctures made by themselves or other insects. Some of the observations reported below would tend to support these suggestions. For example, freshly emerged *Megaselia halterata* are known to imbibe water readily (Binns, 1980).

Some visits of Phoridae to flowers are presumably in order to obtain nectar. Critical studies, however, have yet to be made. Hussey (1960) reported *Megaselia nigra* feeding on 20% sugar solutions in the laboratory. Binns (1980) offered various sugars to *M. halterata* in the laboratory. He reported that these were readily imbibed, and in the case of the females were mainly dispatched to the crop. He did not find the crops of males distended with sugar solutions, so it is likely the sugar entered their midguts straight away. He found that this species accepted all the sugars offered. However, they preferred sucrose, fructose and melezitose (all of which are major honeydew sugars) to lactose, maltose, glucose or D-mannitol. He showed that flies denied sugar meals produced fewer progeny, matured fewer eggs at a time and did not live so long in the laboratory.

Boness (1953) noted that phorids tended to be common at aphid colonies. The honeydew of aphids and other Homoptera is imbibed by phorids. Lindner (1969) reported *Megaselia maranguensis, Megaselia* sp. and *Dohrniphora fraudans* feeding on the honeydew of the white-fly *Trialeurodes vaporariorum* (Aleyrodidae). Gotô (1984b) reported a *Phora* species feeding on honeydew. Brown (1987a) reported *Gymnophora luteiventris* drinking from spots of honeydew on the leaves of low undergrowth. In addition to the the two *Megaselia pleuralis* reported above, I have observed a *M. fuscinervis* male, a *M. pulicaria* female, a male *M. tumida*, a male and three female *M. vernalis*, a *Phora atra* male, a *P. tincta* male and a *Triphleba distinguenda* male imbibing honeydew on leaves below sycamore aphids (*Drepanosiphum platanoidis*). There seems little doubt that detailed study will reveal that honeydew is a major resource for many phorids. Indeed it is probable that the majority obtain their sugars from Homopteran honeydew. It is also probable that this is a primitive habit in the Phoridae.

Downes and Dahlem (1987) have persuasively argued that, in the Diptera, feeding on Homopteran honeydew is a more primitive habit than feeding on flower nectar. The latter resource did not evolve until long after the Homoptera had evolved. The Phoridae, along with other flies, initially taste sugars with their feet (tarsi) and, furthermore, are attracted (visually) to small shiny spots. Such adaptations seem to be primarily related to the location of honeydew on surfaces such as leaves. Furthermore expanded labella of the proboscis, furnished with numerous pseudotracheae, allow flies to ingest dried honeydew spots without risk of significant water loss. Following a secretion of saliva to dissolve the dried honeydew, and a pause, the saliva is re-ingested along with the dissolved sugars. Thus leaf-searching behaviour and expanded labella are probably primitive features within the Phoridae.

If, as is suggested above, feeding on honeydew is the primitive habit in Phoridae and feeding on nectar evolved later in some species only, then we might expect to find opportunistic nectar feeding when there is a dearth of honeydew. Honeydew on leaf surfaces is likely to be drastically reduced after heavy rain. The erratic occurrence of phorids on the flowers of common Umbelliferae (6.2) could be due to fluctuations in the amount of available honeydew on leaf surfaces. Rain may also affect the amount of sugar available in the umbellifer flowers. Comparisons of the numbers visiting different sorts of flower and the numbers searching leaves in relation to the recency of rainfall might be rewarding.

Another anomaly could be explained by the hypothesis that honeydew is

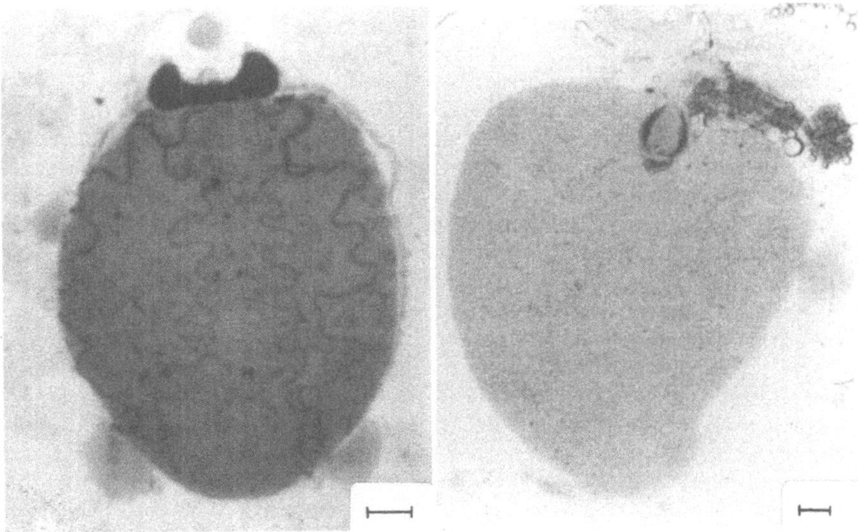

Figure 6.8 Distended crops dissected from females, the dark structures at the anterior (top) ends being Dufour's crop mechanisms. (a) *Woodiphora santoshi*. (b) *Megaselia longicostalis*, distended with carrion fluids from a dead rabbit.

the primary source of sugars for most phorids. In section 6.2 a striking observation is the dearth of phorids visiting the flowers of daffodils (*Narcissus*) in lowland Britain compared with the numbers recorded on the Malham Tarn Estate, at an altitude around 400 m. The timing of the flowering of daffodils seems to be primarily controlled by day length. By contrast temperature is probably more important in determining the spring appearance of many Homoptera. The result is likely to be that some Homoptera are already producing honeydew when the daffodils are flowering in lowland areas. By contrast the daffodils in upland areas probably flower some weeks before honeydew becomes readily available. Any Phoridae seeking sugar meals may therefore be obliged to visit these nectar sources because of the dearth of honeydew.

Males of *Dohrniphora cornuta* in laboratory colonies frequently fed on a mixture of honey, yeast and milk; but females rarely did so (Barnes, 1990a). The females mainly fed on liver, but the males only rarely did so (see below). In a culture of *Megaselia abdita*, whose larvae were reared on a brand of fish food (9.4), females regularly filled their crops with the same protein-rich food (Klaus Sander and Disney, unpublished data). *Wandolleckia achatinae* females are thought to ingest the mucus (slime) of their snail hosts (6.4.1) (Cook, 1897; Wandolleck, 1898).

Dohrniphora incisuralis females have been observed feeding on decaying plant tissues (Brues, 1909b). A male *Diplonevra nitidula* has been observed feeding from a leaf of a runner bean (*Phaseolus*) by piercing it at the junction of two small veins (Disney, 1979f). Binns (1980) observed *Megaselia halterata*

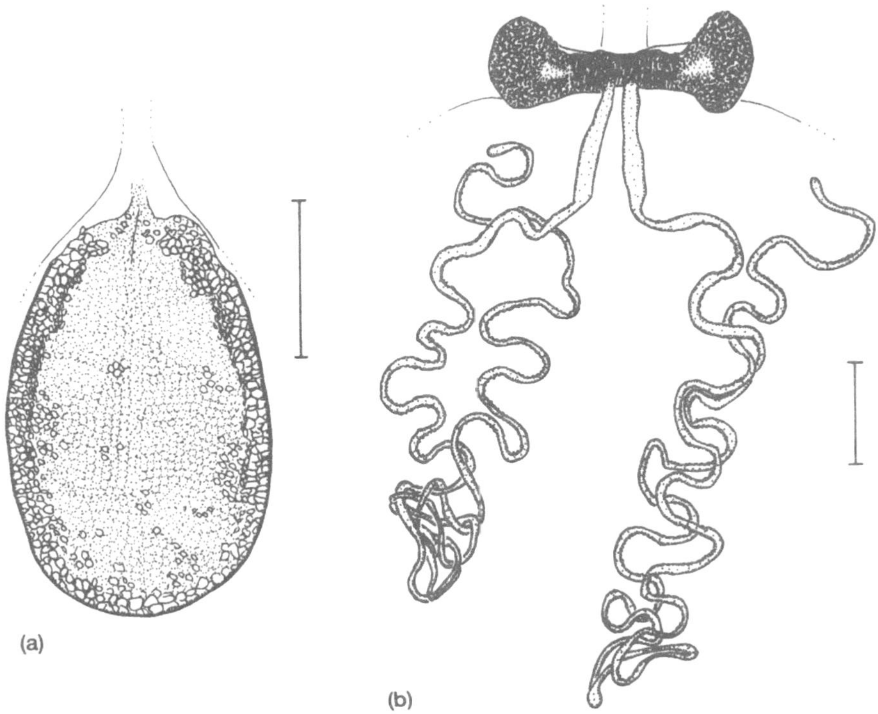

Figure 6.9 Dufour's crop mechanisms of females. (a) *Gymnophora quartomollis* (b) *Woodiphora jacksonae*.

seemingly lacerating the epidermis of the stems of broad bean plants (*Vicia faba*) in order to feed. They also fed at the cut stipes of mushrooms. Indeed Hussey (1960) noted that the females will mature their eggs following the licking of mushrooms. Both males and females of *M. vorata* have been observed feeding on fluid oozing from the syconia of a fig (*Ficus burtt-davyi*) (Compton and Disney, 1991).

In females of the tribe Megaseliini the crop retains meals, usually protein-rich meals. Dufour's crop mechanism (Figs 6.8 and 6.9), in the neck of the crop in these females, probably acts as both a valve and a source of secreted enzymes (Disney, 1987b). In *Woodiphora* the secretory function is probably associated with the tubular glands which extend into the lumen of the crop (Figs 6.8a and 6.9b). The dispatch of protein-rich meals to the crop, not the stomach, is the reverse of the situation in other female Diptera, such as mosquitoes. In the latter the sugar meals are normally dispatched to the crop and the protein-rich blood meals to the stomach (Clements, 1963), but sugar meals are sometimes passed straight to the stomach in female mosquitoes (Culicidae) and blackflies (Simuliidae) (Disney, 1970).

Examination of the crop in the Megaseliini is a valuable means of obtaining information on the types of meals ingested. Some recorded meals are listed below. In many cases the crop contents are too amorphous to allow identification by direct examination alone. However, observation of the fly feeding, followed by immediate examination of the gut, has resolved the identity of some of these meals. The following meals have been recorded in the crops of female phorids (unless otherwise indicated the records are previously unpublished data).

CARRION FLUIDS

Megaselia longicostalis after feeding on a dead rabbit (Disney, 1987b).

FUNGUS SPORES

Megaselia arietina (Čakar and Disney, 1991), *M. brevicostalis, M. dimidia, M. harteni* (Disney, 1991h), *M. latipalpis, M. longicostalis* (Disney, 1987b), *M. nasoni, M. nigra* (Disney, 1987b), *M. ruficornis* and *Woodiphora arunkumari* (Disney, 1989f).

SUGAR SOLUTIONS

Megaselia halterata (Binns, 1980).

YEASTS (*Saccharomyces* species)

Megaselia pulicaria (two records for the variety with 3, not 4, rectal papilae. This variety is currently treated as a morph of a polymorphic species, but some may prove to be a sibling species – Disney, 1989a.)

Direct observation followed by examination of the gut has also provided information on species lacking Dufour's crop mechanism. For example, *Anevrina curvinervis* observed probing dead rabbits with the proboscis were found to have carrion fluids in the midgut, and likewise with *Diplonevra florea* at a dead duck (Disney, 1987b). Bugnion (1913) recorded fungal conidia in the gut of *Clitelloxenia peradeniyae*.

In laboratory colonies *Dohrniphora cornuta* females frequently fed at liver, sinking the proboscis about 0.5 mm into it. In the females of this species there are well-developed tricuspid prestomal teeth (Barnes, 1990a). There are many records in the literature of Phoridae visiting carrion and dung. It has too often been assumed that such visits were for the purpose of oviposition, mainly because the flies were dominantly females. For example, in Cuba Gregor (1977) collected 421 Phoridae visiting baits such as beef, fish, cheese and human faeces. The ratio of females to males was 4.4:1. However, a significant proportion of the species he recorded were of those known to be larval parasitoids of ants. These data suggest that the latter species, at least, were attracted to the baits for the purposes of feeding. In the USA Payne *et al.*

(1968) buried dead piglets 50–100 cm deep in the soil. They observed *Dohrniphora incisuralis* and *Metopina subarcuata* feeding on the carcass fluids at Stage 2 of the decay process. By contrast, *Triphleba subfusca, Megaselia* spp. and *Puliciphora* sp. were reported visiting exposed pig carrion, mainly during the 'advanced decay' stage of decomposition (Payne and Crossley, 1966) (see section 7.5.1 for details of the stages of decomposition). My own observations of slide-mounted females caught visiting carrion have shown that they are rarely gravid, thus ruling out oviposition as the purpose of these visits in most cases. For example, four females of *Triphleba hyalinata* caught (by J. Blasco Zumeta) under a dead garden dormouse (*Eliomys quercinus*) in Spain had the eggs undeveloped. A female *Diplonevra florea* (collected by J. M. Nelson) visiting a dead sheep in Scotland had the eggs only half developed. Other records of phorids visiting corpses of small vertebrates include *Anevrina curvinervis, A. thoracica, A. unispinosa, A. urbana, Conicera tibialis* and *Megaselia pulicaria* (Disney, unpublished data; Disney *et al.*, 1981; Lundbeck, 1922; Wood, 1906). *Diplonevra nitidula* females have been reported swarming over a dead earthworm (Disney, 1979f). In Egypt *Conicera tibialis, C. similis, Diplonevra funebris, Spiniphora bergenstammi* and *Megaselia scalaris* have been recorded visiting dead rabbits buried 10 cm deep in the soil (T. Tantawi and Disney, unpublished data).

Brues (1909b) reported *Dohrniphora incisuralis* females feeding on caterpillar frass and Gotô (1985c) reported *Phora prisca* imbibing bird dung dissolved in water on the surface of leaves.

In laboratory cultures, adult females of *Puliciphora borinquenensis* are reported feeding on dead cockroaches and other insects, supplemented by probing pieces of potato (Miller (1979a). In the laboratory *Dohrniphora cornuta* females were observed feeding on dead adults and larvae of their own kind. One was observed with a living phorid larva suspended from her proboscis; and one was observed feeding on a dead calliphorid (blowfly) larva, which had been offered when still alive (Barnes, 1990a).

Brues (1909b) reported *Dohrniphora incisuralis* females feeding on dead insect larvae. Stuckenberg (1953) reported phorids feeding on the fluid oozing from the honey-bee prey of a mantid (*Polyspilota* sp.) and Sivinski and Stowe (1980) reported *Megaselia* sp. females feeding on the dead-insect prey of a spider. These last two reports represent further cases of kleptoparasitism (see below). A female of the termitophilous *Adenophora pulchella* has been observed dragging a freshly-dead termite from the edge of a foraging column of its host termite (Disney and Kistner, 1989a). Females of *Woodiphora arunkumari* and both sexes of *W. pallidinervis* have been observed feeding from dying termites (Disney, 1989f) and *Dohrniphora caini, D. gilberti* and *D. vorax* doing likewise (Disney, 1990g), as well as numerous *Megaselia* noted (Disney, 1985h) but not yet analysed. Brown (1992b) reports *Dohrniphora* spp. attracted to freshly-dead termites in Costa Rica. Both sexes of *Apocephalus paraponerae* have been reported imbibing fluids from freshly-crushed ants, *Paraponera clavata* (Brown and Feener, 1991b). *Rhyncophoromyia maculineura, R. spinipleura* and *R. trivittata* also feed on the fluids of injured ants (Brown and Feener, 1993).

It seems likely that feeding on freshly-dead insects is commonly practised by many scuttle flies, particularly by females. The systematic study of the importance of this resource is overdue for critical evaluation. Enormous numbers of abandoned, only partly consumed, corpses of the prey of predatory insects, especially the prey of Diptera such as Empididae, Dolichopodidae and Scathophagidae, are left on the leaves of low vegetation or fall to the ground. Where ants abound many of these corpses are quickly removed, so that alternative opportunities such as the sudden exposure of wounded termites are likely to be a valuable resource for other scavenging insects such as Phoridae. Where ants do not abound abandoned insect corpses may remain for many hours, or even days, depending on the situation and weather conditions.

In a preliminary investigation I carried out a survey of the numbers and nature of abandoned insect corpses on the leaves of dog's mercury (*Mercurialis perennis*) in a limestone-scree woodland on the Malham Tarn Nature Reserve, North Yorkshire, on 24 May 1990. This was during a short spell of settled, fine, weather. Fifteen hundred leaves were randomly selected and examined, and the corpses of all insects found were collected into a tube of alcohol, using a moistened paint brush. The score was one corpse per 18 leaves. The 83 corpses collected comprised 70 Diptera (27 Bibionidae, 5 Ceratopogonidae, 9 Chironomidae, 1 Scatopsidae and 28 Sciaridae), 1 Coleoptera (Staphylinidae), 1 Heteroptera (Anthocoridae, nymph), 5 Homoptera (Aphididae), 1 Hymenoptera (Chalcidoidea), 1 Psocoptera and 4 Thysanoptera.

Feeding from freshly-dead or dying insects is only one step away from imbibing fluid from living insects, which in turn is only a step away from actively preying upon other insects. It is now evident that different Phoridae have been observed engaging in one or more of these different activities, no doubt thereby reflecting the course of evolution of obligate predatory behaviour. Wesché (1908) reported a *Diplonevra* species preying upon an acalypterate fly. The females of *Clitelloxenia assmuthi* feed on the blood (haemolymph) of older termite nymphs (Assmuth, 1913).

Phalacrotophora fasciata adults have been reported sucking fluid from the living larvae and pupae of their coccinellid-beetle larval hosts (Martelli, 1914; Lichtenstein, 1920; Richards, 1926). Likewise *P. berolinensis* females have been reported imbibing haemolymph from ladybird beetle pupae (Delucchi, 1953; Colyer, 1952a, 1954g) (3.5.6(b)). Females of both species have been observed making incisions into the dorsal faces of the thorax or abdominal segments 1 and 2 of the beetle pupae. Feeding on haemolymph lasted up to 27 min. From the five hosts observed, phorids were successfully reared in due course (Disney *et al.*, 1994). *Megaselia rubricornis* adults will attack, and suck dry, healthy pupae of chrysomelid beetle pupae, while other adjacent pupae are parasitized by their larvae (3.5.6(b)).

Roger (in Neveu-Lemaire, 1938) reported a *Megaselia* species sucking blood from a mosquito, but no details are given. *Apocephalus paraponerae* females have been observed feeding on fluid oozing from oviposition punctures in the integument of their host-ant *Paraponera clavata* (Brown and Feener, 1991b).

Females of *Megaselia vorata* and *M. parvorata* prey upon fig wasps

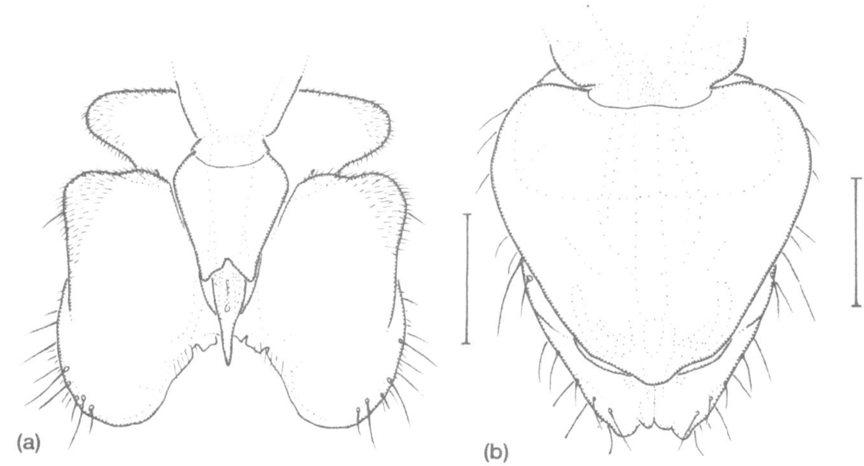

Figure 6.10 *Megaselia vorata* sexual dimorphism in the proboscis. (a) Male. (b) Female, with reduced labella but enlarged labrum.

(Agaonidae), piercing in turn the cuticle of the head, the thorax and the abdomen, and sucking out their contents (Compton and Disney, 1991). The difference in the feeding habits of the males and females of these *Megaselia* species is correlated with a sexual dimorphism in the proboscis (Fig. 6.10). Minute flattened hairs, almost certainly of insect origin, have been reported in the crop of a female *M. jorgensis* (Disney, 1991h). Likewise a variety of hairs and scales, of arthropod origin, have been found in the crop of a female *M. rufipes* (unpublished data).

Phoridae have occasionally been recorded irritating people by attempting to 'bite' them. For example, *Megaselia abdita* has been reported attacking egg-collectors in poultry houses, and it was probably this species reported annoying people in Yorkshire coal mines (Edwards and Schmitz, 1939; Disney, 1987k). *Dohrniphora cornuta* has also been reported attacking man (Schmitz, 1951c), but this appears to be atypical bevaviour (Barnes, 1990a). Large numbers of *M. halterata* have been reported causing annoyance to mushroom pickers (Binns *et al.*, 1979).

Larval kleptoparasitic trophallaxis was described for some Phoridae above (3.2.5). Some adult scuttle flies indulge in similar behaviour also. The habit may be widespread in the family, but the habits of Phoridae whose females reside in the nests of ants, termites and other social insects (see below) have been little studied. Hölldobler (1928) reported that *Metopina formicomendicula* adults solicit food from the ant *Solenopsis fugax*. A fly mounts a worker ant from behind and then strokes the side of the ant's head with its front legs. The ant responds by raising its head, spreading its jaws and regurgitating a drop of food, which the fly then sucks before rapidly departing. Likewise *Allochaeta longiciliata* has been observed soliciting food from the workers of its ant host *Acromyrmex muticinodus* (Borgmeier, 1928b).

Kemner (1922) observed *Javanoxenia punctiventris* females stealing food that was being passed between termite workers. Franssen (1933a) not only confirmed this, but observed *Clitelloxenia hemicyclia* doing the same. The licking of a termite's mouthparts typically lasted 2–3 s (Franssen, 1933b). In addition he observed both species being voluntarily fed by the worker termites. *Termitosagma henningsi* females have been observed inserting the proboscis into the mouthparts of termite nymphs (Malin and Kistner, 1970) and stealing food from nymphs and younger termite workers (Kistner, 1982). *Echidnophora butteli* females have been observed inserting the proboscis into the drop of fluid being passed by a worker termite to its nymph (Disney and Kistner, 1988, 1989b). Likewise *Crasilla setifer* and *Franssenia rocheorum* have been observed indulging in the same behaviour (Disney and Kistner, 1989b).

6.4 SPECIAL ASSOCIATIONS

6.4.1 MOLLUSCA

Ectotrophic phoresy

The peculiar, flightless, females of the Afrotropical *Wandolleckia achatinae* are to be found on the surface of snails, not only running swiftly on the exposed mantle but also under the shell and even entering the pulmonary cavity. It is thought that they feed on mantle secretions. Up to 16 flies have been recorded on one snail (Cook, 1897; Bequaert, 1926). The males were reared by Baer (1953), and found to be fully winged. Only females have been observed on snails, being recorded on the following species: *Achatina achatina* (Baer, 1953), *A. variegata* (Cook, 1897; Schmitz, 1929e, 1958e), *A. ventricosa* (Baer, 1953) and *Pseudotrochus (Perideris)* sp. (Schmitz, 1928a, 1929e). Some of these snails have been spread by man to other biogeographic regions, becoming serious pests in places. Whether *Wandolleckia* has been transported with the snails is not known.

6.4.2 DIPLOPODA (MILLIPEDES)

Schmitz (1939c) reported five males and one female of *Megaselia equitans* on the back of a large black millipede in the Yemen. The only large black millipede known from the Yemen is *Archispirostreptus transmarinus* (A. van Harten, *in litt.*). Males of the closely-related *M. audreyae* have been reported on the backs of the large millipedes *Pachybolus* sp. (Disney, 1978d). The larval biology of these two flies remains unknown.

6.4.3 INSECTA

Large numbers of Phoridae are associated with ants and termites. In most cases these associations are based on the requirements of the larvae (3.5.6(e) and (f)). Some are parasitoids or predators on the adults or brood of their hosts. Others feed on the refuse in the host's nest, while still others exploit the food stores. In many species the females are flightless, frequently being transported by the males while in copula. Such females tend to be resident in the colonies of their hosts. The records of adult Phoridae reported in or at the colonies of social insects are summarized below. Species attracted to ants or termites for the purposes of oviposition only are reviewed in Chapter 3. Species attracted to wounded or moribund termites, and other insects, for feeding purposes are dealt with above.

Borgmeier (1928c) referred to the species whose adult females reside in ant colonies as myrmecophiles and treated them as distinct from the parasitoids. Furthermore he drew attention to the modified ovipositors of the latter group. However, he classed as myrmecophiles some genera, such as *Aenigmatias*, with simple ovipositors and yet with known parasitoid habits. He also subsequently included genera with highly modified ovipositors, such as *Dacnophora*, in his myrmecophile category (Borgmeier, 1961a). It is evident that the degree to which the ovipositor is modified is related more to the placement of the egg than to parasitic habits as such. Where the egg is laid on or near the target host there is no need for a rigid ovipositor. When the egg is placed into the host, whose integument has to be pierced by the ovipositor, the latter may need to be well sclerotized. It seems best, therefore, to regard species whose adult females reside in ant nests as myrmecophiles, regardless of the form of the ovipositor or the habits of their larvae.

Kistner (1982) provides a useful review of the myrmecophilous and termitophilous insects. An essential problem they face is how to avoid being recognized as alien intruders by their hosts. In most phorids avoidance behaviour seems to be the primary adaptation. Thus both *Echidnophora butteli* and *Franssenia rocheorum* tend to associate with the nymphs of their termite hosts. These nymphs are generally sluggish in their reactions to each other and towards intruders. Both flies assiduously avoid the worker termites (Disney and Kistner, 1988). However, other species interact with the termites. For example, *Termitosagma henningsi* females repeatedly rub their front legs over termite nymphs and then rub them on their own saddle regions (at the front of the abdomen). It is thought that by transferring termite scent to these regions they direct the licking behaviour, employed by termites in caring for their own nymphs, towards themselves (Kistner, 1982). They will even groom the antennae of the termite nymphs. Termites have been observed grooming these flies. Likewise *Termitophilomyia gracilis* females palpate their host termites and the latter palpate and lick the flies, especially their abdomens. Furthermore these phorid species possess large pores on the dorsal face of the abdomen. When workers of *Odontotermes culturarum* lick these flies it is postulated that exudates are secreted from these pores (Malin and Kistner, 1970).

(a) Hymenoptera

Diapriidae

Ecitophora comes has been reported associated with the diapriid *Labidopria* sp. (Disney and Kistner, 1989d). This parasitoid wasp is normally associated with the ant *Labidus praedator*, which is also a regular host for this phorid fly (see below).

Formicidae (ants)

The ant classification and nomenclature used here is stated in section 3.5.6(e). The Phoridae are reviewed in two distinct groups, based on what little we know.

A. Female phorids primarily resident in ant nests, and usually flightless (although they may have had a flying stage before shedding their wing membranes on taking up residence with the ants)

***Dorylinae* (driver ants)**

Aenictus sp. – *Aenictacantha setifera* (Disney, 1991m) and *Rhynchomicropteron nudiventer* (Disney, 1992d).

A. aratus – *Aenictacantha* sp. 'A' (Disney, 1991m) and *Maculiphora aenictophila* (Disney and Kistner, 1991).

A. gracilis – *Aenictomyia chapmani* (Brues, 1930).

Dorylus spp. – *Aenigmatopoeus cucullatus* (Colyer, 1958), *Coridophora semicimex*, *Psyllomyia myrmecophila, P. patrizii* and *Puliciphora meneghettii* (Schmitz, 1951d).

D. helvolus – *Psyllomyia braunsi* (Schmitz, 1951d) and *P. testacea* (Brues, 1901).

D. kohli – *Aenigmatopoeus kohli* (Schmitz, 1915d) and *Hexacantherophora cohabitans* (Schmitz, 1914d).

D. sjostedti – *Aenigmatopoeus orbicularis* (Schmitz, 1914d).

***Ecitoninae* (army ants)**

There are 90 species of Phoridae recorded as residents in the colonies of army ants. The records are detailed below.

Unidentified Ecitoninae – *Ecitophora costaricensis* (Brues, 1925a), *Placophorina obtecta* (Borgmeier, 1958), *Thalloptera bisinuosa* (Borgmeier, 1936), *T. schwarzmaieri, Xanionotum bruchi* and *X. spinipes* (Borgmeier, 1928c).

Eciton burchelli – *Acontistoptera brasiliensis* (Schmitz, 1924i), *Adelopteromyia parvipennis* and *A. propinqua* (Disney and Kistner, 1989d), *Apterophora attophila* (Rettenmeyer and Akre, 1968; Prado, 1976), *A. caliginosa* (Brues, 1923c; Prado, 1976), *Chonocephalus mexicanus* and *Ecitophora bruchi* (Rettenmeyer and Akre, 1968), *E. collegiana* (Bristowe, 1925; Borgmeier, 1960b; Rettenmeyer and Akre, 1968; Disney and Kistner, 1989d), *E. comes* (Brues, 1923a), *Ecitoptera centralis* (Borgmeier, 1923d; Rettenmeyer and Akre, 1968), *E. ciliata* (Borgmeier, 1923d), *E. concomitans* (Rettenmeyer and Akre, 1968), *E. schmitzi* (Borgmeier, 1923d), *E. subciliata* (Rettenmeyer and Akre, 1968), *Ecituncula halterata* (Rettenmeyer and Akre, 1968), *E. tarsalis* (Borgmeier, 1925c; Rettenmeyer and Akre, 1968), *Homalophora epichaeta, Pulicimyia triangularis, Puliciphora fenestrata, P. frivola, P. imbecilla* and *P. rata* (Borgmeier, 1960b; Rettenmeyer and Akre, 1968), *P. borinquenensis* and *Thalloptera schwarzmaieri* (Rettenmeyer and Akre, 1968).

E. coecum – Ecitomyia wheeleri (Brues, 1901; Seevers, 1941).

E. dulcius – Commoptera pygmaea (Borgmeier, 1928a), *Ecitophora bruchi* (Schmitz, 1923e), *E. collegiana* (Borgmeier, 1960b; Rettenmeyer and Akre, 1968), *E. parva* (Borgmeier, 1960b), *E. pilosula* (Borgmeier, 1960b; Rettenmeyer and Akre, 1968), *Ecitoptera concomitans* (Borgmeier, 1926a), *E. cordobensis, E. subciliata* (Borgmeier, 1960a), *Ecituncula longipilosa* (Borgmeier, 1960b), *Puliciphora borinquenensis* and *Thalloptera schwarzmaieri* (Rettenmeyer and Akre, 1968).

E. hamatum – Apterophora attophila and *A. caliginosa* (Rettenmeyer and Akre, 1968; Prado, 1976), *Ecitophora collegiana* (Borgmeier, 1960b; Rettenmeyer and Akre, 1968; Disney and Kistner, 1989d), *Ecitoptera subciliata* (Rettenmeyer and Akre, 1968), *Ecituncula halterata* (Borgmeier, 1960b; Rettenmeyer and Akre, 1968), *E. tarsalis* (Borgmeier, 1960b; Rettenmeyer and Akre, 1968; Disney and Kistner, 1989d), *Pachyneurella nitida, Puliciphora borinquenensis* and *Thalloptera schwarzmaieri* (Rettenmeyer and Akre, 1968).

E. lucanoides – Ecitophora collegiana, E. costaricensis and *Ecituncula tarsalis* (Borgmeier, 1960b).

E. mexicanum – Commoptera setiventris (Borgmeier, 1971a), *Ecitophora bruchi, E. collegiana* (Borgmeier, 1960b; Rettenmeyer and Akre, 1968), *E. pilosula* (Rettenmeyer and Akre, 1968), *Ecitoptera concomitans* (Borgmeier, 1960a), *Ecituncula microps* and *Thalloptera comes* (Borgmeier, 1971a), and *T. schwarzmaieri* (Rettenmeyer and Akre, 1968).

E. quadriglume – Acontistoptera brasiliensis and *Adelopteromyia longisetosa* (Schmitz, 1924i), *Chonocephalus ecitophilus* (Borgmeier and Schmitz, 1923a), *Ecitomyia luteola* and *Ecitophora collegiana* (Borgmeier, 1960b), *E. comes* (Schmitz, 1924i), *E. varians* (Borgmeier, 1960b), *Ecitoptera ciliata* (Borgmeier, 1960a), *Ecituncula aptera* (Schmitz, 1924i), *Lepidophoromyia zikani* (Borgmeier, 1925c), *Puliciphora frivola* (Borgmeier, 1960b) and *Thalloptera quadriglumis* (Schmitz, 1924i).

E. schlechtendali – Thalloptera sebastiani (Borgmeier, 1936).

E. schmitti – Ecitomyia wheeleri (Brues, 1901; Seevers, 1941), *Xanionotum hystrix* (Brues, 1903b) and *X. smithii* (Brues, 1936b).

E. vagans – Ecitophora bruchi, E. collegiana, E. varians (Borgmeier, 1960b; Rettenmeyer and Akre, 1968), *Ecitoptera concomitans* (Borgmeier, 1960a), *E. subciliata, Ecituncula halterata* and *Thalloptera schwarzmaieri* (Rettenmeyer and Akre, 1968).

Labidus coecus – Acontistoptera brasiliensis (Brues, 1923a), *Commoptera coeci* (Borgmeier, 1971a), *Cremersia spinicosta* (Borgmeier, 1961a), *Dacnophora setithorax* (Borgmeier, 1961a), *Ecitomyia luteola* (Borgmeier, 1960b), *E. wheeleri* (Borgmeier, 1960b; Rettenmeyer and Akre, 1968), *Ecitophora bruchi* and *E. collegiana* (Borgmeier, 1960b), *E. comes* (Brues, 1923a), *E. costaricensis* (Borgmeier, 1960b), *Ecitoptera centralis, E. humeralis, E. subciliata* and *E. truncatipennis* (Borgmeier, 1960a), *Ecituncula glandularis* (Borgmeier, 1960b), *E. halterata* (Borgmeier, 1960b; Rettenmeyer and Akre, 1968), *E. tarsalis* (Rettenmeyer and Akre, 1968), *Puliciphora lunaris, P. placida* and *P. rufipes* (Borgmeier, 1960b), *Thalloptera imbecilla* and *T. nevermanni* (Borgmeier, 1936) and *Xanionotum tarsale* (Borgmeier, 1968a).

L. praedator – *Acontistoptera brasiliensis* (Schmitz, 1914d; Brues, 1923a; Rettenmeyer and Akre, 1968; Disney and Kistner, 1989d), *Adelopteromyia parvipennis* (Disney and Kistner, 1989d), *A. propinqua* (Borgmeier, 1958; Rettenmeyer and Akre, 1968; Disney and Kistner, 1989d, 1990b), *Apterophora* sp. (Disney and Kistner, 1989d), *Colobomeles ramboi* (Borgmeier, 1927a; Schmitz, 1928a), *Ecitomyia luteola* (Borgmeier, 1960b), *Ecitophora breviptera* (Disney and Kistner, 1989d), *E. collegiana* (Borgmeier, 1960b; Rettenmeyer and Akre, 1968), *E. comes* (Schmitz, 1914d; Brues, 1923c; Rettenmeyer and Akre, 1968; Disney and Kistner, 1989d), *E. fuscipalpis* and *E. luteipalpis* (Schmitz, 1924i), *E. parva* (Schmitz, 1924i; Rettenmeyer and Akre, 1968), *E. pilosula* (Rettenmeyer and Akre, 1968), *E. varians* (Borgmeier, 1960b; Rettenmeyer and Akre, 1968), *Ecitoptera humeralis* (Disney and Kistner, 1989d), *E. maior* (Borgmeier, 1960a), *E. subciliata* (Rettenmeyer and Akre, 1968), *E. truncatipennis* (Borgmeier, 1960a), *Ecituncula halterata* (Borgmeier, 1936), *E. tarsalis* and *Homalophora epichaeta* (Rettenmeyer and Akre, 1968), *Lepidophoromyia labida* (Disney and Kistner, 1989d), *Puliciphora gracilis* and *P. legionis* (Borgmeier, 1960b), *Sphinctomyia aenigmatica* (Borgmeier, 1954), *Thalloptera emarginata* (Schmitz, 1935d), *T. imbecilla* (Borgmeier, 1927d), *T. schwarzmaieri* (Rettenmeyer and Akre, 1968), *Xanionotum scopifer* (Borgmeier, 1968a) and *X. wasmanni* (Borgmeier, 1928c; Schmitz, 1929b).

Neivamymrex alfaroi – *Ecitomyia longipennis* (Borgmeier, 1960b) and *E. wheeleri* (Borgmeier, 1960b; Rettenmeyer and Akre, 1968).

N. densepunctatus – *Commoptera affinis* (Borgmeier, 1971a).

N. diversinodis – *Xanionotum delicatum* (Borgmeier, 1968a).

N. d'orbignyi – *Acontistoptera hirsuta, Commoptera ecitonis* and *Xanionotum spinosius* (Borgmeier, 1926a).

N. goeldii – *Commoptera ecitonis* (Borgmeier, 1971a), *Ecitomyia brevipennis* and *E. longipennis* (Borgmeier, 1960b) and *Xanionotum scopifer* (Borgmeier, 1938b).

N. hetschkoi – *Ecitoptera cordobensis* (Borgmeier, 1925c) and *Xanionotum delicatum* (Borgmeier, 1932b).

N. laevigatus – *Commoptera ecitonis* (Borgmeier, 1971a) and *Xanionotum spinosius* (Borgmeier, 1968a).

N. legionis – *Ceratoconus ecitophilus* (Borgmeier, 1928c; Prado, 1976), *Commoptera affinis* (Borgmeier, 1928a), *Dacnophora legionis* (Borgmeier, 1961a), *Ecitomyia luteola* (Borgmeier, 1960b), *Ecitoptera ciliata* and *E. schmitzi* (Borgmeier, 1960a), *Puliciphora legionis* and *P. rufipes* (Borgmeier, 1960b), *Thalloptera imbecilla* (Borgmeier, 1927d), *Xanionotum bruchi* (Borgmeier, 1928c).

N. minensis – *Commoptera ecitonis* (Borgmeier, 1971a), *Ecitomyia longipennis* (Borgmeier, 1960b) and *Xanionotum spinipes* (Borgmeier, 1932b).

N. nigrescens – *Ecitomyia wheeleri* (Rettenmeyer and Akre, 1968), *Xanionotum hystrix* (Borgmeier, 1963d; Rettenmeyer and Akre, 1968; Disney and Kistner, unpublished data) and *X. scopifer* (Disney and Kistner, unpublished data).

N. opacithorax – *Acontistoptera melanderi* (Brues, 1902b; Rettenmeyer and Akre, 1968), *Ecitomyia longipennis* (Borgmeier, 1960b) and *Xanionotum hystrix* (Rettenmeyer and Akre, 1968).

N. orthonotus – *Xanionotum spiniceps* (Borgmeier, 1968a).

N. pertyi – *Commoptera ecitonis* (Borgmeier, 1971a).

N. pilosus – *Ecitophora parva* (Borgmeier, 1960b), *Xanionotum scopifer* (Borgmeier, 1968a) and *X. wasmanni* (Borgmeier, 1928c).

N. pseudops – *Ecitomyia brevipennis* (Borgmeier, 1960b), *Ecitoptera cordobensis* (Borgmeier, 1960a), *Ecituncula longipilosa* (Borgmeier, 1960b) and *Xanionotum bruchi* (Borgmeier, 1924c, 1932b).

N. punctaticeps – *Aemuliphora reichenspergeri* (Borgmeier, 1938b), *Xanionotum delicatum* (Borgmeier, 1928c) and *X. setulitibia* (Borgmeier, 1938b).

N. raptans – *Acontistoptera hirsuta* (Borgmeier, 1925c; Bruch, 1928), *Commoptera ecitonis* (Bruch, 1928), *Dacnophora legionis* (Borgmeier, 1961a), *Ecitoptera cordobensis* (Borgmeier, 1960a) and *Ecituncula longipilosa* (Borgmeier, 1925c).

Nomamyrmex esenbecki – *Ecitomyia brevipennis* (Borgmeier, 1960b), *Ecitophora breviptera* (Disney and Kistner, 1989d), *Ecitoptera concomitans* and *E. subciliata* (Borgmeier, 1960a), *Ecituncula glandularis* (Borgmeier, 1960b; Rettenmeyer and Akre, 1968), *E. halterata* (Rettenmeyer and Akre, 1968), *E. longipilosa* (Borgmeier, 1960b), *Puliciphora borinquenensis* (Rettenmeyer and Akre, 1968), *Thalloptera socialis* (Borgmeier, 1936), and *Xanionotum bruchi* (Borgmeier, 1932b).

N. hartigi – *Ecitomyia brevipennis, E. ocellata, and Ecitophora bruchi* (Borgmeier, 1960b), *Ecitoptera concomitans* and *E. subciliata* (Borgmeier, 1960a) and *Ecituncula longipilosa* (Borgmeier, 1960b).

These data for army ant colonies are summarized in Table 6.3.

Formicinae

Acropyga acutiventris – *Puliciphora boltoni* (Disney, 1988b; Disney and Kistner, 1988), the larvae being predatory upon the ants (3.5.6(e)).

Brachymyrmex heeri – *Chonocephalus mexicanus* and *Myrmomyia brachymyrmecis* (Silvestri, 1911).

Camponotus rufipes – *Metopina psociformis* (Borgmeier, 1969b).

Prenolepis longicornis – *Rhynchomicropteron caecutiens* (Schmitz, 1914d).

Myrmicinae

Acromyrmex sp. – *Lepidophoromyia zikani* (Schmitz, 1928a).

A. muticinodus – *Allochaeta longiciliata* (Borgmeier, 1926c, 1928b), the flies soliciting food from the ant workers (6.3); *Ecitoptera proboscidalis* (Borgmeier, 1960a).

A. niger – *Allochaeta longiciliata* (Borgmeier, 1928b).

Atta sp. – *Ecitoptera microps* (Borgmeier, 1960a).

A. insularis – *Puliciphora cubensis* (Brues, 1932b).

A. sexdens – *Apterophora attophila* (Borgmeier, 1930a; Prado, 1976) and *Homalophora attae* (Borgmeier, 1958).

Crematogaster impressa – *Euryplatea eidmanni* (Schmitz, 1941e,f).

C. ranavalonae – *Euryophora madagascarensis* (Schmitz, 1915d).

Pheidole spp. – *Ecitomyia manni* (Brues, 1925a; Seevers, 1941), *Pheidolomyia spinosissima* (Borgmeier, 1935d), *Puliciphora myrmecophila, Rhabdomyia setosa* and *R. spinosa* (Brues, 1925a).

Table 6.3 Myrmecophilous Phoridae recorded from nests of army ants (Ecitoninae)

Phorid genus	No fly species (P)	No of ant species (A)	Mean no ant spp per fly sp	Mean no fly spp per ant sp (P/A)
Acontistoptera	3	7	2 3	0 4
Aemuliphora	1	1	1 0	1 0
Adelopteromyia	3	3	1 7	1 0
Apterophora	3	3	1 7	1 0
Ceratoconus	1	1	1 0	1 0
Chonocephalus	2	2	1 0	1 0
Colobomeles	1	1	1 0	1 0
Commoptera	5	11	2 0	0 5
Cremersia	1	1	1 0	1 0
Dacnophora	2	3	1 0	0 7
Ecitomyia	5	14	3 6	0 4
Ecitophora	10	12	3 4	0 8
Ecitoptera	9	14	3 3	0 6
Ecituncula	6	13	3 3	0 5
Homalophora	1	2	2 0	0 5
Lepidophoromyia	2	2	1 0	1 0
Pachyneurella	1	1	1 0	1 0
Placophora	1	1	1 0	1 0
Pulicimyia	1	1	1 0	1 0
Puliciphora	10	8	1 6	1 3
Sphinctomyia	1	1	1 0	1 0
Thalloptera	9	12	1 9	0 8
Xanionotum	12	19	2 1	0 6

The penultimate column is a preliminary index of catholicity of host choice The last column is a preliminary index of niche overlap It is immediately evident that the catholicity of host choice tends to be greater than that of the parasitoid species reviewed in Chapter 3 (3 5 6(e))

P crassipes – *Pheidolomyia alpina* (Borgmeier, 1925c) and *P sodalis* (Borgmeier, 1927c)

P emeryi – *Pheidolomyia alpina* (Schmitz, 1915d)

P fervida – *Hypogeophora macrothrix* (Gotô, 1986b, M Terayama and Disney, unpublished data)

P fimbriata – *Chelidophora baculipennis* (Borgmeier, 1933b)

Pheidologeton sp – *Rhynchomicropteron micraphidiforme* (Disney, 1992d)

P diversus – *Puliciphora rosei* (Disney and Kistner, 1988)

Solenopsis fugax – *Metopina formicomendicula* (Holldobler, 1928), the flies indulging in kleptoparasitic trophallaxis (6 3)

S geminata – *Apterophora attophila* (Prado, 1976), *Commoptera solenopsidis* (Brues, 1901), *Placophorina obtecta* (Borgmeier, 1936), *Puliciphora incerta* (Silvestri, 1911) and *Zikania schmidti* (Borgmeier, 1936),

S saevissima – *Chelidophora solenopsidis* (Borgmeier, 1950b), *Chonocephalus heymonsi* (Borgmeier, 1963d), *Ecitomyia longipennis* (Borgmeier, 1960b), *E*

luteola and *E. minuscula* (Borgmeier and Schmitz, 1923a; Seevers, 1941), *Pulicimyia pauxilla, Puliciphora gracilis*, and *P. interrupta* (Borgmeier, 1960b), *Zikania degenerata* (Borgmeier, 1925a).

Tetramorium aculeatum – Microplatyphora congolensis (Schmitz, 1915d).

Tranopelta gilva – Tranopeltoxenos manni (Brues, 1923a).

It is striking that of the 37 species recorded as residents in the nests of myrmicine ants only 6 are among the 90 species recorded from colonies of Ecitoninae. These are *Apterophora attophila, Ecitomyia longipennis, E. luteola, Lepidophoromyia zikani, Placophorina obtecta* and *Puliciphora gracilis*.

Ponerinae (predatory ants)

Leptogenys spp. – *Puliciphora longipes* (Schmitz and Mjöberg, 1925), *P. rosei* (Maschwitz *et al.*, 1989), *Rhynchomicropteron aphidiforme* (Schmitz and Mjöberg, 1925), *R. necaphidiforme* (Disney, 1990c).

L. diminuta – Puliciphora knighti (Disney, 1988b).

L. distinguenda – Puliciphora rosei, Rhynchomicropteron necaphidiforme and *R. nudiventer* (Disney, 1992d; K. Rosciszewski and Disney, unpublished data).

L. ebenina – Puliciphora microphthalma (Schmitz and Mjöberg, 1924).

L. kittelli – Rhynchomicropteron nudiventer (Brown, 1992a).

L. mutabilis – Puliciphora rosei females following roving bands of ants (Disney, 1988b), *Rhynchomicropteron nudiventer* (Steghaus-Kovac and Disney, 1990).

L. ocellifera – Rhynchomicropteron puliciforme (Annandale, 1912), the females running with a foraging party of the ants.

Myopias maligna – Myopiomyia harmani (Disney, 1987g).

Odontomachus affinis – Lenkoa aurita (Borgmeier, 1969b).

O. rixosi – Stethopathusa corporaali (Schmitz, 1921b).

O. simillimus – Megaselia pagei (Disney, 1987g).

Pachycondylae laevigata – Ecitomyia juxtaposita (Borgmeier, 1960b).

B. Males and flying females attracted to ant colonies, the females probably being at most only temporary residents.

Dolichoderinae

Linepithema (= *Iridomyrmex*) *humile – Ceratoconus setipennis* (Borgmeier, 1928c; Prado, 1976).

Philidris (= *Iridomyrmex*) *cordatus – Woodiphora arunkumari* and *W. pallidinervis* (Disney, 1989f).

Dorylinae (driver ants)

Aenictus cornutus – Dohrniphora kistneri and *Dohrniphora* sp. (K. Rosciszewski and Disney, unpublished data).

A. gracilis – Dohrniphora kistneri and *Dohrniphora* sp. (K. Rosciszewski and Disney, unpublished data).

A. laeviceps – Dohrniphora kistneri (Disney, 1990g).

Ecitoninae (army ants)

Eciton burchelli – Coniceromyia fusca (Borgmeier, 1950a), *Dohrniphora cornuta* and *D. femoralis* (Rettenmeyer and Akre, 1968), *D. ecitophila* and *D. paraguayana* (Borgmeier, 1960b; Rettenmeyer and Akre, 1968), *Megaselia aurea* and *M. enderleini* (Rettenmeyer and Akre, 1968).

E. dulcius – *Dohrniphora ecitophila* (Borgmeier, 1960b; Rettenmeyer and Akre, 1968) and *Megaselia bruchiana* (Borgmeier, 1962c).

E. hamatum – *Apodicrania termitophila* (Rettenmeyer and Akre, 1968).

E. mexicanum – *Dohrniphora ecitophila* (Rettenmeyer and Akre, 1968).

E. quadriglume – *Dohrniphora ecitophila* (Borgmeier, 1960b; Borgmeier and Prado, 1975) and *Megaselia ampullosa* (Borgmeier, 1961a).

E. vagans – *Dohrniphora paraguayana* (Borgmeier, 1960b).

Labidus coecus – *Gymnophora aemula* (Borgmeier, 1960b).

L. praedator – *Holopterina longipalpis* (Borgmeier, 1928c; Rettenmeyer and Akre, 1968), a teneral male of *Pelidnophora schmitzi* (Borgmeier, 1969b).

Neivamyrmex spp. – a male *Trophodeinus spatulatus* (Brown, 1986).

N. opacithorax – *Megaselia* sp. (Rettenmeyer and Akre, 1968).

Formicinae

Camponotus sp. – *Burmophora rostrata* (Borgmeier, 1967c).

C. herculeanus – *Triphleba trinervis* (Schmitz, 1915d).

C. rufipes – *Apodicrania termitophila* (Borgmeier, 1925c).

Plagiolepis pygmaea – *Tubicera lichwardti* (Le Manse, 1941), the larvae being kleptoparasites within the nest (3.2.5).

Myrmicinae

Atta bidens – *Dohrniphora curvispinosa* (Borgmeier, 1925c).

A. sexdens – *Dohrniphora fuscicoxa* (Borgmeier, 1929) and *D. paraguayana* (Borgmeier, 1960b).

Crematogaster ranavalonae – *Hypocerides pterostigma* (Schmitz, 1915d).

Pheidole rufipilis – *Ceratoconus setipennis* (Borgmeier, 1928c).

Pheidologeton sp. – *Woodiphora pallidinervis* (Disney, 1989f).

***Ponerinae* (predatory ants)**

Odontomachus simillimus – *Woodiphora pallidinervis* (Disney, 1989f).

Pachycondyla crenata – *Cataclinusa bucki* (Borgmeier, 1962c).

P. harpax – *Cataclinusa pachycondylae* (Wheeler, 1901, 1910), the larvae are resident in the nest as kleptoparasites (3.2.5).

Pseudomyrmecinae

Tetraponera aethiops – *Plethysmochaeta trinervis* (Schmitz, 1915d).

(b) Isoptera (termites)

The termite nomenclature in the following review follows Snyder (1949), with amendments for *Macrotermes* by Ruelle (1970), and for Neotropical species by Araujo (1977). Some of Silvestri's names for Neotropical termites are not to be found in either Snyder's or Araujo's valuable works. This is because they are names for species which he never got around to describing (they are *nomena nuda*, and therefore must be rejected). *Odontotermes* is overdue for a critical revision. Until this has been accomplished, the locality will be as important as the current name in the records reviewed below, as several of these names are likely to be changed by a future revision.

A. Female flies resident in the termite colony

Females primarily resident in termite nests are usually flightless, at least after entering the host's nest. The Termitoxeniinae, for example, shed their wing membranes on entering the termite mound (Disney and Cumming, 1992).

Unidentified termites – *Chonocephalus elongatus* (Schmitz, 1950a), *Ptochomyia alticola* (Schmitz, 1958e), *Termitosphaera fletcheri* (Wasmann, 1913) and *Thaumatoxena polytricha* (Schmitz, 1951b).

Acanthotermes sp. – *Puliciphora borinquenensis* (Schmitz, 1916c).

A. acanthothorax – *Termitophorina africana* (Borgmeier, 1967b).

Allodontermes giffardi – *Cheiloxenia obesa* and *Termitoscrofa brevis* (Delachambre, 1966) and *Termitostroma ivorensis* (Delachambre, 1965).

A. rhodesiensis – *Paratermitoxenia coatoni* (Beyer, 1965c).

Ancistrotermes cavithorax – *Ptochomyia afra* (W. Rossi and Disney, unpublished data).

A. crucifer – *Ptochomyia afra* (Silvestri, 1920).

Anoplotermes sp. – *Idiophora termitoxena* (Seevers, 1941) and *Metopina psociformis* (Silvestri, 1947a).

Apicotermes angustatus – *Puliciphora* sp. (Schmitz, 1958e).

Bulbitermes constrictiformis – *Puliciphora collinsi* (Disney and Kistner, 1988).

B. singaporiensis – *Puliciphora beaveri* (Disney and Kistner, 1989b).

Coptotermes curvignathus – *Puliciphora coptotermitum* (Disney and Kistner, 1989b).

C. kalshoveni – *Puliciphora collinsi* (Disney and Kistner, 1988).

C. niger – *Chonocephalus heymonsi* (Brues, 1925a).

Cornitermes sp. – *Homalophora reichenspergeri* and *Lepidophoromyia zikani* (Borgmeier, 1923c).

C. cumulans – *Termitophorides heterospinalis* (Bristowe, 1925) and *T. setiger* (Borgmeier, 1923c).

Homallotermes eleanorae – *Puliciphora beaveri* (Disney and Kistner, 1989b).

Hospitalitermes sp. – *Puliciphora beaveri* (Disney and Kistner, 1989b).

H. hospitaloides – *Puliciphora beaveri* and *P. pulex* (Disney and Kistner, 1989b).

H. umbrinus – *Puliciphora beaveri* (Disney and Kistner, 1989b).

Hypotermes obscuriceps – *Ceylonoxenia butteli* (Wasmann, 1913; J. C. F. Fryer and Disney, unpublished data) and *Clitelloxenia peradeniyae* (Bugnion, 1913; Wasmann, 1913).

Kalotermes sp. – *Pachyneurella bicolor* (Borgmeier and Schmitz, 1923a).

Macrotermes sp. – *Aenigmatistes reichenspergeri*, a male and a female being recorded in a mound (Leleup, 1955).

M. barneyi – *Oligophora pallida, Palpiclavina tonkinensis* and *Rhynchomicropteron silvestrii* (Silvestri, 1947b).

M. bellicosus – *Ptochomyia afra* (Brues, 1932b), *Puliciphora borinquenensis* (Schmitz, 1929e; Disney and Kistner, 1989c), *P. decachete* (Disney and

Kistner, 1989c), *Termitophilomyia gracilis* (Reichensperger, 1931; Beyer, 1965c), *Termitosagma henningsi* and *Termitostroma schmitzi* (Reichensperger, 1931), *Thaumatoxena grassei* (Poisson, 1937, 1939; Disney and Kistner, 1992) and *T. pilosior* (Schmitz, 1955d; Disney and Kistner, 1992).

M. carbonarius – *Puliciphora beaveri* (Disney and Kistner, 1989b).

M. falciger – *Thaumatoxena andreinii* and *T. unispina* (Disney and Kistner, 1992).

M. gilvus – *Dicranopteron philotermes* (Schmitz, 1931c; Disney and Kistner, 1990a) and *Palpiclavina kistneri* (Disney and Kistner, 1989b, 1992).

M. malaccensis – *Puliciphora termitum* (Disney and Kistner, 1989b).

M. michaelseni – *Thaumatoxena andreinii* and *T. unispina* (Disney and Kistner, 1992).

M. mulleri – *Penthaplus exilis* (Schmitz, 1958e).

M. natalensis – *Thaumatoxena andreinii* (Beyer, 1965c; Ruelle *et al.*, 1975; Disney and Kistner, 1992) and *T. wasmanni* (Breddin and Borner, 1904; Poisson, 1939; Coaton and Sheasby, 1972; Kistner, 1982).

M. subhyalinus – *Puliciphora decachete* (Disney and Kistner, 1989c), *Thaumatoxena andreinii* (Silvestri, 1906; Disney and Kistner, 1992) and *T. wasmanni* (Disney and Kistner, 1992).

M. ukuzii – *Thaumatoxena nudistriga* (Disney and Kistner, 1992).

Microtermes sp. – *Aenigmatopoeus cucullatus* (Colyer, 1958) and *Mesopathusa hirta* (Borgmeier, 1971a).

Nasutitermes spp. – *Apterophora attophila* and *A. caliginosa* (Prado, 1976), *Ecitoptera schmitzi* (Borgmeier, 1924a; Schmitz, 1924i), *Myrmomyia nigroflava* (Borgmeier, 1958), *Pachyneurella bicolor* (Borgmeier and Schmitz, 1923a) and *Puliciphora anceps* (Schmitz, 1915e).

N. arenarius – *Acontistoptera* sp. (Silvestri, 1947a).

N. corniger – *Adenophora pulchella* (Disney and Kistner, 1989a).

N. costalis – *Puliciphora jacobsonorum* (Disney and Kistner, 1989a).

N. ephratae – *Homalophora metae*, *Pachyneurella bicolor* and *Puliciphora borinquenensis* (Seevers, 1941; Borgmeier, 1958).

N. exitiosus – *Eutermiphora abdominalis* (Lea, 1911, who actually gave the name of the termite as '*Eutermes fumipennis*'. The latter name, however, represents a confusion, which is discussed by Watson and Gay, 1980) and *Diplonevra gynaptera* (Fuller and Lee, 1938).

N. fumigatus – *Eutermiphora abdominalis* (Disney, 1989d).

N. indicola – *Assmutherium rostratum* (Schmitz, 1924d).

N. matangensis – *Puliciphora pulex* (Disney and Kistner, 1989b).

N. meinerti – *Puliciphora tambopata* (Disney and Kistner, 1989a).

N. octopilis – *Puliciphora jacobsonorum* (Disney and Kistner, 1989a).

N. rippertii – *Allochaeta metatarsalis* (Borgmeier, 1924a).

Odontotermes sp. – *Puliciphora termitum* (Schmitz, 1926b).

O. angustatus – *Mesopathusa cloveri* (Disney and Kistner, 1989c).

O. badius – *Alamira termitoxenizans* and *Mesopathusa hirta* (Schmitz, 1951d), *Termitophilomyia gracilis* (Borgmeier, 1959; Beyer, 1965c; Stuckenberg, 1974), *Termitosagma henningsi* (Borgmeier, 1959), *Termitoscrofa pinguissima*, *Termitostroma mergelsbergi* (Schmitz, 1958e) and *T. schmitzi* (Borgmeier, 1959).

O. billitoni – *Echidnophora butteli, Franssenia rocheorum* and *Puliciphora pulex* (Disney and Kistner, 1988).

O. bogoriensis – *Clitelloxenia hemicyclia* (Kemner, 1932b).

O. ceylonicus – *Clitelloxenia peradeniyae* (Wasmann, 1913; Schmitz, 1915b).

O. culturarum – *Termitophilomyia gracilis, Termitosagma henningsi* and *Termitostroma schmitzi* (Kistner, 1969; Malin and Kistner, 1970).

O. dives – *Javanoxenia punctiventris* (Kemner, 1926).

O. djampeensis – *Echidnophora butteli* (Disney and Kistner, 1988).

O. formosanus – *Pseudotermitoxenia nitobei* and *'Termitoxenia'* (? = *Javanoxenia*) *formosana* (Shiraki, 1925).

O. fulleri – *Termitophilomyia gracilis* and *T. leleupi* (Schmitz, 1958e).

O. grandiceps – *Dicranopteron philotermes* (Disney and Kistner, 1990a), *Echidnophora butteli* and *Javanoxenia punctiventris* (Schmitz, 1916a).

O. holmgreni – *Dicranopteron emersoni* and *D. setipennis* (Disney and Kistner, 1990a).

O. horni – *Ceylonoxenia bugnioni* and *Clitelloxenia peradeniyae* (Wasmann, 1913).

O. indrapurensis – *Franssenia termyca* (Disney and Kistner, 1989b).

O. interveniens – *Dundophora fimbriiterga* (Disney and Kistner, 1989c) and *Syntermitoxenia apiogaster* (Schmitz, 1958e).

O. javanicus – *Clitelloxenia hemicyclia* (Schmitz, 1932a; Franssen, 1933b), *Echidnophora butteli* (Schmitz, 1915b; Franssen, 1933b), *Franssenia hirundella* (Schmitz, 1932a), *Javanoxenia fransseni* (Schmitz, 1933e), *J. punctiventris* (Schmitz, 1915b; Franssen, 1933a), *Odontoxenia brevirostris* (Schmitz, 1915b; Franssen, 1933a) and *Puliciphora pallicauda* (Schmitz, 1934g; Beyer, 1967).

O. latericius – *Mesopathusa modesta* (Cumming and Disney, unpublished data), *Syntermitoxenia pseudonanna* (Schmitz, 1958e; Beyer, 1965c; Muesebeck, 1965), *Termitophilomyia braunsi* (Disney and Cumming, 1992), *Termitoxenia havilandi* (Bugnion, 1913; Schmitz, 1914d, 1958e; Beyer, 1965c).

O. longignathus – *Clitelloxenia longiceps* (Schmitz, 1915b).

O. makassarensis – *Echidnophora butteli* and *Franssenia rocheorum* (Disney and Kistner, 1988).

O. montanus – *Mesopathusa hirta* and *Puliciphora decachete* (Disney and Kistner, 1989c).

O. nilensis – *Syntermitoxenia jaegerskioeldi* (Wasmann, 1904b).

O. obesus – *Ceylonoxenia* sp. (Borgmeier, 1970a), *Clitelloxenia assmuthi* (Wasmann, 1902b), *Indoxenia heimi* (Wasmann, 1902b), *I. wasmanni* (Borgmeier, 1964a), *Puliciphora termitum* (Schmitz, 1926b), *P. velocipes* (Schmitz, 1913b) and *Termitosphaera immaculata* (Borgmeier, 1964a).

O. (Hypotermes) obscuriceps – see under *Hypotermes*.

O. patruus – *Termitosagma henningsi* (Borgmeier, 1959; Kistner, 1969; Malin and Kistner, 1970), *Termitostroma reichenspergeri* (Schmitz, 1954c) and *T. schmitzi* (Kistner, 1969).

O. praevalens – *Puliciphora beaveri* (Disney and Kistner, 1989b).

O. proformosanus – *Crasilla setifera* (Disney and Kistner, 1989b).

O. sarawakensis – *Crasilla longiseta* (Disney and Kistner, 1989b), *Dicranopteron brevifurca* (Disney and Kistner, 1990a) and *Rhabdomyia kistnerorum* (Disney and Kistner, 1989b).

O. sundaicus – *Echidnophora butteli* (Disney and Kistner, 1988, 1989b), *Franssenia malaysiae* (Disney and Kistner, 1989b), *F. rocheorum* (Disney and Kistner, 1988) and *Puliciphora collinsi* (Disney, 1988b).

O. takensis – *Echidnophora butteli* (Disney and Kistner, 1989b), *Dicranopteron oligoseta* and *D. setipennis* (Disney and Kistner, 1990a) and *Puliciphora beaveri* (Disney and Kistner, 1989b).

O. taprobanes – *Clitelloxenia peradeniyae* (Bugnion, 1913; Wasmann, 1913).

O. transvaalensis – *Mesopathusa hirta* and *M. modesta* (Disney and Kistner, 1989c), *Termitophilomyia braunsi* (Borgmeier, 1964a; Beyer, 1965c; Disney and Cumming, 1992) and *T. gracilis* (Borgmeier, 1964a; Beyer, 1965c; Stuckenberg, 1974).

O. vulgaris – *Termitomyia mirabilis* (Wasmann, 1900a).

O. wallonensis – *Indoxenia heimi* (Wasmann, 1913).

Pericapritermes sp. –*Puliciphora beaveri* (Disney and Kistner, 1989b).

Procornitermes lespesii – *Termitophorides setiger* (Borgmeier, 1967b).

Protermes prorepens – *Termitostroma ivorensis* (Delachambre, 1965).

Pseudacanthotermes spiniger – *Aenigmatopoeus cucullatus* (Colyer, 1958).

Ruptitermes reconditus - *Paurophora borgmeieri* (Silvestri, 1947a).

Schedorhinotermes brevialatus – *Puliciphora togata* (Borgmeier, 1970a).

Syntermes sp. – *Ecitoptera parallela* (Borgmeier, 1960a) and *Thalloptera bisinuosa* (Borgmeier, 1936).

S. grandis – *Ecituncula inquirenda* and *Puliciphora rufipes* (Silvestri, 1947a).

S. molestus – *Cryptophora coeca* (Silvestri, 1945, 1947a), *C. colombiae* and *Syntermophora microphthalma* (Seevers, 1941).

Termes sp. – *Syntermitoxenia apiogaster* (Schmitz, 1958e).

Termes laticornis – *Puliciphora beaveri* (Disney and Kistner, 1989b).

Trinervitermes trinervius – *Thaumatoxena grassei*, a single record (Poisson, 1937, 1939; Schmitz, 1950b).

B. Species in which the female flies, and sometimes the males, visit the colonies

A number of the phorid species recorded below are primarily attracted to accidental exposures (spillages) of termites, as when a tree crashes to the ground. Such exposures of termites are by no means uncommon, as reported for a tropical forest in Sulawesi (Disney, 1986d). These spillages normally include wounded termites. Those organisms which exploit the opportunity of such spillages, either preying upon the termites or imbibing fluid from those freshly killed, have been described as constituting a 'ghoul guild' (Disney, 1985h). Phorid species reported to be members of this guild and observed imbibing fluids from freshly-killed termites are included below but are also considered in section 6.3.

Some visitors to termite colonies have larvae which are fungivores or saprophage species. Others are probably parasitoids (3.5.6(f)). However, unlike the ant-parasitoid species, I review all the visiting species here, in view of our much greater ignorance of the parasitoids of termites. Those whose larval habits are known are also reviewed in Chapter 3. It seems that some phorids parasitizing termites do not place the eggs directly onto or into the host but scatter them in the termite nest instead (3.5.6(f)).

In our present state of knowledge, most tropical species of *Dohrniphora* and *Megaselia* cannot be named from females taken unaccompanied by their males. I have noted published records of these, as a stimulus to further investigation.

Unidentified termites – *Darwiniphora integricilia, Dicranodeina extravagans, Diplonevra angustipennis, Gymnophora commotria, Haplophleba nigricans, Megaselia confortata, M. paricostalis* and *Tropophleba varians* (Schmitz, 1929i).

Acanthotermes acanthothorax – *Diplostylocera fissicornis* and *Megaselia bilobata* (Borgmeier, 1967b), *M. chaetogaster* (Borgmeier, 1971a) and *M. labiata* (Borgmeier, 1967b).

Bulbitermes germanus – *Dohrniphora vorax* (Disney and Kistner, 1989b).

B. sarawakensis – *Diplonevra mortimeri* (Disney, 1986b) and *Dohrniphora vorax* (Disney, 1990g).

Cornitermes sp. – *Apodicrania termitophila* (Borgmeier, 1923e) and *Metopina zikani* (Borgmeier, 1969b).

Cubitermes spp. – *Dohrniphora diminuens* (Disney and Kistner, 1989c).

Discupiditermes sp. – *Dohrniphora* sp. (Disney and Kistner, 1989b).

Havilanditermes atripennis – *Woodiphora pallidinervis* (Disney, 1989f).

H. monoceros – *Dohrniphora* sp. (Disney and Kistner, 1989b).

Hypotermes xenotermitis – *Dohrniphora* sp. (Disney and Kistner, 1989b).

Kalotermes sp. – *Dohrniphora anterospinalis* (Borgmeier, 1923d), *D. curvispinosa* (Borgmeier, 1923a), *D. fuscicoxa, D. heptacantha* and *D. intrusa* (Borgmeier, 1923d).

Longipeditermes longipes – *Dohrniphora* sp. (Disney and Kistner, 1989b).

Macrotermes bellicosus – *Dohrniphora diminuens* (Disney and Kistner, 1989c).

M. carbonarius – *Dohrniphora* sp. (Disney and Kistner, 1989b).

M. falciger – *Dohrniphora diminuens* (Disney and Kistner, 1989c).

M. gilvus – *Megaselia scalaris* (Schmitz and Delage, 1981) and *Misotermes exenterans* (Kalshoven, 1938).

M. subhyalinus – *Dohrniphora diminuens* (Disney and Kistner, 1989c).

Mastotermes darwiniensis – *Megaselia scalaris* (P. Cranston and Disney, unpublished data).

Microcerotermes crassus – *Woodiphora pallidinervis* (Disney, 1989f).

Nasutitermes spp. – *Diplonevra assmuthi* (Schmitz, 1915e), *Dohrniphora anterospinalis* (Borgmeier, 1923d), *D. cornuta* and *D. curvispinosa* (Borgmeier, 1925c), *D. divaricata* (Borgmeier, 1960b), *D. fuscicoxa, D. heptacantha, D. intrusa* and *D. luteifrons* (Borgmeier, 1923d), *D. myersi* (Brues, 1932a), *D. obscuriventris* (Borgmeier, 1925c; Borgmeier and Prado, 1975), *D. paraguayana* (Borgmeier, 1925c; Brues, 1932a), *D. vorax* and *Hypocerides* sp. (Schmitz, 1915e).

N. bulbiceps – *Dohrniphora* sp. (Disney and Kistner, 1989b).

N. corniger – *Metopina zikani* (Disney and Kistner, 1989a).

N. culasiensis – *Diplonevra mortimeri* and *D. watsoni* (Disney, 1986d), *Dohrniphora caini* and *D. gilberti* (Disney, 1990g) and *Woodiphora pallidinervis* (Disney, 1989f).

N. dendrophilus – *Metopina zikani* (Disney and Kistner, 1989a).

N. ephratae – *Dohrniphora correlata* (Borgmeier, 1932a).

N. latus – *Diplonevra mortimeri* and *D. watsoni* (Disney, 1986d).

N. lujae – *Dohrniphora diminuens* and *D. schmitzi* (Disney and Kistner, 1989c).

N. makasserensis – *Diplonevra mortimeri* and *D. watsoni* (Disney, 1986d) and *Dohrniphora caini* (Disney, 1990g).

N. matangensis – *Dohrniphora* sp. (Disney and Kistner, 1989b).

N. mollis – *Diplonevra mortimeri* (Disney, 1986d), *Dohrniphora gilberti* and *D. vorax* (Disney, 1990g), *Woodiphora arunkumari* and *W. pallidinervis* (Disney, 1989f).

Odontotermes sp. – *Dohrniphora transformata* (Schmitz, 1915e).

O. bangalorensis – *Bolsiusia termitophila* (Schmitz, 1913a).

O. magdalenae – *Megaselia scalaris, M. termitomyca* and *M. zariaensis* (Disney, 1989b).

O. obesus – *Dohrniphora vorax* (Schmitz, 1915e).

O. sundaicus – *Dohrniphora caini, D. gilberti* and *D. vorax* (Disney, 1990g), *Woodiphora arunkumari* and *W. pallidinervis* (Disney, 1989f).

O. transvaalensis – *Dohrniphora diminuens, Megaselia nigricauda and Megaselia* spp. (Cumming and Disney, unpublished data), *Megaselia* spp. and *Metopina obsoleta* (Disney and Kistner, 1989c).

O. (Hypotermes) xenotermitis – see under *Hypotermes.*

Schedorhinotermes sp. – *Dohrniphora rhinotermitis* (Schmitz and Mjöberg, 1924; Fuller and Lee, 1938).

S. javanicus – *Dohrniphora gilberti* and *D. vorax* (Disney, 1990g).

Termes laticornis – *Dohrniphora* sp. (Disney and Kistner, 1989b).

Trinervitermes biformis – *Dohrniphora vorax* (Schmitz, 1915e).

6.5 COURTSHIP AND MATING

The males of some Phoridae form small aerial swarms into which the females are attracted prior to mating. The genus *Phora* is the best-known for this habit. The males generally swarm in sunshine near trees, usually at the tips of the branches (Lundbeck, 1922). Parmenter (1965) describes such swarms of *P. atra*, the flies rising about 2.5 m from the ground as they danced in the air. *Phora stictica* has been observed swarming over a person's head on several occasions and also over a peat hag (Disney, unpublished data). Parmenter (1965) also describes a swarm of *Anevrina urbana* males alighting from time to time on the shaded trunks of poplar trees, less than 1.5 m above the ground. Schmitz (1957a) reported a swarm of *Megaselia palmeni* over the umbellifer *Peucedanum ostruthium. Burmophora angustifrons* also forms swarms over herbaceous plants (Gotô, 1983b). The same author described the swarm of a Japanese *Woodiphora* (Fig. 6.11) (Gotô, 1983a). Both sexes of *Phalacrotophora berolinensis* have been reported around the foliage of Douglas firs (Colyer, 1954g), but these swarms may have served some other purpose than for mating. This species has only been observed mating in the vicinity of the pupae of its ladybird-host pupae (see below). Pairs of *Phalacrotophora halictorum* in copula often fly together (Batra, 1965). Species of *Trineurocephala* form aerial swarms and *Gymnophora* species have been reported mating in flight (Brown, 1992b).

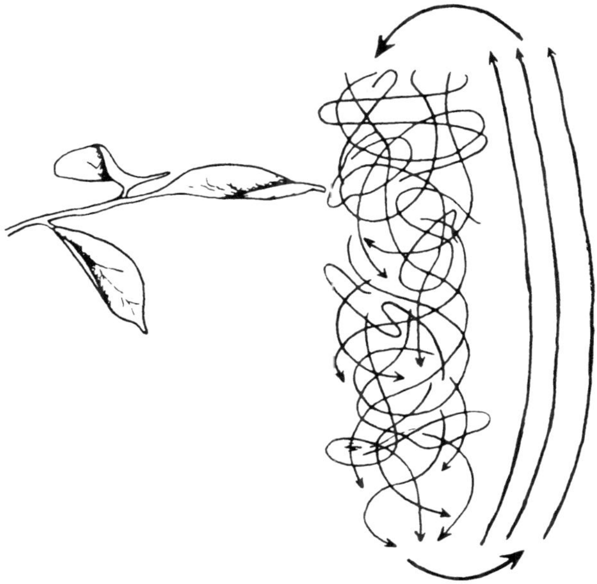

Figure 6.11 The flight paths of flies in a swarm of a Japanese *Woodiphora* species. (After Gotô, 1983a.)

Other species evidently form swarms, as inferred from indirect observations. For example, *Megaselia nigriceps* has been caught in water traps set with their rims just above the top of the herb layer in some woods in Yorkshire. The numbers of this species in four traps swamped the totals for all other phorids and with a marked dominance of males as follows: 77.1% (n = 96), 98.5% (n = 65), 99.3% (n = 135) and 99.7% (n = 291) (Jean Johnson, M. Usher and Disney, unpublished data).

In *Pseudohypocera kerteszi* mating has been observed to take place in the air (Portugal, 1977). The same species will mate in the laboratory on the substratum in culture bottles, mating times recorded being 32–90 s (Chaud-Netto, 1980).

In Sulawesi *Dohrniphora maddisoni* forms swarms, of 1–36 flies, gyrating on the leaves of small saplings less than 30 cm in height. These swarms include both sexes in about equal proportions, and matings are frequent (Disney, 1990g). This species has a well-developed 'sensory complex' at the base of each hind femur (Fig. 6.12a). In *D. cornuta* this complex has fewer spinules (Fig. 6.12b), but it is now known that this species uses these structures during mating. The male grips the females's 7th abdominal segment between the bases of the two hind femora and laterally compresses it while they are in copula, which lasts 2–8 min (Barnes, 1990a). Prior to mating there is a brief courtship. The male pursues a female in a stop–go pattern, pausing to fan his wings at intervals.

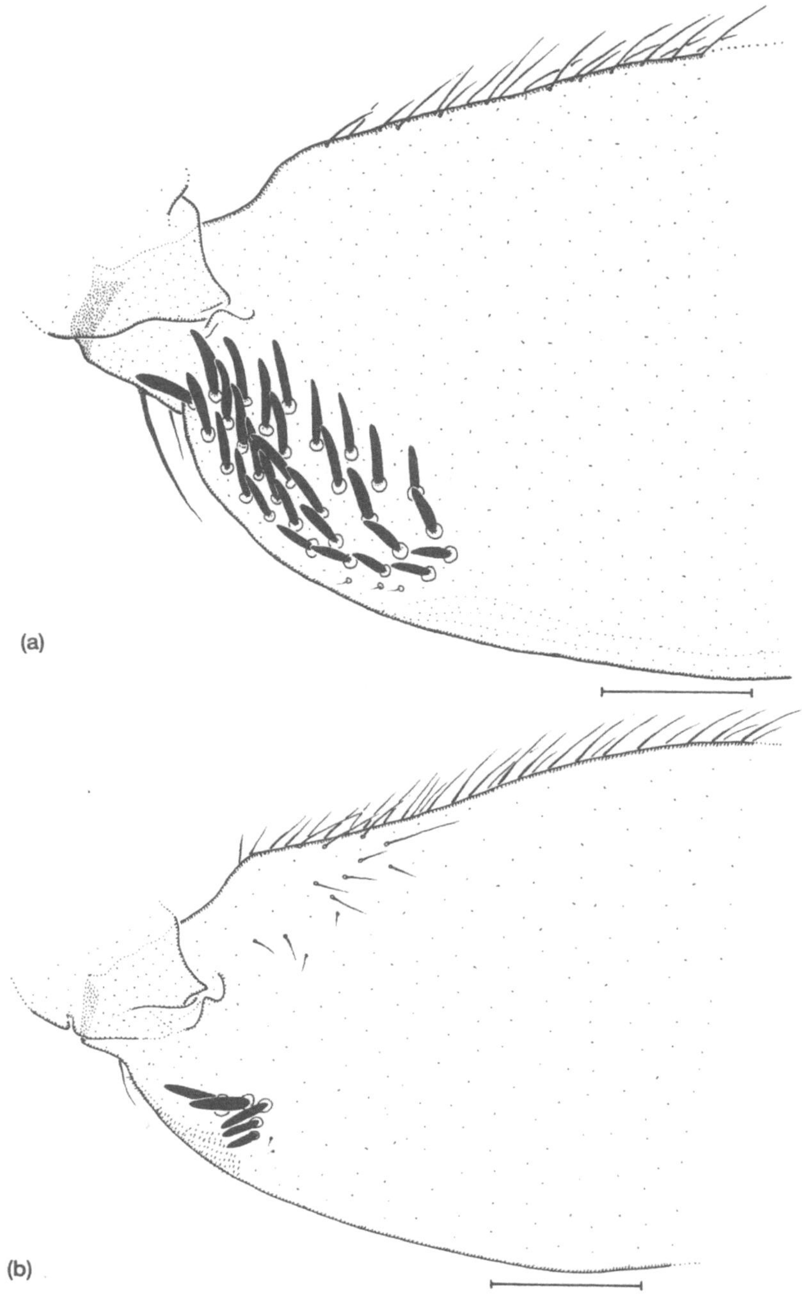

Figure 6.12 The so-called 'sensory complex' at the base of the posterior face of the male hind femur (a) *Dohrniphora maddisoni* (b) *D cornuta*

In Florida, *Rhyncophoromyia conica* forms swarms, of 1–15 flies, on the top surfaces of leaves 1 m or less from the ground. These swarms are made up of all males or all females. In female swarms the flies are very active, investigating each other, and taking short flights. A male encountering a female swarm descends very rapidly, seizes a female, and takes off again in what seems like a single movement (Sivinski, 1988). In Diptera the occurrence of all-female swarms is unusual in relation to mating. However, the same author reports that it appears to be the normal habit in *Megaselia aurea*. He observed only females forming small swarms, of 1–11 individuals (mean 2.5, $n = 125$), on the upper surfaces of horizontally extended leaves. He observed 19 such swarms. The females were observed scurrying in circles several centimetres in diameter. They often approached within a few millimetres of each other, but then rapidly changed direction. Stationary flies were spaced 1–15 cm apart. Three couplings were observed, the males swooping down to mount the females. The pair then instantly took flight. The whole process only took about 1 s. Mated pairs were observed resting on leaves other than the one on which they had coupled.

In Germany *Metopina* (probably *M. oligoneura*) has been observed gyrating on sunlit leaves of lungwort (*Pulmonaria officinalis*, Boraginaceae). They vanished whenever the leaves were in shade (Beyer and Schmitz, 1957). In England *Borophaga incrassata* is to be observed in summer 'drumming' with its hind coxae on the upper surfaces of hedgerow leaves. Whether this serves to attract females or to repel other males is not known. I have records of this species mating on a leaf and on a log. Among other species mating on leaves I have records for *Phora atra*, *P. holosericea* and *P. stictica*, and several of various *Megaselia* species. Many *Megaselia* species, especially in the tropics, can be observed gyrating on leaves. This behaviour probably represents patrolling at mating stations, at which the 'swarms' have been reduced to a single fly. This gyrating should be distinguished from frenetic leaf-searching behaviour by both males and females (6.3). *Diplonevra nitidula* has been observed mating on the ground on several occasions (e.g. Disney, 1991d). I have one record of *Megaselia pleuralis* mating on the surface of a peat pool.

Colyer (1954c) reports a swarm, mainly of males, of *Conicera similis* running in rapid gyratory movements on the leaves of lettuce plants. He also reported *Metopina heselhausi* swarming on a patch of damp ashes, the only females he caught being those caught mating (see Disney, 1979j). Swarms of *Megaselia rubella* males, plus one female, have been reported over piles of pine needles (Colyer, 1952a).

Lundbeck (1922) recorded *Anevrina unispinosa* mating while on the corpse of a sparrow. He reported numbers of *A. thoracica* mating on tree trunks, the peculiar abdominal sacs of this species being inflated at the time. The males of a number of parasitoid species are attracted to the females as they are preparing to oviposit in their hosts. Examples include *Megaselia kodongi* and *M. sembeli* (Disney, 1986b). Likewise *Phalacrotophora berolinensis* and *P. fasciata* mate on or near their beetle-host pupae, or more usually the prepupae. More rarely they mate on a leaf nearby. A female *P. fasciata* has been observed mating just after she had been observed ovipositing, thus indicating that this species will mate on more than one occasion. The duration of copulations

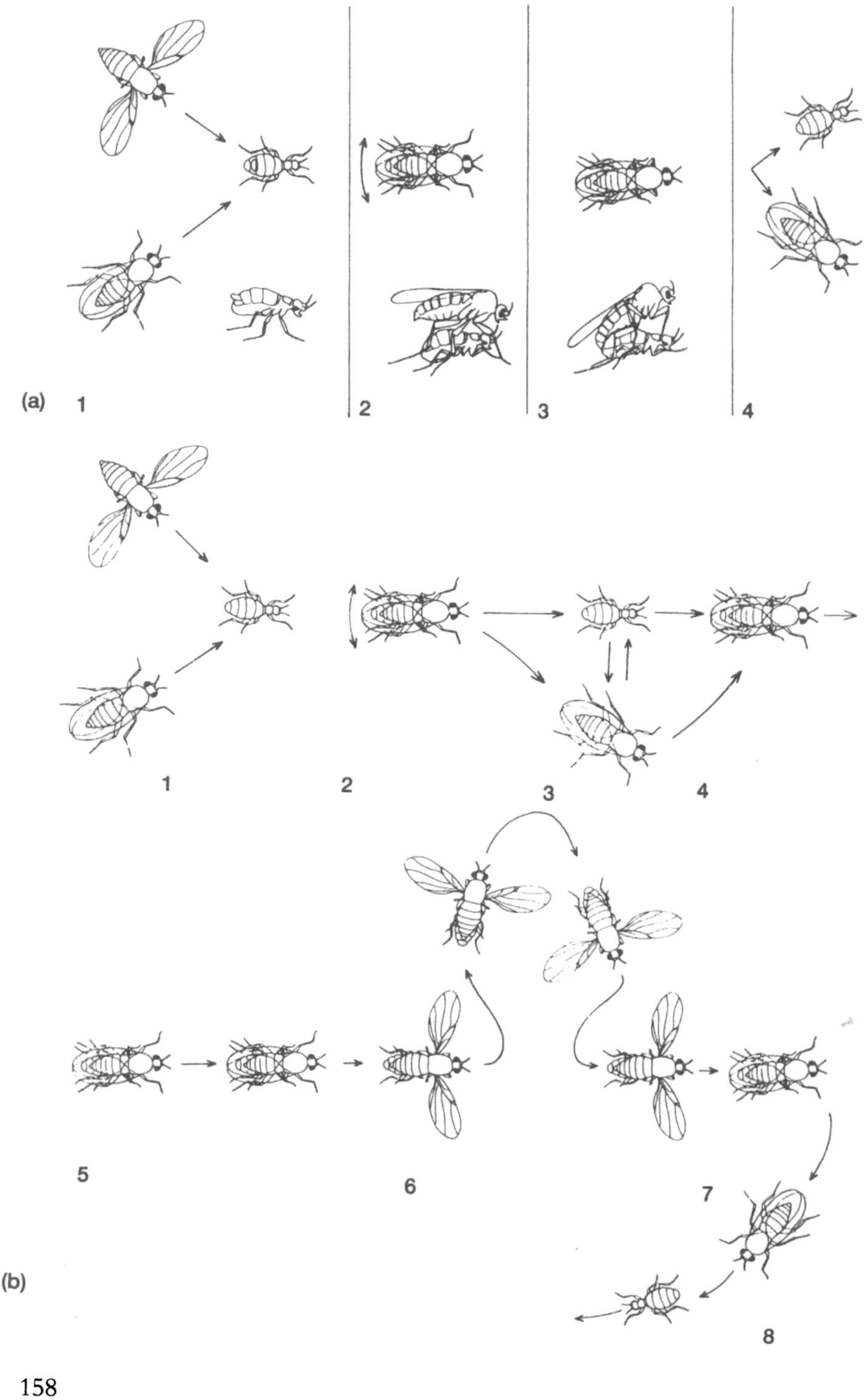
(a)
1
2
3
4
(b)
1
2
3
4
5
6
7
8

observed ranged from 2 min 38 s to 8 min 3 s in *P. fasciata*. In one case, not observed from the initiation of coupling, the duration was 11 min 3 s. A timed mating of *P. berolinensis* lasted 5 min 32 s and a coupling not observed from the start lasted 8 min 54 s (Disney *et al.*, 1994). *Gymnophora* females have been reported mating more than once (Brown, 1992b).

In *M. scalaris* the females have been reported to be receptive to the males 24 h after eclosion, while the males did not mate until 4 days old (Ondraschek, 1953). Miller (1979a) briefly describes the courtship. The male performs a number of complex wing and leg movements. The female responds with lateral wagging movements of the abdomen. The duration of mating is typically about 32 s, but ranging from 16 to 148 s (Benner, 1991). *Megaselia halterata* normally mates within 24 h of emergence (Hussey and Gurney, 1964a), but in some cases not until the second day (Hussey, 1965). The courtship and mating behaviour have been described by Binns (1980), who suggested that the female's habit of pointing the tip of its abdomen into the air is probably associated with the emission of a pheromone to attract the males. The females do indeed attract the males with a pheromone, which has since been shown to be 3,6-dimethylheptan-2,4-dione in a mixture of keto and enol forms (Baker *et al.*, 1982).

Miller (1979a) gives a more detailed account of the courtship of the wingless female of *Puliciphora borinquenensis* by its winged male. The latter holds the female with its middle legs, while keeping his back legs on the ground. With his front legs he makes rapid vibratory movements either side of his partner's head and thorax, in a manner resembling grooming activity. Miller (1984) extends these observations (Fig. 6.13). Mating occurs from around 1600 hours onwards. The males apparently use visual and olfactory cues to locate the females. The latter parade while awaiting copulation, repeatedly 'pumping' the abdomen. With each pumping stroke the anterior flap of abdominal tergite 5 (Fig. 6.14) rises and falls. Possibly this releases a pheromone. Males which have located a suitable medium for oviposition run over it for several minutes. They then make short flights from it and back again. It is evident that they thereby memorize the location of suitable oviposition sites. Having done so they then seek parading females and, while in copula, airlift the females to the oviposition site. A single male may mate with and airlift a succession of females. Indeed one male was recorded airlifting 33 females to the same site. The males may guard a female for a few seconds after mating, thus suggesting the occurrence of sperm competition. Kanmiya (1985) has shown that the males of the closely-related *P. tokyoensis* produce pulses of sound during courtship. Those generated by wing fanning last about 40–60 ms, with an interval of 600–900 ms between pulses. Each pulse is itself made up of 7–27 sinusoidal pulses. The fundamental frequency recorded was 343.7 Hz. Perhaps ultrasound is widely employed during courtship in Phoridae. Furthermore the validity of the claim that *P. tokyoensis*

Figure 6.13 Mating behaviour in *Puliciphora borinquenensis*. (a) Straight-forward sequence of events when mating on the ground. (b) The sequence when mating is followed by airlifting of the female to a site for oviposition. (After Miller, 1984.)

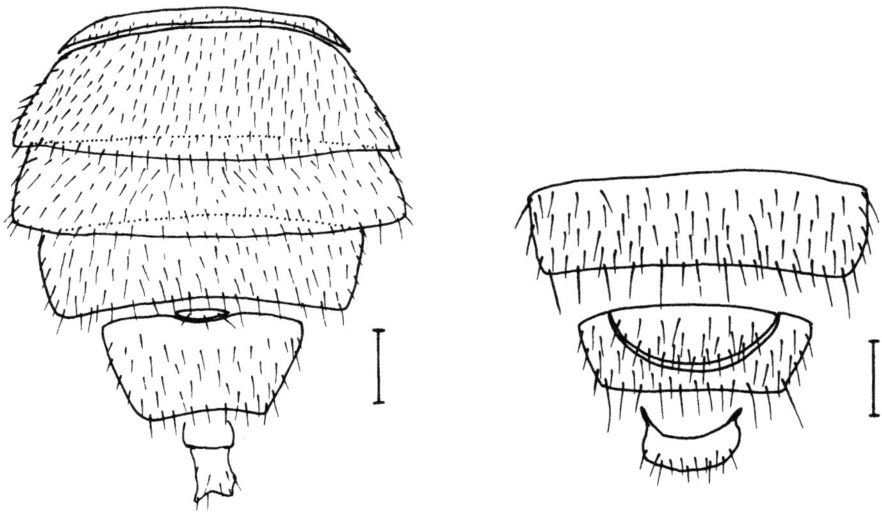

Figure 6.14 Abdominal tergites of *Puliciphora* females. Left, *P. beaveri*, tergites 1–6; right, *P. borinquenensis*, tergites 4–6.

is indeed distinct from *P. borinquenensis* might be evaluated by a comparison of their courtship sounds.

Gotô (1984a) describes the mating behaviour of a Japanese *Beckerina*, in which both sexes are winged (Fig. 6.15).

6.6 LONGEVITY

Data on adult longevity are largely lacking. In laboratory cultures at 23–26 °C *Puliciphora borinquenensis* females live for 2–3 weeks, but males only 2–3 days (Miller, 1979a). Only the females were observed feeding in the laboratory (see above). In laboratory cultures at 22–24 °C, *Dohrniphora cornuta* females survived 7–46 days, but males only 4–14 days (Barnes, 1990a), both sexes feeding in the laboratory, but with different diet preferences (see above).

Megaselia halterata adults lived for more than 3 weeks in the laboratory (Hussey and Wyatt, 1962). They survive best at a relative humidity of 80% (Hussey and Gurney, 1964a). Marked flies have been recovered 14 days after being released (Hussey and Gurney, 1964b). Hussey (1960) reported that *M. nigra* males denied access to sugar, in laboratory cultures at 23–24° C, survived only 2.4 (± 0.8) days and females only 3.0 (± 0.3) days. However, with access to sugar the males lived 9.5 (± 1.5) days, and the females 16.2 (± 3.1) days.

In *Megaselia scalaris* longevity is greater at lower temperatures. Thus at 27° C females survived 7.3 (± 0.25) days and males 17.3 (± 1.35) days (El-Miniawi and Moustafa, 1965). At 25 °C females survived 29.9 (± 9.0) days, and males

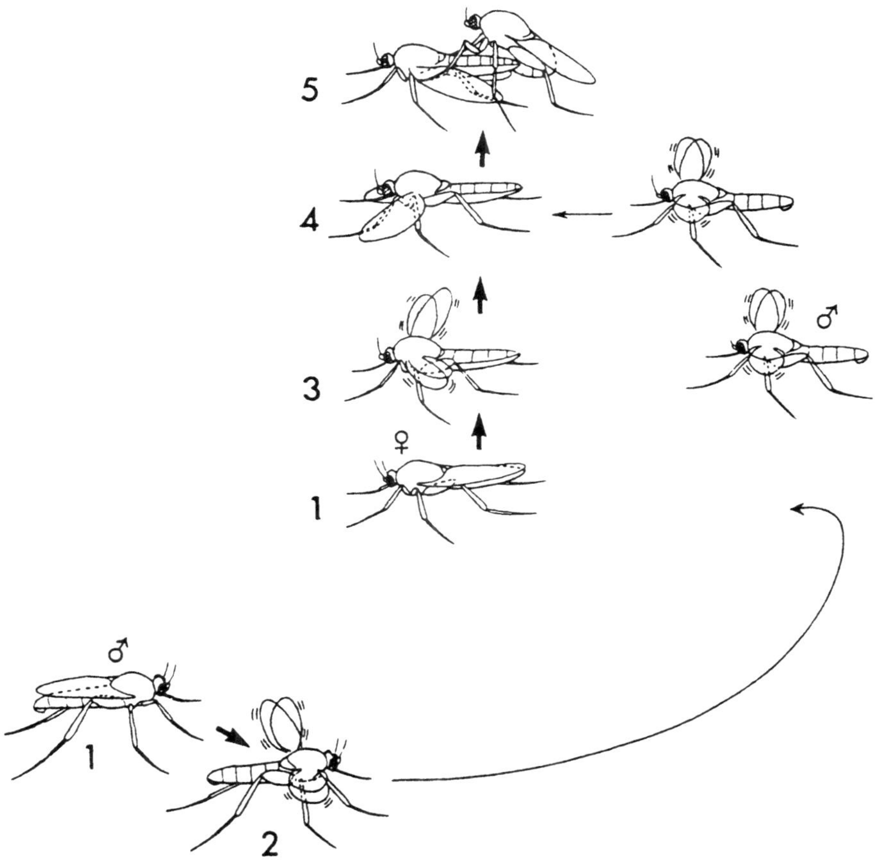

Figure 6.15 Mating behaviour in a Japanese *Beckerina* species. (After Gotô, 1984a.)

24.8 (± 2.4) days. At 15 °C the females lived 51.0 (± 6.9) days and the males 43.0 (± 3.7) days (Prawirodisastro and Benjamin, 1979).

Some species can live several months as resting adults during cold, and possibly during dry, seasons. Data on aestivation are lacking. However, data are available on a number of species overwintering as adults in temperate regions. Not only do specimens caught early in the spring often show evidence of being old (such as extensive fraying of the outer wing margin), but hibernating adults have been collected in the winter. Herbert and Braun (1958) recorded the following species in moss cushions in Europe:

Borophaga subsultans
Megaselia aequalis
Megaselia analis
Megaselia ciliata
Megaselia fusca
Megaselia mixta
Megaselia nasoni
Megaselia parva
Megaselia pulicaria
Megaselia pygmaea

Megaselia glabrifrons	*Megaselia rufipes*
Megaselia largifrontalis	*Megaselia sinuata*
Megaselia lata	*Megaselia subtumida*
Megaselia longicostalis	*Megaselia vernalis*
Megaselia manicata	*Megaselia zonata*

Some of the sporadic records of Phoridae in bird nests (7.5.3) are for hibernating adults. For example, *Megaselia latior* and *M. pumila* have been found in an old nest of a pigeon (*Columba palumbus*) collected, in England, in February (A. G. Irwin and Disney, unpublished data). Schmitz (1934e) reported *M. hibernans* and *M. superciliata* overwintering in rabbit burrows. In the winter in Scotland, *Megaselia ciliata* has been recorded under the bark of decaying fir trees and *M. aequalis* inside rotten fir stumps (Malloch, 1910, 1911). Adults of *M. ciliata* have been found inside rotten larch stumps at 400 m altitude in the Yorkshire Pennines in winter. The specimens were 'greasy' in comparison with the same species caught in the summer months (unpublished data). Leruth (1936) recorded species such as *Megaselia albicaudata* and *M. rufipes* overwintering in caves in Belgium. Also in Europe, Bächli (1969) collected samples of woodland material, often frozen, in winter. The samples were placed in containers in the laboratory and as insects crawled out they were collected for identification. About 40% of the 1610 Diptera collected were Phoridae. These were mainly obtained from leaf mould and bark samples from fir (*Abies alba*) and larch.

Megaselia meconicera adults have been recorded in large numbers in houses in the autumn and winter months (e.g. Scott, 1951; Colyer, 1954c). Such assemblages are frequently females only. However, I was sent a sample of both males and females from a mass invasion, running into thousands, that entered the home of Rev. Bernard Elsdon in north Wales in early October 1982 until well into November. Many of the flies had a 'greasy' appearance. Schmitz (1929g) reported that the fat body of these overwintering flies, at least initially, tends to be greatly enlarged. He also reported overwintering adults of both sexes of *M. ciliata* with enlarged fat bodies. It is clear that both species hibernate as adults. It is probable that they are able to supercool (that is they can survive at temperatures below freezing without ice forming in their tissues), but this remains to be investigated. The 'greasy' appearance of these hibernating flies suggests the presence of the polyhydric alcohols (such as glycerol), which are employed by other insects as a means by which they are able to supercool.

6.7 PHORESY BY MITES ON SCUTTLE FLIES

Mites (Acari) are frequently encountered attached to adult scuttle flies. Many of these are deutonymphs, some are larvae and a few are adults. Where there has been a dearth of adults the identification of these mites has been largely neglected. The only general study is the report of Fain and Greenwood (1991) on some of the mites which I have procured on Phoridae. Otherwise the only detailed studies are on some mites transported by the mushroom pest

Megaselia halterata. The few published records plus some new observations are summarized below.

ASTIGMATA

Anoetidae (ex Acaridae)

Some genera associated with aquatic Diptera were treated by Vitzthum (1932). The following records are for those attached to aquatic Phoridae.

Zwickia spp. deutonymphs attached to legs of *Megaselia schuitemakeri*, the adults of *Z. guentheri* and *Z. nepenthesiana* being present in the *Nepenthes* pitchers in which this fly was breeding (Beaver, 1979).

Myianoetus sp. Two deutonymphs attached to segment seven of the abdominal venter of the pitcher-plant species *M. campylonympha* (Fain, Greenwood, C. Clarke and Disney, unpublished data).

Scutacaridae

Scutacarus baculitarsus – females only, of this species, are carried by *Megaselia halterata*. There is usually one mite per fly (but two and four have been recorded), attached beneath the second abdominal segment of the fly with the front legs. Infestation rates of 2% of the available flies were recorded (Norton and Ide, 1974). Binns (1979, 1981) reported far higher infestation rates, gave details of the natural history of these mites, and provided SEM photographs.

Scutacarus sp. An adult female on the abdomen of a male *Megaselia* sp. (Fain, Greenwood and Disney, unpublished data).

MESOSTIGMATA

Macrochelidae

Macrocheles disneyi adult females are commonly found attached to *Diplonevra nitidula* females, and less commonly to males, in England. Typically they are attached beneath the base of the abdomen facing forwards, but are occasionally found below the thorax of the fly (Fain and Greenwood, 1991). The related *M. muscae* has non-feeding first-stage larvae, but all other stages feed on the eggs of the housefly (*Musca domestica*) (Pereira and Castro, 1945). The apparent specificity of the association of *M. disneyi* with *D. nitidula* suggests that the mite depends either on this fly, perhaps preying on its eggs, or on the fly's earthworm host (3.5.1). If it preys on the eggs of the fly, then this would imply that the latter oviposits into the soil rather than directly into its host.

Macrocheles sp. (cf. *M. matrius*). Nine females, each attached by the proboscis to the abdomen of its host fly, belonging to an undescribed species of *Dohrniphora* caught in a cave in Trinidad (Johanna Darlington, Fain, Greenwood and Disney, unpublished data).

Parasitidae

Gamasodes bispinosus deutonymph on *Conicera dauci* male, under abdomen (Fain and Greenwood, 1991).

Uropodidae

Uroseius acuminatus deutonymphs, often attached by their tails, have been recorded in England on females of *Anevrina thoracica*, both sexes of *Conicera dauci*, a female of *Megaselia rufipes* and a female of *Triphleba distinguenda* (Fain and Greenwood, 1991).

Uroseius sp. (cf. *U. huzikeri*). A female *Triphleba nudipalpis* from Essex (collected by David Henshaw) had 7 deutonymphs attached to the abdomen (K. H. Hyatt and Disney, unpublished data).

Uroseius sp. deutonymph attached by its tail to the abdomen of a female *Megaselia rufipes* (Fain, Greenwood and Disney, unpublished data).

Uroseius sp. Five nymphs, attached by their tail ends to the abdomens of *Conicera megalodus*, caught in a cave in Trinidad (Johanna Darlington, Fain, Greenwood and Disney, unpublished data).

An unidentified larval Mesostigmata. On abdomen of a male *Megaselia pulicaria* from Spain (J. Blasco-Zumeta and Disney, unpublished data).

PROSTIGMATA

Erythraeidae

Leptus sp. larva attached between abdominal tergites 5 and 6 of a male *Megaselia flavicoxa* (Fain and Greenwood, 1991).

Pygmephoridae

Microdispus (= *Brennandania*) *lambi* is carried by *Megaselia halterata* in Australia (Clift and Larsson, 1987).

Tetranychidae

At least some members of this family are known to be plant feeders (Baker and Wharton, 1952). The significance of the following record remains to be discovered.

Petrobia latens female on a male *Phora stictica* (Fain and Greenwood, 1991).

Trombidiidae

The adults of some members of this family are known to feed on insect eggs and early-instar larvae, while larvae are known to parasitize insects (Baker and Wharton, 1952). Details of their interactions with scuttle flies remain to be investigated.

Larval trombidiids have been recorded in Britain and Spain as follows (Fain and Greenwood, 1991, and unpublished data – except where indicated).

Microtrombidium fasciatum larva on a male *Megaselia giraudii*, attached by proboscis just behind abdominal tergite 1 and another larva (also attached by the proboscis) between tergites 2 and 3 of a male *M. pulicaria*, both records from Spain (J. Blasco-Zumeta and Disney, unpublished data).

Microtrombidium sp. 'A' on a male *Megaselia pusilla*.

Microtrombidium sp. 'B' on a male *Megaselia badia*.

Trombidium sp. on a male *Triphleba distinguenda*.

As yet undetermined mites

Conicera dauci – 1 male 1 female with 1 and 2 mites respectively, on abdomen.

C. similis – 1 male with 4 mites on abdomen.

M. brevicostalis – 1 female with 1 mite attached to abdomen by its oral sucker.

M. nigricauda – 1 male from Zimbabwe, with mite attached to side of abdomen.

Among Phoridae collected in Sulawesi in 1985 whose generic placement is still uncertain is a male with 69 mites attached to the base of its abdomen. Also a *Dohrniphora caini* male from Sulawesi had 5 mites attached to the sides of its abdomen.

Much work still needs to done to elucidate the relationships between mites and Phoridae. Taxonomic impediments now lie mainly with the mites, especially as it is frequently larvae and deutonymphs which are found attached to scuttle flies. The above review suggests there is some evidence of host specificity, with the adult females of species such as *Macrocheles disneyi* and *Scutacarus baculitarsus*. However, the deutonymphs of *Uroseius acuminatus* show no evidence of host specificity. These differences need investigation.

6.8 ENEMIES

6.8.1 PREDATORS

(1) Plants

Typically more than 70% of the prey trapped by the leaves of sundews (*Drosera intermedia* and *D. rotundifolia*) are Diptera (Crowder *et al.*, 1990), but most of these are small Nematocera (Van Achterberg (1973). However,

I have collected two male *Megaselia brevicostalis* off the leaves of a butterwort *(Pinguicula vulgaris)* in Yorkshire

(b) Invertebrates

Araneida (spiders)

Phoridae have been recorded as the prey of *Lycosa lugubris* (Lycosidae) (Edgar, 1970) *Phora tincta* (as *P edentata*) has been recorded as the prey of *Misumena vatia* (Thomisidae) (Disney, 1980c) and *Megaselia pulicaria* has been recorded being eaten by juvenile spiders in the laboratory (Evans, 1969, Disney and Evans, 1980) Undoubtedly spiders will prove to be more significant predators of phorids than these few records suggest

Diptera (flies)

The asilid *Leptogaster cylindrica* has been recorded preying on Phoridae (Séguy, 1927, Hobby, 1931a) The aslid *Machimus atricapillus* has been reported preying on phorids (Hobby, 1931a) and on *Borophaga incrassata* in particular (Hobby, 1933) Likewise, *Dioctria rupides* has been reported preying on *Anevrina unispinosa* and on *Diplonevra* sp (Hobby, 1932) The muscid *Coenosia tigrina* has been reported preying on *Borophaga incrassata* and on *Diplonevra* sp (Hobby, 1931b) An empid has been recorded preying on *Megaselia tenebricola* (Schmitz, 1938c) A male *Empis borealis* has been observed presenting a nuptial gift of *Triphleba opaca* to his mate (Svensson *et al* , 1990) *Platypalpus pallidiventris* is recorded preying on *Conicera dauci* (Morley and Atmore, 1915, Aston, 1957) *Rhamphomyia sulcata* is reported feeding on *Diplonevra concinna* (Smith, 1968) Laurence (1951) recorded *Tachypeza nubila* preying on *Megaselia* and *Metopina* species I have collected a freshly-dead *Megaselia pleuralis* male off a leaf of dog's mercury, in a patch where a number of Empididae were hunting at the time Phoridae are among the prey lured to their death by the curious luminous larvae of the mycetophilid gnat *Orfelia fultoni* (Sivinski, 1982) These fragmentary data suggest that other Diptera are major predators of Phoridae

Heteroptera (true bugs)

A reduviid bug has been reported preying on *Phalacrotophora longifrons* (Borgmeier, 1934a)

Hymenoptera

The sphecid wasps *Crossocerus elongatus* and *C nigritus* have been reported preying on *Phora* sp and *Megaselia pulicaria* respectively (Hamm and Richards, 1926) Leaf-cutting ants (*Atta*) sometimes kill phorids attempting to oviposit into them (3 5 6(e))

(c) Vertebrates

Some birds are probably significant predators of Phoridae, but published records are sparse. Borgmeier (1950a) reported *Coniceromyia latimana* in the stomach of a trochilid hummingbird (*Anthracothorax*). Scuttle flies are evidently regularly ingested by swifts (*Apus apus*). Lack and Owen (1955) recorded 15 specimens of *Megaselia* spp. in 12 meals examined on one occasion and 38 in 54 meals examined on another, giving an overall mean of 0.8 per meal. Parmenter (1965) recorded *Diplonevra glabra, Megaselia aequalis, M. rufipes* and *Phora atra* in the meals of this species.

6.8.2 PARASITES AND PARASITOIDS

(1) Hymenoptera – Parasitica

A planidium-type larva has been reported in the base of the abdomen of a male *Megaselia curtinoides* (Disney, 1979a, 1991b). An Opiinae-type (Braconidae) larval exuvium has been reported in the abdomen of a female *Puliciphora malae* (Disney, 1990f). These reports are the result of my habit of slide-mounting Phoridae. Those who pin their specimens, or gum them to a card stage on a pin, tend to miss many interesting data. Lewis (1952), as a result of dissections seeking larval Nematoda, found planidium-type larvae in the heads of Simuliidae. It seems, therefore, that the dearth of records of larval Parasitica in adult flies is probably due to the prevalence of the habit of pinning flies. Knowing that numbers of Phoridae parasitize adult insects (3.5.6) it would be surprising if some parasitic Hymenoptera had not evolved the same habit with respect to adult flies. However, Hymenopterists tend to be reluctant to accept the idea of Parasitica parasitizing adult flies. Clearly further investigations are required.

(b) Nematoda

Adult Phoridae are frequently infested with nematodes, which usually occur in the fly's abdomen. I have recorded them in the following species: *Megaselia curtineura, M. halterata, M. pleuralis, M. pseudobrevior, M. rupides, Woodiphora parvula* and *W. santoshi* (Disney, 1985a, 1988h, 1989f and unpublished data). Only the species parasitizing *M. halterata* has been named and studied in detail.

Howardula husseyi (formerly called *Bradynema*) parasitizes *Megaselia halterata*. All 60 fly specimens examined from a sample trapped in a pasture were found to be infested (Hussey and Gurney, 1964b). Riding and Hague (1974) reported infestation rates up to 0.80 but found about 0.50 was more usual. The fertilized female eelworm probably enters the second-instar fly larva. This female becomes approximately spherical as the host becomes a third instar. At the time of the host's pupation she becomes sausage-shaped, develops a 'heel' at one end and then liberates eggs and newly-hatched eelworm larvae into the host's haemolymph just before the eclosion of the

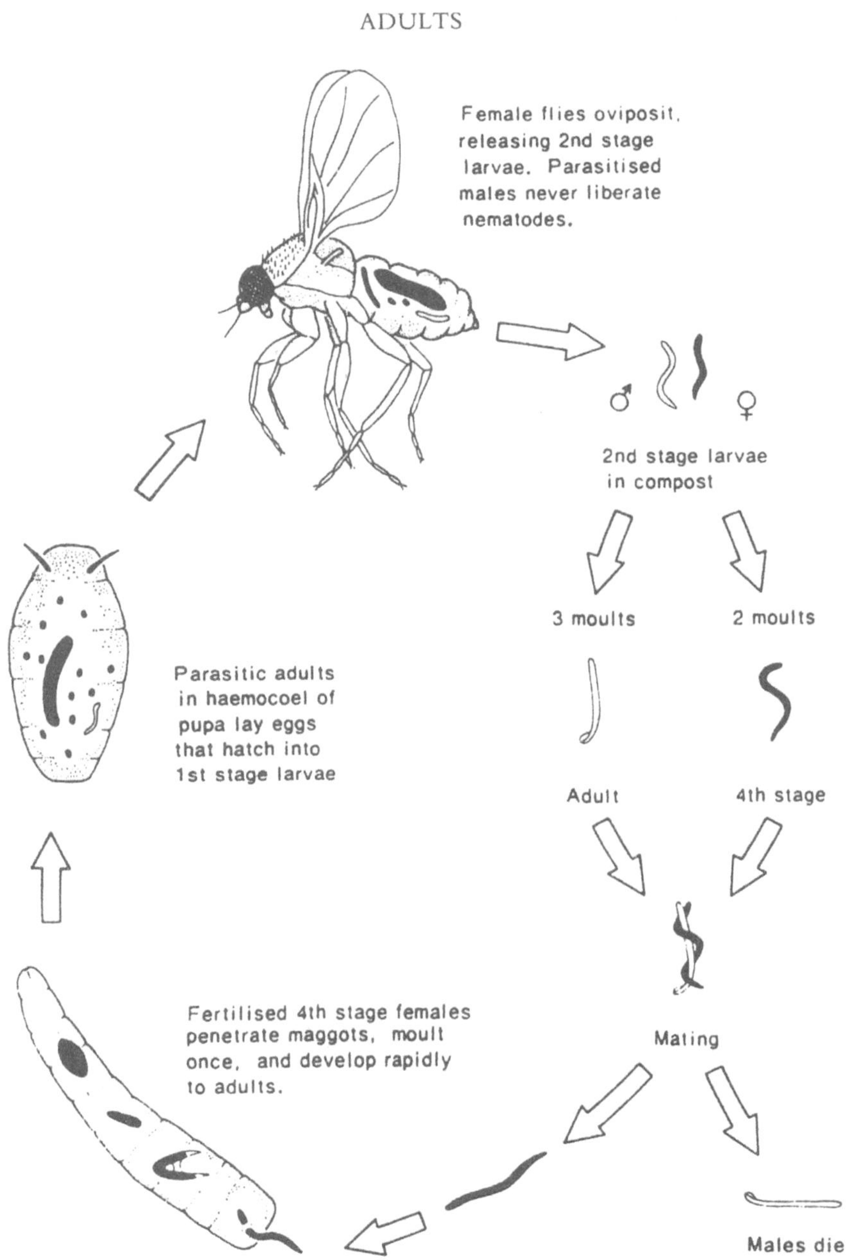

Figure 6.16 Life cycle of *Howardula husseyi* in *Megaselia halterata* (After Richardson *et al* , 1977)

adult fly. The fly's abdomen becomes so packed with young nematodes that in females the number of eggs matured is reduced from around 50 to fewer than 5. Male flies are also parasitized. The nematodes probably leave the host, as second or third instars, either at oviposition or at defecation. Further data on this parasite are published elsewhere (Richardson and Chanter, 1979; Richardson and Hesling, 1977a,b; Richardson *et al.*, 1977). The life cycle is summarized in Fig. 6.16.

Megaselia halterata has also been successfully infected, in the laboratory, with *Heterorhabditis heliothis* and *Steinernema feltiae* (Richardson, 1987).

(c) Fungi

Gol'berg (1979) recorded *Entomophthora culicis* (Entomophthoraceae) attacking and killing phorids found attached to the underside of grapefruit leaves in Israel.

The anamorph states (hyphal bodies) of *Cordyceps* (or a related genus of the Clavicapitales) have been recorded as internal parasites of *Megaselia microcurtineura* and Termitoxeniinae in Zimbabwe (Disney, 1991a). The same group of fungi parasitize cultivated mushrooms, and the mushroom phorid *Megaselia halterata* has been implicated in the transmission of a pest species (7.11).

Undoubtedly other fungi are important pathogens of Phoridae. Samson *et al.* (1988) provide an introductory atlas of entomophagous fungi. These mycopathogens of insects are reviewed by Evans (1989) and Keller and Zimmermann (1989).

7

Ecology

All biologists are more impressed by what we don't know than they are by what we know, and such an emphasis is thus a true reflection of our state of knowledge and our state of mind (Bates, 1951)

7.1 INTRODUCTION

Ecology is concerned with the economy of the biosphere. Its currency is energy, nutrients and information. Its primary focus is the patterns of the interrelationships between the different organisms and between them and the environment. Many of the patterns and processes that define these interrelationships are only revealed by the collection of quantitative data. In common with other large and taxonomically complex families of small insects, such data are sparse for the Phoridae. This has caused ecologists to underestimate the ecological significance of such families. The fragmentary data reviewed below constitute no more than a rough sketch of the ecological significance of the scuttle flies. Our current knowledge is also unbalanced by being largely derived from studies in temperate latitudes.

7.2 THE PHORID FAUNAS OF DIFFERENT HABITATS

7.2.1 AQUATIC HABITATS AND MICROHABITATS

A large number of terrestrial plants provide containers which retain water. These container habitats are termed phytotelmata and their characteristics are reviewed by Frank and Lounibos (1983). The arthropod community of bamboo internodes is reviewed by Kovac and Streit (in press).

The aquatic phorids of container habitats, whose larval habits are known, feed on drowned insects and the fungi associated with them (3.2.1). The larval habits of the intertidal *Megaselia minutior* (Cheng and Hogue, 1974) are unknown. The pioneering studies of freshwater Phoridae are those of Meijere (1910), Schmitz (1931d, 1955c), Thienemann (1932, 1934) and Beaver (1979, 1983). Their findings have been critically reviewed (Disney,

1991i) and are summarized below, based on this recent review supplemented by Disney (1993b) and further data (indicated by the citation of additional publications).

Axil waters

Cocos nucifera (coconut palms) – *Megaselia* sp. (Haworth, 1924).
Zingiber macradenium – *M. bivesicata*.

Bromeliads

An epiphytic species – *Dohrniphora* sp. (Winder, 1977).

Decaying coconuts

– *M. scalaris*.

Bamboo stems (waters held in internodes)

Undetermined species – *Megaselia bivesicata*.
Gigantochloa scortechinii – *M. anomaloterga*, *M. deningi*, *M. gombakensis*, *M. humida*, *M. kovaci*, *M. nepenthina*, *M. scalaris* and *Plastophorides gigantochloae*.

Pitcher plant pitchers

Darlingtonia californica – *M. orestes*.
Nepenthes albomarginata – *M. nepenthina*, *M. schuitemakeri*.
N. ampullaria – *M. bivesicata*.
N. bicalcarata – *M. campylonympha*.
N. distillatoria – *M. decipiens*, *M. deningi*, *M. gregalis*.
N. gracilis – *M. nepenthina*, *M. schuitemakeri*.
N. gymnamphora – *M. decipiens*, *M. gregalis*.
N. mirabilis – *M. bivesicata*, *M. campylonympha*, *M. nepenthina*, *M. schuitemakeri*.
N. rafflesiana (probably) – *M. deningi*.
N. tobaica – *M. tobaica*.
Nepenthes spp. – *M. cambodiae*.
Sarracenia flava – *Dohrniphora cornuta*.

Most of the *Megaselia* species larvae live fully submerged, but *M. decipiens*, *M. gombakensis*, *M. gregalis* and *M. humida* tend to live in the meniscus or suspended from the water surface. Indeed *M. decipiens* has a pair of long caudal processes, with which it can suspend itself from the water surface when not engaged in creeping about, leech-like, in the meniscus. The *M. scalaris* larvae recorded in the bamboo internode were in the meniscus, and bore many colonies of aquatic peritrich protozoans (Ciliophora, Peritricha). The larva of *M. humida* (Fig. 3.2a) is especially characteristic of the meniscus region, and in life tends to be camouflaged by a covering of debris. The poorly-known *M. gombakensis* seems to have the same habits. The *Dohrniphora cornuta* larvae tend to favour pitchers that are starting to dry out. This species also invades trickling filter sewage beds, as does *Megaselia rufipes* in certain circumstances. The larva of the former species (Fig. 3.1a) more resembles that of strict hygrophilous species, such as *M. humida*, than does that of *M. rufipes* (Fig. 3.1b), which is more rarely recorded in aquatic habitats.

The water in the pitchers of *Nepenthes* presents a challenging environment in terms of water chemistry. Before the pitchers open the water has a very low pH. However, this steadily rises to around neutrality after the pitcher opens and the accumulation of prey debris takes place. In addition digestive enzymes accumulate, such as ribonuclease, lipase, esterase and protease (Heslop-Harrison, 1976). It is probably significant, therefore, that larvae that live fully submerged, such as *Megaselia bivesicata* and *M. deningi*, have enlarged anal gills (Fig. 3.3c). These are presumably involved in osmoregulation. Indeed *M. deningi* seems to be thereby pre-adapted for invading osmotically challenging environments, having been reported in steepers of takuwan (Japanese garden radishes pickled with rice bran and salt) (Kaneko and Furukawa, 1983) and in a case of urinary myiasis (3.5.7).

Until the discovery of the intertidal breeding habits of *Megaselia minutior* by Cheng and Hogue (1974) it had not been realized that there were any marine Phoridae. These authors obtained 20 specimens in an emergence trap set over mud, and 2 when set over sand, in intertidal mangrove swamps in the Gulf of California. The larval biology of this species clearly invites investigation. The probability of other intertidal scuttle flies turning up would seem to be high. Indeed Saunders (1928) reported that the larvae of a *Megaselia* species (referred to as '*M. ursina*' at the time) 'can withstand short immersion in sea-water', but gives no further details of this fly of the Canadian Pacific coast.

7.2.2 CAVES

Leruth (1936, 1939) reviews species found in caves. However, all the species he lists as being exclusive to caves are now known from other habitats as well. Indeed the species he recorded most commonly in caves, *Megaselia rufipes*, is far commoner outside caves. But caves are evidently an overwintering habitat for species such as *M. rufipes* and *M. albicaudata*. Franz (1949) and Weber (1989) also list Phoridae recorded in caves and Matile (1970) covers scuttle flies in his review of cave Diptera. Baumann (1976a) provided a detailed bibliography for published records of Phoridae in caves, especially for Europe. The principal species characteristic of caves in Europe are *Triphleba antricola* and *T. aptina*. These species breed in bat dung. *Megaselia melanocephala* will also breed in caves, the larvae preying on the eggs of the cave spider (3.5.3(b)). At high altitudes (i.e. above 2000 m) species normally associated with other habitats may more frequently turn up in caves. *Megaselia manicata, M. teneripes, M. xanthozona, Metopina heselhausi* and *M. oligoneura* have been reported from such caves in the Sierra Nevada of Spain and *M. teneripes* from the Dolomites. Furthermore, *M. xanthozona* is also known from the north-eastern Alps in the Plesitocene (Mani, 1968). In North America *M. cavernicola* is especially associated with caves.

The phorids of caves in the tropics are virtually unexplored. *Diploneura peregrina* inhabits caves in Malaysia, where its larvae feed on freshly-dead bats (McClure *et al.*, 1967). However, collections by Johanna Darlington from caves in Trinidad indicate that such caves merit investigation. Her collections include an undescribed *Chonocephalus* species, *Conicera megalodus*, *Diplonevra impressa*, *Dohrniphora cornuta*, *D. divaricata*, *D. lugens*, an undescribed *Dohrniphora*, *Megaselia saprophaga*, *M. scalaris*, an undescribed *Megaselia* in abundance and small numbers of other undescribed species of *Megaselia*, *Puliciphora borinquenensis* and an undescribed *Puliciphora* species (Disney, unpublished data).

7.2.3 CROPS

(a) Soils

Some preliminary data have been published on scuttle flies obtained in emergence traps operated in a winter wheat/fallow rotation system in England (Disney, 1980f). The species obtained were *Conicera floricola*, *Diplonevra nitidula*, *D. pilosella*, *Megaselia aequalis*, *M. ciliata*, *M. nigriceps*, *M. pleuralis*, *M. pulicaria and Triphleba nudipalpis*. A number of males of a previously undescribed species, *Conicera crassicosta*, were recently obtained in emergence traps in a winter wheat field in Germany (Disney, 1991l).

Prescher (1992) obtained 12441 Diptera (an estimate of 6220.5 m^{-2}), of which 21.5% (or 1339.5 m^{-2}) were Phoridae, in emergence traps operated in a summer wheat field in Germany, from March to November 1988. *Metopina oligoneura* represented 72.8% of the phorid catch; *Megaselia* spp. 13.7%; *Diplonevra nitidula* 5.4%, *D. funebris* 2.2% and *D. pilosella* 0.04%; *Conicera similis* at least 3% (the females not being distinguished from *C. tibialis*); and *Triphleba nudipalpis* 1.7%. The principal *Megaselia* species were *M. aculeata*, *M. brevicostalis*, *M. pleuralis*, *M. pulicaria* complex and *M. pusilla*.

Prescher (1992) also operated emergence traps in a maize field in 1989 and a potato field in 1990, from March to November in each case. She obtained 14545 (an estimated 7272.5 m^{-2}) Diptera in the maize field, the phorids being 35.2% (or 2558 m^{-2}). *Metopina oligoneura* represented 61.5% of the phorid catch, *Megaselia* spp. 22.9%, *Conicera similis* at least 6%, *Diplonevra funebris* 2.7%, but *D. nitidula* only 0.9% and *D. pilosella* 0.8%, and *Triphleba nudipalpis* 2.2%. In the potato field she obtained 8484 (an estimated 4242 m^{-2}) Diptera, of which 27.3% (or 1158.5 m^{-2}) were Phoridae. *Metopina oligoneura* represented 50.5% of the phorid catch, *Megaselia* spp. 15.4%, *Conicera similis* at least 10%, *Diplonevra nitidula* 8.9%, but *D. funebris* only 1.2% and *D. pilosella* 0.8%, and *Triphleba nudipalpis* 6.3%.

Some emergence traps operated over cultivated raspberry stools in

Scotland (Disney and Gunn, 1980; Disney, 1989a) caught 195 scuttle flies. The percentages for the different species were as follows:

Triphleba nudipalpis	33.3	*Megaselia altifrons*	1.5
Megaselia dimidia	17.4	*Phora holosericea*	1.0
Metopina oligoneura	13.3	*Anevrina thoracica*	0.5
Megaselia aequalis	10.3	*Diplonevra pilosella*	0.5
Diplonevra nitidula	9.2	*Megaselia giraudii*	0.5
Megaselia ciliata	4.6	*Megaselia hirtiventris*	0.5
Conicera similis	4.1	*Megaselia pleuralis*	0.5
Megaselia brevicostalis	2.6		

Emergence-trap collections of Phoridae in Poland were 1.8% (0.7% of the biomass) of the Diptera specimens collected from rye fields, and 1.4% (0.4% of biomass) from potato fields (Dabrowska-Prot and Karg, 1974). In Germany Heynen and Wübbeler (1990) compared two arable plots subjected to different types of management. They sampled adult insects with emergence traps, obtaining 3876 Diptera, and sampled larvae with a flotation/washing procedure, obtaining 3979 Diptera. Adult Phoridae represented 17.6% of the Diptera at one site (compared with 25.1% Sciaridae, 25.8% Cecidomyiidae and 10.5% Drosophilidae) and 23.6% at the other site (compared with 23.9% Sciaridae, 27.2% Cecidomyiidae and 13.2% Chironomidae). No larval Phoridae were obtained at either site. Morris (1927) extracted insects from the soil of a mangold (*Beta vulgaris*) field in England, by washing and sieving soil samples, but obtained only insignificant numbers of larval Phoridae.

Froese (1992a–c), using emergence traps, obtained 9507 phorids from the soils of arable fields in Germany. Of these 79.7% were *Metopina oligoneura*; 9.4% were *Megaselia* species, representing 17 species, notably *M. aequalis, M. altifrons, M. brevicostalis, M. ciliata, M. longicostalis, M. pectoralis, M. posticata, M. pulicaria* and *M. pusilla*. Other species included *Conicera similis* (0.1%), *Diplonevra funebris* (0.7%), *D. nitidula* (0.6%), *D. pilosella* (4.5%), *Metopina pileata* (1.9%) and *Triphleba nudipalpis* (3.1%). He reported (1992b) up to 35 m^{-2} *T. nudipalpis* adults at the peak of emergence in May. He also (1992c) reported 250–600 m^{-2} emerging adults of *Metopina oligoneura*, and (using a sieve/flotation technique) procured larval estimates of 2900 m^{-2} for this species.

Most workers, except Froese, have encountered difficulty in extracting larval phorids from soil samples. By contrast, in the compost in which cultivated mushrooms are grown the densities of the larvae of *Megaselia halterata* can reach 5–45 per 30 gm of compost and in extreme cases up to 400 per 30 gm. Damage to the mushroom crop is likely to occur when the levels exceed 100/30 gm (Moreton and John, 1955; Hussey, 1960, 1961b).

(b) The crop

Brasse (1975) sampled Diptera in cereal fields in Germany, using pitfall traps. Of the 41 233 Diptera specimens obtained 10.9% were Phoridae. However, in vacuum-net samples from cereal crops in Sussex, England, the Phoridae represented only 0.34–1.27% of the Diptera caught (Potts and Vickerman, 1974).

Boness (1958) provided some preliminary data on Phoridae in clover and lucerne fields in Germany. The most abundant species were *Megaselia aequalis, M. brevicostalis, M. longicostalis, Metopina* spp., *Diplonevra* spp. and *Conicera* spp.

Dabrowska-Prot and Karg (1974) reported that Phoridae represented 0.5% (0.2% of the biomass) of the Diptera specimens caught in rye fields in Poland, and 0.9% (0.3% of the biomass) of Diptera in potato fields. These are significantly lower proportions than the numbers of Phoridae emerging from the soil in the same fields (see above), presumably because many of the Diptera species found on the crop foliage do not breed in the soil.

Goos (1975) employed yellow traps in sugar-beet crops in Poland and found Phoridae to represent from 8.4% to 40.6% of the Diptera specimens caught (n = 182 422).

7.2.4 MEADOWS AND PASTURES

(a) Soils

In Poland Olechowicz (1971) used emergence traps in uncultivated, unharvested meadows (*Stellario-Deschampsietum*) and compared the returns with similar trap data for fertilized, cultivated meadows (*Arrenatheretum*), which were harvested twice a year. In the former the yield of phorid specimens was estimated to be 39.9 m^{-2} per season, representing 5.7% of the Diptera specimens. In the cultivated meadow the yield was 107.9 m^{-2} or 15.6% of the Diptera. The families of Diptera showing comparable or greater yields were Sciaridae (165.3 m^{-2} or 23.8% in the uncultivated and 136.7 m^{-2} or 19.7% in the cultivated), Cecidomyiidae (83.7 m^{-2} or 12.0% uncultivated, and 165.6 m^{-2} or 23.9% cultivated) and Ceratopogonidae (57.2 m^{-2} or 8.2% uncultivated, and 12.0 m^{-2} or 1.7% cultivated).

In England emergence traps were operated in resown *Lolium* pastures at ten different sites simultaneously (Disney, Henderson *et al.*, 1981). These traps were operated for eight periods of 20 days each spread over a year. The estimated numbers of scuttle flies emerging per square metre in 20 days were as given in Table 7.1.

Table 7.1 Estimated numbers of Phoridae m^{-2} emerging from the soil of ten pastures in England in 20 days

		20-day period ending on:							
Sites	n	6 May	29 May	17 Jun	29 Jul	20 Aug	1 Oct	20 Oct	13 Nov
All	14 467	3.5	22.4	25.7	75.2	103.3	18.7	12.4	6.5
Least	43	7.4	18.5	5.6	31.5	7.4	0	3.7	5.6
Most	354	0	20.4	124.1	177.8	257.4	42.6	25.9	7.4

The highest individual estimate was 261.1 m^{-2} Phoridae in 20 days, or about 13 m^{-2} per day, emerging from a Shropshire pasture in August.

The interesting feature of these estimates, of the numbers of scuttle flies breeding in pasture soils, is that they imply population densities of larvae in excess of those indicated by extractions of Diptera larvae from such soils. However, precise estimates of larval densities are sparse in the literature. Salt *et al.* (1948) extracted 109 Diptera larvae from 20 small cores of soil (0.30 m deep) taken from an English pasture soil. From their data one can estimate a density of 86.3 m^{-2} Cyclorrhapha larvae. This compares with an estimate of 382.4 m^{-2} larval Sciaridae, the Cyclorrhapha representing 12.8% of the Diptera larvae and the Sciaridae 56.9%. Subsequent studies indicate that Phoridae are often found to be the dominant family of Cyclorrhapha in emergence-trap collections. Thus an estimate of about 80 m^{-2} phorid larvae would seem likely. In North America Menhinick (1962) was a little more successful in extracting phorid larvae from soil samples taken from a mown-grass roadside verge. He obtained 29 phorid larvae, representing 4.2% of the 691 Diptera larvae and an estimate of 7.8 m^{-2} for the phorid larvae. These are, however, low figures in comparison with the other Cyclorrhapha larvae he obtained (e.g. 213 Agromyzidae and 72 Drosophilidae). Part of the explanation of this apparent deficiency of larval Phoridae may lie in the parasitoid habits of many phorids. Probably many of the 'missing' larvae are concealed within other invertebrates obtained by extractions using standard methods.

The principal species breeding in the ten pasture soils in England (Disney, Henderson *et al.*, 1981) were, in ranked order of abundance, as follows:

1. *Diplonevra nitudula*
2. *Megaselia brevicostalis*
3. *Diplonevra pilosella*
4. *M. ciliata*
5. *M. aequalis*
6. *M. pulicaria* complex
7. *Metopina oligoneura*
8. *Megaselia pusilla*
9. *Triphleba nudipalpis*
10. *Diplonevra funebris*
11. *M. propinqua*
12. *M. pectoralis*
13. *M. lucifrons*
14. *Conicera floricola*
15. *Megaselia giraudii*
16. *M. pleuralis*

Neun and Weber (1985) investigated the soil fauna of reclaimed rubble tips in Germany, using emergence traps. Out of 18 403 Diptera obtained in the traps 50.4% were Phoridae, the most numerous family by far. The next most abundant families were Cecidomyiidae (21.1%) and Sciaridae (11.9%). Their more detailed analysis of the Phoridae (Weber and Prescher [neé Neun], 1990) provides some information on the succession of species following the reclamation. Two plots were about 3 years old, one (Plot 3N) being left to natural succession (*Tanaceto-Artemesietum* with abundant grass and patches of *Arctium*) and another (Plot 3S) being sown with grasses (*Lolium perenne* and *Festuca rubra*) and regularly mown. A third plot (8N) underwent a natural succession for about 8 years (*Rubus fruticosus, Solidago, Betula, Salix* and *Rosa canina* being well established). The results for selected species were as given in Table 7.2.

Table 7.2 Relative abundances of species of phorid obtained in emergence traps operated on three different reclaimed rubble tips in Germany (see text for details)

Species	Plot 3S (%) ($n = 4156$)	Plot 3N (%) ($n = 2362$)	Plot 8N (%) ($n = 2803$)
Anevrina urbana	0.02	–	2.9
Diplonevra funebris	4.7	7.4	4.9
D. pilosella	–	–	4.3
Metopina galeata	68.6	59.1	25.1
M. oligoneura	12.3	13.9	44.7
Triphleba intermedia	2.4	0.04	–
T. nudipalpis	–	1.3	4.6

Of the 4949 *Metopina galeata* obtained in the three plots 57.6% were from Plot 3S, 28.2% from Plot 3N and only 14.2% from 8N. This species clearly preferred the cultivated plot and diminished during the course of succession. Of the 2092 *M. oligoneura* 24.4% were from Plot 3S, 15.7% from Plot 3N and 59.8% from Plot 8N. Thus this species also preferred the cultivated plot to the natural succession plot, but it greatly increased during the course of natural succession. By contrast the 509 *Diplonevra funebris* were more evenly distributed among the three plots, with 38.7% in 3S, 34.2% in 3N and 27.1% in 8N. All 121 specimens of *D. pilosella* occurred in Plot 8N. Furthermore *D. nitidula* represented 0.04% of the phorids in plot 8N, but was not obtained in the younger plots. Both these *Diplonevra* species parasitize earthworms (3.5.1), which take several years to invade reclaimed sites (Dunger, 1989). With *Triphleba nudipalpis* 2.3% were from Plot 3N and 97.7% from Plot 8N ($n = 133$).

(b) The vegetation

Boness (1953) sampled insects in the vegetation of meadows in Germany and reported abundant Phoridae. Common taxa were *Megaselia* spp., *Diplonevra funebris*, *Conicera dauci* and *Metopina* spp. Bährmann (1987) reported that the Phoridae represented from 16.2% ($n = 1173$) to 37.7% ($n = 3678$) of the Diptera, less the Nematocera, obtained in pitfall traps in two grassland sites in Germany. In the first site only Sphaeroceridae were more abundant (at 20.1%) and Chloropidae (at 14.2%) were the next most abundant family. In the second site the Phoridae were the most frequent, followed by the Sphaeroceridae (22.1%) and Chloropidae (7.6%). By contrast a grassland site polluted by a phosphate factory produced only 9.5% Phoridae ($n = 6927$), with the Tethinidae (normally associated with the seashore) being the most abundant family (at 17.2%), followed by Opomyzidae (13.5%), Empididae (10.0%) and Chloropidae (9.7%). However, this relatively low percentage seems to have reflected more the increase in other families, especially the Tethinidae, rather than any adverse effect of the pollution on the Phoridae. Ricou (1978) reported Phoridae to represent 3.5% of the Brachycera + Cyclorrhapha in some limited samples from grasslands in Britain.

7.2.5 MOORLANDS

A survey of moorland sites in northern England (Disney *et al.*, 1981), using pitfall traps, obtained more than 7000 specimens of Phoridae. However, the selectivity of pitfall traps (9.2.4) means relatively few species were caught at each site. The mean number of species was 10.4 (range 1–18). Overall *Megaselia longicostalis* represented more than 60% of the specimens caught. A comparison for eight pairs of sites on peat and mineral soils gave a mean of 6.9 species on the peat soils and 11.0 on the mineral soils (the difference being significant, $p < 0.05$).

Also in northern England Nelson (1965) sampled the aerial insects above a moorland stream and above its banks. The estimated densities of Phoridae were only $0.09\,m^{-2}$ ($\pm$ 0.07) over the stream and $0.47\,m^{-2}$ ($\pm$ 0.10) over the bank in a period of about 9 months. The low figures overall are probably mainly a reflection of the type of trap (horizontal sticky traps – see 9.2.2) being inappropriate for phorids.

7.2.6 SCRUB

Bährmann (1984) compared the Diptera in the herb layer and the shrub layer of a site in Germany, using net samples. He procured 6.7 times as many phorid specimens in the herb layer as in the shrub layer.

7.2.7 FORESTS AND WOODLANDS

(a) Soils

Bächli (1970) reared insects from soil samples taken from mixed woodland habitats in Switzerland. The 2877 Diptera comprised 82.2% of the emerging insects. The 217 Phoridae represented 9.2% of these Diptera, being exceeded only by Sciaridae (39.3%) and Cecidomyiidae (22.1%). Szadziewska (1977) obtained 720 phorids in emergence traps over soil in forest habitats in Poland. These represented 13.1% of the Diptera, with only the Cecidomyiidae (32.0%) and Sciaridae (30.8%) being more abundant. In France Mollon (1982) used emergence traps in mixed woodland. She obtained about 8% Phoridae out of all the Diptera, exceeded in abundance only by Cecidomyiidae (about 67%) and Chironomidae (about 11%). The Sciaridae (about 3.7%) and Empididae (about 2.6%) were less abundant. Her actual estimates of the numbers of Phoridae emerging were $126\,m^{-2}$ in 14 days in an oak/hornbeam plot and $115.4\,m^{-2}$ in an oak/pine plot. Baumann (1976b) obtained about 3000 phorids in 20 emergence traps operated for about a year in riverine woodland bordering the Upper Rhine in Germany. The yield of phorid specimens worked out at about $1000\,m^{-2}$ per year.

Altmuller (1976, 1979) reported Phoridae to be only 1.5% of the Diptera in emergence traps set over acid humus in a beech forest in Germany, and the production of adults to be $62.3\,m^{-2}$ in 1973. The families more abundant than phorids were Sciaridae (77.9%), Mycetophilidae-Sciophilinae (9.5%), Cecidomyiidae (7.1%) and Empididae (1.8%). In contrast to these families, his extractions of larval Diptera from the organic layer of the soil (horizons L, F and H) revealed a dearth of phorid larvae. Heynen (1988) carried out a nine-year study of the Diptera larvae in the soil of a beech wood in Germany. Despite an average estimate of Diptera larvae of 700–1900 m^{-2} he obtained no phorid larvae. Likewise Healey and Russell-Smith (1970) extracted 733 Diptera larvae, using five different methods, from soil samples from an English wood on clay. Of these only one larva was a phorid. These findings confirm the situation reported in other habitats (see above). Using emergence traps in a less acid German beech forest on limestone, Hövemeyer (1984) reported that Phoridae represented 9.9% of the Diptera. Furthermore he subsequently (1989) carried out an intensive investigation of larval Diptera in the leaf litter, comparing normal litter with litter artificially increased fivefold. With normal leaf litter phorids occurred with a mean density of 12 and maximum of 68 larvae m^{-2}. Under conditions of increased litter the densities were a mean of 16 and a maximum of 135 larvae m^{-2}. While these recorded densities are exceptionally high for phorid larvae, they are low compared with other Diptera. For example, the figures for larval Sciaridae were a mean of 481 and maximum of 1460 larvae m^{-2} under normal litter and a mean of 2579 and maximum of 14 704 larvae m^{-2} under increased litter conditions. The increased densities of larval Diptera, including Phoridae, under increased litter probably resulted from an increased curtailment of the movements of larger predatory invertebrates.

Using emergence traps in spruce (*Picea*) plantations in Germany, Thiede (1977) obtained estimates for Phoridae of 11.4–29.8 m^{-2} per year. This was out of an estimate of total Diptera of 2100–3700 m^{-2}.

A provisional list of species obtained in emergence traps set in mixed woodland habitats in Europe is given in Table 7.3 (based on data in Baumann, 1976b, 1979; Disney, 1978a, 1982h; Disney and Szadziewski, 1979).

Adult Phoridae are generally well represented in the litter layer in woodlands in temperate regions (e.g. Cole, 1946) and also in tropical forests (e.g. Williams, 1941).

(b) The groundstorey and understorey

Phoridae appear to be abundant in tropical forests, but few quantitative data are available. Analysis of the catch in a Malaise trap operated for a few days in Sulawesi yielded 62.2% Diptera out of 1239 insects caught. Phoridae represented 11.4% of the total catch, contrasting with 5.0% for ants and 9.0% for Parasitica (Hymenoptera). An identical trap operated for a shorter time in cleared land adjacent to the forest yielded 32.9% Diptera (out of a total of 167 insects). The Phoridae were only 2.4%, ants 6.6% and Parasitica 19.2% of the catch (Disney, 1985h).

Dead wood is undoubtedly an important microhabitat for many Phoridae. For example, in America Blackman and Stage (1924) and Rhoads (1924) reported numerous unidentified phorids reared from 4- and 5-year-old

Table 7.3 Phoridae recorded in emergence traps in mixed woodland in Europe

Anevrina thoracica	*Megaselia altifrons*
A. unispinosa	*M. breviterga*
A. urbana	*M. cinerea*
Borophaga agilis	*M. flava*
B. incrassata	*M. fuscinervis*
Chaetopleurophora erythronota	*M. lutea*
C. pygidialis	*M. picta*
C. spinosissima	*M. pleuralis*
Conicera floricola	*M. pulicaria*
C. tarsalis	*M. rufa*
Diplonevra abbreviata	*M. ruficornis*
D. concinna	*M. tumida*
D. florea	*Phora edentata*
D. nitidula	*Spiniphora bergenstammi*
D. pilosella	*S. dorsalis*
D. unisetalis	*Triphleba autumnalis*
Gymnophora arcuata	*T. gracilis*
G. integralis	*T. nudipalpis*
G. nigripennis	*T. papillata*
G. quartomollis	*T. tumidula*
Hypocera mordellaria	

samples of dead and decaying hickory (*Hicoria glabra*) wood placed in rearing cages. Likewise Elton (1966) reported *Megaselia* species, along with Sciaridae, to be abundant in rotting beech wood in Wytham Woods, England. Standing dead timber appears to be favoured by some Phoridae. For example, a study of standing dead pines, in the USA, recorded numbers of *Megaselia* larvae under the bark of trunks which had died 4–6 years previously (Howden and Vogt, 1951). Some of these species will be parasitoids of other insects breeding in these situations, for example, the *Megaselia* species parasitizing sciarid larvae (3.5.6(c)). On several occasions *M. fungivora* and *Triphleba gracilis* have been reared from pupae found under the bark of fallen trees (Disney, 1979f and unpublished data). There is considerable scope for rearing experiments aimed at determination of the larval habits of the various species of Phoridae breeding in dead wood of different ages.

(c) Tree canopies

Durska (1981) published some preliminary data on Phoridae in the crowns of trees in urban green areas in Warsaw. Some quantified data (obtained by Sir Richard Southwood and identified by myself) for some oak trees near Oxford (England) are presented in Table 7.4. Of particular note is the prevalence of species previously considered to be rare, notably *Megaselia correlata* and *Obscuriphora sheppardi*. Other 'rare' species reported from tree canopies in Europe are *Megaselia durskae*, *Microselia southwoodi* and *Phora greenwoodi* (Disney, 1988c, 1989c,e).

Table 7.4 Phoridae obtained from the canopies of oak trees (*Quercus* spp.) in woods near Oxford (1980–1984), using the insecticide knockdown method

Species	May/June				July/August			
	Wytham Woods		Bagley Wood		Wytham Woods		Bagley Wood	
	No	%	No	%	No	%	No	%
Borophaga carinifrons	0	–	0	–	1	0.7	0	–
B. incrassata	0	–	0	–	1	0.7	2	2.0
Diplonevra pilosella	1	1.8	0	–	0	–	0	–
Gymnophora quartomollis	5	9.1	16	10.0	3	2.1	2	2.0
Megaselia ciliata	4	7.3	1	0.6	27	18.5	6	5.9
M. correlata	6	10.9	11	6.9	0	–	1	1.0
M. diversa	0	–	1	0.6	4	2.7	4	3.9
M. pulicaria	1	1.8	1	0.6	7	4.7	0	–
Other *Megaselia* spp.	17	30.9	16	10.0	46	31.5	37	36.2
Metopina ulrichi	0	–	0	–	1	0.7	0	–
Obscuriphora sheppardi	0	–	3	1.9	3	2.1	2	2.0
Phalacrotophora berolinensis	0	–	1	0.6	0	–	0	–
P. fasciata	0	–	0	–	0	–	5	4.9
Phora atra	2	3.6	2	1.3	3	2.1	2	2.0
P. edentata	0	–	0	–	0	–	3	2.9
P. tincta	18	32.8	108	67.5	50	34.2	38	37.2
Spiniphora bergenstammi	1	1.8	0	–	0	–	0	–
Totals	55	100	160	100	146	100	102	100

In Germany Buchs (1983, 1988) has studied the fauna obtained in emergence traps attached to the trunks and branches of woodland trees. Phoridae proved to be the dominant family of Diptera. Those emerging from the branches represented 21.8% of the Diptera, with an overall estimate of 242 phorids m^{-2} of bark area. However, the percentage of phorids varied with the tree species, being 12% on ash (*Fraxinus excelsior*), 14% on oak (*Quercus robur*), 26% on poplar (*Populus nigra*) and 48% on elm (*Ulmus laevis*). The species obtained included *Megaselia aequalis*, *M. annulipes*★, *M. ciliata*★, *M. pulicaria*★, *M. giraudii*, *M. lata*, *M. nasoni*, *M. tama*, *M. tarsalis*, *Metopina oligoneura* and *Phalacrotophora berolinensis*. Those marked with an asterisk (★) each represented more than 1% of the Diptera obtained in the traps, with *M. annulipes* being the most prolific at 6.7% (or an estimated density of 75 flies m^{-2}).

Tree canopies are clearly rich in Phoridae in the tropics and elsewhere, but quantitative data are lacking (mainly due to the lumping of Phoridae with other Cyclorrhapha). Numbers of species have been described from specimens obtained in tree canopies (e.g. Disney, 1983h).

7.3 THE DEARTH OF LARVAE IN SOIL EXTRACTIONS

Estimates of the numbers of Diptera larvae from soil extractions tend to lie within the broad range of 100–20 000 m^{-2} (Persson and Lohn, 1977). In much of the literature on extractions of insects from the soil the Phoridae are not distinguished from other Cyclorrhapha larvae. However, a common perception, that emerges from the few quantitative data available, is that the density estimates of Phoridae breeding in soil obtained with emergence traps greatly exceed those indicated by the extraction of Diptera larvae from soil samples. The data of Heynen and Wübbeler (1990) (7.2.3(a)) are particularly striking. Overall they found Phoridae represented 20.5% of the 3876 adult Diptera in their emergence trap samples, but 0% of the 3979 larval Diptera extracted from their soil samples from the same sites. This may be partly a methodological problem with regard to the extraction procedures used, as Menhinick (1962) (7.2.4(a)) was more successful, obtaining an estimate for phorid larvae of 7.8 m^{-2}. Likewise Hövemeyer (1989) procured larvae in densities up to 135 m^{-2}, but mostly 12–16 m^{-2} in beech-wood leaf litter, but still proportionally far less than he obtained with adults in emergence traps (1984). Thus these better figures for larvae are still far lower than most estimates based on emergence trap data for adults, which imply that the larval densities range from 10 to 300 m^{-2} with no allowance being made for any larval or pupal mortality. The only estimate that seems at all likely to reflect reality is that of 2900 m^{-2} for the larvae of *Metopina oligoneura* reported by Froese (1992c). When compared with his estimates of adult emergences (the species being almost 80% of the phorids obtained in the emergence traps), of 250–600 m^{-2}, there is clearly significant mortality during the pre-adult stages.

It has been suggested that much of the explanation is that many of the 'missing' phorid larvae are parasites inside other invertebrates (Disney, Henderson *et al.*, 1981). Other larvae may be inside plants. For example, more than 60% of the available root nodules of the chickpea (*Cicer arietinum*) may be infested with the larvae of *Metopina ciceri* (3.4). Further investigations are called for.

7.4 HABITAT PREFERENCES

Largely because of the effort involved, apart from the problem of ensuring the comparability of one's sampling procedures in different habitats (see Disney, 1986g), there have been few studies involving simultaneous quantitative sampling in different habitats in the same locality. Some data exist for *Megaselia pleuralis* in the USA, which were obtained using 1-phenylethanol-baited traps (see Chapter 9) in four habitats simultaneously (Kamm *et al.*, 1987). The percentage breakdown for the 1899 specimens of *M. pleuralis* trapped was as follows: red clover grown for seed 27.2, blue grass grown for seed 4.7, mixed grass pasture 32.9 and grove of oak woodland 35.2. Froese (1992b) employed emergence traps in a transect from a hedge into an arable field. The results indicated that *Triphleba distinguenda* was more abundant at, or in the vicinity of, the hedge. *T. nudipalpis* tended to be more abundant away from the hedge, especially when the numbers of *T. distinguenda* were relatively high. Likewise the numbers of *Metopina oligoneura* increased with the distance from the hedge, most being obtained more than 5 m from the hedge (Froese, 1992c).

Tropical forests appear to be especially rich in Phoridae. Quantitative data, however, are lacking, apart from the preliminary comparison between tropical forest and adjacent land cleared for agricultural purposes given in section 7.2.6(b).

7.5 SPECIAL MICROHABITATS AND RESOURCES

7.5.1 CARRION

The decomposition of an exposed carcass passes through four principal stages (microseral stages), which have been variously subdivided by different authors. The total process lasts from less than 1 to more than 9 months, depending on the environmental conditions (especially temperature). The four stages are:

1. FRESH – this stage seldom exceeds 2 days.
2. BLOAT – this stage begins when the build up of gases inside the carcass, as a result of anaerobic protein decomposition (putrefaction), causes bloating. It typically lasts 2–5 days in warm weather. In cold weather it may last several weeks, and may involve alternations of inflation and deflation as a result of thawing and freezing.
3. DECAY – this stage begins with the onset of aerobic protein decomposition (decay). It is frequently initiated by the burrowing activities of insect larvae allowing air to enter the carcass, and then accelerated by their digestive enzymes. During decay the carcass may lose up to 90% of its original weight.
4. DRY – this stage begins when no stage-three maggots remain on the carcass (apart from subsequent invasions of pieces of skin moistened by rain).

Adult Phoridae visit carcasses to feed (see Chapter 6) or, more rarely, to oviposit (see section 3.2.3, for carrion breeding species). Johnson (1975) sampled scuttle flies visiting mammal carcasses (squirrel, rabbit, cat and opossum), in Illinois, throughout all the micoseral stages. His results were as follows:

Stage	*Fresh*	*Bloat*	*Decay*	*Dry*
%	0	19	74	7

In Europe Lundt (1964) buried pieces of meat at different depths. He recorded adult Phoridae up to depths of 0.5 m as early as 4 days from the time of burial. The species recorded in his study included *Anevrina unispinosa, Conicera tibialis, Megaselia rufipes, Metopina heselhausi, Triphleba autumnalis, T. dudai*

Table 7.5 Phoridae recorded at buried meat by Lundt (1964)

Depth (m)	Phorid larvae		Adult flies			
			C. tibialis		*M. heselhausi*	
	No.	%	No.	%	No.	%
0.025	141	12.1	53	44.2	81	20.8
0.10	283	24.3	58	48.3	106	27.2
0.25	574	49.4	5	4.2	145	37.3
0.50	165	14.2	4	3.3	57	14.7
Totals	1163	100%	120	100%	389	100%

and *T. nudipalpis*. As well as adult flies, he recorded phorid larvae but did not identify them. Some figures are given in Table 7.5.

The relative sparsity of *Conicera tibialis* (the coffin fly), especially at the lower depths, probably reflects the fact that the buried meat was somewhat fresh for this species. Whether the *Metopina* was breeding in the meat or just feeding on it is not known. Also there must be an element of uncertainty over the specific identification (by Beyer), as much misidentification occurred in this genus prior to a revision of the European species (Disney, 1979j). *Triphleba nudipalpis* has been reared from corpses of white mice buried about 0.3 m deep (Disney, 1991d).

Forensic Entomology and Phoridae

Forensic entomological evidence is frequently employed in order to estimate the minimum time since the death of a murder victim. Hence, the faunal succession in human corpses is of especial interest in forensic entomology. The details of the phorid species breeding in human corpses are briefly considered in Chapter 3. Here we are concerned with general principles. These have been reviewed by a number of authors, and these reviews are collated by Smith (1986). A protocol for collecting entomological evidence from a corpse is also provided by Erzinçlioğlu (1989).

Exposed corpses are generally considered to experience eight waves of faunal invasions, in relation to the state of the corpse, over the first 3 years. It should be noted that 'exposed' corpses include those which are concealed but still exposed to the air. Thus murder victims have frequently been concealed under floor boards, but are still exposed to flies on the wing via the air vents. In buried corpses four waves of invasion have been proposed. Phoridae are considered characteristic of the fifth wave of invasion in exposed corpses, at 4–8 months since the onset; and characteristic of the third wave in buried corpses, at about 1 year from the time of burial. However, these generalizations probably never apply in their entirety in the case of the succession in a particular corpse. Indeed Erzinçlioğlu (1983b) concluded that 'it is impossible to generalize about the insect succession on corpses'. Furthermore there are records of phorids breeding in corpses exposed for far shorter periods of time. For example, unidentified scuttle fly pupae have been reported in a corpse only about 3 weeks old (Leclercq and Lerclercq, 1948) and other unidentified larvae and pupae were obtained in corpses in Belgium ranging in age from 3 to 4 months (M. Leclercq, *in litt.*). Some cases with named Phoridae (3.2) involve corpses less than 4 months old. Many variables, such as temperature, time of year, habitat context (especially the type of soil in the case of buried corpses), the weather pattern before and after the time of exposure, all conspire to render the succession in every corpse unique.

It seems unlikely that the refinement of interpretations of the significance of any particular phorid species being present in a corpse will come primarily from attempts to fine-tune the definition of the successive waves of faunal succession. It seems more likely that experimental studies on the development times of individual species (5.3), and the key ecological factors regulating their population dynamics and ecological niches, will provide more useful information. The detailed documentation of particular cases (3.2) is most likely to provide the empirical data which will lead to the elucidation of the critical phorid species and ecological factors for forensic entomology.

Webster's (1890) report of *Conicera* (presumably *C. tibialis*) breeding in a corpse of a man found to have been poisoned with arsenic (3.2.3) points to another use of Phoridae in forensic cases. Smith (1986) cites the analysis of other fly maggots, from a badly decomposed corpse, to identify the presence of a drug when insufficient of the original soft tissues of the corpse were available for study. Perhaps the larvae and pupae of *C. tibialis* could be used in similar cases.

Other uses of Phoridae in forensic cases are suggested by the report of a *Megaselia* species associated with cannabis in a security area in London (Smith, 1986). Determination of the species in such cases might provide clues as to the place of origin of an illegal importation of drugs, as reported by Crosby *et al.* (1986), who used beetles and ants to infer the origin of a batch of cannabis illegally imported into New Zealand.

7.5.2 FUNGI

Scuttle flies are a significant component of the Dipteran fauna of fungus sporophores (e.g. Hackman and Meinander, 1979). In a study in Switzerland Phoridae represented 7.1% of 6960 Diptera reared from 1246 fungi belonging to 121 species (Burla and Bächli, 1968). They came third in the list, after Mycetophilidae (51.2%) and Drosophilidae (35.3%). In a study in Hungary the Phoridae represented 6.7% of 2931 Diptera reared from a range of fungi (Dely-Draskovits and Babos, 1976). They came fifth in the list, after Mycetophilidae (33.1%), Drosophilidae (22.7%), Muscidae (14.1%) and Heleomyzidae (10.8%). The importance of the Phoridae varied with the family of fungi, ranging from 2.2% of the Diptera from Cortinariaceae (n = 406) to 24.1% reared from Agaricaceae (n = 212). The species recorded breeding in different species of fungi are listed in section 3.3.

Cultivated mushrooms are attacked by a number of species (3.3), which have therefore attracted much study. However, although more than a dozen genera of fungi are cultivated (or at least semi-cultivated) in the world (Chang, 1991), it is only the phorid pests of the two cultivated *Agaricus* species that have attracted much attention. Binns (1978) provided a review of outbreaks in England, and emphasized the change from *Megaselia nigra* being the principal pest to *M. halterata* becoming the main problem species. Part of the explanation lies in the fact that *M. nigra* will only oviposit in the light, so that the simple expedient of blacking out the doors and windows of mushroom houses has largely eliminated this species as a pest.

The population dynamics of *Megaselia halterata* in mushroom houses have been studied (Binns *et al.*, 1979). Comparable data for this species in the wild are lacking. The relationship between the larval density of *M. halterata* and yield reductions of cultivated mushrooms is discussed by White (1987). The results of a recent questionnaire survey of mushroom growers in Britain suggested that about half the farms had experienced infestations of phorids (White, 1993).

Clift (1979) has reported the occurrence of this species in Australia, mainly in abandoned railway tunnels used for the cultivation of mushrooms but also in growing sheds. However, its numbers are rarely sufficient to cause economic damage. It is now known from New Zealand as well (Disney, 1994b). In India the introduced *M. agarici* has become a pest of *Agaricus bisporus* (Sandhu and Bhattal, 1986; Disney, 1993d).

In America *Megaselia halterata* larvae attacking the mycelium have become the dominant pest in the USA and *M. agarici* larvae attacking the sporophore are now generally kept under control (White, 1985). *M. agarici* used to be the main phorid pest of cultivated mushrooms (Robinson, 1977). Out of 333 pest problems investigated by Pennsylvania State University in the period 1961–1970, 51 (15.3%) implicated *M. agarici* (as '*M. bovista*'). Only pygmy mites (18.6%) and sciarid midges (18.0%) were more frequently reported (Snetsinger, 1972).

The report of *Megaselia halterata* and *M. nigra* as pests of *Pleurotus sajor-caju* in India (Somchoudhury *et al.*, 1988) is almost certainly based on misidentifications. An undescribed species of *Megaselia* and *M. scalaris* are now known to attack this fungus. It is possible that *M. scalaris* is normally a secondary invader of mushrooms already spoiled by sciarid and phorid attack (Johal and Disney, in press).

7.5.3 NESTS AND BURROWS MADE BY BIRDS AND MAMMALS

In a survey of Diptera in the burrows of small mammals in Finland, Hackman (1963) recorded numbers of Phoridae, especially *Anevrina thoracica* and *Conicera floricola*. Baumann (1977b) carried out a similar survey in upland meadows (670–715 m altitude) in Central Europe. Phoridae were again one of the more abundant families, with *Conicera floricola* and *Metopina oligoneura* being the most frequent species identified. In the Hope Collections at Oxford there are specimens of *Anevrina curvinervis* from rabbit burrows. Jansson (1919) reported *Megaselia rufipes* and other *Megaselia* species in old squirrel nests. Brown (1992b) reported *Anevrina variabilis* in burrows of gophers (*Thomomys talpoides*) and Borgmeier (1963d) reported *A. sulcatifemur* in a gopher burrow. Brown also reported *A. spinipes* in the burrows of *Marmota monax* (Sciuridae).

E. B. Basden's extensive rearings of Diptera from bird and mammal nests in Britain did not show the Phoridae to be a significant component of the fauna (Rotheray, 1989). There are, however, a number of records in the literature (e.g. Collin, 1939). In some cases of rearings from bird nests these are from resources such as dead nestlings. The commonest such saprophage encountered in Britain is *Megaselia giraudii*. Lundbeck (1922) reared the same species from a chaffinch nest in Denmark. A specimen of the saprophage *Spiniphora bergenstammi*, along with its puparium, from a blackbird's nest (*Turdus merula*) is preserved in the Hope Collections (Oxford University). Colyer (unpublished notebooks) reared *Megaselia frontalis* from the nest of a swan (*Cygnus olor*). Armstrong's (1953) records of Phoridae in the nests of wrens (*Troglodytes troglodytes*) include the saprophage *Megaselia giraudii*, along with *M. aequalis, M. ciliata and M. minuta*. Hicks (1959) summarizes a number of records of these, mostly sporadic, occurrences of Phoridae in bird nests. In many cases these appear to be merely hibernation records for adult flies, that happen to be from bird nests (6.6). It is evident, however, that the nests of sand martins (*Riparia riparia*) frequently, and probably habitually, harbour Phoridae. The following have been recorded: *Conicera tibialis, Gymnoptera vitripennis, Megaselia rufipes* (Stadler, 1948) and *Diplonevra* sp. (Buttiker, 1969). Schmitz (1928f) also recorded *C. tibialis* from several old nests. It is probable that Phoridae habitually utilize nests located in burrows, in contrast to their sporadic occurrence in nests elsewhere.

It seems that burrows are more regularly utilized by Phoridae than are nests, and the available data suggest that burrows are of greater importance where the site is more exposed and the climate is more extreme.

The following species have been recorded in the nests of moles (*Talpa europaea*) in Europe: *Anevrina thoracica* (Malloch, 1908a; Lundbeck, 1922), *Anevrina urbana* (Falcoz, 1912b; Malloch, 1908a; Lundbeck, 1922), *Conicera floricola* (Schmitz, 1953a; Smith, 1977), *Megaselia pleuralis* (Colyer, in Disney, 1978a) and *Megaselia rufipes* (Falcoz, 1912a, b).

7.5.4 INSECT NESTS

The nests of termites and Hymenoptera are utilized by numerous Phoridae (see Chapters 3 and 6). Quantitative data, however, are sparse. Franssen (1933b) estimated that for around 20 000 termite eggs, in a nest of *Odontotermes javanicus* in Java, there were about 500 Termitoxeniinae eggs along with about 150 *Javanoxenia puntiventris* and 40 *Odontoxenia brevirostris* females. Biegel (1953) recorded the numbers of phorid larvae (of *Triphleba lugubris* plus *Gymnoptera vitripennis*) in the nests of wasps (*Vespula* spp.) as ranging from 56 to 21 933 (mean = 7513.5) per nest.

In their field studies of Phoridae utilizing the colonies of army ants Rettenmeyer and Akre (1968) estimated the numbers of females, most of which are flightless, running with ants during their nomadic phase. Most of the flies follow the tail-end of these emigration columns. In laboratory studies, however, the phorids (unlike some other myrmecophilous insects) did not follow ant-scent trails (Akre and Rettenmeyer, 1968). The reasons for this finding require investigation. In the field the same authors observed densities of 100 phorids per metre of trail, and they estimated that *Eciton vagans* and *Labidus praedator* were harbouring more than 4000 phorids per colony. The larval natural histories of these flies are considered in section 3.2.4(c).

7.6 DIURNAL PATTERNS

The diurnal patterns of activity by insects tend to be correlated with the daily fluctuations in temperature, light, and other environmental variables. These have been little studied in Phoridae. A few observations are summarized below.

In *Megaselia halterata* the temperature threshold for flight was reported to be 12–13° C, but most flies only took off when the air temperature had reached 15–16° C (Hussey, 1965). However, this species has subsequently been observed flying when there was an airfrost (Binns *et al.*, 1979). Clearly further studies are required. A characteristic curve for flight activity is depicted in Fig. 7.1. Flight activity was found to be curtailed by a light threshold, but wind and rain tended to curtail flight at any time. In mushroom cropping-houses flight can be initiated by turning on the lights. Girard *et al.* (1972) investigated the responses of this species (under the name '*M. bovista*': see Robinson, 1977) to light, temperature and gravity. The adults showed a significant tendency to move upwards and towards the light. Given a choice of temperature zones the majority preferred the range 15.6–25 °C, some favoured 25–28.3 °C and only a few 28.3–44.4 °C. Lewis and Taylor (1965) report two peaks of flight activity in this species. The mean flight times, in three experiments, were 10.43 hours and 15.03 hours (GMT), for both sexes together; 10.48 hours and 15.06 hours for males; and 10.51 hours and 14.50 hours for females.

In the tropics some Phoridae are evidently on the wing at night as much as they are by day (Young, 1986) (6.2). However, Orr (1992) found that *Neodohrniphora curvinervis* will normally only attack the foraging workers of its ant host in daylight.

Unique events in the life of a fly, such as pupation or adult eclosion, frequently occur at a time of day which is characteristic for a particular species. For example, mosquitoes (Culicidae) tend to pupate at a particular time of day. As pupal period is largely a function of temperature, the subsequent eclosion of adults may occur at any time of the day or night (Clements, 1963). In blackflies (Simuliidae) the same applies, but with the modification that if the time for adult eclosion falls in the night then the eclosion is delayed until morning (Disney, 1969). Studies of the timing of

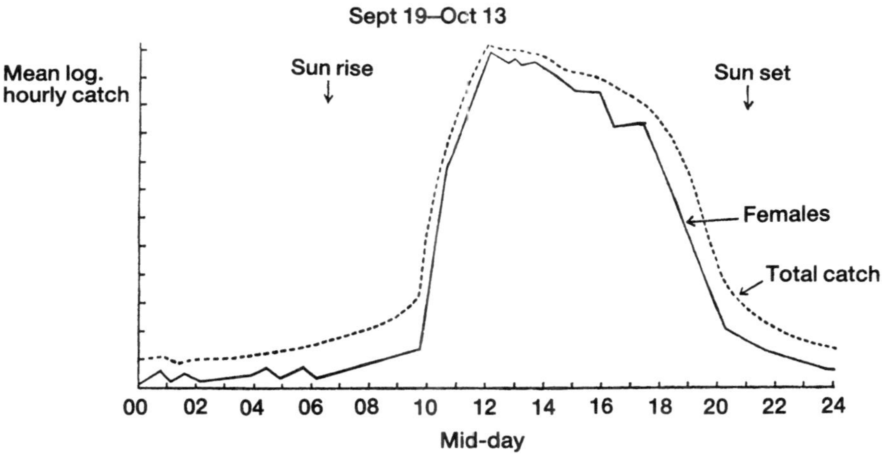

Figure 7.1 Diurnal flight activity of *Megaselia halterata*. (From Hussey, 1965.) Vertical axis is number of flies caught; horizontal axis is time of day (British Summer Time, i.e. GMT + 1 h).

such events in Phoridae are almost totally lacking. However, Franssen (1933b) reported that in *Clitelloxenia hemicyclia*, *Javanoxenia punctiventris* and *Odontoxenia brevirostris* the peak time of the hatching of the larvae from the eggs was 0600–0700 hours. With the relatively constant conditions prevailing in the fungus gardens inside the nests of their termite hosts, this might reflect a restriction of oviposition to a particular time of day. On the other hand the very long incubation period for the eggs of these species (5.3) would tend to suggest that this is not the explanation for this precise timing of the hatching of the eggs.

It has already been noted (2.5) that *Megaselia nigra* will only oviposit in the light. In Zimbabwe the arrival of mating pairs of *Termitophilomyia braunsi* at the nest mounds of their host termites is restricted to the morning between 0500 and 1000 hours, with the peak between 0630 and 0930 hours (Disney and Cumming, 1992).

7.7 PHENOLOGY

Detailed, quantified, studies of the seasonal fluctuations in the numbers of adult scuttle flies are lacking for most species. The factors regulating these phenological patterns are largely unexplored. While most studies are unlikely to reveal interesting patterns without identification to the species level, a few published data based on family-level identification only are not without interest. For example, Johnson's (1975) data for scuttle flies visiting carcasses in Illinois showed clear spring and autumn peaks:

Month	*Mar.*	*Apr.*	*May*	*Jun.*	*Jul.*	*Aug.*	*Sept.*	*Oct.*	*Nov.*
%	5	43	8	9	5	5	21	4	0

Table 7.6 Averaged phenology data for emergence traps operated in ten different pastures in England (Disney, *et al* , 1981)

Species	*n*	Percentages of totals for each species							
		16 Apr –6 May	9–29 May	28 May–17 Jun	9–29 Jul	1–20 Aug	11 Sept –1 Oct	1–20 Oct	24 Oct –13 Nov
Diplonevra nitidula	313	0	0 3	16 6	35 5	39 3	5 8	2 2	0 3
D pilosella	132	0	0	20 5	6 1	64 4	8 3	0 7	0
Megaselia aequalis	117	3 4	28 2	4 3	9 4	48 7	2 6	2 6	0 8
M brevicostalis	216	0 5	1 4	4 2	56 0	31 0	2 3	2 8	1 8
M ciliata	137	1 5	7 3	2 9	22 6	44 5	5 1	11 0	5 1
M pusilla	84	6 0	45 2	1 2	2 4	13 1	14 3	7 1	10 7
Metopina oligoneura	63	0	3 2	4 8	33 3	30 1	25 4	3 2	0
Triphleba nudipalpis	50	6 0	45 2	1 2	2 4	13 1	14 3	7 1	10 7

Table 7.7 Pitfall trap returns for moorland habitats in northern England in a normal year (1978) and a drought year (1976)

Species	Year	*n*	A	M	J	J	A	S	O	N
Diplonevra	1976	68	–	–	1 5	–	30 9	67 6	–	–
funebris	1978	31	–	–	9 7	58 0	19 4	12 9	–	–
Gymnophora	1976	12	–	–	–	–	–	100 0	–	–
healeyae	1978	32	–	–	–	84 4	15 6	–	–	–
Megaselia	1976	90	–	1 2	2 5	3 8	28 8	51 3	1 2	11 2
brevicostalis	1978	107	0 9	–	2 8	7 4	11 2	25 2	46 0	6 5
M flavicoxa	1976	57	–	–	–	5 3	80 7	14 0	–	–
	1978	163	–	–	–	1 8	34 6	63 6	–	–
M pleuralis	1976	133	–	1 5	6 8	44 3	15 8	23 3	1 5	6 8
	1978	1133	–	0 4	36 3	32 3	2 0	13 1	11 8	4 1
Triphleba	1976	8	–	12 5	–	–	12 5	25 0	–	50 0
nudipalpis	1978	44	–	4 6	9 1	13 6	13 6	6 8	50 0	2 3

Perhaps alternative sources of protein-rich meals are more frequently exploited in the summer months (6 3)

The emergence-trap data for species breeding in the soils of 10 English pasture soils (7 2 4(a)) allows averaged phenology patterns for some of the commoner species (Table 7 6)

In Germany Froese (1992b) reported a similar pattern for *Triphleba nudipalpis* His first peak of emergence in May yielded up to 35 flies m^{-2}, and his second peak in September gave a peak of up to 11 flies m^{-2} He also reported (Froese, 1992c) data for *Metopina oligoneura*, which show a first peak of emergence in April/May, sometimes a minor peak in June, a major peak in July or July/August and a peak in August/September

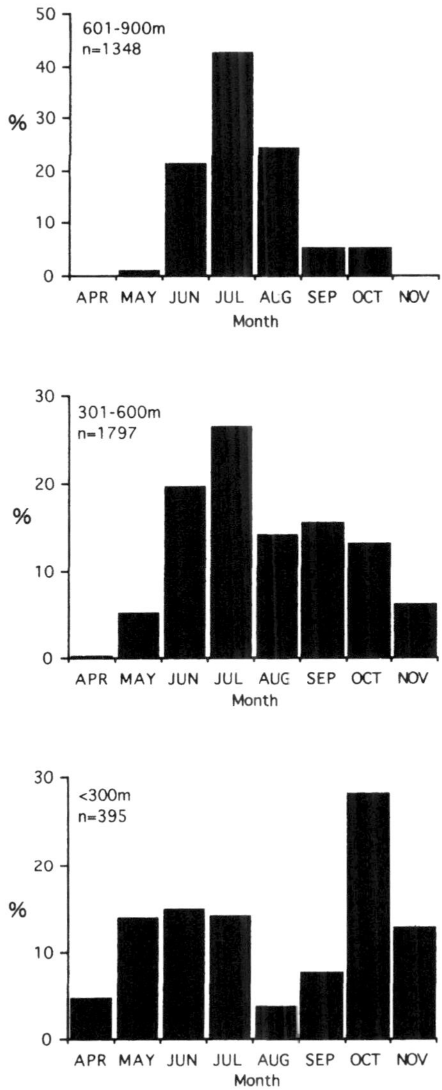

Figure 7.2 The phenology of *Megaselia longicostalis* in moorland sites in northern England in a normal year (1978) at different altitudes.

A survey of moorland sites in northern England (Disney, Coulson, *et al.*, 1981) yielded some quantitative data on phenology, and was carried out in two contrasting years. 1976 was a year of severe drought throughout the summer; 1978 was a more normal year. The data obtained for some of the common species are presented in Table 7.7. The figures are percentages of the total catch for the year, for the species in question.

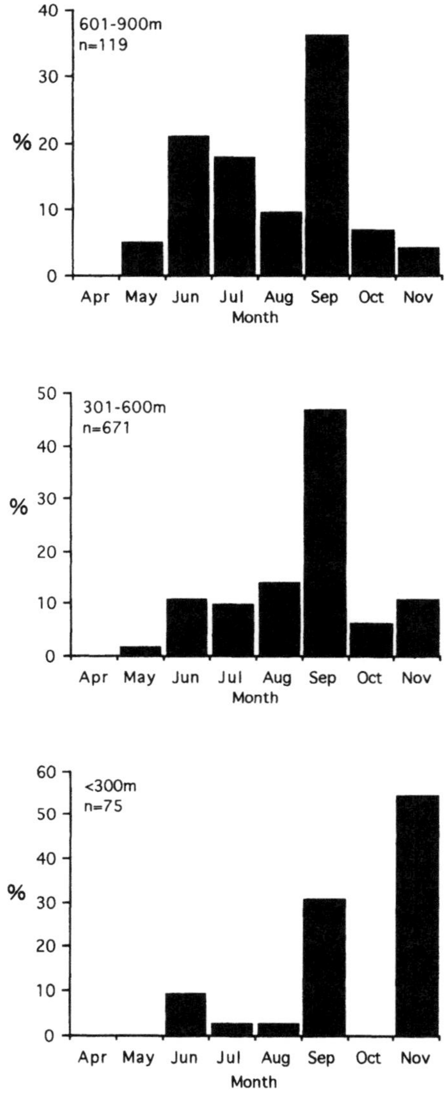

Figure 7.3 The phenology of *Megaselia longicostalis* in moorland sites in northern England in a drought year (1976) at different altitudes.

The data in Table 7.7 suggest that for most species the drought delayed the main peak of emergence. However, in *M. flavicoxa* and *M. brevicostalis* emergence peaked earlier in the drought year. The drought seems to have reduced the populations of most species. *Diplonevra funebris*, however, was more abundant in the drought year and the numbers of *M. brevicostalis* were unaffected.

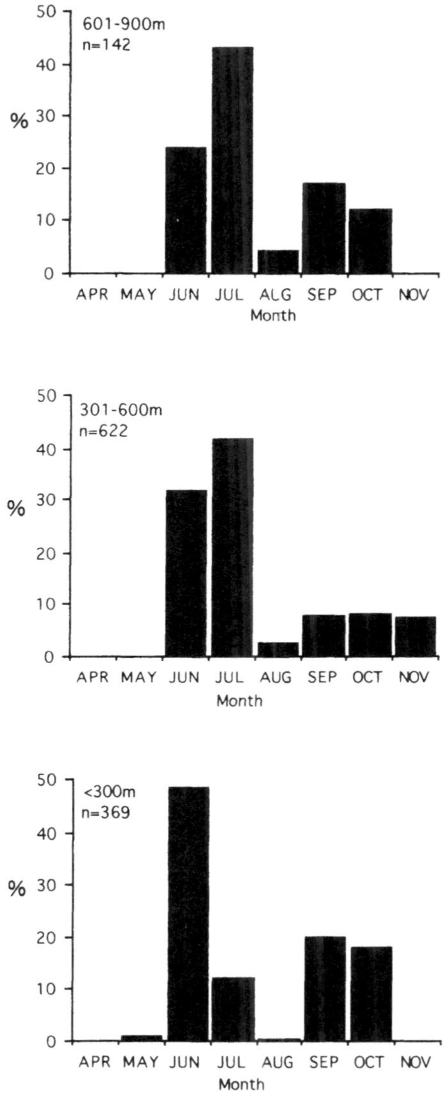

Figure 7.4 The phenology of *Megaselia pleuralis* in moorland sites in northern England in a normal year (1978) at different altitudes.

Sufficient data were obtained for *M. longicostalis* and *M. pleuralis* to permit examination of the effect of differences in altitude. These are presented in Figs 7.2–7.4.

The plots for 1978 (Figs 7.2 and 7.4) indicated a spring and an autumn peak of emergence for both species. With *M. longicostalis* at lower altitudes the spring peak of emergence started earlier and was not as sharp as it was at

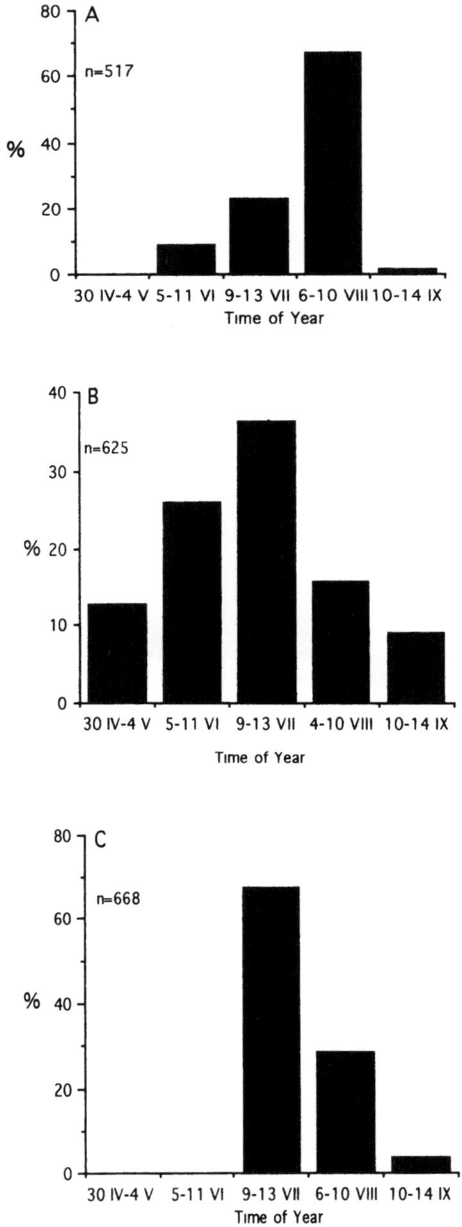

Figure 7.5 The phenology of *Megaselia pleuralis* near Corvallis, Oregon, in three lowland (<100 m) habitat types A = red clover seed crop, B = mixed grass pasture, C = an oak grove (woodland) (From Kamm *et al* , 1987)

higher altitudes. The autumn peak was marked at lower altitudes but not evident at higher altitudes. In the drought year 1976 (Fig. 7.3) the spring peak was greatly reduced at lower altitudes and there was a marked autumn peak at all altitudes. The actual numbers in the drought year (1976) were only a quarter of the 1978 levels. Thus the relative heights of these autumn peaks of *M. longicostalis* were essentially the result of the drastic reduction of the spring/summer peaks. In the case of *M. pleuralis* in 1978 (Fig. 7.4) the spring peaks were sharper, and the first started earlier and the second ended sooner, at the lower altitudes. At the highest altitudes the autumnal peak was cut short sooner than at intermediate altitudes. There is, therefore, an overall trend from being bivoltine at lower altitudes to being univoltine at higher altitudes. In the drought year (1976) the overall numbers were only about a tenth of those in the normal year (1978).

These differences in phenology patterns evidently reflect differences in climate. This suggests that the monitoring of these patterns at different altitudes might provide a sensitive biological indicator of global climatic change. The problem with some current attempts to use the presence or absence of indicator species of insects for this purpose is that it is never easy to prove an absence for a small insect species. Shifts in the altitudinal zone in which a species, such as *M. longicostalis*, changes from being bivoltine to univoltine might be easier to monitor and interpret.

Data are available on the phenology of the holarctic *Megaselia pleuralis* in a lowland (<100 m altitude) region in Oregon (Kamm *et al.*, 1987). The overall pattern for a pasture habitat (Fig. 7.5B) is comparable with the higher altitude data for England (Fig. 7.4), and an interesting finding was that the pattern was partly a function of habitat type (cf. Figs 7.5A,C). In the tropics the combination of short generation times (of a few weeks only, rather than months), a lack of marked seasonal changes in the climate, and the often restricted extent of torrential rain storms, means that phenological patterns for insects may vary over short distances. For example, simuliid blackflies may show different patterns for each tributary of a river system (Disney, 1976a). The evidence of habitat type affecting the phenology of a temperate-region scuttle fly merits further investigation.

In Finland Hackman (1963) trapped Phoridae in vole burrows. He obtained 96 specimens of *Anevrina thoracica*, distributed seasonally as follows:

	May	*Jun.*	*Jul.*	*Aug.*	*Sept.*	*Oct.*
%	16.7	47.9	28.1	7.3	0	0

Likewise Baumann's (1977b) data for Diptera from vole burrows, at 670–715 m altitude in central Europe, yields sufficient data (Table 7.8) to reveal phenology patterns for two species.

The pattern for *Metopina oligoneura* shows a much sharper late-summer peak, without the tail of autumn emergence found in the lowland pastures in England (see above). Weber and Prescher (1990) provide further phenological data on this and other species, from a reclaimed rubble dump in Germany, which both confirm and extend the findings of these other workers. Likewise Prescher (1992) provides further phenological data for

Table 7.8 Seasonal variation in catches of two phorid species in upland vole burrows in Central Europe

Species		Feb.	Mar.	Apr.	May	Jun.	Jul.	Aug.	Sept.	Oct.	*n*
Conicera floricola	%	–	–	–	33.8	5.2	10.4	37.6	13.0	–	77
Metopina oligoneura	%	–	–	–	12.4	9.3	76.3	2.0	–	–	97

phorids from arable fields in Germany (7.2.3(a)) and Weber (1989) provides data for *Megaselia rufipes* and *Triphleba antricola* in German caves.

Data for tropical species are largely lacking. Gregor (1977) gives some preliminary data for *Megaselia* species in Cuba.

7.8 COLD-ADAPTED ADULTS

It is evident from collecting data that the adults of some temperate-region species are only active in the winter months. This is especially characteristic of some *Triphleba* species. For example, in America Brues (1924c) collected numbers of both sexes of *T. microcephala* in winter, following a period of several weeks when the temperature had remained below 8 °C. I have collected *T. autumnalis* in January in England, including gravid females. Leclercq and Watrin (1973) reported *T. hyalinata* to be primarily on the wing in winter in Belgium. Some of the cave species (7.2.2) spend their entire lives at nearly constant, cool temperatures. For example, *T. aptina* larvae and adults have been reported from an Austrian cave whose temperature is always about 8 °C (Weissmair and Hauser, 1993). Many phorid species overwinter as hibernating adults (6.6).

7.9 NICHE SEPARATION AND COMPETITION

Competition is reduced when co-existing species exploit different resources, different parts of the same resource, the same resource in different microhabitats, or the same resource at different times of the day or the year. Many data in this chapter, along with data summarized in Chapters 3 and 6, provide examples of these mechanisms. Here attention is drawn to some particular cases of niche separation and also to evidence of severe competition when such mechanisms are lacking.

The larvae of *Megaselia halterata* and *M. nigra* both feed on mushrooms, especially *Agaricus campestris* and *A. bisporus*, at the same time. Whereas *M. halterata* larvae feed on the mycelium, the *M. nigra* larvae feed on the gills and cap tissues of the developing sporophores (Hussey, 1960). In the USA *M. longipennis* also largely restricts itself to feeding on the mycelium (Hussey, 1960), whereas *M. agarici* larvae attack the sporophores (White, 1985).

Megaselia scalaris coexisting with the larger *Fannia howardi* (Fanniidae), and competing with it for the same limited resource, seems to owe its success to its more rapid development. Pupation occurs 2 days earlier than that of *F. howardi* (Kneidel, 1985). Kneidel (1983) had already shown that *M. scalaris* survived better on slug carcasses (of *Limax maximus*) when it was the first colonizer, and that in these circumstances it had the greatest impact on other species of Diptera. Similarly the presence of the blowfly *Lucilia* (= *Phaenicia*) *coeruleiviridis* prevented the successful development of *Puliciphora occidentalis* in small mammal carcasses. Otherwise this scuttle fly successfully completes its development in such carcasses (Kneidel, 1984b). Similarly Beaver (1987) exposed dead snails (*Achatina fulica*) in Malaysia and reared a mean of 12.8 (range 1–61) individuals of a *Puliciphora* species. When, however, he exposed a snail which was protected from oviposition by larger flies and from predation by ants he reared 887 specimens. He also reported that most of the Phoridae (such as *Conicera formosensis*) oviposited from the first day onwards. However, *Spiniphora genitalis* will only oviposit from the second day of exposure onwards. The reason for this delay invites investigation.

Studies in England of *Phalacrotophora berolinensis* and *P. fasciata*, parasitizing the same species of ladybird (coccinellid) pupae in the same localities at the same time (3.5.6(b)), produced no evidence of niche separation (Disney *et al.*, 1994). In the field it is probable that different species are frequently able to exploit the same resource simultaneously, if the resource in question is patchily distributed and the eggs of each of the competing species are distributed in an aggregated manner. Indeed studies on *Drosophila* have shown that, in these circumstances, there may be little evidence of traditional resource partitioning or of competition leading to the extinction of populations of inferior competitors (Shorrocks, 1991).

The evolutionary trends of accelerated larval development and reduction in the size of adults may be, in part, adaptations allowing phorids to exploit small, patchily distributed, ephemeral resources as opposed to more obvious resources such as large carcasses. Certainly the Phoridae provide a particularly challenging opportunity to the ecologist interested in resource partitioning or the lack of it. The sheer diversity of phorids, especially in the giant genus *Megaselia*, would suggest that every possible theoretical scenario is likely to operate in the field, the particular model that is applicable depending upon the particular ecological situation prevailing. Until the basic natural histories of co-existing species are known in detail, we can only speculate.

7.10 DISPERSAL

Detailed studies of dispersal by Phoridae have yet to be carried out. In America Felt (1935) procured phorids on the roof of a building 40 m above street level and on top of a tower, 18 m in height, situated on a hill of nearly 600 m altitude. Using sticky screens mounted on an aeroplane, Collins and Baker (1934) caught Phoridae at an altitude of more than 300 m above Massachusetts. In Yorkshire Hardy and Milne (1938) procured *Megaselia* spp. in nets attached to kites, up to 150 m (or more) from the ground. In his classic study, using insect traps mounted on a small aeroplane, Glick (1939) collected

121 Phoridae over the Mississippi Valley (USA). Fourteen of these were collected on night flights, and 45 (37.2%) belonged to the giant genus *Megaselia*. He collected 42 below 100 m altitude, 64 between 100 and 1000 m, and 15 specimens above 1000 m altitude. Altogether the Phoridae represented 1.1% of the Diptera (n = 11 304), and this percentage was the same for both day and night catches. The Diptera were the largest group in his samples (being 35.8% of all the insects caught; and being almost three times as numerous as Coleoptera, the beetles, in second place). In England Lack and Owen (1955) recorded a mean of 0.8 phorids per meal (n = 66 meals) in swifts (*Apus apus*), which confirms that scuttle flies are commonly encountered some distance above the ground.

Some scuttle flies are clearly transported long distances by the wind, as evidenced by their occurrence on high-altitude snowpatches. For example, a collection of 1678 Diptera from such snowpatches in Alaska included 0.42% Phoridae, immobilized by the cold on the surface of the snow (Edwards and Banko, 1976). It is evident, however, that most Phoridae tend to stay close to the ground in exposed situations. For example, Lewis (1967) investigated the horizontal and vertical distribution of insects near artificial windbreaks. Phoridae were more abundant in the lee of a fence 2.4 m high than in an exposed situation. In the latter situation the numbers rapidly declined above 1 m from the ground, but in the sheltered areas the numbers peaked just below 2 m.

Wind is not the only agency of passive transport. Lundbeck (1922) recorded five species reared from pupae found in flood refuse in Denmark. Boness (1975) sampled flood refuse deposited beside rivers in Germany, and regularly obtained viable pupae of Phoridae.

I have regularly observed Phoridae being transported in vehicles, from cars to trains. Lever (1944) recorded *Megaselia scalaris* in an aeroplane just arrived in Fiji from the New Hebrides, 800 km to the West. Britain has acquired a number of exotic species, including *Chonocephalus heymonsi*, *Dohrniphora cornuta*, *Megaselia scalaris* and *Puliciphora borinquenensis*, through the agency of man (Disney, 1983a, 1989a). In the case of *D. cornuta* and *M. scalaris* I have been sent several samples of each found in various cargoes unloaded from ships. An even more intriguing case concerns the details of a specimen from Spain, sent to me by Javier Blasco-Zumeta. His bird-ringer friend R. López reported that small flies frequently emerged from among the feathers of his trapped birds while he was handling them. The flies tend to fly away rapidly. However, he managed to capture one which had emerged from the plumage of a rock sparrow (*Passer* (*Petronia*) *petronia*). It proved to be a female *M. scalaris*. Clearly this observation needs following up.

Wingless females are transported by the males while in copula. Miller (1979a) recorded males of *Puliciphora borinquenensis* transporting females at least 3.6 m in a single flight, these airlifts only occurring in the light. The effectiveness of such airlift-dispersal is shown by the studies of Wilton (1961), who examined the colonization of 68 refuse containers by flies in Honolulu. *Megaselia scalaris* successfully colonized 51.6% of the containers, *P. borinquenensis* reached 50.0% and *Dohrniphora cornuta* 17.6%. Thus *P. borinquenensis* was evidently no less successful at locating containers than the two species with fully-winged females.

Females which are fully winged may also be transported by the males, as reported for *Megaselia scalaris* by Miller (1979a).

Data on sex ratios suggest that in many species the males are more prone to dispersal than the females. Thus catches of *Megaselia halterata* show a remarkable predominance of males away from mushroom cropping houses. Hussey (1965) reported a ratio of 1.06:1 (n = 42 000) in the cropping houses, but 62:1 (n = 68 000) outside the cropping houses, in suction trap returns. He also reported a swarm in a house to be 87:1 (n = 200) and a sticky-trap catch in a field to be 27.9:1 (n = 433).

In emergence-trap returns in Germany the ratio of males to every female caught was 5.05:1 (n = 121) for *Gymnophora arcuata* and 13.9:1 (n = 401) for *G. quartomollis* (Baumann, 1979). Likewise Froese (1982b) reported *Triphleba nudipalpis* males to be considerably more abundant in emergence traps than were the females.

The figures above contrast with the sex ratios of flies caught at the larval resource. Thus the ratio for *Diplonevra mortimeri* males at exposures of termites, to which the females are attracted for oviposition (2.5), is only 0.20:1 (n = 229), and 0.03:1 (n = 99) for *D. watsoni* (Disney, 1986d). In *D. mortimeri* the female has somewhat reduced wings, and so the higher ratio of males for this species may be because some females are being transported to the scene by the males. The ratio of males for every female for the ant-parasitoid *Megaselia sembeli* caught at exposures of its host is 0.23:1 (n = 89) (Disney, 1986b).

The hibernating aggregations of some species tend to be dominated by females. Thus Scott (1951) reported that 'swarms' of *Megaselia meconicera* in houses in October, in England, were exclusively females. However, both males and females were recorded in March.

7.11 AGRICULTURAL AND MEDICAL SIGNIFICANCE

Economic damage to crops caused by Phoridae has mainly been reported for cultivated mushrooms. Some of the species known to attack crop plants may also cause economic damage. For example, in Texas the ears of green corn (maize) have been reported suffering economic damage due to attack by larvae of *Megaselia scalaris* (Walter and Wene, 1951). Phoridae are reported as economically significant pests of bee hives, especially in the Neotropical Region. These cases are detailed above in the relevant sections (3.2.5, 3.4 and 7.5.2). The following review is restricted to phorids and disease.

7.11.1 PHORIDAE AND DISEASES OF CROPS

Chickpeas (*Cicer arietinum*) are attacked by the rust fungus *Uromyces ciceri arietina*. Both sexes of the leaf-mining pest of chickpeas, *Megaselia arietina*, pick up the spores of this fungus on the labella of the proboscis. In addition the females ingest spores, taking them into the crop (Čakar and Disney,

1991). It seems highly probable, therefore, that this fungus is transmitted from plant to plant by this fly species. Indeed *M. arietina* possibly only mines leaves already attacked by *Uromyces* (3.4).

The fungus disease of cultivated mushrooms (*Agaricus bisporus*) caused by the parasitic fungus '*Verticillium fungicola*' (previously called *V. malthousei*, but probably in reality the anamorph state of a *Cordyceps* species) is readily transmitted in laboratory cultures by *Megaselia halterata* (Cross and Jacobs, 1968; White, 1981). Indeed *M. halterata* may cause a greater economic threat as a vector of the sticky spores of *V. fungicola* than by the damage to the mushroom mycelium caused by its larvae (White, 1987). It is perhaps worth noting here that the anamorph states of a *Cordyceps* species, or a related genus, have been recorded as internal parasites of adult phorids in Zimbabwe (Disney, 1991a; see 6.8.2(c)). Clearly the interactions of these Clavicapitales fungi with Phoridae require further investigation.

The results of a recent survey in Britain suggest that there may be a correlation between phorid infestation and the mushroom disease caused by the growers' so-called '*Dactylium* fungus' (White, 1993). This disease is normally caused by *Hypomyces rosellus* (whose conidial stage is called *Cladobotryum dendroides*, which was previously called by the synonym *Dactylium dendroides*), but is occasionally due to the related *H. aurenteus*.

7.11.2 PHORIDAE AND DISEASES OF PEOPLE AND LIVESTOCK

Apart from cases of myiasis in man and his livestock (3.5.7) phorids have been implicated both in the transmission of disease and more directly in causing disease.

Megaselia agarici sometimes occurs in large numbers in the vicinity of mushroom farms. Sensitization to masses of flies may cause or aggravate cases of bronchial asthma (Kern, 1938; Trüitt, 1951).

Roberg (1915) showed, by means of laboratory experiments, that *Megaselia scalaris* can transmit the cholera *Vibrio* (*V. cholerae*). Flies mechanically transmit the bacteria from faeces to food and he showed that '*V. cholerae*' will pass from the mature larvae through the pupal stage to the adult fly. While there must be some doubt, in the light of our present knowledge of *V. cholerae*, that Roberg's bacterium in the latter experiment was indeed this species, the survival of a bacterium through the metamorphosis of the fly is not without significance. Furthermore the flies were able to escape through eight layers of stretched gauze (of 8 meshes per centimetre). These laboratory findings need to be set alongside the unhygienic habits of this fly outside the laboratory. Thus I have been sent a series of this species (by Dr Brian Laurence) from a population breeding in dry-composting toilets in Dar-es-Salaam, Tanzania. Another sample I was asked to examine was from a population infesting sanitary-towel disposal bins in Barbados. Other samples of this species, along with *Puliciphora borinquenensis*, have been sent to me from populations breeding in laboratory bacteria cultures, in some cases from blood agar plates before inoculation with bacteria. I have examined a sample of *Dohrniphora cornuta* from an infestation of a supermarket in England. Almost certainly these flies originated from a larval population in a trickling-filter sewage bed nearby. The species has been reported breeding in sewage beds in North America (Kloter *et al.*, 1977).

Saprophage species in general are liable to invade food products. Discarded food provides an opportunity for these species to oviposit. Even the dregs of milk left in discarded bottles have been reported as a larval pabulum for *Megaselia giraudii, M. rufipes* and *Spiniphora bergenstammi* (Oldroyd, 1964; Disney, 1979f, 1983a). More serious is the invasion of food before it is discarded, especially before it is purchased by the customer. I have examined samples of *Dohrniphora cornuta* from a packet of bacon and from a packet of flour, both purchased in England. Barnes (1990a) reported the same species in barbecue sauce imported into America from China. I have examined samples of *Megaselia rufipes* from a freshly-purchased oven-ready rice-based meal and *M. scalaris* from packets of food shipped to England, as follows: in vermicelli from Singapore, spaghetti from Italy, dates from Iraq and soya flour from Singapore. I have also seen a sample of this species breeding in newly-purchased cheese in Indonesia and another from dried fish in Malawi. The involvement of these phorids in the transmission of alimentary infections of man is a probability, and should not be dismissed out of hand. The species reported from these food contamination cases are also reported breeding in a range of dung types (3.2.2).

7.11.3 PHORIDAE AS AGENTS OF BIOLOGICAL CONTROL

Weber (1972) suggested that parasitoid phorids may be helpful in decreasing populations of the leaf cutter ants (*Acromyrmex* and *Atta*), which cause considerable economic damage to some crops. Clearly the data on larval biology (summarized in Chapter 3) indicate that some phorids are strong candidates for being useful agents for the biological control of pests. In particular, species attacking slug eggs (3.5.2), larval craneflies (3.5.6(c)), scale insects (3.5.6(d)), ants (3.5.6(e)), termites (3.5.6(f)) or moth larvae (3.5.6(g)) merit further investigation. In the case of ants, e.g. the fire-ant problem in the USA, some authors have doubted whether mortality due to Phoridae could ever reach a high enough level to make it worth trying to utilize these phorids as control agents. However, Feener and Brown (1992) have shown that the mere presence of phorids significantly disrupts the rate of food retrieval by the ants. Orr (1992) has reported significant reduction of daytime foraging by leaf-cutter ants in the presence of attacking phorids and also an influence on the caste size of the foraging ants. The effects of such disruption need to be fully explored in any evaluation of these phorids as potential biological control agents.

The effectiveness of phorids as potential biological control agents is indicated by an unfortunate experience in New Caledonia. The ladybird beetle *Olla v-nigrum* was introduced as a biological control agent against a psyllid pest, *Heteropsylla cubana*, of the important fodder crop *Leucaena leucocephala*. Unfortunately the ladybird pupae experienced attack by an indigenous parasitoid phorid, *Phalacrotophora quadrimaculata*, at rates ranging from 0.15 to 0.79 (Chazeau *et al.*, 1989; Disney and Chazeau, 1990). Several phorid parasitoids of ladybird pupae, although apparently restricted to this family of beetles, are evidently not very host specific (3.5.6(b)).

7.12 CONSERVATION EVALUATION USING PHORIDAE

The prevalence of scuttle flies and their biological diversity make them an excellent group for the evaluation of the faunal diversity of different sites, as a measure of conservation value in terms of species richness (diversity). The evaluation of sites for this purpose, using quantitative sample surveys, has been discussed elsewhere (Disney, 1986g).

Preliminary indications of the species richness of individual sites are available for a few sites only. A limited survey for Hayley Wood (Cambridgeshire, England) produced a list of 82 species (Disney, 1987f, 1988e). I have recorded more than 60 species in my suburban garden in Cambridge and more than 120 species on the upland nature reserve at Malham Tarn in the Yorkshire Pennines (England).

The enormous ignorance of the phorid faunas of different regions and habitats, along with the lack of an agreed phylogenetic classification, precludes the employment of other measures of conservation value, as discussed by Vane-Wright (1992). However, we know enough of the diversity of phorid natural history to suggest that simple measures of species richness of the phorid faunas of different sites are likely to be useful indicators of overall conservation value. It could be argued that a weighting factor for species with known, or inferred, parasitoid or predatory larvae might make such measures an even better indicator of overall conservation value; but the mass of *Megaselia* species whose larval habits are unknown would be likely to reduce the value of such an exercise. Suffice it to say that a site or habitat type shown to have a high score for Phoridae, in terms of species richness, is likely to be of high conservation value in terms of numerous other taxa.

8

Identification

Concepts differ from facts by continuing to change over time (Mayr, 1991)

8.1 INTRODUCTION

The principal impediment to the study of the natural history of scuttle flies has been the difficulty of identifying specimens with confidence. For several regions the lack of a key to genera means it is difficult to even start. The use of a generic key for another region can leave the user wondering whether an error has been made at an earlier stage. The aim of this chapter, therefore, is to provide a key to the genera of the world currently recognised in the literature and to indicate which keys or papers are required for identification to the species level.

8.1.1 SYSTEMATICS

The most recent key to the genera of the world is that by Schmitz (1929g). This dealt with nearly 100 genera. Section 8.3 lists 229 genera, allowing for some reinterpretations proposed below. Not all of these genera will prove to be valid. Some will prove to be the 'missing' sex of another genus currently only known in one sex. More than 50% of the described genera are still only known in one sex. It seems likely that some modern molecular techniques could be used to establish the conspecificity of sexually dimorphic species whose two sexes are currently assigned to separate genera, each of which is currently considered to be only known in one sex. Some genera will disappear as they are amalgamated with others, following the description of new species which erode the gaps between two genera. Others will be split up as it is discovered that they are polyphyletic. New genera will continue to be described, particularly as tropical faunas become better known.

Implicit in these few comments is the recognition that a genus embodies a set of opinions about the affinities of species and the status of the clusters of species we have constructed to reflect those opinions. The process of clustering related species, **systematization**, is in principle a process of scientific discovery through the erection of testable hypotheses. The assignment of a rank (genus, subgenus or whatever) to a cluster of species, **categorization**, owes as much to historical accident and prejudice as to scientific insight. A **classification**, therefore, is partly a reflection of the results of scientific discovery and partly the result of arbitrary decisions. The latter introduce an element of historical constraint, which needs to be respected in the interests of nomenclatural stability. It is too often forgotten that a classification is meant to be a system of reference to aid communication. Continual changes in names, or the concepts signified by them, impair effective communication.

I conclude that even if we were confident that all species had been correctly assigned to monophyletic clusters, the status of any such monophylum is to a large degree a matter of opinion. My own prejudice is in favour of minimizing unnecessary nomenclatural changes. Thus I will change a name if I discover evidence that a species has been assigned to the wrong genus. I object, however, to the practice of raising subgenera to the generic level in a futile attempt to render different genera 'equivalent' (whatever that might mean – see Disney, 1983d). Likewise I recognize no obligation to reduce a genus to the subgeneric level because it is found to be a distinct clade within a larger monophyletic 'genus'. For example, the genera *Gymnophora* and *Woodiphora* are essentially distinct clades within a monophylum comprising these two genera plus the genus *Megaselia* (Disney, 1989f). The purist will argue that this means the 'genus' *Megaselia* is, therefore, an incomplete monophylum (a paraphyletic taxon) and that this is unacceptable. The solution, it is argued, is to resolve the paraphyly either by reducing the excluded 'genera' to subgenera or by raising the subgenera of *Megaselia* to generic level. If we were in the position to adopt the second option (which we are not, in our present state of knowledge), the result would be to change the names of over 1300 species at a stroke. If we pursued the first option, then all the species of *Gymnophora* and *Woodiphora* would have to change to become species of *Megaselia*. It is for this reason that I have proposed (Disney, 1989j) that we accept that we will be stuck with numerous unresolved paraphyletic genera as far as we can see into the future. If we wish to indicate that we have identified cases of this, as with *Megaselia*, we can do so by calling these paraphyletic genera 'Paragenera'. In these cases the first word of the species binomial, e.g. *Megaselia* in the case of *Megaselia scalaris*, is then a paragenus rather than a genus in the narrow cladistic sense. In the interests of nomenclatural stability I make no apologies for this conscious adoption of non-equivalent notions of genera being applied to different clusters of phorid species. I repeat that a major function of a formal classification is as a system for the ordering and retrieval of biological data. The purist's abhorrence of paragenera is a recipe for continuous name changing, to the detriment of unambiguous communication.

The notion of a genus may be contentious, but the suprageneric classification is even more so. In section 8.4 I indicate the sort of problems currently being debated. Because there is, as yet, no consensus I offer no key to subfamilies and tribes, but go straight into keys to genera, one for males and one for females. These keys are followed by a guide to the literature for each genus. A section on fossil Phoridae then follows.

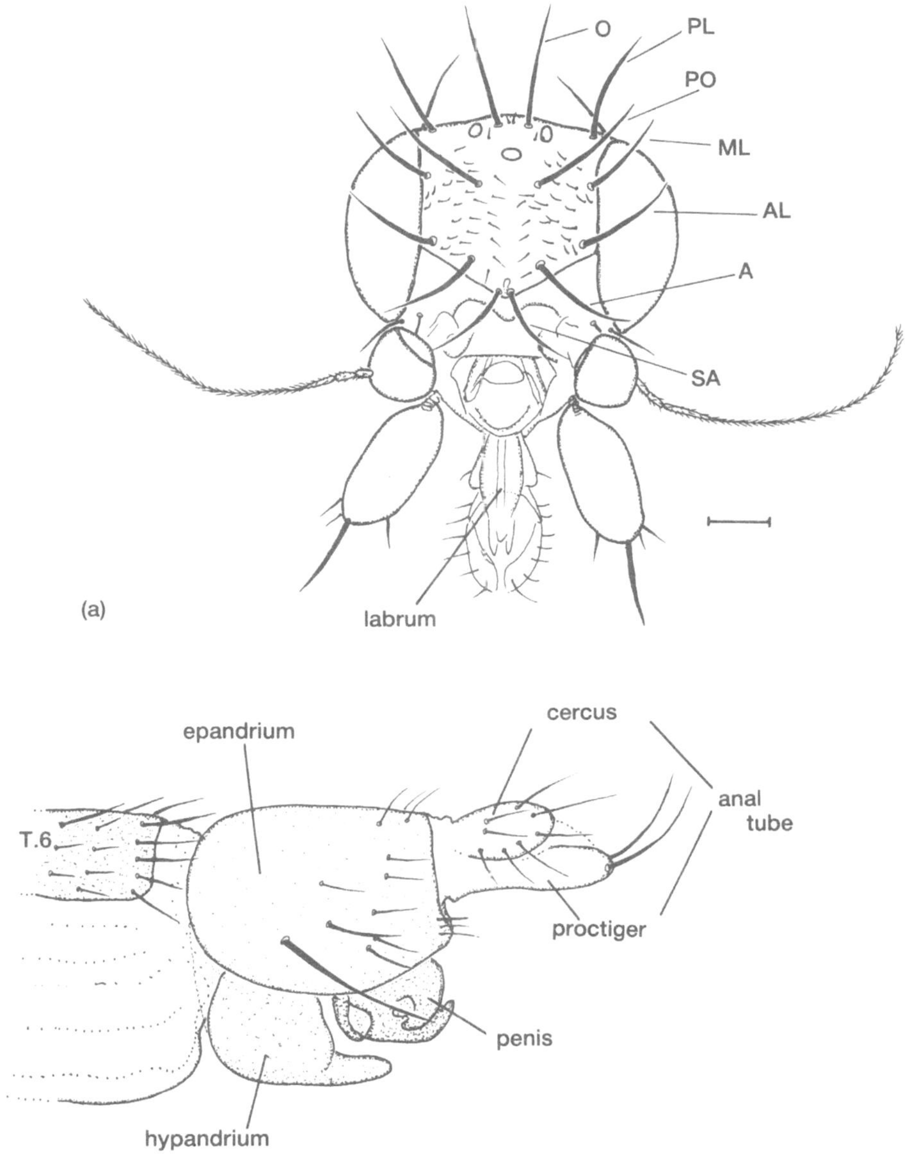

Figure 8.1 (a) Frontal view of head of *Triphleba nudipalpis* male, showing bristles (O = ocellars, PL = postero-lateral, PO = pre-ocellar, ML = medio-lateral, AL = antero-lateral, A = antial, SA = supra-antennal). (b) Side view of male hypopygium of a generalized *Megaselia*-type (T6 = abdominal tergite 6). *(Scale bars = 0.1 mm, as in all other figures.)*

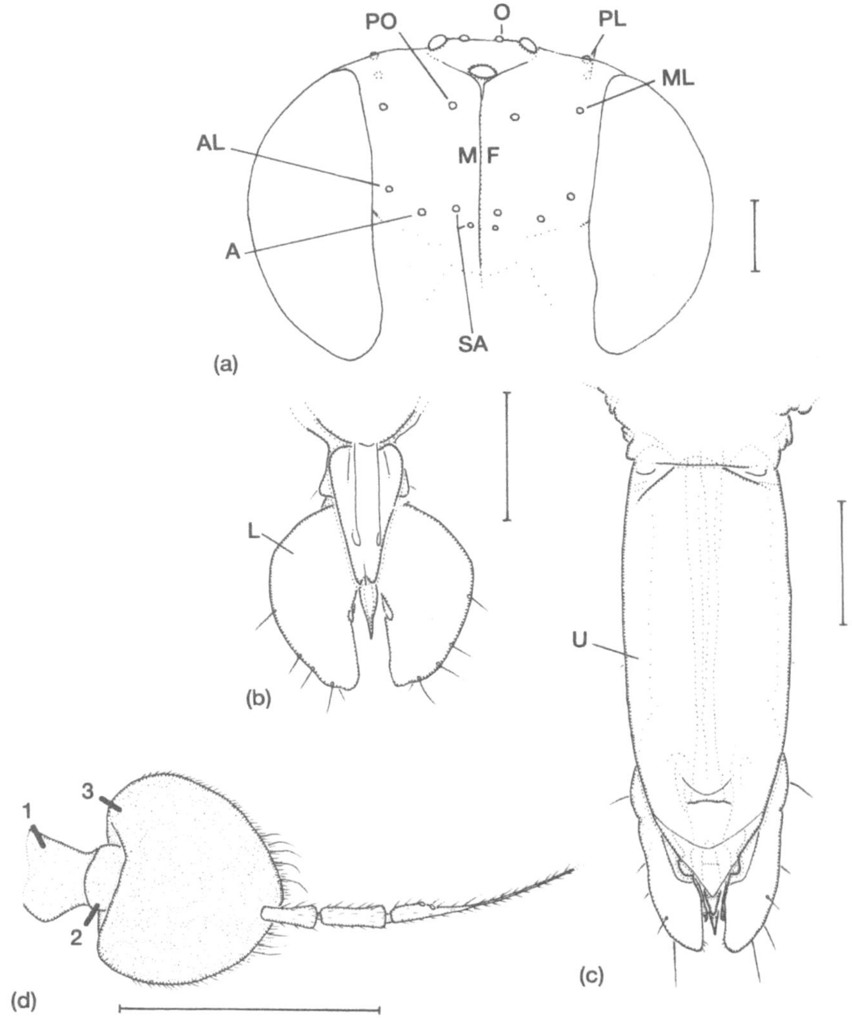

Figure 8.2 Details of head. (a) Frons of *Megaselia deningi* male, showing median furrow (MF) and bristle sockets, labelled as in Fig. 8.1(a). (b) and (c) Proboscis of male (b) and female (c) of *M. chlumetiae* (L = labellum, U = labrum). (d) Antenna of male *M. microcurtineura* (1–3 = segments 1–3).

8.1.2 MORPHOLOGY

The principal details of the morphology of a scuttle fly are indicated in Figs 8.1–8.15. The **head** is typically as in Fig. 8.1(a). In many genera a median furrow (MF in Fig. 8.2a) runs from the anterior ocellus to the front margin. The 'standard' set of bristles on the frons is labelled in Fig. 8.1(a). Bristles are liable to be knocked off, but their positions are always indicated by their basal sockets (Fig. 8.2a). In many genera a pair of the 'standard' bristles may be absent. Sometimes there are additional bristles, especially more than four supra-antennals. The proboscis is frequently sexually dimorphic (e.g. Figs 8.2b,c), mainly in terms of the breadths and lengths of the labrum and labella. The antennae (Fig. 8.2d) typically comprise three basal segments and a three-segmented arista, but the arista may have a reduced number of segments or be absent. The first (basal) segment (the scape) (1 in Fig. 8.2d) is fully exposed; the second (the pedicel, labelled 2) is largely enveloped by the enlarged third segment (the first flagellar segment, labelled 3). The third segment may be elongated (e.g. Fig. 8.31a).

The **thorax** viewed from above mainly comprises the enlarged scutum (mesonotum) and the scutellum behind. In flightless females the scutellum is frequently reduced or absent. Details of the sides of the thorax are useful in both classification and identification. The principal features are labelled in Fig. 8.3. There is sometimes a gland opening at the front end of the notopleuron (the edge of the scutum bordering the mesopleuron), which may be partly or completely concealed in a notopleural cleft (Fig. 8.28b and 8.93b,c).

The **legs** are typical in having a coxa, trochanter, femur, tibia and tarsus, the latter normally having five segments (Fig. 8.4a). Sometimes the front tarsus has only four or two segments. The fifth tarsal segment may be longer than the fourth, and is sometimes tapered (e.g. Fig. 8.4b). The presence of isolated bristles on the tibiae (Fig. 8.4c) is a useful character. Likewise the presence of, usually near-dorsal, longitudinal hair palisades (Figs 8.4d,e), as opposed to spaced spines (e.g. Fig. 8.4c), is a useful character.

The **wings** have a surprisingly uniform venation within the family. Despite this they exhibit a number of variations useful in identification. The anterior series of veins, the so-called 'thick veins', comprise the following. The marginal costal vein (C in Fig. 8.5b) bears costal cilia (c in Fig. 8.5a), usually in two rows. In published descriptions lengths of costal cilia refer to those on the anterior row of costal section 3. Typically the costa extends about half the wing length (measured from the level of the basal bristle – B in Fig. 8.5b – along the mid line to the wing tip). It may be longer or shorter. In the traditional phorid literature the Costal Index is much used. It is the length of the costa (measured from the basal bristle) divided by the wing length. The subcosta may extend to vein 1 and be fused to it (S in Fig. 8.5b), be free (e.g. Fig. 8.5a) or be absent. Vein 1 (R1) occasionally has hairs along its upper face. Vein Rs (R in Fig. 8.5b) may also have hairs along its upper face, but typically has only one at the base (e.g. Fig. 8.5b) or none. It frequently divides into vein 2 (R2+3) and vein 3 (R4+5). When vein 2 is missing, the single vein Rs/R4+5 is still referred to as vein 3, and is said to be unforked. In the traditional literature the costal ratios are much used. These are the ratios of the lengths of the costal sections 1 to 3, or 1 to 2 only if vein 3 is unforked, to each other (S1, S2 and S3 in Fig. 8.5b). They are best expressed by dividing the lengths of 1 and 2 by the length of 3, so that section 3 always scores as 1.

The **abdomen** typically has tergites on the first six segments, and typically no sternites. Rarely there is a sternite on segment 6 only. Even more rarely

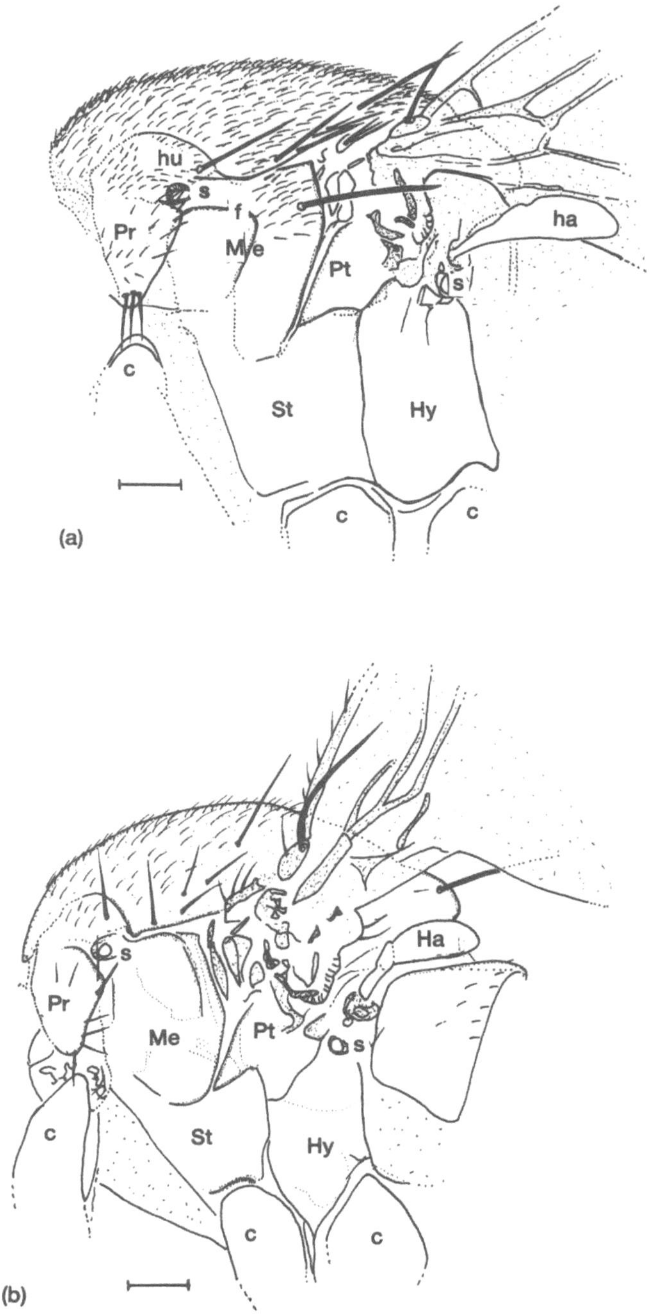

Figure 8.3 Left sides of thorax of *Megaselia ciliata* (a) and *Conicera dauci* (b). s = spiracle, Pr = propleuron, Me = mesopleuron, f = mesopleural furrow, hu = humerus, Pt = pteropleuron, St = sternopleuron, Hy = hypopleuron, ha = haltere, c = coxa.

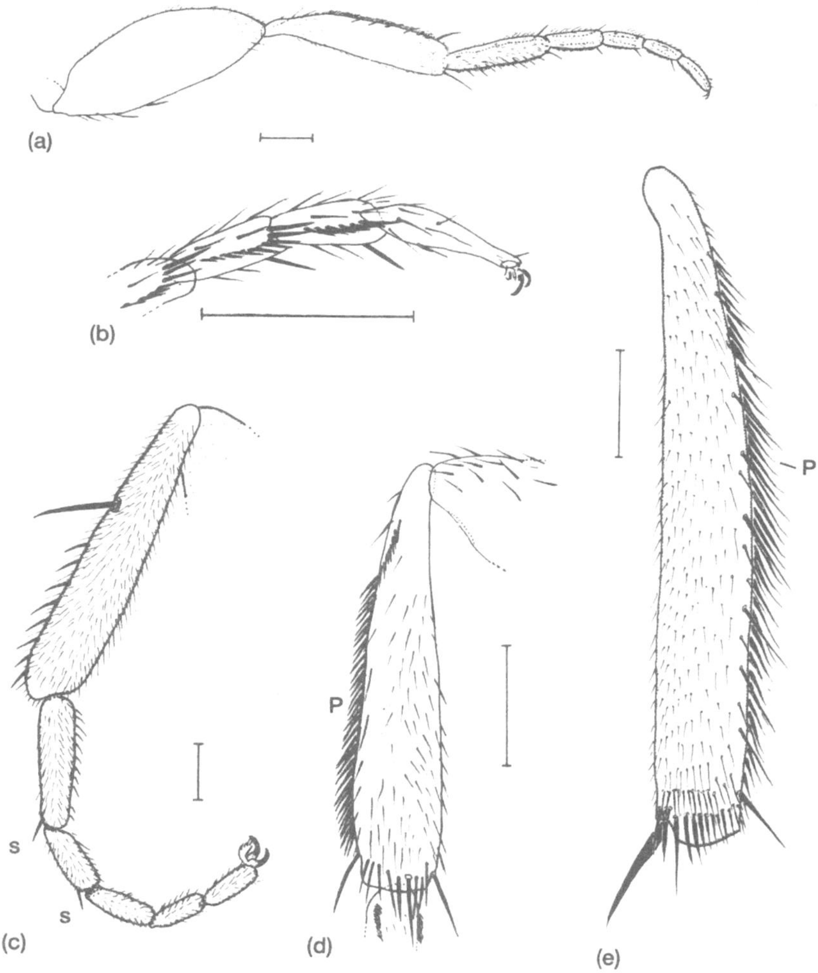

Figure 8.4 Parts of legs. (a) Hind leg of *Megaselia beaveri* female, omitting coxa and base of trochanter. (b) *M. beaveri* female, last three tarsal segments of middle leg. (c) Front tibia and tarsus of *Spiniphora bergenstammi*, showing an isolated bristle and spaced spines on tibia and single spines (s) at tips of first two tarsal segments. (d) and (e) posterior faces of hind tibiae bearing a near-dorsal hair palisade (P), of *Pseudacteon brevicauda* (d) and *Megaselia biarticulata* (e).

there is a sternite on segment 4 and vestiges on segment 3. Tergite 1 is typically shorter than 2 and frequently the tergites are progressively narrower from 2 onwards (e.g. Figs 8.6a and 8.7c). In females the tergites of the first six segments may be reduced in both size and number. The terminal segments of the abdomen are discussed in section 8.2(A) below.

The terms used for the parts are those most widely used in the literature on the family. In recent years there has been a tendency to modify the

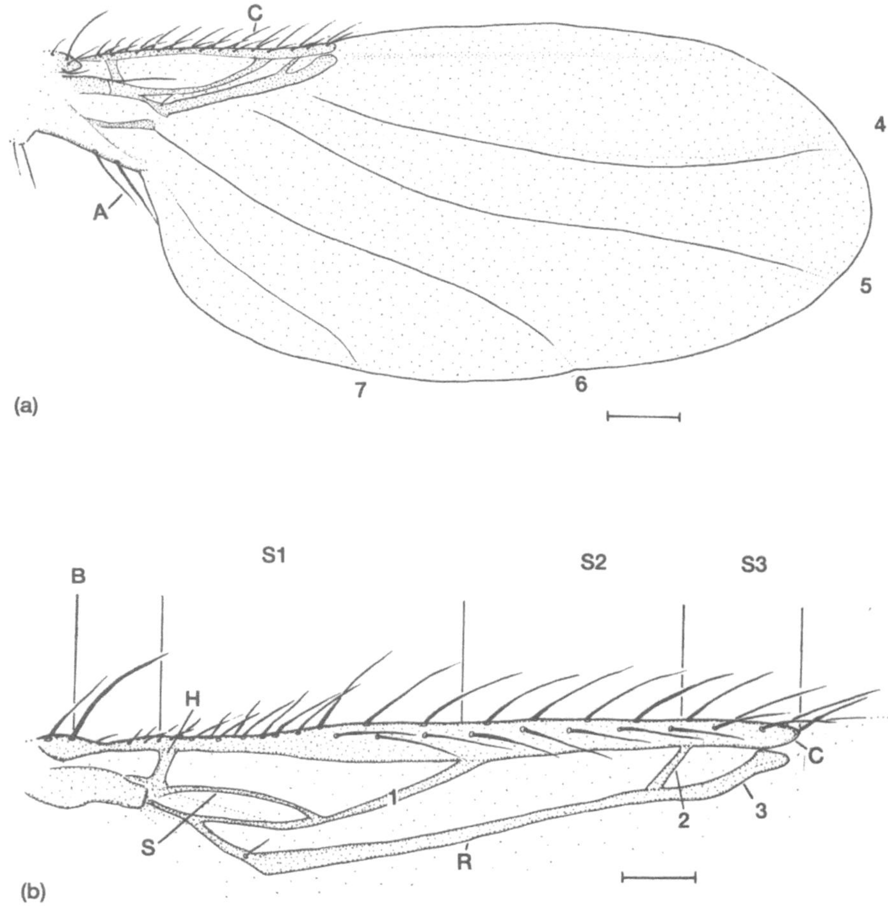

Figure 8.5 Details of wing (a) Right wing of *Megaselia stenoterga* male (A = axillary bristles, c = costal cilia on costal vein, 4–7 = tips of veins 4–7). (b) Anterior basal region of right wing of *M. subtumida* female (B = basal bristle of costa, H = humeral vein, C = tip of costal vein, 1–3 = veins 1–3, S = subcostal vein (Sc), S1–S3 = costal sections 1–3).

morphological terms in every new major text on the Diptera of a particular biogeographic region. There is little sign of any stability being achieved in the literature of the second half of the twentieth century. In view of this it is considered more useful to stick with the traditional labels. The passion to avoid identical terms in different orders of animals, in case these are taken to imply homology of the structures so labelled in the two orders, is seen to be absurd if taken to its logical conclusion. Thus the labelling of a leg bone in man as the 'femur' certainly has precedence over the labelling of the section of an insect leg with the same term. The logic of the purist's argument, therefore, is that we should coin new terms for such parts of the insect leg! I consider it most useful to employ the terms most frequently employed in the literature required to identify specimens to the species level.

8.2 KEYS TO GENERA

Many specimens can be identified to the genus by examining them in a watchglass of 70% alcohol under a stereo, dissecting, microscope with a good light. I use a relatively cheap instrument but with a separate fibre-optic lighting system. I mainly use this, however, for making slide mounts of the flies. Details of slide mounting are given in section 9.7. Slides are examined under a compound microscope using ×10, ×20 and ×40 objectives. Success in seeing fine detail tends to be achieved more often by adjustment of the lighting than by switching to a higher magnification!

In the following keys variation between species in a genus has been dealt with by keying out some genera at more than one couplet. Also some couplets can be reached by more than one route. The effect of this device is that by one route the user may only confront a particular choice once, but by another route a particular choice may be encountered twice.

A number of genera, particularly several small Neotropical genera, I have encountered only in the literature. There is, therefore, an increased risk of misapprehension in my perception of these genera. I would welcome notification of any errors in these keys. Comparison with other published keys to phorid genera will quickly reveal that those I offer below represent a radical new approach. Having been repeatedly frustrated when trying to use the existing keys, I have adopted an entirely novel arrangement. The most radical difference is in my total disregard of the supposed subfamilies or tribes. The result is that quite unrelated genera may end up in the same couplet, simply because they resemble each other in terms of the recognition characters I have chosen to emphasize.

The presence or absence of certain bristles are very useful in the recognition of certain genera. The basal sockets of detached bristles remain intact and can be readily observed. Traditional keys have given great emphasize to the inclination of the 'supra-antennal' bristles as well as of the 'antials'. I have disregarded this, as it is evident that this is not as significant, taxonomically, as has been supposed. Furthermore some of the differences given in the literature are almost certainly based on erroneous homologies of the bristles in question. Thus some cases of upswept 'supra-antennals' in conjunction with 'missing antials' could equally plausibly be interpreted as missing supra-antennals and approximated antials! In view of this sort of possibility for confusion, it seemed wiser to use other characters.

(A) DETERMINATION OF THE SEX OF A SCUTTLE FLY

The first task is to distinguish between the males and the females. In some females there may be a reduction in the size and number of tergites on the first

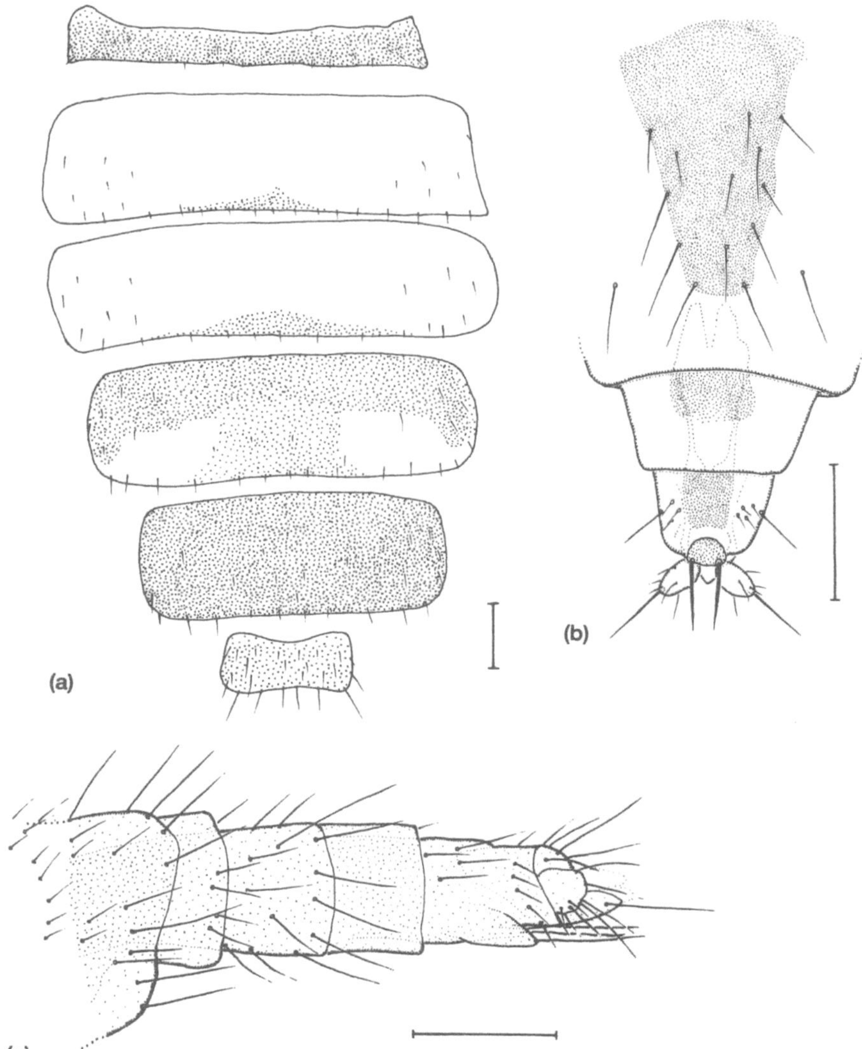

Figure 8.6 Parts of female abdomen. (a) Tergites 1–6 of *Megaselia humida*. (b) Tergite 7 and terminal segments of *M. humida* from above. (c) Terminal segments of *M. termitomyca* in side view.

six abdominal segments. The main external differences between the sexes, however, are in the terminal segments (those beyond segment 6).

In *females* the two cerci are separated (Figs 8.6b and 8.7a). In some they are reduced (e.g. Fig. 8.8a), or absent, but in these cases the abdomen ends in an obvious ovipositor-like structure (e.g. Figs 2.5–2.7). The basic abdominal

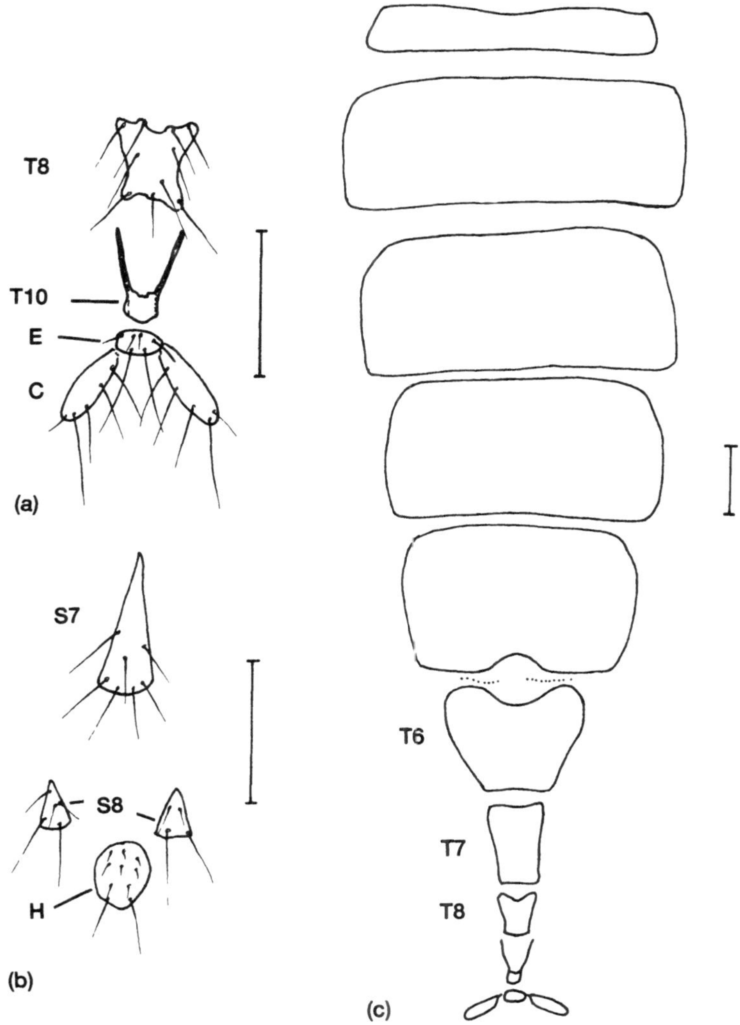

Figure 8.7 Female abdominal tergites and sternites of *Megaselia subnudipennis*. (a) Tergite to epiproct. (b) Sternite 7 to hypoproct. (c) All tergites and epiproct. (T6–T10 = tergites 6–10, S7 and S8 = sternites 7 and 8, C = cercus, E = epiproct, H = hypoproct.)

terminalia are as in Figs 8.6b,c and 8.7. Tergite 9 is normally fused to the anterior apodemes of tergite 10, thus appearing to be absent (Fig. 8.7a). Only rarely is it distinct (e.g. Fig. 8.8a). Sternite 8 may be simple (e.g. Fig. 8.8c), modified (e.g. Fig. 8.8b) or divided into two (e.g. Fig. 8.7b). Sternite 9 is either absent or internalized. In the latter case it is typically ring-shaped (e.g. Fig. 8.8a), but is occasionally a more elaborate structure bearing hairs.

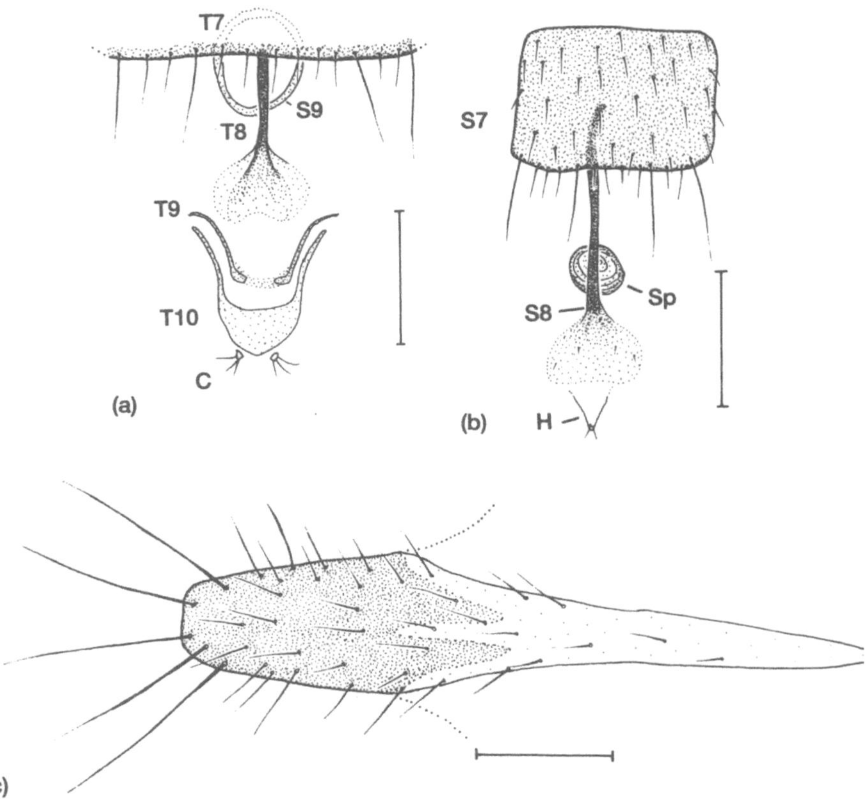

Figure 8.8 Female abdominal tergites and sternites (a) Terminal tergites of *Chonocephalus blackithorum* (b) Sternites of *C blackithorum* (c) Sternite 8 of *Chaetopleurophora erythronota* T7–10 = tergites 7–10, S7–9 = sternites 7–9, sp = spermatheca, H = hypoproct, c = cercus

In *males* the cerci are closely approximated and lie along the dorso-lateral regions of the proctiger, frequently being completely fused to the latter, and the three structures form the **anal tube** (Fig 8 1b) This anal tube emerges from the posterior region of the **epandrium** (= the enlarged Tergite 9) The latter embraces the penis complex, which may be completely withdrawn (Fig 8 9a) or be fully extruded (Fig 8 9b) Beneath the penis complex lies the **hypandrium** (Fig 8 1b), but in some genera an asymmetric development of the epandrium may cause the hypandrium to be displaced to one side Sometimes the hypandrium and epandrium are fused to form a single structure Adjacent to, and below, the anal tube there may be a pair of **surstyli** (Fig 8 10) These may be fused to the epandrium (e g Fig 8 11a,b), to form surstylar lobes A surstylus may be present on one side only (e g Figs 8 11c and 8 12a,b) Sometimes the tergite and sternite of segment 10 are still retained and fused to form a ring (Fig 8 12c) or cylinder (Fig 8 13b) at the

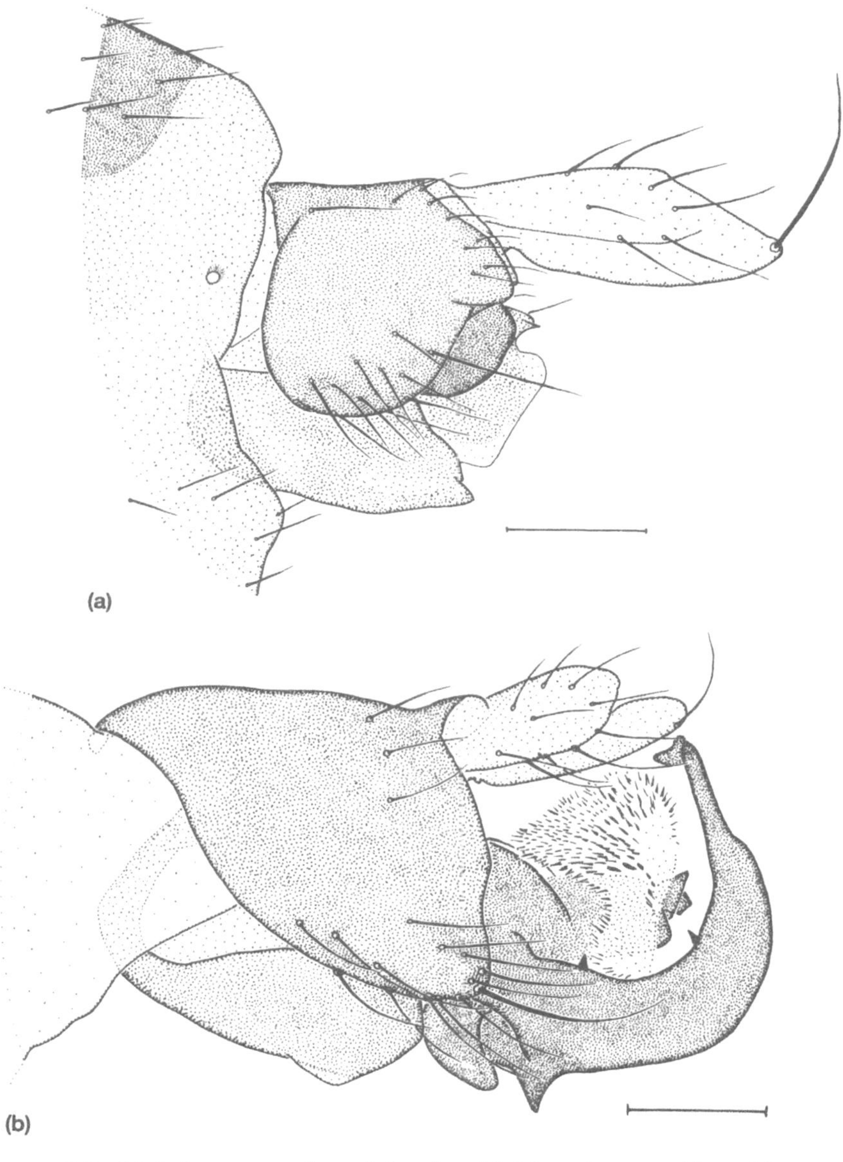

Figure 8.9 Male hypopygia from left sides. (a) *Megaselia nepenthina* with penis withdrawn. (b) *M barbulata* with penis extruded.

base of the anal tube. In the latter case any surstyli may be fused to this cylinder (e.g. Fig. 8.13a). Normally the remnants of the sclerites of segment 10 are lost or fused, at least in part, to the rear of the epandrium. The **hypandrium** frequently has one or two posterior lobes (e.g. Fig. 8.14b) and

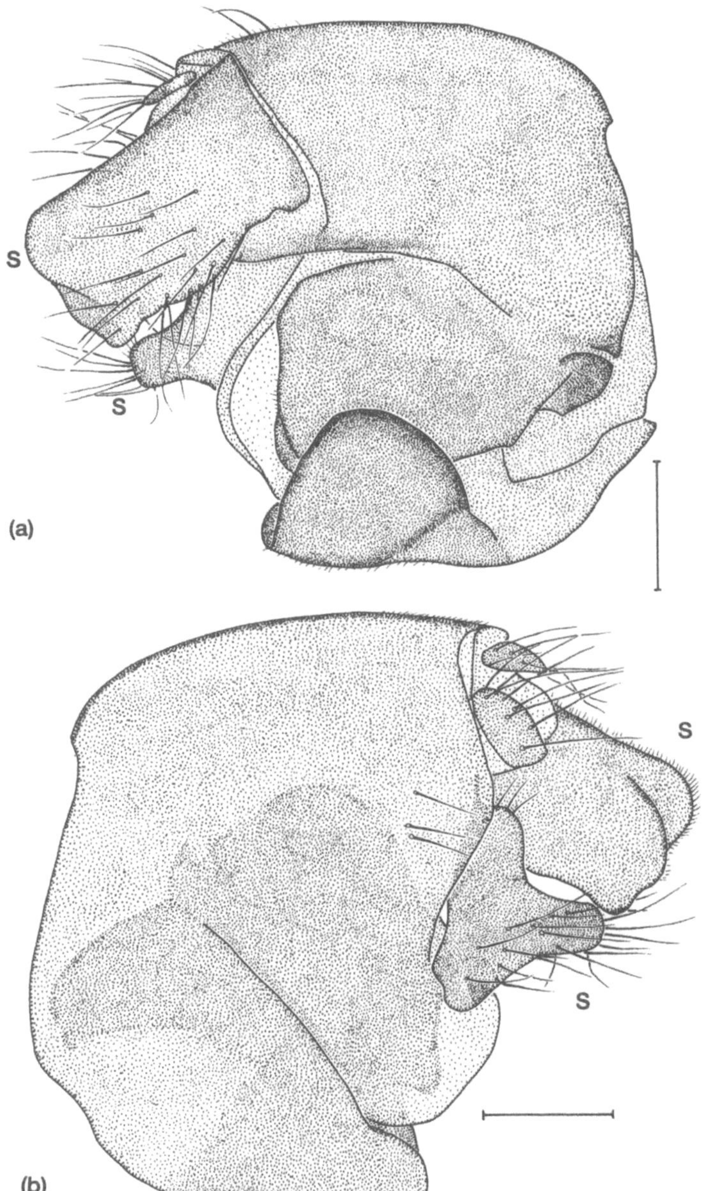

Figure 8.10 Male hypopygium of *Beyermyia spinulicosta* from right (a) and left (b) sides. (S = surstyli.)

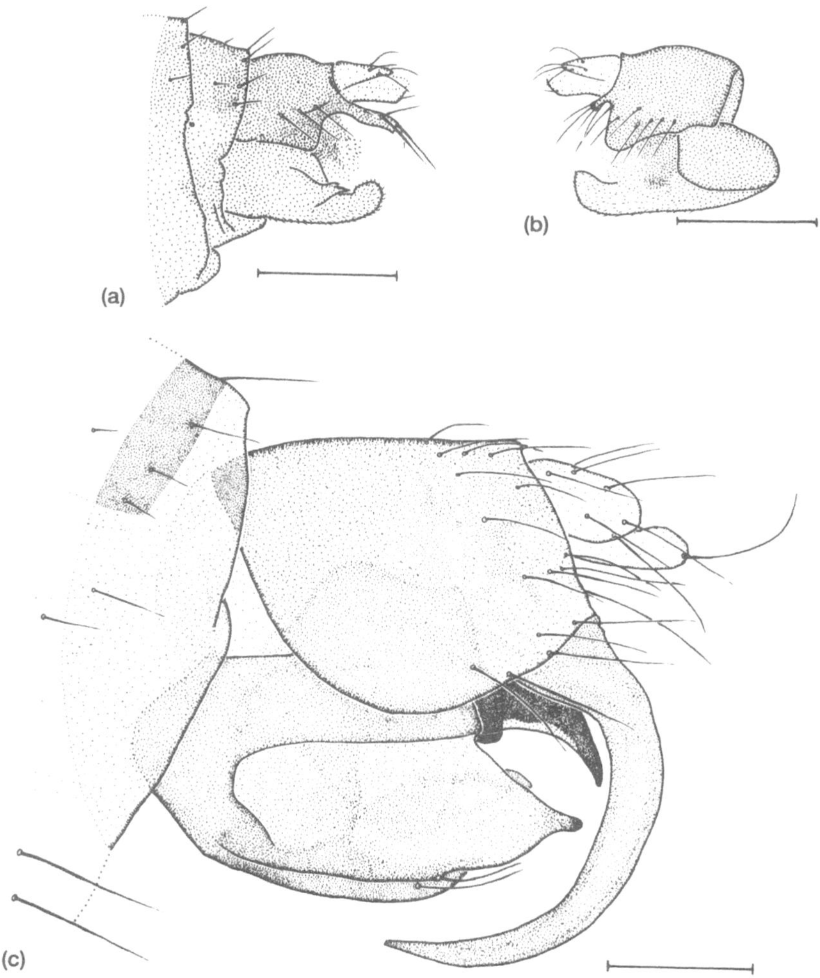

Figure 8.11 Male hypopygia. (a) *Maculiphora aenictophila* left face and (b) the same, right face. (c) *Megaselia biarticulata* left face.

may rarely carry a pair of gonopods (e.g. Fig. 8.15a). The latter, however, are usually absent or represented by a pair of bristles only (Fig. 8.14a). The entire terminal complex (i.e. segment 9 and all beyond) is called the **hypopygium**.

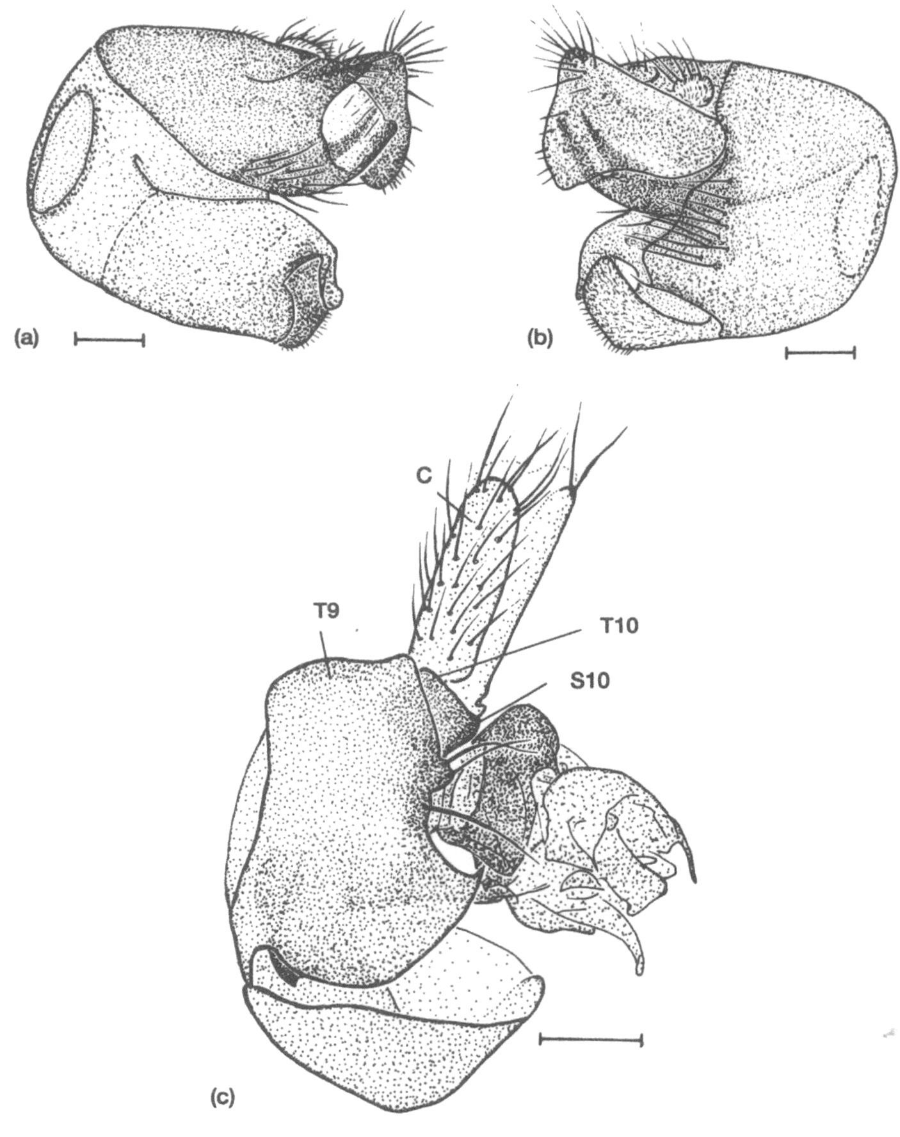

Figure 8.12 Male hypopygia. (a) and (b) left and right faces of *Phora penicillata*. (c) left face of *Dohrniphora cornuta* (T9 = tergite 9, C = cercus, T10 + S10 = fused tergite 10 and sternite 10 to form a ring).

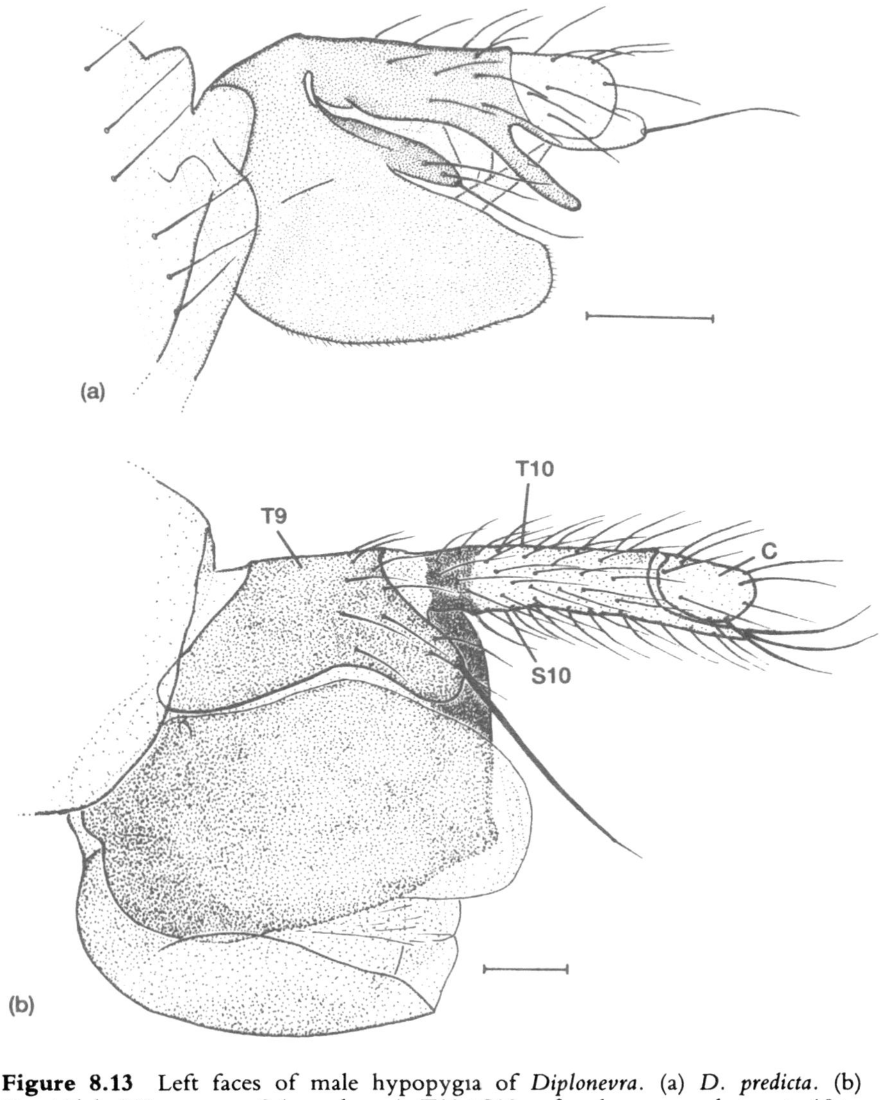

Figure 8.13 Left faces of male hypopygia of *Diplonevra*. (a) *D. predicta*. (b) *D. nitidula* (T9 = tergite 9 (epandrium), T10+S10 = fused tergite and sternite 10 to form cylinder, C = cercus).

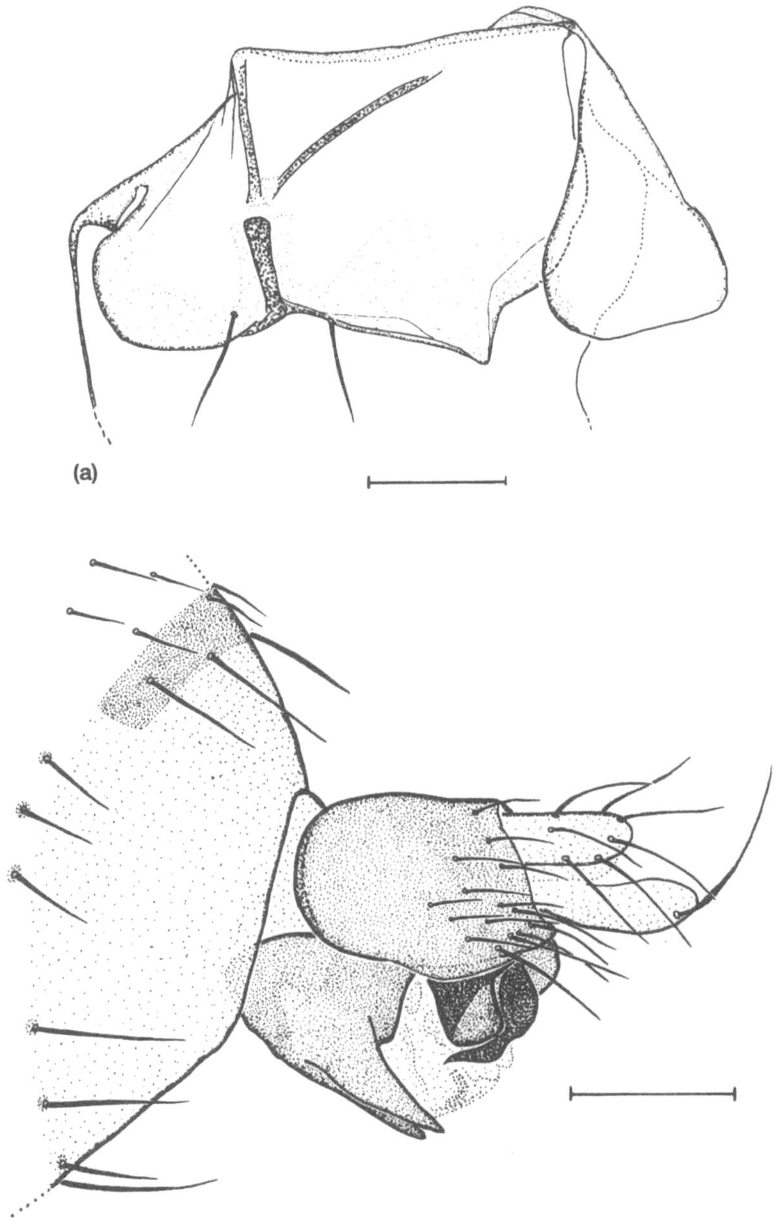

Figure 8.14 Male hypopygia. (a) *Gymnophora healeyae* hypandrium from below. (b) *Megaselia badia*, left face of hypopygium.

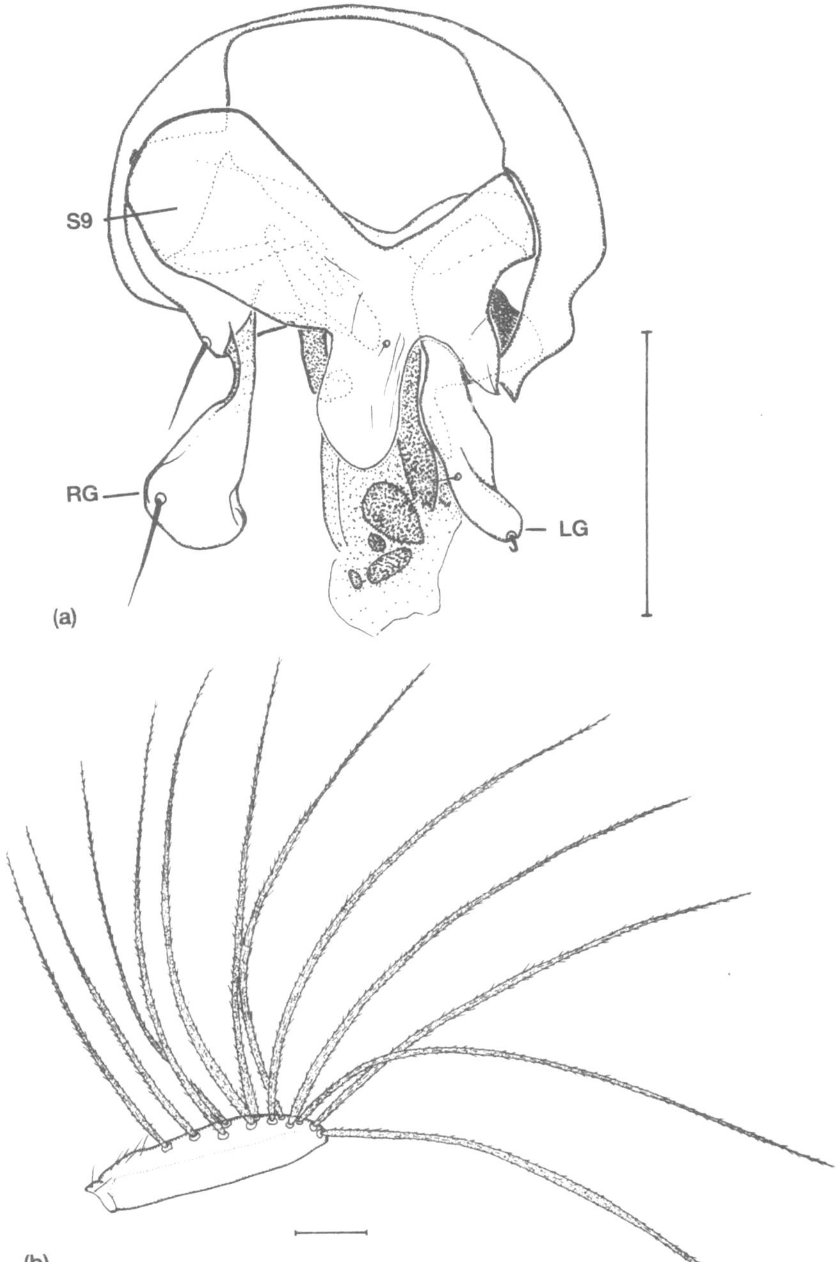

Figure 8.15 (a) Ventral view of male hypopygium of *Chonocephalus blackithorum* (S9 = sternite 9 (hypandrium), RG and LG = right and left gonopods). (b) *Acontistoptera brasiliensis*, right wing of female.

(B) KEY TO GENERA OF MALES

1. Wings absent — 2
– Wings present — 3
2. Fewer than 15 ommatidia in each eye — *Arrenaptenus*
– More than 40 ommatidia in each eye — *Aptinandria*
3. Vein Rs (i.e vein 3 before any apical fork) with several hairs along the dorsal face (e.g. Fig. 8.21a). These hairs may be short and fine (or at least shorter and finer than a more bristle-like hair at base of vein) — 4
– Apart from a few hairs (usually only one) in the basal quarter, the dorsal face of vein Rs is bare (apart from microsetae) — **13**
4. Vein 1 with hairs along its dorsal face — *Phlebothrix*
– Vein 1 bare on its dorsal face — 5
5. Hind tibia with at least one dorsal, or near-dorsal, longitudinal hair palisade (e.g. Figs 8.4d,e) — 6
– Hind tibia without dorsal hair palisades, but there may be spaced hairs (as in Fig. 8.4c) — 10
6. A longitudinal hair palisade present on front tibia — 7
– Front tibia without a dorsal, longitudinal hair palisade, but there may be a row of spaced bristles or spines — 8
7. Distance between anterior ocellus and front of frons much greater than distance between two posterior ocelli. Vein 3 forked or unforked (Fig. 8.21a) — *Aenigmatias*
– Distance between the anterior ocellus and front of frons less than distance between posterior ocelli. Vein 3 unforked — *Misotermes*
8. Height of frons much greater than width — *Trineurocephala* (part)
– Frons width about equal to height (from anterior ocellus to front margin) or greater — 9
9. Posterior ocelli close to eye margin and ocellar triangle strongly demarcated at front by a sinuous furrow (Fig. 8.94d) — *Stichillus*
– Posterior ocelli well removed from eye margin and ocellar triangle not demarcated in this way at front — *Borophaga*
10. Third antennal segment apically elongated and tapered, with apical arista — *Burmophora*
– Third antennal segment more globose; arista frequently pre-apical and dorsal in position — 11
11. Costa less than two fifths of wing length. Medio-lateral and pre-ocellar bristles absent — *Euryophora*
– Costa extends about half wing length, or more. Medio-lateral and pre-ocellar bristles present — 12
12. More than 10 hairs on vein Rs — *Anevrina*
– Fewer than 10 hairs on vein Rs — **13**
13. Apex of third antennal segment drawn out into long, narrow, long-haired process; with longer, short-haired, arista originating near base of this apical process — *Diplostylocera*
– Antenna not of this form — 14
14. Apex of third antennal segment drawn out into long, tapered extension bearing apical arista (e.g. Figs 8.16b and 8.31a) — 15

- – Apex of third antennal segment not drawn out in this way, except in some cases where arista is pre-apical in position or is absent. Sometimes arista replaced by apical pseudo-arista (which is an extension of the third antennal segment itself, rather than articulating with it) **39**
15. Arista with only one or two segments and mesopleuron bare 16
- – Without this combination (i.e. arista normally with two short basal segments and long apical segment, but if with only one or two segments, then mesopleuron has hairs) 17
16. Arista with single long segment. Veins 1 and 3 well separated apically (Fig. 8.36b). Only fine hairs below basal half of hind femur *Dahliphora* (part)
- – Arista short and with two segments (Fig. 8.49c). Veins 1 and 3 closely approximated (as Fig. 8.50d). Base of hind femur with spinules (Fig. 8.49b). (Hypopygium as Fig. 49e) *Iridophora*
17. Mid tibia with one or two differentiated dorsal or near-dorsal bristles in basal half 18
- – Mid tibia without such bristles, although a row of postero-dorsal differentiated hairs may be present along the length of tibia **25**
18. Only one bristle in basal half of mid tibia. Vein 3 forked or unforked 19
- – Two bristles in basal half of mid tibia. Vein 3 unforked 20
19. Vein 3 unforked **83**
- – Vein 3 forked **84**
20. Hind tibia with two longitudinal hair palisades *Antipodiphora*
- – Hind tibia without any longitudinal hair palisades 21
21. Hind tibia with isolated bristles including at least onè pair (one dorsal and one antero-dorsal) in basal half (e.g. Fig. 8.32a) 22
- – Hind tibia without any isolated bristles situated close enough to another to form such a pair in basal half 24
22. Hind tibia with at least eight bristles before apical quarter *Contopteryx*
- – Hind tibia with only four bristles in upper three-quarters 23
23. Hind tibia with lower dorsal bristle beyond middle (Fig. 8.32a). Vein Sc does not reach vein 1 (Fig. 8.31c). (Antenna and claspers (surstyli) as Fig. 8.31b) *Conicera* (part)
- – Hind tibia with lower dorsal bristle just before middle. Tip of vein Sc reaches vein 1 and fuses with it *Darwiniphora* (part)
24. At least one strong spine at postero-lateral extremity of left side of epandrium (Fig. 8.47c) *Gymnoptera* (part)
- – No such spines on epandrium (see also couplet 71) *Coniceromyia* (part)

25. Hind tibia with a longitudinal dorsal hair palisade 26
- – Hind tibia without a dorsal hair palisade, but there may be a row of spaced hairs 34

26. Vein 3 forked 27
- – Vein 3 unforked 31
27. Frons breadth equal to or clearly greater than height 28
- – Frons clearly with height greater than breadth **39**
28. Anal tube not straight, narrow and long (not exceeding mid-line length of epandrium) 29

– Anal tube strikingly long (exceeding mid-line length of epandrium), straight and narrow **39**
29. Mesopleuron with hairs and a bristle 30
– Mesopleuron bare **39**
30. Antero-lateral bristles absent from frons *Acanthophorides*
– Antero-laterals present *Macrocerides* (part)
31. Mesopleuron with hairs. (Arista very short.) *Tubicera* (part)
– Mesopleuron bare 32
32. Tip of vein 3 does not reach margin of wing but runs to tip of the costa, which is deflected rearwards through a right angle bend *Stenoneurellys*
– Tip of costa not deflected in this way 33
33. Anal tube long and slender, being clearly longer than length of epandrium **39**
– Anal tube only about as long as epandrium and not obviously slender *Ceratoconus*
34. Vein 3 forked *Rhopica*
– Vein 3 unforked 35
35. Costa extends less than one-third of wing length. Tapered extremity of third antennal segment bears hairs which are clearly much longer than those on segment 3 of arista *Leptilla*
– Costa extends for at least half wing length. Hairs on tapered extremity of third antennal segment little, if any, longer than those on segment 3 of arista 36

Note: If neither applies try couplet 196.

36. Vein six deflected anteriorly in middle section, so that space embraced by it and vein 5 has characteristic 'waist' (Fig. 8.56a) *Metopina* (part)
– Vein six not deflected in this manner 37
37. Frons clearly broader than high 38
– Frons with height at least as great as breadth or greater. *Aristocerina*
38. Frons bearing conspicuous, strongly developed, bristles (Fig. 8.16b). Mesopleuron bare (see also Figs 8.16a, 8.17a) *Adelopteromyia*
– Frons devoid of bristles. Mesopleuron with a few hairs dorsally *Cyphocephalus*
39. Front part of frons at a right angle to the top face of frons, the transition being marked by an arch-shaped ridge (Fig. 1.4) *Aenigmatistes* (most)
– Frons not of this form 40
40. At least two differentiated dorsal or near-dorsal bristles in basal two-thirds of mid tibia 41
– Only one bristle or none in basal two thirds of mid tibia, but there may be a longitudinal row of dorsal, differentiated hairs **78**
41. Hind tibia with dorsal, longitudinal hair palisade (Fig. 8.4d,e). Sometimes with two or three such palisades (Fig. 8.39f) 42
– Hind tibia without longitudinal hair palisade, but there may be a row of spaced hairs or a series of overlapping transverse combs **53**
42. Vein 3 unforked 43
– Vein 3 forked 49
43. Arista absent *Abaristophora*
– Arista present 44

44. Frons narrow, the height being clearly greater than width. *Trineurocephala* (part)
 – Frons with width equal to or greater than height 45
45. Mid tibia with one or two dorsal or near-dorsal longitudinal hair palisades. Veins 1 and 3 divergent apically 46
 – Mid tibia without longitudinal hair palisades. Veins 1 and 3 subparallel throughout 47
46. Mid tibia with two longitudinal hair palisades. Supra-antennal bristles absent. Anal tube relatively short *Aenigmatistes* (part)
 – Mid tibia with only one longitudinal hair palisade. A well-developed pair of supra-antennal bristles. Anal tube relatively long (Fig. 8.45a) *Epicnemis*

Note: The unknown male of one *Diplonevra* species may key out here, but will be distinguished by its cerci being well separated from the epandrium (see couplet 52), and by having both supra-antennals and two palisades.

47. Frontal bristles strongly developed. Form of hypopygium as Fig. 8.48c. (Typically palp as Fig. 8.48b and tip of vein 1 as Fig. 8.49a.) *Hypocerides*
 – Frontal bristles greatly reduced (Figs 8.67a and 8.89a). Hypopygium relatively short but high (e.g. Fig. 8.66c, d) 48
48. Palp with several bristles (Fig. 8.66b) and supra-antennal bristles distinct (Fig. 8.67a). (Hypopygium as Fig. 8.66c, d, mid tibia as Fig. 8.66a.) *Palpiclavina*
 – Palp with only one bristle (Fig. 8.89b) and supra-antennal bristles vestigial or absent (Fig. 8.89a) *Thaumatoxena*
49. Frons lacks supra-antennals and other bristles at front. Base of broad anal tube projects above level of dorsal face of epandrium. Scutellum with at least ten bristly hairs *Cyrtophorina* (part)
 – Supra-antennals and other bristles present at front of frons. Base of anal tube not projecting above dorsal face of epandrium. Scutellum usually with fewer than ten hairs, but when ten present then at least two clearly differentiated as strong bristles 50
50. Costal cilia very reduced and coloured as costa, so that at low magnifications latter appears to be bare *Rhynchomicropteron* (part)
 – Black costal cilia clearly evident at low magnifications 51
51. Mesopleuron with hairs in upper part. Axillary ridge with single, generally pale, bristle which is broadest in middle (Fig. 8.40b). Hypopygium characterized by short but tall epandrium and hairless sclerotized ring (the fused tergite and sternite 10) at base of anal tube (Fig. 8.12c). (Hind femur typically with patch of spinules at base of posterior face – Fig. 6.12, and sometimes with sensory pit as well – Fig. 8.40a, or other modifications) *Dohrniphora*
 – Mesopleuron bare. Axillary ridge with at least two bristles, which are generally dark brown or black and evenly tapered. Epandrium generally longer than high and sclerites of segment 10 either form hairy cylinder, with or without paired surstyli fused to the sides (Fig. 8.13), or entirely absent 52
52. Sclerites of segment 10 form cylinder bearing hairs only (Fig. 8.13b) or

hairs and surstyli (Fig. 8.13a). In either case bases of cerci are well separated from epandrium. (Bases of hind femora frequently with spinules and/or processes – Fig. 8.39.) *Diplonevra*

– Sclerites of segment 10 not evident, so that bases of elongated cerci lie immediately behind rear margin of epandrium *Psyllomyia*

53. Vein 3 unforked 54

– Vein 3 forked **68**

54. Axillary ridge with single bristle which is very long plumose (Fig. 8.29a). (Tip of costa swollen – Fig. 8.29b. Hypopygium as Fig. 8.10.) *Beyermyia*

– Axillary ridge without such a long-plumose bristle, and frequently with more than one bristle 55

55. No isolated bristles on upper two-thirds of hind tibia 56

– At least one bristle on upper two-thirds of hind tibia 57

56. Arista apical *Billotia*

– Arista in dorsal, pre-apical position 57

57. Mesopleuron with hairs 58

– Mesopleuron bare 61

58. Third antennal segment globose 59

– Third antennal segment more elongated, being more pear-shaped **68**

59. Hind tibia with series of overlapping transverse combs along length of dorsal face (Fig. 8.46c). (Hypopygium as Fig. 8.48a.) *Hypocera*

– No such combs 60

60. Arista in dorsal, pre-apical position. Costa at least two-fifths of wing length *Plethysmochaeta*

– Arista apical. Costa less than two-fifths of wing length *Mannheimsia*

61. Width of frons clearly less than height. Mid tibia with one or more true dorsal bristles below one in basal pair of bristles *Phora*

– Width of frons subequal to or greater than height. Only true dorsal bristle in upper two-thirds of mid tibia is one in basal pair of bristles 62

62. Hind tibia with two or more isolated bristles in upper two-thirds (e.g. Fig. 8.32a) 63

– Hind tibia at most with only one isolated bristle in upper two-thirds **68**

63. Veins 1 and 3 apically divergent (Fig. 8.31c). Arista apical in position 64

– Veins 1 and 3 subparallel (Fig. 8.65c). Arista dorsal in position. (Hypopygium as Fig. 8.65d, e; hind leg as Fig. 8.65b.) *Obscuriphora*

64. Palps longer than breadth of frons, strongly tapered, devoid of bristles, and directed upwards in front of broad hairless band running from anterior ocellus to front of frons *Bothroprosopa*

– Neither palps nor frons of this form 65

65. The most basal anterior bristle of hind tibia is paired with a dorsal bristle 66

– Without dorsal bristle adjacent to most basal anterior bristle of hind tibia 67

66. One pair of supra-antennal and eight other bristles on frons in front of anterior ocellus. (Antenna as Fig. 8.32b, claspers (surstyli) of hypopygium as Fig. 8.31b.) *Conicera* (part)

– Both supra-antennals and antero-laterals absent, leaving only six bristles in front of anterior ocellus *Darwiniphora* (part)
67. Apical ventral region of front metatarsus developed into an elongated process **68**
– Front metatarsus simple *Gymnoptera* (part)
68. Costal cilia very reduced and coloured as costa, so that at low magnifications latter appears to be bare *Rhynchomicropteron* (part)
– Dark costal cilia clearly evident at low magnifications 69
69. Third antennal segment conspicuously elongated apically and tapered from swollen base 70
– Third antennal segment globose, pear-shaped or oval, but without strongly tapered apical portion 71
70. Arista clearly pre-apical in position *Cyphometopis* (part)
– Arista clearly apical 71
71. Front metatarsus with apical-ventral region developed into elongated process. (Front tibia sometimes with strong spine, which may be twisted, and/or notched) *Coniceromyia* (part)
– Front metatarsus without such a process 72
72. Palps greatly elongated, being longer than breadth of frons, and devoid of bristles *Palpocrates*
– Palps shorter, or if enlarged there are small bristles at tip at least 73
73. Mesopleuron with hairs, and usually with strong bristle at rear as well *Chaetopleurophora*
– Mesopleuron bare 74
74. Medio-lateral bristles absent *Tarsocrates*
– Medio-lateral bristles present (but antero-laterals sometimes reduced or absent) 75
75. Mid tibia with anterior bristle in third quarter clearly longer than width of tibia at point of insertion *Spiniphora*
– Mid tibia with anterior bristle in third quarter greatly reduced or absent 76
76. Fork of vein 3 complete *Triphleba* (part)
– Base of inner branch (R2+3) of 'fork' of vein 3 obliterated, so there is no actual forking of vein 3 as such 77
77. Restricted to New Zealand *Kierania*
– Restricted to Holarctic Region *Triphleba* (part)

Note: See comment under review of *Kierania* below.

78. One differentiated bristle in basal half of mid tibia 79
– No differentiated bristle in basal half of mid tibia, but there may be longitudinal row of near-dorsal or dorsal hairs **85**
79. Hind tibia with one or more longitudinal hair palisades 80
– Hind tibia without longitudinal hair palisades 81
80. Hind tibia with only one, dorsal, hair palisade. Scutellum with at least ten hairs *Cyrtophorina* (part)
– Hind tibia with several longitudinal hair palisades (Fig. 8.60a). Scutellum with only four hairs or bristles. (Frons and hypopygium as Fig. 8.61, wing as Fig. 8.60d.) *Multinevra* (part)
81. Third antennal segment conspicuously elongated apically and tapered

from swollen base. Arista clearly pre-apical in position *Cyphometopis* (part)
- Third antennal segment otherwise. If tapered apically then arista is apical; otherwise arista is dorsal 82

82. Vein 3 forked *Triphleba* (part)
- Vein 3 unforked **83**

83. Arista dorsal, in pre-apical position. Subcosta well developed *Triphleba* (part)
- Arista apical. Subcosta weakly developed or absent *Xanionotum*

84. Frons with antial and antero-lateral bristles absent. A weak anterior bristle in basal half of hind tibia *Ceratoplatus*
- Antial bristles present above the upper pair of supra- antennal bristles in a pre-ocellar position. No isolated bristle in basal half of hind tibia *Ceratusa*

85. Hind tibia with at least one dorsal, or near-dorsal, longitudinal hair palisade (e.g. Fig. 8.4d, e) 86
- Hind tibia without longitudinal hair palisades, but there may be row of spaced hairs **175**

86. Front tibia with longitudinal dorsal, or near-dorsal, hair palisade 87
- Front tibia at most with spaced hairs or spinules along dorsal face 89

87. Hind tibia with several longitudinal hair palisades (Fig. 8.60a). (Frons and hypopygium as Fig. 8.61; wing as Fig. 8.60d.) *Multinevra* (part)
- Hind tibia with a single, dorsal, hair palisade 88

88. Vein 3 forked. Costa at most only half length of wing, and ending level with tip of vein 3 89
- Vein 3 unforked. Costa extends well past middle of wing and overreaching tip of vein 3 *Eremophora*

89. Mesopleuron with hairs, and sometimes with one or more bristles also. Rarely with a single strong bristle only 90
- Mesopleuron bare **109**

90. Vein 3 unforked 91
- Vein 3 forked 95

91. Scutellum with at least six bristles. With six or more supra-antennal bristles, but otherwise no bristles on frons between these bristles and anterior ocellus *Aenigmatopoeus*
- Scutellum with only four bristles or two bristles and two hairs. With only four supra-antennal bristles, but with several strong bristles between these and anterior ocellus 92

92. Abdominal venter with paired pigmented patches on segment 5 (Fig. 8.71). (Front tarsus as Fig. 8.72b.) *Plastophorides*
- Abdominal venter without these pigmented patches 93

93. Veins 1 and 3 subparallel (as Fig. 8.65c) 94
- Veins 1 and 3 apically divergent (as Fig. 8.31c) *Megaselia* (part)

94. Anal tube very short, with tip of proctiger only extending as far as rear margin of epandrium *Syneura* (part)
- Anal tube long, extending well beyond rear of epandrium *Lepta*

Note: Since completion of this key, Brown (1993a) has published a description of the hitherto unknown male of *Colyeria*. It runs to this lead. The anal tube extends beyond the rear margin of epandrium but is clearly shorter than it. The mesopleuron has a single strong bristle only. The body and hypopygium are somewhat flattened dorso-ventrally.

95. Eight bristles on scutellum 96
– At most six bristles on scutellum, usually with four bristles or with two bristles and two hairs 97
96. Epandrium with a pair of long surstylar lobes, extending from rear margins below anal tube. Terminal hairs of proctiger clearly longer and stronger than those on cerci *Egregiophora*
– Epandrium without surstylar lobes. Hairs at tip of proctiger not longer or stronger than those on cerci *Johowia* (part)
97. Hind tibia with antero-dorsal row of differentiated hairs, as well as postero-dorsal row (the two rows being separated by hair palisade) 98
– Hind tibia without an antero-dorsal row of differentiated hairs 100
98. Hind tibia with antero-ventral row of differentiated hairs as well *Physoptera* (part)
– Without this antero-ventral row of hairs 99
99. Palp bristles and supra-antennal bristles only about size of upper occipitals, but the former sometimes a little longer *Phalacrotophora* (part)
– Either palp bristles or supra-antennals, or both, clearly longer than upper occipitals *Megaselia* (part)
100. Frons narrow, with height clearly much greater than breadth 101
– Frons not obviously narrowed or with breadth greater than height 102
101. Costa only extending about one-third of wing length. Third antennal segment globose, but with pointed apex. *Stenodicrania*
– Costa extends about half wing length. Third antennal segment about twice as long as broad *Ceratophoromyia*
102. Frons about three times as broad as long. Third antennal segment with apex drawn out so that it is an elongated pear shape. Anal tube very long and thin. Costa about half length of wing *Brachycephaloptera*
– Frons at most twice as broad as long. Without other characters in combination 103
103. Anal tube clearly longer than length of dorsal face of epandrium. (Mesopleuron with a long bristle at rear in addition to hairs.) 104
– Anal tube little, if any, longer than dorsal face of epandrium 106
104. Section 1 of costa with a swelling. Only two minute supra-antennal bristles. (Anal tube slender.) *Pleurophorina*
– Costal section 1 without a swelling. With two or four supra-antennal bristles, of which at least two are not obviously reduced in size 105
105. Anal tube slender and very long relative to length of epandrium (Fig. 8.54a) *Menozziola*
– Anal tube not so long and slender *Megaselia* (part) and *Rhyncophoromyia* (part)

Note: These two genera can only be distinguished in the female sex. *Rhyncophoromyia* is only known from the Nearctic and Neotropical Regions.

106. Third antennal segment elongated *Macrocerides* (part)
– Third antennal segment globose 107

107. More than four supra-antennal bristles *Johowia* (part)
– Only two or four supra-antennal bristles 108
108. Anal tube very short (both cerci and proctiger being only about as long as high). Epandrium with well-developed paired surstylar lobes (at least twice as long as width at base) which bear more conspicuous hairs than rest of epandrium. *Megaselia* (part, including subgenus *Kerophora*)
– Anal tube usually longer. Surstylar lobes absent or if developed they are shorter and the hairs of rest of epandrium are usually more conspicuous *Diocophora* and *Megaselia* (part)

Note: It is not clear how to separate these two genera in the male sex. The hypopygium of a *Diocophora* is shown in Fig. 8.38(c). This genus is only known from the Nearctic and Neotropical Regions.

109. Arista absent, but third antennal segment drawn out into a long, tapered, apical extension 110
– Arista present 112
110. Vein 3 forked. (The narrowed extension of third antennal segment at least two thirds of total length.) *Macrocerides* (part)
– Vein 3 unforked 111
111. Anal tube with inflated cerci (third antennal segment short-haired) *Tubicera* (part)
– Cerci not inflated (third antennal segment with at least some long hairs in apical half) **218**
112. Vein 3 unforked 113
– Vein 3 forked **136**
113. Two longitudinal hair palisades on hind tibia *Distichophora* (part)
– Only one palisade on hind tibia 114
114. More than four supra-antennal bristles 115
– At most with four supra-antennal bristles 116
115. Part of frons bearing supra-antennals is extended forwards so that front margin is almost V-shaped. Anal tube shorter than mid-line length of dorsal face of epandrium *Cataclinusa*
– Front margin of frons only imperceptibly produced in middle. Anal tube longer than epandrium *Anticofimbria*
116. Anal tube at most as long as mid-line length of dorsal face of epandrium 117
– Anal tube longer than mid-line length of epandrium 120
117. Notopleural cleft present (Fig. 8.93b,c) *Woodiphora* (part)
– No notopleural cleft 118
118. Veins 1 and 3 almost subparallel, only slightly divergent apically 119
– Veins 1 and 3 clearly divergent in apical third or more 120
119. Frons clearly broader than high. Third antennal segment globose *Syneurina*
– Frons higher than broad. Third antennal segment more oval *Syneura* (part)
120. Third antennal segment globose 121
– Third antennal segment more oval, conical, or at least always longer than greatest breadth 124

121. Anal tube relatively short 122
– Anal tube relatively long (e.g. Fig. 8.25) 123
122. Last third of costal section 1 distinctly narrower than first two-thirds *Pheidolomyia*
– Costal section 1 of even thickness throughout or gradually thickening from base *Megaselia* (part)
123. Anal tube with narrow cerci (Fig. 8.91b) (Costa about one-third of wing length) *Trucidiphora*
– Anal tube with greatly inflated cerci *Dinocercus*

Note: Since I completed this key, Brown (1993a) has published a description of the hitherto unknown male of *Borgmeieriphora*. It runs to this couplet. Its anal tube is similar to that of *Trucidiphora*, but its costa is at least half the wing length and the tip of vein 4 is deflected rearwards close to wing margin.

124. Entire third antennal segment long-haired, these hairs being much longer than those on arista *Dichocerina*
– Third antennal segment with any hairs longer than those on arista restricted to apical region 125
125. Without supra-antennal bristles in a median position at front of frons 126
– At least two supra-antennal bristles in a median position at front of frons 131
126. Antero-lateral and/or medio-lateral bristles absent 127
– Both antero-lateral and medio-lateral bristles present 128
127. Pre-ocellar and antial bristles also absent *Melaloncha* (part)
– Pre-ocellar and antial bristles present 130
128. A strong pair of supra-antennal bristles present, but each standing in a triad with an antial and antero-lateral bristle and thus much further apart than pre-ocellars (e.g. Fig. 8.27b) *Auxanommatidia* (part)
– Supra-antennal bristles entirely absent 129
129. Costa only about one-third of wing length. Hairs at tip of proctiger much longer than those on cerci *Cremersia* (part)
– Costa about half wing length. Hairs at end of proctiger only about as long as those on cerci. (Antennae close together.) *Melaloncha* (part)

Note: Since I completed this key, Brown (1993a) has published a description of the hitherto unknown male of *Melittophora*. It runs to this couplet. The costa is about half the wing length. The widely-separated antennae and somewhat dorso-ventrally flattened body will distinguish it from *Melaloncha*.

130. Antero-lateral bristles of frons absent *Apocephalus* (part)
– Antero-lateral bristles present *Melaloncha* (part)
131. Antero-lateral and/or medio-lateral bristles of frons absent 132
– Both antero-lateral and medio-lateral bristles present 133
132. Epandrium symmetrical and with broad, hairy, postero-lateral lobes *Styletta*
– Epandrium tends to be asymmetrical and at least one postero-lateral lobe tends to be narrow and tapered (e.g. Fig. 8.25a) *Apocephalus* (part)
133. Anal tube relatively short and thick 134

– Anal tube long and slender (e.g. Fig. 8.74a) 135
134. More than four bristles on axillary ridge of wing *Apodicrania*
– Fewer than three axillary bristles **173**
135. Frons more than 1.5 times as broad as high *Veruanus*
– Frons less than 1.5 times as broad as high *Pseudacteon*
136. With at least eight bristles on scutellum *Chaetaspidia*
– With at most only six bristles on scutellum 137
137. Hind tibia with more than one longitudinal hair palisade, even if some are irregular (e.g. Fig. 8.92c, d and 8.73a) 138
– Hind tibia with a single, dorsal, hair palisade 141
138. Mesopleuron with hairs *Beckerina* (part)
– Mesopleuron bare 139
139. At anterior end of notopleuron the orifice of the duct from notopleural gland is fully exposed and its diameter is less than diameter of anterior thoracic spiracle (Fig. 8.28b) *Beckerina* (part)
– Orifice of duct from notopleural gland at least partly concealed within a notopleural cleft (Fig. 8.93b, c). The length of this cleft is greater than diameter of anterior thoracic spiracle 140
140 Hind tibia as Fig. 8.73a. (Hypopygium as Fig. 8.70b) *Plectanocnema*
– Hind tibia with fewer and less complex hair palisades (as Fig. 8.92c, d) *Woodiphora* (part)
141. Postero-dorsal hairs of hind tibia set in enlarged asymmetric pits and with broadened tips. Wings with a distinct elongated dark mark between bases of veins 5 and 7. Costa extends well beyond apex of vein 3. *Pelidnophora*
– Without this combination 142
142. At least six supra-antennal bristles *Johowia* (part)
– At most only four supra-antennal bristles 143
143. Six short bristles on scutellum. With four subequal supra-antennal bristles and eight other bristles on frons below pre-ocellars. Costa extending almost two-thirds of wing length *Clinochaeta*
– Without this combination 144
144. Frons breadth more than five times mid-line height, the anterior ocellus being at the front margin. Third antennal segment concave on external face and with greatest breadth about the same as frons width *Lecanocerus*
– Frons not so broad. Third antennal segment otherwise 145
145. Outer two-thirds of hind femur postero-ventrally hollowed to accommodate tibia, and this region with a distinctive oblique dark mark at its proximal end (Fig. 8.74d). A pre-apical tooth present on all tarsal claws (Fig. 8.74b). Anal tube very short. More than eight bristles on axillary ridge of wing *Pseudohypocera*
– Without this combination 146
146. Basal half of hind margin of wing fringed with fine hairs a little longer than costal cilia on costal section 3. Costa more than half wing length. Third antennal segment globose. Anal tube clearly much shorter than epandrium *Woodiphorides*
– Not with this combination 147

147. Hind femur with triangular projection at base (Fig. 8.78c). Anal tube shorter than epandrium (Fig. 8.78d). Third antennal segment clearly longer than broad (Fig. 8.78e) *Razorfemora*
– With a different combination 148
148. Eye with lower facets (ommatidia) at least twice diameter of those uppermost (e.g. Fig. 8.27b). Anal tube long relative to epandrium (e.g. Fig. 8.27a). Third antennal segment longer than greatest breadth *Auxanommatidia* (part)
– Without these features combined 149
149. Supra-antennal bristles absent, but antials may be in middle of front margin of frons 150
– At least two small supra-antennal bristles are differentiated from adjacent hairs on frons 152
150. Anal tube clearly longer than epandrium and slender 151
– Anal tube either shorter than mid-line length of epandrium or if a little longer it is not slender. (Third antennal segment oval or pear-shaped.) *Neodohrniphora* and *Cremersia* (part)

Note: The males of both genera are poorly known. The known *Neodohrniphora* have the costa at least two-fifths of wing length, whereas it is only about one-third of length in the *Cremersia* species keying out here.

151. Pre-ocellar bristles absent. (Costa about half of wing length.) *Melaloncha* (part)
– Pre-ocellar bristles present 152
152. The last quarter of vein 6 (third thin vein) absent. Frons more than 2.5 times as broad as high. Antial bristles close to antero-laterals. Third antennal segment globose, but with small apical point *Exochaeta*
– Without this combination 153
153. Hind tibia with a clearly differentiated antero-dorsal row, as well as a postero-dorsal row, of hairs (the two rows separated by hair palisade) 154
– At most with a postero-dorsal row of differentiated hairs on dorsal face of hind tibia 155
154. Palp bristles and supra-antennals only about size of upper occipitals, but former sometimes slightly longer *Phalacrotophora* (part)
– Palp bristles and at least upper pair of supra-antennals clearly longer than upper occipitals 155
155. Third antennal segment globose 156
– Length of third antennal segment clearly greater than greatest breadth 162
156. At front end of notopleuron there is the opening of duct from notopleural gland, either fully exposed (Fig. 8.28b) or concealed (in part at least) within a notopleural cleft (Fig. 8.93b, c). (Anal tube at most only about as long as epandrium.) 157
– No such opening, or cleft, at front of notopleuron 158
157. Opening of duct from notopleural gland at least partly concealed within a notopleural cleft (Fig. 8.93b, c). Length of this cleft is greater than diameter of anterior thoracic spiracle *Woodiphora* (part)
– Opening of duct from notopleuron gland fully exposed to view and its

diameter is less than that of anterior spiracle (Fig. 8.28b) *Beckerina* (part)

158. Costa at least half wing length 159
 – Costa less than half wing length *Megaselia* (part)
159. Anal tube at most as long as mid-line length of epandrium 160
 – Anal tube clearly longer than epandrium 161
160. Costal at most only half wing length *Megaselia* (part)
 – Costa more than half wing length *Megaselia* (part) and *Trophithauma* (part)

Note: These two genera can only be separated in the female sex; indeed the latter may only be aberrant species of *Megaselia*.

161. With at most only three bristles on axillary ridge of wing *Megaselia* (part)
 – With at least four axillary bristles *Megaselia* (part), *Physoptera* (part) and *Rhyncophoromyia* (part)

Note: The males of these three genera are difficult to separate at the generic level. The last two are restricted to the Nearctic and Neotropical Regions, and both have obviously elongated anal tubes.

162. Anal tube at most as long as mid-line length of epandrium 163
 – Anal tube clearly longer than epandrium (e.g. Fig. 8.25) 166
163. Costa extends less than half wing length 164
 – Costa extends more than half wing length 165
164. Inner branch of fork of vein 3 either disconnected at base or clearly much thinner than outer branch (R4+5). Third antennal segment oval *Syneura* (part)
 – Inner branch of fork of vein 3 intact and strong, even if a little thinner than outer branch. Third antennal segment oval or pear-shaped *Myrmosicarius* (part)
165. Third antennal segment bean-shaped. Only two supra-antennal bristles *Diaclinella*
 – Third antennal segment more nearly globose. With four supra-antennals *Megaselia* (part)
166. Costa extends to middle of wing or beyond 167
 – Costa less than half wing length 171
167. With at least three bristles on axillary ridge of wing 168
 – With fewer than three axillary bristles 170
168. Third antennal segment conical (tapering towards apex) 169
 – Third antennal segment more oval *Megaselia* (part)
169. Conical third antennal segment with long, tapered, apical portion. With four supra-antennal bristles. With at least four axillary bristles *Macrocerides* (part)
 – Without this combination *Apocephalus* (part)
170. Vein Sc obscure. Third antennal segment pear-shaped. Only two supra-antennal bristles. Antial and antero-lateral bristles present *Anaclinusa*
 – Without these characters in combination *Apocephalus* (part)
171. Anal tube relatively long and slender (e.g. Fig. 8.25) 172

– Anal tube relatively short and broad (e.g. Fig. 8.57b) **173**
172. Supra-antennal bristles as strong as antero-laterals *Neophora*
– Supra-antennal bristles either absent or clearly shorter than antero-laterals or, if latter are absent, than antials *Apocephalus* (part) and *Dacnophora*

Note: It is not clear how to distinguish these two genera in the male sex.

173. Epandrium longer than high and postero-lateral lobes well developed below anal tube (e.g. Fig. 8.57b) *Microselia* (part)
– Epandrium as high as high as long, or a little higher, without obvious postero-lateral lobes 174
174. Vein 3 forked *Synclinusa*
– Vein 3 unforked *Pradea*
175. Arista absent, but apex of third antennal segment drawn out into a long pseudo-arista (e.g. Fig. 8.36a) 176
– Arista present and articulated with third antennal segment 177
176. Both hairs and bristles on frons, in front of anterior ocellus *Rhopica* (part)
– Only hairs on frons, in front of anterior ocellus *Dahliphora* (part)
177. Vein 3 forked 178
– Vein 3 unforked **187**
178. Mesopleuron with fine hairs in postero-dorsal region and notopleuron with a cleft (Fig. 8.47a,b). Frons with only small hairs and no bristles in front of ocelli. Costal cilia only about as long as breadth of costal section 3 and costal section 1 clearly longer than sections 2+3 *Gymnophora*
– Without this combination and mesopleuron always bare 179
179. Third antennal segment with elongated, tapered, apical part 180
– Third antennal segment essentially globose, even if a little pointed apically 181
180. Frons much wider than high *Corynusa*
– Frons exceptionally narrow, with height very much greater than breadth *Cyphometopis* (part)
181. The bend in basal third of vein 5 is opposed to forward deflection in middle of vein 6, so that space between the two veins is characteristically narrowed just before middle (as in Fig. 8.56a) *Metopina* (part)
– Without this feature 182
182. Apart from supra-antennals, there are fewer than eight bristles in front of pair on ocellar triangle. (Frons clearly broader than high.) 183
– The normal eight bristles are present on frons in front of ocellars (as in Fig. 8.1a) 184
183. Medio-lateral bristles absent *Macroselia*
– Medio-lateral bristles present, even if somewhat higher on frons than usual *Aphiura*
184. Costa clearly extends a little way beyond tip of vein 3 *Allochaeta*
– Costa ends at tip of vein 3 185
185. Vein Sc obscure or absent *Phymatopterella* (part)
– Vein Sc strongly developed and its tip merging with vein 1 186
186. Opening of duct from notopleural gland fully exposed to view and its

diameter less than diameter of the anterior thoracic spiracle (Fig. 8.28b) *Beckerina* (part)

– Opening of duct from notopleural gland at least partly concealed in a notopleural cleft (Fig. 8.93b,c), the length of latter being greater than diameter of anterior thoracic spiracle *Woodiphora* (part)

187. Eyes greatly enlarged and almost meeting behind ocelli, and frons narrow (Figs 8.70c and 8.73b). Wing with very short costa and weakly developed vein 1 (Fig. 8.70d) *Postoptica*

– Without this combination 188

188. Front tibia with apical ventral process and/or pre-apical triangular notch in ventral margin *Apterophora*

– Front tibia without apical ventral process or ventral notch 189

189. Front metatarsus with notch in middle of ventral margin. Scutellum with two bristles and four hairs (two between the bristles and one either side) *Eurycnemis*

– Without these features 190

190. Axillary ridge of wing with at least one bristle 191

– Axillary ridge of wing without bristles 195

191. Subcosta strongly developed. Anal tube at most subequal to length of epandrium, usually clearly shorter 192

– Subcosta obscure or absent. Anal tube clearly longer than epandrium *Phymatopterella* (part)

192. With the posterior pair of bristles on scutellum shorter than the anterior pair. (Frons about twice as broad as long.) *Paraphiura*

– The four scutellar bristles either subequal or the anterior pair shorter, and occasionally reduced to fine hairs 193

193. Front tarsi modified as in Fig. 8.92a,b *Woodiphora* (part)

– Front tarsi not modified in this way 194

194. Third antennal segment oval. Right side of epandrium with tapered lobe from hind margin extending almost as far as tip of anal tube, which is about as long as epandrium. With four bristles on scutellum *Distichophora* (part)

– Third antennal segment globose. Epandrium without tapered lobe from hind margin. If anal tube is almost as long as epandrium then scutellum with posterior pair of bristles and anterior pair of minute hairs *Woodiphora* (part)

195. Wings somewhat reduced (Fig. 8.59a,b). Posterior ocelli absent. No frontal bristles apart from one reduced pair of supra-antennals and ocellar pair *Metopina* (part)

– Without this combination 196

196. Front femur with at least six long, subventral, hairs in basal half (these hairs at least as long as greatest breadth of front tibia) *Homalophora*

– Without these long hairs in basal half of front femur 197

197. Front femur somewhat swollen and with a darkened patch just beyond middle (Fig. 8.53a,b). Costa clearly less than half wing length and veins 5 and 6 much weaker than veins 4 and 7 (Fig. 8.53c,d). (Hypopygium as Fig. 8.11a,b.) *Maculiphora*

– Without these features in combination 198

198. Anal lobe of wing large, so that inner margin of left wing viewed from above is strongly S-shaped. Vein 1 with at least distal half missing. Costa only about two-fifths of wing length. Third antennal segment conical, with an apical arista. Both antero-lateral and medio-lateral bristles of frons absent *Travassophora*
– Not with this combination 199
199. Costa at most only one-third of wing length. Anal tube clearly shorter than epandrium 200
– Costa at least two-fifths of wing length. Anal tube shorter or longer than epandrium **208**
200. Vein 1 distinct. Vein 3 straight and ending at tip of costa. Vein 4 originates well beyond tip of vein 3. Veins 4–6 end very close to wing margin *Brachycosta*
– Without this combination 201
201. Arista long-haired (Figs 8.76b and 8.84b) 202
– Arista short-haired (e.g. Fig. 8.63a) or a little longer (Fig. 8.84c) 203
202. Second segment of arista longer than broad (Fig. 8.84b). Veins 4–6 end very close to wing margin (Fig. 8.85b) *Termitophilomyia*
– Second segment of arista broader than long (Fig. 8.76b). Veins 4–6 end well clear of wing margin (Fig. 8.76a). (Front leg as Fig. 8.76c.) *Ptochomyia*
203. Vein 4 originates beyond apex of vein 3 (e.g. Fig. 8.82a, d) 204
– Vein 4 originates at or before apex of of vein 3 (e.g. Figs 8.63c and 8.88a) 205
204. A clearly differentiated pair of supra-antennal bristles present. Vein 1 distinct. Veins 4–6 almost reach wing margin. A faint vein 7 present *Brachyphlebina*
– Any hairs in position of supra-antennals are at most only as strong as other small hairs on frons. Vein 1 obscure, veins 4–6 ending somewhat to considerably short of wing margin, and no indication at all of vein 7 (Fig. 8.82a, d) *Syntermitoxenia*
205. Left surstylus with jagged antero-ventral margin (Fig. 8.33a) *Perissa*
– Left surstylus with smooth, or only slightly irregular, antero-ventral margin (Figs 8.87c and 8.88b) 206
206. Vein 5 directed towards wing tip (Fig. 8.88a). Second segment of arista with breadth greater than or equal to length (Fig. 8.88c). (Hypopygium as Fig. 8.88b.) *Javanoxenia*
– Vein 5 directed to point behind wing tip (Fig. 8.87d). Second segment of arista longer than broad (Figs 8.63a and 8.84c) 207
207. Fore tarsal segments relatively long (Fig. 8.63b). With two fine bristles in lower half of propleuron. (Antenna as Fig. 8.63a, wing as Fig. 8.63c, hypopygium as Fig. 8.69a.) *Necperissa*
– Fore tarsal segments relatively short (Fig. 8.84d). Propleuron bare. (Hypopygium as Fig. 8.87c.) *Termitoxenia*
208. Outer half of costal section 1 with sudden swelling *Pachyneurella*
– Costal section 1 either of uniform thickness throughout or only gradually increasing in breadth 209
209. Ocelli absent. (Vein Sc strongly developed and its tip joining

R1.) *Wandolleckia*
- Ocelli present 210

210. Anal tube clearly shorter than epandrium and cerci little, if any, longer than high (e.g. Fig. 8.30.) 211
- Anal tube subequal or longer than epandrium. When a little shorter, then cerci are clearly longer than high 214

211. Vein 6 (third thin vein) deflected forwards in middle third, or just beyond, so that space between veins 5 and 6 is a little to very markedly (Fig. 8.56a) constricted. The microsetae fringing the rear margin of wing only about as long as the distance between adjacent microsetae 212
- Veins 5 and 6 almost straight or only slightly sinuous, so there is no obvious constriction of space between these two veins (e.g. Fig. 8.29c). The microsetae fringing rear margin of wing about three times as long as distance between adjacent microsetae 213

212. Vein 6 strongly deflected and anal region (i.e. part of wing bearing anal vein – vein 7) with inner border at a steep angle to axillary ridge (Fig. 8.56a) *Metopina* (part)
- Vein 6 less strongly deflected and anal/axillary angle shallow *Penthaplus*

213. Wing as Fig. 8.29(c), the base of Rs joining base of vein 1 at an oblique angle. (Hypopygium compact and often complex – Figs 8.15a and 8.30.) *Chonocephalus*
- Wing with base of Rs joining base of vein 1 at an angle of about 90 *Epichonocephalus*

214. Vein Sc strongly developed and its tip fusing with vein 1 (as in Fig. 8.5b) 215
- Vein Sc less strongly developed, and its tip not reaching vein 1, or absent 217

215. Anal tube a little shorter than mid-line length of epandrium. Lower margin of left side of epandrium concavely emarginate in middle. Palps large (breadth being greater than diameter of third antennal segment) with only short fine hairs. At most only pre-ocellar bristles on frons between supra-antennals and ocelli *Trophodeinus*
- Without this combination 216

216. Both antial and antero-lateral bristles absent Pre-ocellars and medio-laterals present. (Hypopygium as Fig. 8.54b,c, wing as Fig. 8.59d, front leg as Fig. 8.59d.) *Mesopathusa*
- Antero-lateral and/or medio-lateral bristles absent. Antials present *Commoptera* and *Puliciphora*

Note: It is not clear how to distinguish these two genera in the male sex. Borgmeier (1926a) erroneously stated that the Sc vein is absent in *Commoptera*, despite it being illustrated as being present. The error was then perpetuated in his Nearctic monograph (Borgmeier, 1963d).

217. Costa just under half wing length. Vein Sc evident, and sometimes with 1–2 small bristles at base *Acontistoptera*
- Costa extending just over half wing length. Vein Sc without bristles and frequently absent altogether *Ecitomyia*,

Ecitophora,
Ecitoptera
and *Ecituncula*

Note: These genera are only reliably separated in the female sex, in our present state of knowledge.

218. Vein 4 runs straight to wing margin and vein 3 has a slight angle near origin of vein 4 — *Pseudacteon* (part)
– Vein 4 almost reaches wing margin but then turns sharply rearwards and gradually approaches wing margin at a narrow angle. Vein 3 straight in vicinity of origin of vein 4 — *Trichocerina*

(C) KEY TO GENERA OF FEMALES

1. Wings and halteres entirely absent (e.g. Figs 6.7, 8.20b, 8.22a). Sometimes wing rudiments fused to thorax and not articulated with it (e.g. Fig. 8.44a) — 2
– Wings present, even if reduced to rudiments or stumps (e.g. Figs 6.4–6.6, 8.26b, 8.55a, 8.77a, 8.81d and 8.90). These rudiments articulate with thorax. Halteres usually present — **47**

Note: In some cases the wing rudiments are partly fused to the thorax. Sometimes developments of the anterior segments of the abdomen may tend to conceal wing stumps (e.g. Fig. 8.46a). These cases are keyed both ways.

2. End segments of abdomen directed rearwards — 3
– End segments of abdomen deflected ventrally, even to the extent of the anus being directed forwards (e.g. Fig. 5.1) — **87**
3. Dorsal-basal region of abdomen extended forwards over top of thorax (Fig. 8.46a). First antennal segment as long as tapered third segment (Fig. 8.46b) — *Eutermiphora* (part)
– Without this combination — 4
4. Proboscis more than four times length of head, with distal section elongated. (Abdominal tergites 1–6 present, but 6 reduced, and 5 with antero-median gland complex. Only two ocelli present, the anterior one being absent.) — *Apterophora*
– Proboscis little, if any, longer than head and usually shorter — 5
5. Abdominal tergites entirely absent — 6
– At least tergite 2 is present, even if weakly pigmented — 9

Note: Variable genera are keyed both ways.

6. A pair of bristles in basal half of mid tibia. Three ocelli present — *Diplonevra* (part)
– Mid tibia without pair of bristles in basal half, but there may be some long hairs along dorsal face. Ocelli absent — 7
7. Diameter of eye clearly less than diameter of third antennal segment — 8
– Diameter of eye subequal to diameter of third antennal segment — 9
8. All tibiae bearing dorsal, or near-dorsal, hairs which are longer than greatest breadth of tibia. Similar hairs on femora and most tarsal segments. Dorsal faces of abdominal segments 2–5 also with conspicuous hairs — *Oligophora*

– Hairing of legs and abdomen much shorter 9
9. Abdominal tergites greatly reduced in number, with only 2 (which is sometimes very pale), and sometimes 5 as well, present 10
– Abdominal tergites 2–5 present, and sometimes 1 and 6 also **19**
10. Ocelli present 11
– Ocelli absent 12
11. Thorax (viewed from above) more than twice as wide as long. Abdomen with no trace of tergite or associated gland opening on segment 5 *Apterella*
– Thorax at most only twice as wide as long. Abdominal segment 5 with a reduced tergite associated with a gland opening 12
12. Diameter of eye clearly less than diameter of third antennal segment 13
– Diameter of eye subequal to, or greater than, diameter of third antennal segment 16
13. Abdominal tergite 2 absent *Trophodeinus* (part)
– Abdominal tergite 2 present 14
14. Abdominal tergite 2 with front margin broader than hind margin *Trophodeinus* (part)
– Abdominal tergite 2 with hind margin broader than front margin 15
15. Hairs on dorsal faces of abdominal segments 1–5 short, being clearly shorter than bristles on thorax. Sometimes with long hairs on flanks of segments 3 and 4 *Ecituncula* (part)
– Hairs of dorsal faces of abdominal segments 1–5 long, some being almost as long as bristles on thorax, contrasting with short hairs on tergite 2 (apart from longer ones on hind margin) *Rhabdomyia* (part)
16. Abdominal tergite 2 extends full width of segment, but sometimes largely represented by patch of longer hairs contrasting with shorter hairs on rest of abdominal dorsal face 17
– Abdominal tergite 2 reduced, and clearly not extending full width of segment 18
17. Abdominal tergite 2 mainly represented by longer hairs contrasting with rest of hairs on dorsal face of abdomen. No gland complex evident on segment 5 *Morahania*
– Abdominal tergite 2 distinctly demarcated and bearing short hairs (no longer than hairs, as opposed to bristles, on thorax). A distinct gland opening on segment 5 *Zikania*
18. Abdominal tergite 2 at most as broad as long, and usually longer than broad. General hairing of dorsal face of abdomen shorter than hairs on tarsi *Ecituncula* (part)
– Abdominal tergite 2 more than twice as wide as long. General hairing of dorsal face of abdomen clearly longer than hairs of tarsi *Wandolleckia*
19. Not strongly dorso-ventrally flattened; the anterior thoracic spiracles in normal, lateral, position; head usually not very much wider than long; antennae usually closely approximated; the postero-lateral corners of thoracic notum not forming backwardly-directed tapered processes 20
– Strongly dorso-ventrally flattened; head very much wider than long; antennae widely separated; anterior thoracic spiracle in dorsal position; postero-lateral extremities of thoracic notum may form tapered

processes (representing wing rudiments fused to notum) (e.g. Fig. 8.44) **54**

20. Hind femur with dorsal, pre-apical, feathered bristle and all tibiae with similar dorsal or near-dorsal bristles (Fig. 8.19). Antennae widely separated and frons with long hairs (Fig. 8.20a). Thorax and abdominal tergites also long-haired (Fig. 8.20b) *Aenictacantha*
– Without these features in combination 21
21. Both eyes and ocelli absent. Third antennal segment pear-shaped. Abdominal tergites 2 and 3 short but wide, with long bristles on their concave hind margins. Long bristles along dorsal faces of mid and hind tibiae *Cryptophora*
– Without this combination 22
22. No supra-antennal bristles in middle at front of frons, as they are displaced laterally in association with the wide separation of antennae (e.g. Fig. 6.7). Apart from bristles on vertex, which are either reduced or absent, there are no other bristles on frons 23
– Supra-antennal bristles, usually two pairs, situated in middle at front of frons. Antennae usually more closely approximated. There may be other bristles on frons, apart from supra-antennals and those on vertex 25
23. Abdomen with six tergites extending full widths of segments (Fig. 6.7). (Terminal abdominal segments as Fig. 8.8a, b) *Chonocephalus*
– Abdomen with at most only five tergites extending full widths of segments 24
24. Only three abdominal tergites extending full widths of segments. (Apart from antero-laterally situated supra-antennals, with no bristles on frons. Ocelli absent. Eyes much reduced. Long bristles on flanks of abdominal segments 3–6.) *Paurophora*
– Abdomen with five tergites extending full widths of segments 25
25. Dorsal face of thorax at least four times as wide as long (e.g. Fig. 8.55d) 26
– Dorsal face of thorax less than four times as wide as long (but if almost four times as wide as long, then arista is shorter than palp) 28
26. At least four abdominal tergites extending full widths of segments (e.g. tergites 4–6 as Figs 6.14 and 8.55b, c) 27
– Abdominal tergites 3–5 progressively much narrower than widths of segments *Stethopathusa*
27. At rear of abdominal tergites 2–5 at least some of the bristles are as strong as those on thorax (e.g. Fig. 8.55c) *Mesopathusa* (part)
– Bristles at rear of abdominal tergites 2–5 not as strong as those on thorax *Puliciphora* (part)
28. Abdominal tergites 1 and 2 tapered laterally and extending well down flanks (Fig. 8.41a) *Dundophora*
– Abdominal tergites 1 and 2 not tapered in this way 29
29. A pair of ocellar bristles present (i.e. a pair of closely approximated bristles in middle of vertex) even though ocelli may be absent. If these bristles are somewhat small, then diameter of eye at least equal to that of third antennal segment 30

– Ocellar bristles absent. (Ocelli absent; antennae widely separated; dorsal face of thorax more than three times as wide as long; four abdominal tergites as wide as segments; diameter of eye less than that of third antennal segment; arista shorter than palp.) *Tranopeltoxenos*

30. Ocelli absent 31
– Ocelli present **40**

Note: Variable genera are keyed both ways.

31. At least four bristles on frons between supra-antennals and those on vertex 32
– At most only two bristles on frons between supra-antennals and those on vertex 33

32. Abdominal tergites 2–4 each with two transverse rows of strong bristles (as strong as those on thorax), one row in anterior half and one at rear margin. Tergite 5 with a large semicircular flap occupying a wide emargination of the front margin (similar to Fig. 6.14 right) *Hexacantherophora*
– Abdominal tergites 2–4 each with at most a single transverse row of, weaker, bristles at rear margin only. Tergite 5 with only a small emargination of middle of front margin (less than one third of tergite width) and without a semicircular flap *Aptinandria*

Note: Since I completed this key, Brown (1993a) has described new species of *Borgmeieriphora* whose females will run to this lead (rather than couplet 69), as they lack wing rudiments. Their down-curved, pointed ovipositors, with cerci vestigial or absent, will distinguish them from *Aptinandria*. If none of these options apply, proceed to couplet 40.

33. Eyes greatly reduced, their diameters being at most only one-third of that of third antennal segment 34
– Diameter of an eye at least two-thirds, and usually greater than, that of third antennal segment 36

34. Frons without bristles between supra-antennals and those on vertex. Without any indication of a gland opening in middle of front margin of abdominal tergite 5 *Syntermophora*
– Frons with pair of bristles between supra-antennals and those on vertex. A small semicircular flap occupying indentation in middle of front margin of abdominal tergite 5 (e.g. Fig. 8.78a, b) 35

35. Abdominal tergite 4 much wider than abdominal tergite 5 (e.g. Fig. 8.78a) *Puliciphora* (part)
– Abdominal tergite 4 scarcely any wider than tergite 5. 36

36. Abdominal tergite 4 only as wide as tergite 5. Frons with small pair of antial bristles, four supra-antennals, and pair of ocellars on vertex *Hylophora*
– Abdominal tergite 4 wider than tergite 5. With no bristles on frons between supra-antennals and those on vertex 37

37. Abdominal tergites 2–6 extend full widths of segments 38
– At least tergite 5 is much narrower than width of segment (e.g. Fig. 8.78b) *Puliciphora* (part)

38. Abdominal tergite 5 with semicircular flap (associated with a gland opening), which may be small (e.g. Fig. 8.78a, b), in middle of front

margin 39

– No such flap (or gland opening) at front of abdominal tergite 5 *Coridophora* (part)

39. Anterior flap at front of abdominal tergite 5 less than one quarter of width of tergite (e.g. Fig. 8.78b) *Puliciphora* (part)

– Anterior flap at front of abdominal tergite 5 more than half width of tergite (Fig. 8.18a) *Adenophora*

40. Frons without bristles between supra-antennals and those on vertex 41

– Frons with at least two bristles between supra-antennals and those on vertex. (With gland complex in middle of front margin of abdominal tergite 5, e.g. Fig. 6.14.) 44

41. Gland opening in middle of front margin of abdominal tergite 5, usually associated with small semicircular flap (e.g. Fig. 6.14) *Pachyneurella* and *Puliciphora* (part)

Note: These two genera can only be separated in the male sex, in our present state of knowledge).

– No such gland complex at front of abdominal tergite 5 42

42. Top of thorax with only four strong bristles near rear margin, there being none on middle third *Coridophora* (part)

– With six strong bristles near rear of thorax, there being a pair in middle third 43

43. Viewed from above, back of head concave *Homalophora*

– Back of head not concave *Penthaplus*

44. Top of thorax with 4–6 bristles near hind margin 45

– Top of thorax with only a single pair of bristles at rear *Placophorina*

45. Eyes greatly reduced, their greatest diameter being less than half that of third antennal segment *Arrenaptenus*

– Eyes with greatest diameter clearly more than half that of third antennal segment 46

46. Viewed from above, diameter of eye at most only slightly greater than that of third antennal segment, usually subequal or less. There is a bristle, sometimes on a distinct papilla, on side of thorax towards rear (representing the last vestige of the ancestral wing). (Abdominal tergites as Fig. 6.14.) *Puliciphora* (part)

– Viewed from above, diameter of eye clearly greater than that of third antennal segment. There is no bristle on side of thorax towards rear (i.e. in the position that would have been occupied by the wing in the ancestral state) *Myrmomyia*

47. Wings with at least membranous area (part bearing the thin veins) reduced 48

– Wings fully developed (e.g. Fig. 1.1) **84**

Note: Borderline cases, where the membranous area is only reduced a little (e.g. Fig. 6.2b) are keyed both ways.

48. Wings represented by intact, but reduced, wings or rudiments (e.g. Figs 6.2c, d, 6.3–6.6, 8.26, 8.42d, 8.52a, 8.55a, 8.79d and 8.90) 49

– Wings represented by stumps, the membranous area having been largely shed, leaving a ragged outer margin (e.g. Figs 6.1, 8.77a) **84**

49. Wing membrane greatly narrowed, but not obviously shortened (e.g. Fig. 6.3) 50
– Wing otherwise 51
50. Costa narrow (no wider than vein 3) with at most two rows of costal cilia, which are clearly longer than width of costa in outer half. Veins 4–6 (first three thin veins) evident. The tergite rudiments associated with gland opening on abdominal segment 5 relatively small *Chelidophora*
– Costa broad, with at least three irregular rows of hairs which are relatively short. Veins 4–6 not evident (e.g. Fig 6.3). Semicircular flap associated with gland opening on abdominal segment 5 relatively large (Fig. 8.45b) *Franssenia*
51. Wing rudimentary in that the membrane (i.e. part beyond end of costa and behind vein 3) is greatly reduced or absent. When some of membrane present costal index exceeds 0.9 52
– Wing complete, even if membrane is shorter than usual. Costal index always less than 0.9 **84**
52. Strongly dorso-ventrally flattened and abdominal segments 1–6 largely covered by single large tergite; antennae in pits (Figs 8.26 and 8.90) 53
– If strongly dorso-ventrally flattened, then abdominal tergites 1–6 largely covered by six separate tergites or 3–6 reduced so that much of dorsal face is membranous. Antennae usually clearly not in pits **54**
53. As Fig. 8.90, with head very much broader than long and proboscis shorter than head. Hind femur with at least one (usually two) pre-apical anterior spines (Fig. 8.91a) *Thaumatoxena*
– Head and thorax as Fig. 8.26, with proboscis longer than head. Hind femur without pre-apical anterior spines *Assmutherium*
54. Six tergites evident and entirely covering dorsal face of abdomen. Frons with anterior face at right angle to top face (Fig. 8.23b, c) and sometimes latter greatly shortened (Fig. 8.24). Wing rudiments simple (Fig. 6.4) *Aenigmatistes*
– Without this combination 55
55. Without wing rudiments or they are fused to postero-lateral corners of thorax, and do not articulate with it (e.g. Figs 8.22a and 8.44) 56
– Wing rudiments present, and articulating with thorax 58
56. Diameter of eye greater than that of third antennal segment (Fig. 8.22a). At rear of abdominal segment 7 with a series of strong hooks each side (Fig. 8.21b, c) (Hind tibia as Fig. 8.60b.) *Aenigmatias*
– Diameter of eye less than that of third antennal segment. No strong hooks at rear of abdominal segment 7 57
57. Palps globose (Fig. 8.44b). Mid femur with antero-ventral row of strong, dark, spine-like bristles in outer half (Fig. 8.43b, c) *Epicnemis*
– Palps flattened (Fig. 8.23a). Antero-ventral row of bristles on outer half of mid tibia weaker and no darker than adjacent hairs *Aenigmatopoeus*
58. Wing rudiment with very long, strong, feathered bristles on costa (e.g. Figs 8.15b and 8.17b) 59
– No such bristles on costa 61
59. Wing with radial veins distinct and thin veins also evident (Fig. 8.17b).

Normal halteres present. Ocelli present *Adelopteromyia* (part)
- Radial veins obscure and thin veins not discernible (e.g. Fig. 8.15b). Normal halteres may be present, but usually strongly reduced or absent. Ocelli absent 60

60. Frons with only four supra-antennal bristles, arranged two in front and two behind, at front of frons between moderately separated antennae. Scutellum with pair of hairs, weaker than pre-scutellar dorso-central bristles, or bare *Xanionotum*
- Frons with at least six bristles at front, arranged in a single transverse row, between widely separated antennae. Scutellum with pair of bristles at least as strong as pre-scutellar pair of dorso-centrals *Acontistoptera*

61. Wing rudiment with costa and radial veins distinct 62
- Wing rudiment more reduced, the radial veins (at least) being not discernible 67

Note: Variable cases are keyed both ways.

62. Proboscis clearly longer than head. Ocelli absent. A long bristle at apex of costa and two similar bristles on vein 3 (Fig. 8.75a). (Hind tibia as Fig. 8.60c.) *Psyllomyia*
- Without this combination 63

63. Mesopleuron with one or more hairs and/or bristles 64
- Mesopleuron bare 65

64. Mesopleuron with very strong, very long, bristle (about as long as costa), and sometimes a few minute hairs as well. Wing somewhat rounded and with very long costal bristles (Fig. 6.2d) *Hypogeophora*
- Bristles on mesopleuron much shorter and finer. Wing more than 2.5 times as long as greatest breadth and with short costal hairs only *Gymnophora* (part)

Note: If palps are unusually large and two-segmented (Fig. 8.37a,b) and abdominal tergites are reduced to lateral remnants of 1 and 2 only (Fig. 8.37c), go to couplet 248.

65. Abdominal tergite 2 at least 1.5 times as broad as long, and covering at least three-quarters breadth of segment 66
- Abdominal tergite 2 with breadth at most only 1.2 times length, and covering at most only half breadth of segment *Commoptera*

66. Wing rudiment relatively long and narrow (greatest breadth being about one-third of length) and veins 1 and 3 clearly defined *Thalloptera*
- Wing rudiment usually relatively short and broad (greatest breadth being at least two-fifths of length) and veins 1 and 3 poorly defined (e.g. Figs 6.5a, 8.55a) 67

67. *Note:* This couplet is reached by two different routes to allow for differences in the wing rudiments:
- Wing rudiment long and narrow, with narrow membrane behind hairy, basally broadened, costa (Figs 6.5a, 8.81d). Proboscis clearly longer than head (e.g. Figs 8.80b, 8.81a) *Rhynchomicropteron*
- Wing rudiment and proboscis otherwise 68

68. Cerci normally developed. Abdominal tergites 1–5 all broader than long and extending full widths of segments (e.g. tergites 4–6 as Fig.

8.55b,c). Thorax much broader than long (e.g. Fig. 8.55d). Wing rudiment simple, with long terminal bristle (Fig. 8.55a) *Mesopathusa* (part)

– Without this combination 69

69. Abdominal tergites 1–5 wider than long and extending full widths of segments. Tergite 6 a little narrower, but still wider than long. Ocelli absent. Diameter of eye sometimes less than that of third antennal segment. Wing rudiment a short lobe with up to six costal cilia, the end one being as long as rudiment itself. Sometimes only this bristle is present. Abdomen ending in sharply pointed, sclerotized, ovipositor *Borgmeieriphora*

– With a different combination 70

70. Wing rudiment long and rod-shaped (being longer than hind femur). Mid tibia with dorsal and antero-dorsal bristle in basal third and second dorsal bristle near middle. Abdomen with tergites 1–6 present, but progressively narrower from 2 to 6. Scutellum absent *Contopteryx*

– Not with this combination 71

71. Wing rudiment an elongated irregular triangle (e.g. Figs 6.5b and 8.52a). Abdominal tergite 2 clearly broader than long and extending most, if not all, width of segment (e.g. Figs 6.6 and 8.52b) 72

– Wing rudiment otherwise. Abdominal tergite 2 usually extending much less than full width of segment and sometimes longer than broad 73

Note: Since completion of this key, Brown (1993a) has described a new genus *Cootiphora*, which runs to this couplet. It has no ocelli; widely separated antennae; a dorso-ventrally flattened body, with dorsally-situated anterior thoracic spiracles; a transversely-triangular wing rudiment tapering rearwards, whose greatest breadth is more than wing length; and abdominal tergites 2–6 extending full widths of segments.

72. Ocelli absent and eyes greatly reduced; abdominal tergite 2 extends full width of segment; no other tergites on abdomen (Fig. 6.6). Mid and hind femora each with a near-dorsal pre-apical spine (Fig. 8.51c). Mid and hind tibiae with near-dorsal, spine-like, bristles (Fig. 8.51b). Wing rudiment as Fig. 6.5b *Maculiphora*

– Ocelli present. Eye with diameter at least as great as that of third antennal segment. Abdominal tergites as Fig. 8.52b, or with tergites 3,4 present or absent. Mid and hind femora without pre-apical spines. Mid and hind tibiae without near-dorsal, spine-like, bristles. Wing rudiment as Fig. 8.52(a) *Lepidophoromyia* (part)

73. Wing rudiment paddle-shaped (Fig. 8.79d). Abdomen with strong bristles arising from sclerotized discs, which may be large (e.g. Fig. 8.80a) *Rhabdomyia* (part)

– Wing rudiment not paddle-shaped. Abdominal bristles usually much weaker and only very rarely arising from sclerotized discs 74

74. Ocelli absent 75

– Ocelli present 79

Note: Variable cases are keyed both ways.

75. Thorax clearly longer than broad and sharply narrowed behind, before

reaching slightly broader (confluent) scutellum. Postero-laterally rear margin of head each side is extended rearwards as a rounded lobe *Lenkoa*

– Thorax usually a little broader than long or length and breadth subequal. No rearwardly-directed lobe each side of head 76

76. Eyes reduced, so that diameter of third antennal segment is clearly greater than diameter of eye. (Wing rudiment a long, narrow, hairy appendage.) 77

– Diameter of eye subequal to or greater than that of third antennal segment 78

77. Abdominal tergite 2 clearly longer than wide *Aemulophora*

– Abdominal tergite 2 clearly wider than long *Idiophora*

78. Wing rudiment very short (at most only four times as long as broad). Abdominal segments 3 and 4 devoid of tergites *Lepidophoromyia* (part)

– Wing rudiment clearly longer. At least tergite 4 is present, even if small 79

79. Abdominal tergite 2 clearly broader than long 80

– Abdominal tergite 2 as long as broad or longer (e.g. Fig. 8.42a), but sometimes very pale 82

80. Abdominal tergites 3–5 well developed, but progressively narrower, and 5 with gland complex (Fig. 6.14) 81

– Abdominal tergites 3–5 absent. (Tergite 2 narrower at rear than at front. Wing rudiment at most only four times as long as wide.) *Colobomeles*

81. Wing rudiment a very small finger-like structure (at most only three times as long as broad) bearing minute hairs only. *Pulicimyia*

– Wing rudiment a rounded, membranous lobe bearing hairs and at least one bristle at, or near, tip *Puliciphora* (part)

82. Wing rudiment reduced to haltere-like structure (W in Fig. 8.42d) *Ecitophora* (part)

– Wing rudiment a long, narrow, hairy appendage (e.g. Fig. 8.42b) 83

83. A pair of scutellar hairs present at rear of thorax (S in Fig. 8.42d) *Ecitophora* (part)

– No scutellar hairs present *Ecitomyia*

84. Wing with only two thin veins (e.g. Fig. 8.63d); or, if membranous region bearing thin veins has been shed, tip of abdomen is deflected ventrally (e.g. Fig. 5.1) 85

– With at least three thin veins; or, if membranous region has been shed, tip of abdomen directed rearwards **108**

85. First antennal segment not elongated 86

– First antennal segment elongated (e.g. Fig. 8.46b) **108**

86. No gland opening evident in middle of front margin of abdominal segment 5 **87**

– A distinct gland opening evident in middle of front margin of abdominal segment 5 (as Fig. 8.45b, but smaller) **108**

87. Palp enlarged apically and without bristles at tip *Pseudotermitoxenia*

– If palp a little enlarged apically, then at least one bristle at tip 88

88. At least 40 fine bristles on thoracic scutum. (Tip of costa extensively overreaches tip of costa.) *Termitoscrofa*
– With far fewer fine bristles on thoracic scutum 89
89. Arista long, but sparsely, haired (e.g. Fig. 8.84a) to very long-haired (e.g. Fig. 8.64a) 90
– Arista short, and more densely, haired (e.g. Fig. 8.64b) 95
90. Abdomen with tube-like appendages in anterior half 91
– Abdomen without such appendages 92
91. Some of hairs on sides of abdomen arising from discs (as in Figs 8.86a and 8.87b) whose rims are made up of a series of outwardly projecting lobes (appearing as dashes in plan view). Proboscis normal *Termitosphaera*
– No abdominal hairs arising from such discs. Ventral face of base of proboscis (postmentum) grossly inflated *Cheiloxenia*
92. Tip of costa clearly overreaching tip of vein 3 (e.g. Fig. 8.85a). Anterior thoracic spiracles not enlarged 93
– Tip of costa little, if any, overeaching tip of vein 3 (e.g. Fig. 8.77a). Anterior thoracic spiracles enlarged (e.g. Figs 8.64d and 8.77c). (Antenna as Fig. 8.64a, front metatarsus as Fig. 8.64c, proboscis as Fig. 8.77d.) *Ptochomyia*
93. Head as broad as long *Termitomyia*
– Head longer than broad 94
94. Abdominal tergite 2 present *Alamira*
– Abdominal tergite 2 absent (Fig. 5.1) *Termitophilomyia*
95. Costa extensively overreaches tip of Vein 3, and latter normal (i.e. almost parallel-sided) (e.g. Fig. 8.86c,d) *Termitostroma*
– Costa ends at or only a little beyond tip of vein 3, or if somewhat overreaching tip then latter is more egg-shaped (e.g. Fig. 8.88d) 96
96. Proboscis relatively short (as in Fig. 8.77d). Mesopleuron always with some fine bristles 97
– Proboscis relatively long (e.g. Fig. 8.64e). Mesopleuron frequently, but not always, bare 99
97. Palps long and club-shaped (Fig. 8.69b). (Front leg and wing as Fig. 8.69c,d.) *Perittophora*
– Palps short and more leaf-like 98
98. Vein 3 near normal (i.e. almost parallel sided), even if somewhat obscure (Fig. 8.84e) *Termitosagma*
– Vein 3 more egg-shaped and distinct (as Fig. 8.83a). '*Perissa*' (part)
99. Palps short and rounded and basiproboscis long (Fig. 8.64e). (Antenna as Fig. 8.64b, wing as Fig. 8.63d.) *Necperissa*
– Palps longer and more tapered apically (as in Fig. 8.77d) and basiproboscis usually shorter 100
100. Vein 3 almost parallel-sided. With four bristles on scutellum. Abdominal segment 2 not inflated to form a 'collar' *Paratermitoxenia*
– Without this combination 101
101. Propleuron bare 102
– Propleuron with at least one bristle in lower half and usually a second in upper half 103

102. Many fine bristles on basal segments of abdomen situated on saucer-like discs (as Figs 8.86a and 8.87b), whose rims are made up of a series of projecting lobes (appearing as dashes in plan view) *Termitoxenia*
 – No saucer-like discs present at bases of any of fine bristles on abdomen *Odontoxenia*
103. Apical section of hind femur inflated (Fig. 8.33c) 104
 – Apical section of hind femur normal 107
104. Some of fine bristles on basal segments of abdomen situated on saucer-like discs (e.g. Fig. 8.33b), whose rims are a series of outwardly-projecting lobes (appearing as dashes in plan view) 105
 – No abdominal bristles situated on such discs 106
105. Mesopleuron bare *Javanoxenia*
 – Mesopleuron with one or more (usually two) fine bristles. (Head as Fig. 8.33d.) *Clitelloxenia*
106. Back of head extended rearwards so that it faces upwards (as in Fig. 8.33d) *Indoxenia*
 – Back of head normal (i.e. facing rearwards) *Ceylonoxenia*
107. Some of fine bristles on basal segments of abdomen are situated on circular discs of crowded micropapillae (Fig. 8.82e) *Syntermitoxenia*
 – Some of fine bristles on basal segments of abdomen set on saucer-like discs (as Figs 8.86a and 8.87b), whose rims are made up of a series of outwardly-projecting lobes (appearing as dashes in plan view) '*Perissa*' (part)

108. Vein 3, apart from 1 to 3 (usually 1) hairs at base, with a row of, usually finer, hairs along upper face 109
 – Vein 3 with 0–3 hairs at base only **126**
109. Vein 1 also with row of hairs along upper face. (With pair of bristles, a dorsal and a near-dorsal, in basal third of mid tibia; mesopleuron with hairs.) *Phlebothrix*
 – With at most a single hair on vein 1 110
110. Veins 4–6 (first three thin veins) also with rows of hairs (e.g. Fig. 8.62a) 111
 – Veins 4–6 without rows of hairs 112
111. Vein 1 with single hair; scattered erect hairs on membrane between veins; vein 3 unforked. Hind tibia without dorsal longitudinal hair palisade. Mid tibia without differentiated bristles in basal third *Dorsochaeta*
 – Vein 1 bare; no scattered hairs on membrane; vein 3 forked (Fig. 8.62a). Hind tibia with dorsal longitudinal hair palisade. Mid tibia with pair of bristles (one dorsal and one near-dorsal) in basal third. (Head as Fig. 8.62b.) *Myopiomyia*
112. A pair of bristles (one dorsal and one near-dorsal) in basal half of mid tibia 113
 – No isolated bristles in basal two-thirds of mid tibia **123**
113. Vein 3 forked 114
 – Vein 3 unforked 117
114. Mesopleuron bare. Hind tibia without longitudinal hair palisades 115

– Mesopleuron with hairs. Hind tibia with two longitudinal, dorsal or near-dorsal, hair palisades *Borophaga* (part)
115. Labrum more than three times as long as greatest breadth *Burmophora*
– Labrum much shorter 116
116. With more than ten hairs on vein 3 *Anevrina*
– With fewer than ten hairs on vein 3 127
117. Hind tibia with one or two dorsal, or near-dorsal, longitudinal hair palisades (as Fig. 8.4d,e) 118
– Hind tibia without longitudinal hair palisades, but there may be row of spaced, differentiated, hairs (as Fig. 8.4c) 120
118. Frons as broad as, or broader than, high (long) 119
– Frons clearly higher (longer) than broad *Trineurocephala*
119. Posterior ocelli clearly closer to eyes than to anterior ocellus, and the ocellar region usually clearly bounded in front by a sinuous furrow (as Fig. 8.94d) *Stichillus*
– Posterior ocelli clearly closer to anterior ocellus than to eyes, and ocellar region not clearly demarcated in this way in front *Borophaga* (part)
120. Frons with one pair of bristles, the supra-antennals only. Maximum breadth of wing subequal to length of costa *Euryophora*
– Frons with at least 14 bristles. Maximum breadth of wing less than length of costa 121
121. Hind tibia with transverse combs (Fig. 8.46c). Mesopleuron with a strong bristle as well as hairs *Hypocera*
– Hind tibia without transverse combs. Mesopleuron without a strong bristle and sometimes without hairs also 122
122. Mesopleuron with hairs. Abdominal tergites 5 and 6 fused, to give only five tergites on segments 1–6 *Plethysmochaeta*
– Mesopleuron bare. With six tergites on abdominal segments 1–6 *Kuenburgia*
123. Costa clearly less than half length of wing. Eight bristles on frons between supra-antennals and ocelli. Body not obviously flattened dorso-ventrally, so anterior thoracic spiracles are on sides *Brachyselia*
– Costa clearly more than half wing length. With at most a single pair of bristles on frons below (in front of) anterior ocellus. Body somewhat flattened dorso-ventrally, so that anterior thoracic spiracles are on top 124
124. Costa clearly longer than greatest breadth of wing. Anterior ocellus as close to front margin of frons as to hind margin. Some differentiated fine bristly hairs on frons. All tibiae with a dorsal longitudinal hair palisade *Misotermes*
– Costa subequal to or shorter than greatest breadth of wing. Anterior ocellus clearly closer to hind margin than to front margin of frons. No differentiated bristly hairs on frons. All tibiae lack dorsal longitudinal hair palisades 125
125. Greatest diameter (height) of eye at least twice length of third antennal segment, and viewed from above eyes extend full length of frons *Microplatyphora*
– Greatest diameter of eye less than 1.5 times length of third antennal

segment, and viewed from above eye is less than half length of frons *Euryplatea*

126. Antennae partly concealed in pits adjacent to eyes (Fig. 8.68b). Palps short with long, unpigmented, bristles with frayed tips in addition to normal black bristles (Fig. 8.67b). Costa with short hairs and long bristles (Fig. 8.68a). (Wing membrane may be shed.) *Palpiclavina*

– Without this combination 127

127. Mid tibia with a pair of bristles (one dorsal and one near-dorsal) in basal half (as in Fig. 8.32a) 128

– Mid tibia without a pair of bristles in basal half, but occasionally a single bristle is present **154**

128. Hind tibia with at least one dorsal, or near-dorsal, longitudinal hair palisade (e.g. Fig. 8.39f) 129

– Hind tibia without longitudinal hair palisades, but there may be row of spaced, differentiated, hairs (as Fig. 8.4c) 134

129. Vein 3 forked 130

– Vein 3 unforked 132

130. Mesopleuron with hairs. (Only one longitudinal hair palisade on hind tibia. Wing with single axillary bristle. Abdominal tergite 7 typically well developed laterally, but middle third, or more, absent – e.g. Fig. 8.40c.) 131

– Mesopleuron bare **281**

131. Scutellum with six bristles, of which middle one each side is more than twice as long as other two. Bristle at base of vein 3 longer than distance between point of insertion and the fork of vein 3 *Aenictomyia*

– These bristles much shorter, and usually only four on scutellum. When there are six the middle one each side is only as long as the one behind *Dohrniphora*

132. Mesopleuron with hairs *Abaristophora* and *Antipodiphora*

Note. It is not possible, in our present state of knowledge, to distinguish these two genera in the female sex.

– Mesopleuron bare 133

133. Hind tibia with two, dorsal or near-dorsal, longitudinal hair palisades (Fig. 8.39f). Palp two-segmented (as Fig. 8.41b,c) *Diplonevra* (part)

– Hind tibia with single, dorsal, longitudinal hair palisade. Palp single-segmented. (Typically vein 1 as Fig. 8.49a.) *Hypocerides*

134. Vein 3 forked 135

– Vein 3 unforked 141

135. Mesopleuron with hairs 136

– Mesopleuron bare 137

136. Median furrow present on frons. Vein 2 and Sc weakly developed and latter usually not reaching vein 1 *Coniceromyia* (part)

– Median furrow absent. Vein 2 and Sc well developed and tip of latter confluent with vein 1 *Chaetopleurophora*

137. Frons without a median furrow 138

– Median furrow present on frons *Coniceromyia* (part)

138. Bristle at tip of each fifth tarsal segment dorso-ventrally flattened (Fig.

8.94c). Abdominal tergite and sternite of segment 7 as Figs 8.94(a, b) *Spiniphora*

– Bristle at tip of each fifth tarsal segment normal (i.e. round in cross-section). Tergite and sternite of abdominal segment 7 otherwise 139

139. The base of vein 2 missing, so that it is not attached to vein 3 to form a complete fork *Kierania* (part) and *Triphleba* (part)

Note: It is not possible to separate these two genera in our present state of knowledge. *Kierania* is only known from New Zealand. All *Triphleba* species with this feature are only known from the Holarctic Region.

– Vein 2 complete 140

140. Frons with medio-lateral bristles absent but antero-laterals present *Tarsocrates*

– Medio-laterals present and usually antero-laterals also (e.g. Fig. 8.1a) *Triphleba* (part)

141. The single bristle on axillary ridge of wing has very long, branched, hairs (Fig. 8.29a) *Beyermyia*

– Axillary bristle or bristles with only short hairs or none 142

142. Mesopleuron with hairs 143

– Mesopleuron bare 144

143. Frons without median furrow *Mannheimsia*

– Frons with distinct median furrow *Coniceromyia* (part)

144. Median furrow of frons present 145

– Median furrow absent 147

145. One or more dorsal bristles on mid tibia beyond one in basal pair of bristles. Usually more than one bristle on axillary ridge of wing. (Very black species.) *Phora*

– Only dorsal bristle on mid tibia is one in basal pair of bristles. Only one axillary bristle on wing 146

146. Antial, antero-lateral, pre-ocellar and medio-lateral bristles present on frons *Coniceromyia* (part)

– Antero-lateral bristles missing; other bristles present 147

147. Hind tibia with two or more dorsal or antero-dorsal bristles in upper two-thirds 148

– Hind tibia with at most only a single near-dorsal or antero-dorsal bristle in upper two-thirds 151

148. Hind tibia with two dorsal and one or two antero-dorsal bristles in upper two-thirds (e.g. Fig. 8.32a) 149

– Hind tibia at most with only one dorsal and with one to three antero-dorsal bristles in upper two-thirds 150

149. Hind tibia with one (Fig. 8.32a) or (rarely) two antero-dorsal bristles (the second being near lower dorsal bristle). Vein Sc usually clearly not reaching vein 1 (Fig. 8.31c), and always free when there are two antero-dorsals on hind tibia *Conicera*

– Hind tibia with two antero-dorsal bristles in upper two-thirds. Tip of vein Sc reaches vein 1 and fuses with it *Darwiniphora*

150. Costal section 1 much longer than section 2, as veins 1 and 3 are closely

approximated and subparallel (Fig. 8.65c). Hind tibia as Fig.8.65b. (Frons strongly prolongated in front – Fig. 8.65a.) *Obscuriphora*

– Costal section 1 subequal to or clearly shorter than section 2, as the tips of veins 1 and 3 are widely separated. The two to three antero-dorsal bristles on hind tibia are more widely spaced *Gymnoptera*

151. A single near-dorsal bristle in upper two-thirds of hind tibia 152

– No differentiated bristle in upper two-thirds of hind tibia 153

152. With only six bristles on frons between the two supra-antennals and ocelli, the antero-laterals being absent *Bothroprosopa*

– With eight bristles on frons between the two supra-antennals and ocelli (e.g. Fig. 8.1a) *Kierania* (part) and *Triphleba* (part)

Note: See note in couplet 139.

153. Arista apical in position *Billotia*

– Arista in dorsal, pre-apical position *Triphleba* (part)

154. Hind tibia with at least one dorsal, or near-dorsal, longitudinal hair palisade (as Fig. 8.4d,e) 155

– Hind tibia without longitudinal hair palisades, but there may be a row of spaced, differentiated, hairs (as Fig. 8.4c) **247**

155. Vein 3 unforked 156

– Vein 3 forked **182**

Note: Some genera are keyed both ways, especially when the base of vein 2 is obscure.

156. A clearly demarcated, darkened, oval patch on wing between bases of veins 4 and 5 (the first two thin veins). Veins 1 and 3 subparallel or even convergent, in basal third or more. *Phymatopterella* (part)

– Without this combination 157

157. Mid and hind femur each with pair of strong spines on anterior face in pre-apical position (similar to Fig. 8.91a) *Colyeria*

– No such spines on mid and hind femora 158

158. Front tarsal segments 1–4 or 1–3 asymetrically widened (Fig. 8.72a) and tarsal segment 5 of all legs tapered (Fig. 8.72c,d). Mesopleuron with a few hairs *Plastophorides*

– Front tarsi without asymetrically widened segments. Fifth tarsal segments rarely tapered; but when they are, mesopleuron is bare 159

159. Eyes greatly enlarged and almost meeting below, and lower facets (ommatidia) markedly larger than those above (Fig. 8.28a). Legs slender (Fig. 8.18b). Ovipositor of elaborate design (e.g. Fig. 2.5) *Auxanommatidia* (part)

– Without these features combined 160

160. Body somewhat dorso-ventrally flattened, so that anterior thoracic spiracles are dorsal in position 161

– Body not modified in this way; anterior thoracic spiracles on sides 162

161. Apical section of third antennal segment elongated, tapered and bearing apical arista. Scutellum with four strong bristles *Platydipteron*

– Third antennal segment more oval, not tapered apically, and arista just pre-apical. Scutellum with ten hairs on margin. (Hair palisade of hind tibia complete in basal part, but more intermittent distally as tranverse combs more developed.) *Melittophora*

162. Costa less than third of wing length and distal half or more of section 3 turned rearwards, through a right angle, to meet tip of vein 3 well behind wing margin *Stenoneurellys*
– Costa not modified in this way 163
163. Mesopleuron with hairs 164
– Mesopleuron bare 165
164. Veins 1 and 3 subparallel (section 1 being about three times length of section 2). Abdominal tergite 6 at least eight times as broad as long (being only about as long as tergite 1) and with small median emargination in front margin *Syneura* (part)
– Veins 1 and 3 divergent (section 1 being clearly less than three times length of section 2). Abdominal tergite 6 longer and narrower *Megaselia* (part)
165. Abdominal tergite 6 with an embayment of front margin which embraces a roughly pear-shaped, hairless, sclerite (Fig. 8.93a). Internally Dufour's crop mechanism not only very short and wide, and usually darkly sclerotized, but also with pair of long tubules extending from middle into lumen of crop (Figs 6.8a and 6.9b) *Woodiphora* (part)
– Abdominal tergite 6 and Dufour's crop mechanism (when present) not modified in these ways 166
166. Costa less than two-fifths of wing length 167
– Costa more than two-fifths of wing length **174**

Note: Variable genera are keyed both ways.

167. Only two bristles on front (lower) half of frons. (Third antennal segment elongate with pre-apical, dorsal arista; veins 1 and 3 subparallel.) *Styletta*
– At least four bristles on front half of frons 168
168. Third antennal segment more than twice as long as broad, with dorsal arista clearly before tip *Tubiceroides* (part)
– Third antennal segment usually clearly less than twice as long as broad; but if nearly twice as long, then arista is apical or, occasionally, absent 169
169. Supra-antennal bristles absent. Ovipositor asymmetrical and complex (usually with one or two strong black spines; venter of abdominal segment 6 often with well-developed sternite and frequently with conspicuous bristles and/or spines) *Cremersia* (part)
– At least one pair of supra-antennal bristles present. Ovipositor symmetrical 170
170. Abdominal tergites 5 and 6 both very short, and 6 completely encircling segment *Trucidophora*
– Not thus 171
171. Abdominal sternite 6 present and bearing bristles; modified ovipositor well developed and sometimes bizarre (e.g. Fig. 2.6) *Pseudacteon* (part)
– No sternite on venter of abdominal segment 6 172
172. Terminal segments of abdomen developed as a somewhat dorsoventrally flattened ovipositor 173

– Terminal abdominal segments not so modified *Megaselia* (part)

173. Costa at most only one-third of wing length *Pradea*

– Costa more than one-third of wing length *Pseudacteon* (part)

174. Epistoma (clypeus) inflated between bases of antennae and extending to front of frons *Apodicrania*

– Epistoma not unusually inflated 175

175. Hind tibia, in addition to dorsal hair palisade, with irregular antero-dorsal palisades in basal third. No supra-antennal bristles. Vein Sc obscure or absent. Median furrow of frons obscure or absent. Third antennal segment elongated and tapered, with an apical or subapical arista. Abdominal tergites wide, extending onto upper parts of flanks *Melaloncha* (part)

– Without this combination 176

176. Third antennal segment about twice as long as greatest breadth, strongly tapered, with apical arista. With two to four supra-antennal bristle *Ceratoconus*

– Without both characters in combination 177

177. Third antennal segment oval to conical 178

– Third antennal segment globose, but sometimes slightly pointed (in latter case tip of vein 4 deflected rearwards) 179

178. A brown sternite bearing hairs present on abdominal segment 6 (e.g. Fig. 2.6) *Pseudacteon* (part)

– No sternite on abdominal segment 6 *Apocephalus* (part) and *Pseudacteon* (part)

Note: A few *Apocephalus* species have an unforked vein 3, as is typical of *Pseudacteon* species lacking a sternite on segment 6. The former cannot be distinguished with confidence, in our present state of knowledge, unless associated with their males.

179. Costal index exceeds 0.7. Veins 4 and 5 widely separated, vein 4 being very pale and originating near apex of vein 3; base of vein 5 originating well before apex of vein 1. Abdominal tergites 3 and 4 fused to form a single elongated tergite *Parasyneura*

– Without this combination 180

180. Third antennal segment slightly pointed; costa more than half length of wing; with elongated ovipositor with long narrow cerci; tip of vein 4 deflected rearwards and running parallel to wing margin for short distance *Ectochaeta*

– Not with this combination 181

181. Without row of differentiated postero-dorsal hairs behind hair palisade of hind tibia *Beckerina* (part)

– With differentiated row of postero-dorsal hairs on hind tibia *Megaselia* (part)

182. Mesopleuron bare 183

– Mesopleuron with hairs, and sometimes bristles as well **224**

183. Front tibia with dorsal longitudinal hair palisade, as on mid and hind tibiae. Fore tarsus with only four segments *Neodohrniphora*

– Without these two features in combination 184

184. Front tarsus with only two, long, segments, the second usually being

tapered 185

– Front tarsus with five segments 186

185. Venter of abdominal segment 6 with distinct, semicircular or horseshoe-shaped, sternite bearing bristles on hind margin. (Costal index at least 0.32, usually 0.34 or more.) *Dacnophora* (part)

– Venter of abdominal segment 6 without distinct sternite. (Costal index at most 0.36, usually 0.33 or less. Lower ommatidia of each eye a little larger than upper ones.) *Myrmosicarius*

186. Eyes greatly enlarged and almost meeting below, and lower facets (ommatidia) larger than those above (e.g. Fig. 8.28a). Legs slender (Fig. 8.18b). Ovipositor elaborate (e.g. Fig. 2.5) *Auxanommatidia* (part)

– Not with these features combined 187

187. Fork of vein 3 very large, so that costal section 3 is about four times as long as section 2. Vein Sc strongly developed and its tip fusing with vein 1. Axillary ridge with at least five bristles. Abdominal tergite 6 almost divided in two by anterior V-shaped notch. On each flank at boundary of abdominal segments 5 and 6 there is an eversible finger-like process *Dicranodeina*

– Without this combination 188

188. First half to two-thirds of costal section 1 clearly wider than rest of costa, and with bristle situated dorsally at end of wide part at least as long as rest of costa. Fork of vein 3 large, with vein 2 as strongly developed as veins 1 and 3. Costal cilia of section 3 longer than vein 2. (The wing membrane may be shed.) *Pheidolomyia*

– Not thus 189

189. With 16–20 supra-antennal bristles; otherwise frontal bristles normal. Frons broader than long. (Costa only about one-third of wing length. Ovipositor short, broad, with narrowly rounded tip.) *Apocephalus* (part)

– Usually with fewer than 14 supra-antennals (usually 2–4), but if as many as 16 the frons is as long as broad 190

190. Hind tibia with single, dorsal, hair palisade 191

– Hind tibia with several, irregular, hair palisades (Fig. 8.73a) 194

191. Six or more supra-antennal bristles 192

– None to four supra-antennal bristles 193

192. With differentiated row of postero-dorsal hairs behind hair palisade on hind tibia *Johowia* (part)

– Without differentiated postero-dorsals on hind tibia *Beckerina* (part)

193. With dark oval patch on wing between tip of vein 3 and base of vein 4 *Phymatopterella* (part)

– Any dark marks on wing either are traversed by veins or are between bases of veins 4 and 5 194

194. With more than one longitudinal hair palisade on hind tibia, even if those on anterior face only present on basal third 195

– With only one, dorsal, hair palisade on hind tibia 197

195. Several irregular hair palisades on anterior face of mid and hind tibiae, those on hind leg running full length of tibia (Fig. 8.73a) *Plectanocnema*

– Any irregular hair palisades on anterior faces of mid and hind tibiae are restricted to basal halves or less; but regular, even if intermittent, palisades may be present beyond basal third (as Fig. 8.92c,d) 196

196. Irregular hair palisades of anterior face of hind tibia restricted to basal third. Without supra-antennal bristles. Vein Sc obscure or absent. Abdominal tergites wide, extending onto upper parts of flanks. Terminal segments obviously developed as an ovipositor *Melaloncha* (part)

– Without this combination. Abdominal tergite 6 as Fig. 8.93a, and no obviously developed ovipositor behind. (Internally, Dufour's crop mechanism not only very short and heavily sclerotized, but also with pair of long tubules extending from middle into lumen of crop – Figs 6.8a and 6.9b) *Woodiphora* (part)

Note: One Neotropical species assigned to *Beckerina* (*B. nudipleura*) will also key out here. However, it seems likely that it is really a *Woodiphora* species. Its description differs from that of a typical *Woodiphora* by its more acute angle between veins 2 and 3. Its transference to the latter genus would require reporting of a notopleural cleft in the male and the above type of Dufour's crop mechanism in the female.

197. The distal section of proboscis (distiproboscis) long and narrow (being longer than length of head) *Rhyncophoromyia* (part)

– If proboscis longer than head, then this not primarily due to elongation of distproboscis (labrum plus labella) 198

198. Epistoma (clypeus) longer than broad. Membranous region in front bearing palps also elongated, with result that latter originate well in front of (below) head. Abdominal tergite 2, and sometimes 3 also, elongated; tergites 3–5 or 3–6 shortened and of complex structure associated with gland opening. Segments 3–6 usually with one or more finger-like processes on flanks *Trophithauma*

– Without this combination 199

199. Third antennal segment more than twice as long as greatest breadth and with pre-apical, dorsal, arista. Frontal bristles all close to margins (six at front; two each side, one near middle and one near top of each eye; and four on vertex). Base of vein 2 detached from vein 3 *Tubiceroides* (part)

– Not so 200

200. Wing with triangular dark patch between bases of veins 4 and 5 201

– With no such dark patch 202

201. Base of triangular dark patch on wing against hind margin of vein 3. There is a small oval swelling near base of vein 3 at base of vein 6 *Physoptera* (part)

– Base of triangular dark patch clearly separated from vein 3. Without oval swelling near bases of veins 3 and 6 *Phymatopterella* (part)

202. Wing with small oval swelling near base of vein 3 at base of vein 6. Basal half of hind tibia with postero-ventral row of differentiated hairs, which are either shorter or longer than adjacent hairs. Costa more than half wing length *Physoptera* (part)

– Not with this combination 203

203. Hind tibia with differentiated row of antero-dorsal hairs as well as postero-dorsal row (the two rows separated by dorsal hair

palisade) 204
- Hind tibia at most with postero-dorsal row of differentiated hairs only 205

204. Palp bristles and supra-antennal bristles only about as strong as upper occipital bristles. Abdominal tergite 3 frequently with a pair of oval to round papillae near front margin (Fig. 8.70a) *Phalacrotophora* (part)
- Either the palp or supra-antennal bristles or both clearly longer than upper occipitals. Without such papillae on abdominal tergite 3 *Megaselia* (part)

205. Last tarsal segments on both front and middle legs, at least, longer than fourth segments and usually tapered (Figs 8.4b and 8.56b, c) 206
- Last tarsal segments shorter than fourth segments on all legs or else longer on front pair only 208

206. Palps with bristles, even if they are short. Costa at most only two-fifths of wing length 207
- Palps without bristles. Costa just over two-fifths of wing length. (Supra-antennal bristles absent. Abdominal tergites, ovipositor, and fifth tarsal segments as Fig. 8.79a–c.) *Razorfemora*

207. Third antennal segment pear-shaped, with apical arista and conus (apical bulb) of second segment inserted into side, not base, of third segment *Epacteon*
- Third antennal segment bean-shaped with dorsal, pre-apical, arista and conus of second segment inserted into base of segment in normal way (e.g. Figs 8.56d and 8.57a). (Ovipositor well developed – e.g. Fig. 8.58.) *Microselia*

208. Abdominal segment 6 with distinct sternite bearing bristles 209
- Abdominal segment 6 without sternite 212

209. Only four bristles at front margin of frons. Fifth tarsal segment of front leg frequently longer than fourth and tapered (as Fig. 8.56b) 210
- At least six bristles at front margin of frons. Fifth tarsal segment not elongated or tapered *Synclinusa*

210. Ovipositor symmetrical 211
- Ovipositor asymmetrical and of complex structure (usually with one or two strong black spines) *Cremersia* (part)

211. Costa clearly less than half wing length *Dacnophora* (part)
- Costa at least half wing length *Macrocerides* (part)

212. Hind tibia without postero-dorsal row of differentiated hairs, the hairs in this position being no different from adjacent ones of posterior face 213
- Hind tibia with row of distinctly differentiated postero-dorsal hairs, even if they are fine 214

213. Abdominal tergite 6 as Fig. 8.93(a). Internally Dufour's crop mechanism not only very short and heavily sclerotized, but also with pair of long tubules extending from middle into lumen of crop (Figs 6.8a and 6.9b) *Woodiphora* (part)
- Abdominal tergite 6 otherwise. Without Dufour's crop mechanism *Beckerina* (part)

214. With only four bristles at front margin of frons. Ovipositor asymmetri-

cal and complex (usually with one or two strong black spines) *Cremersia* (part)
- Without this combination 215

215. Third antennal segment conical, with apical arista. With only one bristle on axillary ridge of wing *Procliniella*
- Without these two features in combination 216

216. Costa less than half length of wing 217
- Costa at least half length of wing 220

Note: Borderline cases are keyed both ways.

217. Terminal abdominal segments (beyond tergite 6) clearly modified into dorso-ventrally flattened basal part embracing pointed ovipositor beyond 218
- Terminal abdominal segments not obviously modified as specialized ovipositor 219

218. Supra-antennal bristles as strong as other bristles at front of frons *Neophora*
- Supra-antennals clearly shorter and weaker than other bristles at front of frons *Apocephalus* (part)

219. Only two supra-antennal bristles. Costa only about one-third of wing length. Abdominal tergite 6 reduced and much wider than long *Syneura* (part)
- Without these three features in combination *Megaselia* (part)

220. The two supra-antennal bristles as strong as antials. Third antennal segment conical with apical, or slightly subapical, arista. Base of ovipositor broader than long, embracing the pointed distal portion (ovipositor proper). With long bristles at rear of segment 6, on sides and below *Anaclinusa*
- Without this combination 221

221. Terminal abdominal segments (beyond tergite 6) clearly modified into dorso-ventrally flattened basal part embracing pointed ovipositor (e.g. Fig. 2.5) 222
- Terminal abdominal segments not obviously modified to form specialized ovipositor 223

222. Frons clearly broader than long; with only two, strong, supra-antennal bristles in nearly straight transverse row with antials and antero-laterals at front of frons. Antials close to antero-laterals *Lecanocerus*
- Frons usually as long as broad or longer; with antials near mid line and two to four reduced supra-antennals below these or absent *Apocephalus* (part)

223. With only two supra-antennal bristles. Postero-dorsal hairs of hind tibia only weakly differentiated from adjacent hairs. Internally without Dufour's crop mechanism *Beckerina* (part)
- Without first two features in combination. Internally Dufour's crop mechanism is present (e.g. Fig. 6.8b) *Megaselia* (part)

Note: If there is a small swelling near base of vein 3 at base of vein 6 and costa exceeds half wing length, it may be a *Physoptera* species without ventral hairs on hind tibia (see couplet 202).

224. With tube-like elongation of basiproboscis (largely embracing distiproboscis) so that reduced palps are borne at tip of proboscis, which is clearly longer than head. (Third antennal segment globose. Costa at least half wing length. Abdominal tergite 6 longer than any other tergites and with small notch in front margin.) *Calamiscus*

– Proboscis usually much shorter and palps always at base 225

225. Wing margin incised at tip of vein 5, and the posterior lobe bordering incision with dense patch of minute hairs. Veins 4 to 6 not thin (as in most phorids), but as thick as vein 1 *Menozziola* (part)

– Without these features in combination 226

226. Eight bristles on scutellum 227

– Two to six bristles on scutellum 228

227. In addition to rest of normal bristles on frons, with only four to six differentiated supra-antennals *Egregiophora*

– In addition to rest of normal bristles on frons, with four median supra-antennals and several additional differentiated bristles either side of these *Johowia* (part)

228. Distiproboscis (labrum plus narrow labella) elongated, exceeding length of head. Labrum almost parallel-sided 229

– Distiproboscis shorter than head, or if longer then labrum is much broader basally and tapers towards its tip 230

229. Labrum broader (e.g. Fig. 8.2c) and cerci well developed (e.g. Fig. 8.6b) *Megaselia* (part)

– Labrum narrower. Cerci vestigial *Rhyncophoromyia* (part)

230. At least six differentiated supra-antennal bristles 231

– At most only four supra-antennals 232

231. With six of eight supra-antennals on sides of a narrow forward projection of middle of front margin of frons (between bases of antennae). Only two bristles on axillary ridge of wing. Abdominal tergites 1–5 very wide but short, but 3 being narrower than others. Tergite 6 reduced to small semicircle associated with gland opening *Lobochaeta*

– Additional supra-antennals either side of normal ones. At least four axillary bristles. Abdominal tergites otherwise *Johowia* (part)

232. Hind tibia with row of differentiated antero-dorsal hairs as well as postero-dorsals (the two rows being separated by hair palisade) 233

– Hind tibia without differentiated row of antero-dorsals 235

233. Basal half of hind tibia with differentiated row of very short antero-ventral hairs. Wing with small oval swelling near base of vein 3 at base of vein 6 *Physoptera* (part)

– Without this combination 234

234. Palp bristles and supra-antennals only about as strong as upper-occipital bristles. Abdominal tergite 3 frequently with pair of oval to round papillae near front margin (Fig. 8.70a) *Phalacrotophora* (part)

– Either palp or supra-antennal bristles or both clearly longer than upper occipitals. Without such papillae on abdominal tergite 3 *Megaselia* (part)

235. Second quarter of hind tibia with differentiated near-ventral row of longer hairs. Wing with small oval swelling near base of vein 3 at base of

vein 6. Abdominal tergites 3 and 4 greatly reduced, so that dorsal faces of these two segments largely membranous *Physoptera* (part)
– Not with this combination 236

236. More than 12 bristles on axillary ridge of wing. Arista with very short and widely spaced hairs (Fig. 8.74c), appearing to be bare at low magnifications. Three bristles on humerus, the middle one being above anterior thoracic spiracle. All tarsal claws with pre-apical tooth (Fig. 8.74b) *Pseudohypocera*
– Without these features in combination 237

237. Frons clearly longer than broad; with five bristles each side, one above the other and close to eye, the only other bristles being pre-ocellars and ocellars. Scutellum with six bristles, the middle one each side being longest and strongest. Costa only about one-third of wing length. Terminal abdominal segments dorso-ventrally flattened and embracing a sharply-pointed ovipositor *Stenodicrania*
– Not so 238

238. Fifth tarsal segments on all legs longer than fourth segments and strongly tapered (as Fig. 8.56b). With sharply-pointed ovipositor and vestigial cerci. Abdominal tergite 6 with conspicuous anterior median notch. Mesopleuron with fine hairs but no bristles. Costa less than two-fifths of wing length *Lepta*
– Without this combination 239

239. Frons clearly broader than long and without median furrow; with four median bristles near front and three each side (one above the other) near eye. Third antennal segment pear-shaped with apical arista. Terminal segments of abdomen long and narrow with sharply-pointed ovipositor. Mesopleuron with hairs and one long bristle *Acanthophorides*
– Without these features in combination 240

240. Mesopleuron with strong bristle as well as hairs 241
– Mesopleuron with hairs only 243

241. Abdominal tergites 5 and 6 clearly shorter and narrower than tergite 4. A pair of finger-like processes on sides of segment 5 (e.g. Fig. 2.6a) *Menozziola* (part)
– Not with both these features 242

242. Ovipositor not well developed but cerci, bearing hairs, well developed (e.g. Fig. 8.6b) *Megaselia* (part)
– Ovipositor well developed but cerci vestigial and without hairs (e.g. Fig. 8.38b) *Diocophora*, *Macrocerides* (part) and *Megaselia* (part)

Note: The females of *Macrocerides* species, whose males have a bristle on the mesopleuron, have not yet been described. A few *Megaselia* species with modified ovipositors cannot be distinguished from *Diocophora*, in our present state of knowledge, apart from one *Diocophora* with elongated, tapered, fifth tarsal segments, one with bifid tarsal claws and one whose vein 4 ends behind the wing tip.

243. Ovipositor blade-like (laterally compressed) with well developed cerci below (Fig. 8.51a) *Megaselia* (part, subgenus *Kerophora*)

– Ovipositor not thus 244
244. Terminal segments of abdomen (behind semicircular tegite 6) dorso-ventrally flattened, broadening apically and with concave hind margin. Ovipositor much narrower and normally withdrawn *Ceratophoromyia*
– Terminal abdominal segments otherwise 245
245. Frons clearly wider than long. Costa just over half wing length. Only two bristles on axillary ridge of wing. Third antennal segment oval, with pre-apical arista. With well developed ovipositor embraced by modified terminal segments *Brachycephaloptera*
– Without this combination 246
246. With well-developed ovipositor and cerci vestigial or absent. Third antennal segment short-conical with pre-apical arista. Wing with vein Sc well developed, more than six bristles on axillary ridge and costa more than half length *Macrocerides* (part)
– Not with this combination *Megaselia* (part)
247. Mesopleuron with hairs and wing vein 3 forked 248
– Mesopleuron bare and vein 3 forked or unforked 250
248. Palps very large, two-segmented, and with scattered hairs but no strong bristles (Fig. 8.37a,b). Fork of vein 3 large; wing membrane may be shed (Fig. 6.1a–c). Remnants of tergites present on abdominal segments 1 and 2 only (Fig. 8.37c) *Dicranopteron*
– Not thus 249
249. Without supra-antennals, antials, antero-laterals and pre-ocellar bristles, and usually medio-laterals as well. A notopleural cleft present (Fig. 8.47a,b) *Gymnophora* (part)
– With four supra-antennal bristles and other normal bristles on frons. Without notopleural cleft. (Fork of vein 3 large.) *Tropophleba*
250. Wing vein 3 forked 251
– Vein 3 unforked **260**
251. An isolated bristle in basal third of mid tibia *Triphleba* (part)
– Without such a bristle 252
252. Wing with vein 6 deflected forwards opposite basal bend of vein 5, so that space between two veins is constricted (as Fig. 8.56a). No bristles on axillary ridge *Metopina* (part)
– Vein 6 not deflected in this way. At least one bristle on axillary ridge 253
253. Third antennal segment clearly longer than greatest breadth, strongly tapered from broad base (the tapered part being at least one-third of length), with apical arista *Corynusa*
– Third antennal segment either globose, with pre-apical arista, or if somewhat pear-shaped with apical arista it is scarcely, if any, longer than greatest breadth 254
254. End of costa clearly overreaches tip of vein 3 a short distance. With only two axillary bristles, which are shorter than costal cilia on costal section 3 *Allochaeta*
– Without this combination 255
255. Apart from bristles on vertex, with only eight or fewer bristles on frons. Median furrow of frons absent 256

– Apart from bristles on vertex, with 10 or 12 bristles on frons. Median furrow present, but sometimes obscure 258

256. Only four or six bristles on frons. Breadth of frons clearly more than twice length from anterior ocellus to front margin. *Macroselia*

– Eight bristles on frons. Frons breadth not twice midline length 257

257. With only one or two bristles on axillary ridge of wing *Rhopica* (part)

– With at least three bristles on axillary ridge *Aphiura*

258. Proboscis long and slender (the distal part being longer than frons). Fork of wing vein 3 large (costal section 3 being as long as, or longer than, section 2); axillary ridge with only two bristles *Pericyclocerina*

– Without this combination 259

259. Abdominal tergites 6 and 7 as Fig. 8.93(a). Internally Dufour's crop mechanism not only very short and heavily sclerotized, but also with pair of long tubules extending into lumen of crop (Figs 6.8a and 6.9b) *Woodiphora* (part)

– Abdominal tergites 6 and 7 otherwise. Without Dufour's crop mechanism *Beckerina* (part)

260. Wing with very long bristles on costa (Fig. 8.17b) *Adelopteromyia* (part)

– Costa with shorter, finer, bristles or hairs. When some long bristles present in outer half, then section 1 (at least) with hairs 261

261. Fifth tarsal segment on all legs longer than fourth and a little to strongly tapered (e.g. Fig. 8.50b, c) 262

– Fifth tarsal segments not longer than fourth segments 263

262. All tibiae with near-dorsal pre-apical bristle in last quarter (and sometimes one or two additional near-dorsals above this one). Outer half of costa with 4–5 long bristles (some being longer than vein 3). (Wing membrane may be shed.) Third antennal segment globose with apical arista. Abdominal tergites 1–6 absent. Ovipositor not obviously modified *Termitophorides*

– Tibiae without pre-apical near-dorsal bristles. Wing (Fig. 8.50d) with short costal cilia only. Third antennal segment elongated and without arista (Fig. 8.50a). Abdominal tergites 1–6 present. Ovipositor as Fig. 2.7 *Iridophora*

263. First antennal segment elongated and third segment conical, with pre-apical arista; proboscis narrow and elongated (Fig. 8.46b). Without tergites on abdominal segments 1–6 (Fig. 8.46a). (Wing membrane may be shed – Fig. 6.1d.) *Eutermiphora* (part)

– Without this combination 264

264. Third antennal segment globose, with apical arista. More than 30 bristles on thoracic scutum (mesonotum). No tergites on abdominal segments 1–6. Costa clearly less than one-third of wing length and with two longer bristles in outer half, preceded by shorter hairs *Sphinctomyia*

– Not with these features combined 265

265. Third antennal segment with apical half developed as long tubular extension bearing widely-spaced long hairs, contrasting with very short hairs of apical arista. (Only four bristles on frons in front of

anterior ocellus, two being supra-antennals; costa at least half wing length.) *Dahliphora* (part)

– Third antennal segment globose, without apical extension bearing long hairs 266

266. At least two bristles on axillary ridge of wing 267

– Without axillary bristles 269

267. At most only ten bristles on frons in front of anterior ocellus, including two supra-antennals 268

– Twelve bristles on frons in front of anterior ocellus, including four supra-antennals. (Dufour's crop mechanism wider than long, and with pair of long tubules extending into lumen of crop – e.g. Fig. 6.9b.) *Woodiphora* (part)

268. Only eight bristles on frons in front of anterior ocellus. Fewer than four bristles on axillary ridge of wing *Rhopica* (part)

– Ten bristles on frons in front of anterior ocellus. More than five axillary bristles *Haplophleba*

269. Thoracic scutum (mesonotum) more than 2.5 times as long as broad and tapering towards distinctly separated scutellum. Frons about twice as long as wide, with four supra-antennals, a pair of reduced pre-ocellars and bristles on vertex only. Abdominal tergites 1–5 present, with front half of 5 developed as semicircular flap *Stenophorina*

– Without these features combined 270

270. At least abdominal tergite 2 present 271

– Abdominal tergites 1–6 absent, apart from rudiment of 5 associated with gland opening. (Without bristles on frons between anterior ocellus and two supra-antennals. Thoracic scutum clearly wider at front than at rear. Wing membrane may be shed.) *Termitophorina*

271. With only tergite 2 and rudiment of 5, associated with gland opening, present on abdominal segments 1–6 272

– With at least four tergites present on abdominal segments 1–6 273

272. Abdomen with numerous long bristles (longer than tergite 2). Wing membrane may be shed; when present costa at most only one-third of wing length and only two thin veins present. Costal cilia on section 2 longer than vein 1 *Echidnophora*

– Abdomen with short hairs only, which are only a little longer at hind margins of segments. Wing membrane not shed; costa more than half wing length and three thin veins present. All costal cilia shorter than vein 1 *Bolsiusia*

273. Costa less than one-third of wing length. Only one bristle on palp. Only two supra-antennal bristles and two pre-ocellars on frons, apart from those on vertex. Terminal segments of abdomen developed as down-curved beak-shaped ovipositor *Brachyphlebina*

– Costa about half, or more, of wing length. Without other characters in combination 274

274. Wing as Fig. 5.2(b). Eyes absent or represented by one or two reduced ommatidia only. Ocelli absent *Metopina* (part)

– Wing with thin veins evident. Eyes and ocelli present 275

275. Only four bristles, including supra-antennals, on frons in front of anterior ocellus 276
– At least six bristles on frons in front of anterior ocellus 278
276. Vein 6 bent towards basal curve of 5, so that space between veins is constricted in middle (e.g. Fig. 8.56a) 277
– Veins 5 and 6 not thus (e.g. Figs 8.34a and 8.35a). (With four supra-antennals as only bristles on frons, apart from those on vertex. Abdominal tergites as Figs 8.34b and 8.35b.) *Crasilla*
277. Two supra-antennals and two pre-ocellar bristles on frons, apart from six bristles on vertex *Dahliphora* (part)
– Frons with four supra-antennals only, apart from those on vertex *Metopina* (part)
278. Frons at least three times as broad as long. ('pre-ocellar' bristles level with posterior ocelli. 'Thin' veins 4–6 may be almost as thick as outer third of costa.) *Holopterina*
– Frons less than three times as broad as long 279
279. Vein 6 bent towards basal curve of vein 5, so that space between veins is constricted in middle (e.g. Fig. 8.56a) *Metopina* (part)
– Vein 6 not bent in this way (indeed vein 6 may be abbreviated or absent, in association with narrowing of basal half of wing) 280
280. Wing broadest at level of tip of costa or beyond (Fig. 8.43a). Abdominal tergite 2 at most only twice as wide as long *Ecitoptera*
– Wing broadest before level of tip of vein 1. Abdominal tergite 2 more than four times as wide as long *Cataclinusa*
281. Head, thorax and abdomen somewhat flattened dorso-ventrally, with anterior thoracic spiracles dorsal in position. Eyes somewhat reduced. Hind tibia with antero-ventral row of spines before apical spur *Synaptophora*
– Without this combination. (With two or more longitudinal hair palisades on hind tibia – Fig. 8.39f. Wing with two or more axillary bristles.) *Diplonevra* (part)

8.3 GUIDE TO LITERATURE FOR SPECIES IDENTIFICATION

The following review gives the number of known species in each genus and cites the keys or descriptions that will allow identification of the species. It has been estimated that perhaps as few as one-tenth of the world's scuttle fly species have been described and named. Furthermore the limits of distribution for the described species are poorly known. There is therefore a high probability of turning up species new to science or new to a region.

Abaristophora Schmitz
Two Palaearctic and one Nearctic species are known (Schmitz, 1951c; Michailovskaya, 1988; Brown, 1988).

Acanthophorides Borgmeier
The seven Neotropical species are keyed by Borgmeier (1967h).

Acontistoptera Brues
The one Nearctic and three Neotropical species are keyed by Borgmeier (1963d).

Adelopteromyia Schmitz
The three or four Neotropical species (one species may be merely the short-winged morph of another) are keyed by Disney and Kistner (1990b).

Adenophora Borgmeier
The single Neotropical species, known only in the female sex, is covered by Borgmeier (1960b).

Aemulophora Borgmeier
The single Neotropical species, known only in the female sex, is covered by Borgmeier (1938b).

Aenictacantha Disney
Two species from Malaysia and Sulawesi, known only as females, are covered by Disney (1991m).

Aenictomyia Brues
The single Oriental species, known only in the female sex, is covered by Borgmeier (1967a).

Aenigmatias Meinert
A dozen species are known from the Nearctic and Palaearctic Regions. The former are keyed by Borgmeier (1963d) and the latter by Schmitz (1955b), supplemented by Disney (1983a) for males and (1993c) for females.

Aenigmatistes Shelford
The 20 Afrotropical species are keyed by Disney (1991k).

Aenigmatopoeus Schmitz
The females of the six Afrotropical species are keyed by Borgmeier (1963e). Colyer (1958) describes the only known male.

Alamira Schmitz
The two Afrotropical species, known only in the female sex, are keyed by Disney (1990j). A third species, *A. peckorum*, is a species of *Termitostroma* (see below).

Allochaeta Borgmeier
The four Brazilian species are keyed by Borgmeier and Prado (1975).

Anaclinusa Borgmeier
The single Neotropical species is covered by Borgmeier (1969b, 1971a).

Anevrina Lioy
The five Palaearctic species are covered by Schmitz (1941c) and Disney (1983a). The nine Nearctic species are keyed by Borgmeier (1963d). Beyer (1958a) describes three species from Burma.

Anticofimbria Schmitz
The male of the single known species from Japan is described by Schmitz (1951e). The female is unknown.

Antipodiphora Schmitz
Six species from New Zealand are keyed by Schmitz (1939a). Undescribed species occur in the Oriental Region.

Aphiura Schmitz
Three Australasian species are discussed by Borgmeier (1967a), along with an unnamed female that probably belongs to some other genus.

Apocephalus Coquillett
There are certainly over 120 species in this genus, which is known from the Australasian, Oriental, Nearctic and Neotropical Regions. Brown is currently carrying out revisionary studies. Until these are complete, specimens can be keyed with Borgmeier (1971a), supplemented by Prado (1976) and Disney (1980i, 1981d, 1982k). A Nearctic species assigned to a new genus, *Zyziphora* (Peterson and Robinson, 1976), has been transferred to *Apocephalus* by Brown (1992b). It is also figured by Peterson (1987). Brown (1993b) provides a revised key to Nearctic and Neotropical species of the subgenus *Mesophora*.

Apodicrania Borgmeier
The three Neotropical species are keyed by Borgmeier (1963e).

Apterella Borgmeier
A single species from Costa Rica is known only in the female sex (Borgmeier, 1935a).

Apterophora Brues
The five Neotropical species are discussed by Prado (1976), allowing identification of the males. In our present state of knowledge, females can only be named when associated with their males.

Aptinandria Schmitz
Schmitz (1921f) describes the female of the single Afrotropical species. The male has not been described, but it was included in a key by Beyer (1965b).

Aristocerina Borgmeier and Prado
A single Neotropical species, known only in the male sex, is described by Borgmeier and Prado (1975).

Arrenaptenus Schmitz
Two Afrotropical species are known (Schmitz, 1958e; Borgmeier, 1959).

Assmutherium Schmitz
A single Oriental species, known only in the female sex, is covered by Disney (1991m).

Auxanommatidia Borgmeier
The validity of this genus is called into question by observations on the significance of ommatidial size differentiation (Disney and Schroth, 1989). The single Nearctic species is covered by Beyer (1963), Borgmeier, (1963d) and Barnes (1991b). Females of four Neotropical species are keyed by Borgmeier and Prado (1975) and the males of two species by Borgmeier (1971a), in which the generic assignment of a further three Neotropical species and the Nearctic species is called into question. An Afrotropical species is described by Disney (1981h).

Beckerina Malloch
This genus is recorded from all regions except the Afrotropical. The 19 known species are keyed by Borgmeier (1971a) with a Neotropical addition by Borgmeier and Prado (1975). The Nearctic species are keyed by Borgmeier (1963d) with the addition of a 'Neotropical' species by Barnes (1991a). The sole European species is covered by Disney (1983a) and the sole New Zealand species by Schmitz (1939a).

Beyermyia Disney
The female of the single Afrotropical species is dealt with by Beyer (1965b) and the male by Disney (1991m), who adds to the description of the female.

Billotia Schmitz
The single European species is covered by Schmitz (1953a).

Bolsiusia Schmitz
The single Oriental species, known only in the female sex, is covered by Schmitz (1913a).

Borgmeieriphora Prado
The four Neotropical species are keyed by Brown (1993a).

Borophaga Enderlein
This genus is known from every region except the Australasian. Twenty-two of the 28 known species are keyed by Beyer (1959a). The key to Nearctic species (Borgmeier, 1963d) includes two of the omitted species. Beyer (1965b) adds an Afrotropical species and Borgmeier and Prado (1975) add a Neotropical species. Colyer (1966) added an eastern Palaearctic species. Otherwise the Palaearctic species are keyed by Schmitz (1951c), with the British species keyed by Disney (1983a). Brown (1992b) has proposed splitting up this genus into a number of separate genera (*Godavaria,*

Latiborophaga and *Peromitra*, in addition to *Borophaga* in a more restricted sense). As the polarities of a number of the postulated transformation series advanced in support of this proposal are in contention, I consider it more sensible to treat these new taxa as subgenera (at least until a consensus is achieved). Brown adds a new Oriental species.

Bothroprosopa Schmitz
A single species from New Zealand is described by Schmitz (1939a).

Brachycephaloptera Borgmeier
A single Brazilian species is described by Borgmeier (1924a).

Brachycosta Prado
This genus is known from a single male from Brazil (Prado, 1976).

Brachyphlebina Borgmeier and Prado
A single species from Brazil is described by Borgmeier and Prado (1975).

Brachyselia Schmitz
A single Afrotropical species, known only in the female sex, is covered by Schmitz (1929g).

Burmophora Beyer
This genus, known from Burma, Japan and North America, is sexually dimorphic to the extent that the female was placed in a separate genus (*Crinophleba*) from the male. There are four described species, but one may be the opposite sex of another. The species are dealt with by Beyer (1958a), Borgmeier (1967c), Gotô (1983b) and Brown (1990a).

Calamiscus Borgmeier
The single Neotropical species, known only in the female sex, is covered by Borgmeier (1960b).

Cataclinusa Schmitz
It is doubtful that the single Nearctic and single Neotropical species really belong to the same genus. The former is dealt with by Borgmeier (1963d) and the latter by Schmitz (1929g).

Ceratoconus Borgmeier
The five Brazilian species are covered by Borgmeier (1928c) and Prado (1976).

Ceratophoromyia Beyer
A single Afrotropical species is described by Beyer (1959d).

Ceratoplatus Schmitz
The male of a single New Zealand species is described by Schmitz (1939a).

Ceratusa Borgmeier
The male of the single Neotropical species is described by Borgmeier (1967f).

Ceylonoxenia Schmitz
There are two described species from the Oriental Region, known only as females; but I have at least three new species awaiting description. In the meantime reference should be made to Borgmeier (1964a).

Chaetaspidia Borgmeier
The single species, known only from males from Costa Rica, is dealt with by Borgmeier (1961a).

Chaetopleurophora Schmitz
Six Palaearctic species are keyed by Schmitz (1941c), with the four British species also keyed by Disney (1983a). Six Nearctic species (including one Holarctic species) are keyed by Borgmeier (1963d). Seven Neotropical species are keyed by Borgmeier (1969b) and three Oriental species are covered by Beyer (1958a) and Borgmeier (1962c). Brown (1992b) has proposed raising two segregates to the generic level (*Chaetocnemistoptera* and *Neopleurophora*). Until the grounds for doing so command consent, I consider it wiser to treat these segregates as subgenera.

Cheiloxenia Delachambre
The single Afrotropical species, known only in the female sex, is covered by Delachambre (1966).

Chelidophora Borgmeier
Two Neotropical species, known only in the female sex, is covered by Borgmeier (1950b).

Chonocephalus Wandolleck
More than 40 species have been described and named and several are known but not named. The marked sexual dimorphism, along with the naming of species based on the description of one sex only, has caused much confusion. The recognition of both sexes of two cosmopolitan species, transported around the world by man, has been recently clarified (Disney, 1991h). Borgmeier (1963d) keys the Nearctic species. Paulian (1958) deals with species from Madagascar. Australasian and Oriental species are covered by Schmitz (1929g), Bohart (1947), Borgmeier (1967f) and Disney (1986c). Neotropical species are covered by Borgmeier (1967f), Borgmeier and Prado (1975) and Disney (1980h). Pairs caught in copula or in reared series are needed for the resolution of taxonomic problems in this genus. New species should only be named when both sexes are available for description.

Clinochaeta Borgmeier
A single species, known only from males collected in Brazil, is covered by Borgmeier (1923c).

Clitelloxenia Kemner
Borgmeier (1964a) refers to six described species, known only as females, from the Oriental Region. The genus is currently undergoing revision. At least one synonym is suspected and three undescribed species are awaiting description.

Colobomeles Borgmeier
The single Neotropical species, known only in the female sex, is covered by Borgmeier (1927a).

Colyeria Borgmeier and Prado
The female of the single Neotropical species is covered by Borgmeier and Prado (1975) and the male by Brown (1993a).

Commoptera Brues
The females of the one Nearctic and five Neotropical species are keyed by Borgmeier (1971a) and former also covered by Borgmeier (1963d). The male of only one species is known (Borgmeier, 1926a); its characters are briefly summarized by Borgmeier (1963d), but with the perpetuated error that the subcosta ('mediastinal vein') is said to be absent, despite its being illustrated in the original description.

Conicera Meigen
Twenty-five species are known. The Palaearctic species are covered by Schmitz (1953a) and Disney (1983a, 1991l). The Australasian and Oriental species are keyed by Disney (1990k). Beyer (1965b) records the genus in the Afrotropical Region, but no species has been described. The Nearctic species are keyed by Borgmeier (1963d), but see also Disney (1983a). The Neotropical species are covered by Schmitz (1929g) and Borgmeier (1968a, 1969a).

Coniceromyia Borgmeier
More than 30 species are known, mainly from the Neotropical Region but with at least three in the Nearctic. Borgmeier (1963e) provides a key, which needs supplementing with Borgmeier (1950a, 1969a, 1969b), Borgmeier and Prado (1975), Prado (1976), Peterson and Arntfield (1971), Peterson (1982) and Barnes (1991a).

Contopteryx Schmitz
A single Neotropical species is known (Schmitz, 1929i; Borgmeier and Prado, 1975).

Cootiphora Brown
The single Neotropical species, known only in the female sex, is covered by Brown (1993a).

Coridophora Schmitz
There are two Afrotropical species, known only as females, and covered by

Borgmeier (1959) (under *Edaphomyia*) and Disney (1982e). Since the latter paper further material, indicating greater variation than previously supposed, has allowed the following synonymy:

Coridophora convexa Schmitz, 1954b: 14.
Edaphomyia nigriventris Borgmeier, 1959: 194. **Syn. nov.**

Corynusa Schmitz
The male of the single Neotropical species was originally assigned to *Cyphometopis*. It is described by Schmitz (1929i) and the female (as a separate species) by Schmitz (1931b). Borgmeier (1967f) discusses the genus.

Crasilla Borgmeier
Three Oriental species, known only as females, are covered by Borgmeier (1967a) and Disney and Kistner (1989b). It is possible that this genus is a polyphyletic assemblage.

Cremersia Schmitz
Twelve species are Neotropical and two of these, plus an additional species, are also Nearctic. The females have been keyed by Borgmeier (1961a). Males can be named only when associated with their females.

Cryptophora Borgmeier
Two Neotropical species, known only as females, are covered by Borgmeier (1935d) or Silvestri (1945), and Seevers (1941).

Cyphocephalus Borgmeier
The only known species was described from a single male from Costa Rica (Borgmeier, 1967f).

Cyphometopis Borgmeier
Three species are known from males only, all from Brazil. Borgmeier and Prado (1975) provide a key.

Cyrtophorina Borgmeier and Prado
A single species from Brazil is known only in the male sex (Borgmeier and Prado, 1975).

Dacnophora Borgmeier
A key to the females of five Brazilian species is provided by Borgmeier (1961a). The female of a Nearctic species is described by Brown (1988). Males can be named only when procured in association with their females.

Dahliphora Schmitz
The males of the three Neotropical species and one Australasian species are keyed by Borgmeier and Prado (1975). Females can only be named by association with their males.

Darwiniphora Schmitz
The two species, both from Chile, are covered by Borgmeier (1963e).

Diaclinella Borgmeier
The single species from Cuba, known only in the male sex, is covered by Borgmeier (1962c).

Dichocerina Borgmeier
A single species from Brazil, known only in the male sex, is covered by Borgmeier (1971a).

Dicranodeina Schmitz
Schmitz (1929i) assigned a single Neotropical species, only known in the female sex, to a separate subgenus of *Tropophleba*. The differences, however, are such that I am herewith raising this subgenus to the generic rank.

Dicranopteron Schmitz
The five Oriental species, known only as females, are keyed by Disney and Kistner (1990a).

Dinocercus Borgmeier
A single species from Brazil, known only in the the male sex, is covered by Borgmeier (1971a).

Diocophora Borgmeier
The validity of this genus is doubtful (Disney and Kistner, 1989d). Two Nearctic species are dealt with by Borgmeier's (1963d) key to females, which also includes seven of the eight Neotropical species. The omitted species is known only in the male sex. It is included in Borgmeier's (1959) key to the five Neotropical species known as males.

Diplonevra Lioy
About 70 species have been described in this cosmopolitan genus. The Palaearctic species are keyed by Schmitz (1949c), augmented by the key to British species (Disney, 1983a). Borgmeier (1963d) keys the Nearctic species and also (1969b) the Neotropical species. The Australasian and Oriental species are keyed by Disney (1990e) and the Afrotropical species are covered by Beyer (1959g, 1960a, 1965b).

Diplostylocera Borgmeier
The single Afrotropical species, known only in the male sex, is covered by Borgmeier (1967b).

Distichophora Schmitz
The males only are known for one species from Australia (Borgmeier, 1963b) and one from New Zealand (Schmitz, 1939a). It is doubtful, however, that these really belong to the same genus.

Dohrniphora Dahl
More than 100 species are known. The single Palaearctic species is virtually cosmopolitan, through the agency of man. It is covered by Disney (1983a), and every other key cited below. The Nearctic species are keyed by Borgmeier (1963d), supplemented by Khalaf (1971) and Barnes (1991a). The Neotropical species are keyed by Borgmeier (1960b), substantially augmented by Borgmeier (1961a, 1969a,b), Borgmeier and Prado (1975), Prado (1976) and Disney (1983h). The Afrotropical species are keyed by Beyer (1959a), supplemented by Disney and Kistner (1989c) and Disney (1991a). The Australasian and Oriental species are keyed by Disney (1990g).

Dorsochaeta Borgmeier and Prado
The single Neotropical species is known only in the female sex (Borgmeier and Prado, 1975).

Dundophora Schmitz
A single Afrotropical species, known only in the female sex, is described by Schmitz (1955a).

Echidnophora Schmitz
The single species, known only in the female sex, is known from the Oriental Region and Sulawesi. Specimens from Sulawesi differ slightly from the Oriental material, and may represent a distinct species (Disney and Kistner, 1988). The species is described by Schmitz (1916a) supplemented by Kemner (1932a).

Ecitomyia Brues
Females of the eight Neotropical and one Nearctic species are keyed by Borgmeier (1960b). The Nearctic species is also covered by Borgmeier (1963d), who includes a description of the only known male.

Ecitophora Schmitz
The females of the ten Neotropical species are keyed by Borgmeier (1960b), supplemented by Disney and Kistner (1989d).

Ecitoptera Borgmeier and Schmitz
The females of the 12 Neotropical species are keyed, along with the five known males, by Borgmeier (1960a).

Ecituncula Schmitz
Borgmeier (1960b) keys the females of five of the seven known Neotropical species. Silvestri (1947a) and Borgmeier (1971a) cover the other two. The only known male is described by Silvestri (1947a) as a '*Puliciphora*'.

Ectochaeta Borgmeier
The single Neotropical species, known only in the female sex, is probably only an aberrant species of *Megaselia*. It is described by Borgmeier (1958).

Egregiophora Beyer
The single known species from Burma is covered by Beyer (1958a).

Epacteon Borgmeier
A single Australian species, known only in the female sex, is described by Borgmeier (1963b).

Epichonocephalus Schmitz
A single species from the Bismarck Archipelago, and known only in the male sex, is covered by Schmitz (1929g).

Epicnemis Borgmeier
Three species, from the Oriental Region, are covered by Disney (1991m).

Eremophora Borgmeier
The single species, from Panama, is known only in the male sex (Borgmeier, 1961a).

Eurycnemis Borgmeier
A single species, from Brazil, is known only in the male sex (Borgmeier, 1971a).

Euryophora Schmitz
A single species is known, from Madagascar. The female is described by Schmitz (1915d) and the male by Schmitz (1929g).

Euryplatea Schmitz
The single Afrotropical species, known only in the female sex, is covered by Schmitz (1941e).

Eutermiphora Lea
The single Australian species, known only in the female sex, is redescribed by Disney (1989d).

Exochaeta Schmitz
A single species, from Argentina, is known only in the male sex (Schmitz, 1931b).

Franssenia Schmitz
Four species, from the Australasian and Oriental Regions, are known only in the female sex. They are keyed by Disney and Kistner (1989b).

Gymnophora Macquart
More than 50 species are known. Both sexes of the British species are keyed by Disney (1983a). Otherwise only males can be named in our present state of knowledge. The Neotropical species are keyed by Brown (1987b). The Nearctic species and many Palaearctic species are keyed by Brown (1987a). Further Palaearctic species are covered by Brown (1989a), along with some Oriental species.

Gymnoptera Lioy
Two Palaearctic species are keyed by Disney (1983a). Three Australasian/ Oriental species are dealt with by Colyer (1956a). The single Neotropical species, which only doubtfully belongs in this genus, is covered by Borgmeier (1958).

Haplophleba Schmitz
A single Neotropical species, known only in the female sex, is covered by Schmitz (1929i).

Hexacantherophora Schmitz
The single Afrotropical species, known only in the female sex, is described by Schmitz (1914d).

Holopterina Borgmeier
A single Neotropical species, known only in the female sex, is covered by Borgmeier (1928c, 1960b).

Homalophora Borgmeier
Six Neotropical species are known. Four are described by Borgmeier (1958) and two are described by Borgmeier (1960b, 1969a).

Hylophora Borgmeier
Two Neotropical species are known only as females, and are covered by Borgmeier (1960b).

Hypocera Lioy
The single Palaearctic species is covered by Lundbeck (1922), Schmitz (1951c) or Disney (1983a), and two Nearctic species by Borgmeier (1963d). The three Oriental species are covered by Beyer (1958a) and Borgmeier (1967a).

Hypocerides Schmitz
Seven species are known. Three are Afrotropical, one Australasian, two Neotropical and one from the latter two regions plus the Nearctic. A preliminary key is provided by Disney (1991m).

Hypogeophora Gotô
A single Japanese species, known only in the female sex, is described by Gotô (1986b) (but it should be noted that he overlooked the reduced gland opening on abdominal segment 5).

Idiophora Borgmeier
The single Neotropical species, known only in the female sex, is covered by Borgmeier (1960b).

Indoxenia Schmitz
Three Oriental species, known only as females, are referred to by Borgmeier (1964a). The genus is currently undergoing revision.

Iridophora Disney
A single species is known from Sulawesi (Disney, 1986b).

Javanoxenia Schmitz
This genus is currently undergoing revision. Two Oriental species, known only as females, are referred to by Borgmeier (1964a). Both sexes of a third species assigned to *Perissa* are covered by Disney (1990j). It is transferred herewith:

Javanoxenia georgei (Disney, 1989g: 111.) **Comb. nov.**

Johowia Silva Figueroa
Seven Neotropical species are known. Six are keyed by Borgmeier (1960b) and a seventh species added by Borgmeier (1971a).

Kierania Schmitz
It is probable that this genus is merely a slightly aberrant *Triphleba*. The single species from New Zealand is covered by Schmitz (1939a).

Kuenbergia Schmitz
A single Palaearctic species, known only in the female sex, is covered by Schmitz (1953a).

Lecanocerus Borgmeier
The single Nearctic species is possibly only an aberrant *Megaselia*. The male is described by Borgmeier (1963d) and the female by Peterson (1987).

Lenkoa Borgmeier
A single Neotropical species, known only in the female sex, is covered by Borgmeier (1969b).

Lepidophoromyia Borgmeier
The three Neotropical species, known only in the female sex, are covered by Borgmeier (1925c, 1971a) and Disney and Kistner (1989d). The genus is probably polyphyletic, as presently conceived.

Lepta Schmitz
The single European species is described by Schmitz (1938m).

Leptilla Borgmeier
The single Australian species is known only in the male sex (Borgmeier, 1963b).

Lobochaeta Borgmeier
A single species from the Philippines, known only in the female sex, is covered by Borgmeier (1967a).

Macrocerides Borgmeier
Borgmeier (1969b) provides a key to the males of the seven Neotropical and

one Afrotropical species. He commented that 'this genus is rather heterogeneous. It is quite possible that some of the species . . . do not belong'. Indeed the 'genus' is almost certainly polyphyletic and may only represent an assemblage of *Megaselia* species with aberrant antennae. Beyer (1965b) covers the Afrotropical species.

Macroselia Schmitz
The validity of this genus is doubtful. The three species are from Burma, Chile and New Zealand. They are keyed by Beyer (1966).

Maculiphora Disney
The two species, from the Philippines and Australia, are covered by Disney and Kistner (1991).

Mannheimsia Beyer
The single Afrotropical species is covered by Beyer (1965b).

Megaselia Rondani
Smith (1984) has described this genus as 'the Diptera enfant terrible'! At least 45% of the known Phoridae belong to this genus, with nearly 1400 described species currently recognized. However, the complexities of this genus means that they are often put to one side when new taxa are being described. The best-known phorid fauna is that of the British Isles (Disney, 1983a, 1989a, 1991j, 1993c), in which about 70% of the species belong to *Megaselia*. Furthermore the boundaries of the genus are by no means clear. This has been discussed with reference to the Palaearctic fauna, with the return of some species to the genus (Disney, 1990b).

It remains the case that a number of described genera of the tribe Megaseliini are probably no more than somewhat atypical *Megaselia*. The temptation to raise distinctive species groups to the generic level needs to be strongly opposed. The type species of the genus belongs to a distinctive segregate. Thus a policy of raising such groups to the generic level would be likely to change the names of more than 1300 species in due course. Such a programme of nomenclatural upheaval would serve the interests of nobody but the pedant. The affinities of clusters of species within the genus can be adequately expressed by means of subgenera and species groups, in due course. Some poorly-characterized 'genera' currently excluded from *Megaselia* may be best transferred to (in many cases returned to) the genus. Other well-defined 'genera', such as *Gymnophora* and *Woodiphora*, belong to the same monophylum as *Megaselia*. Their exclusion, therefore, renders *Megaselia* paraphyletic. One could reduce these genera to subgenera of *Megaselia*. While this would resolve the paraphyly, the nomenclatural upheaval is certainly not justified (as argued in section 8.1 above). The above considerations lead me to reject Brown's (1992b) proposal to designate a group of *Megaselia* species as a new genus *Myriophora*. Even if there were consensus regarding the validity and boundaries of this taxon, he makes no convincing case for not treating it as a species group or a subgenus.

In preparing the keys to genera, it became apparent that the genus

Pericyclocera Schmitz could not be characterized. The sole character, uniting an otherwise disparate assemblage of *Megaselia* species, is the dorsal deflection of the median part of the epistoma (clypeus) in the females. This deflection is associated with a wider than usual separation of the antennae. Borgmeier (1966a) added that the proboscis is short, the buccal opening large and the hypostoma somewhat produced. The hypostomal character was illustrated by Schmitz (1929g), but in fact is not found in all the species. Indeed these additional features are found in many *Megaselia* species. Furthermore, apart from the epistomal feature, the other characters are not found in all the *Pericyclocera* species. For example, *P. diptychogastra* has a long proboscis, very similar to that of *Megaselia chlumetiae* (Fig. 8.2c). However, despite inclusion of *P. diptychogastra* in his key to world species, Borgmeier (1967e) added the short-proboscis character to his diagnosis of the genus. He then (1969b) added three further species primarily on the basis of a short, stout, labrum, even though the degree of deflection of the epistoma was well within the normal range for *Megaselia*. The evidence is that *Pericyclocera* is a polyphyletic assemblage of *Megaselia* species in which convergent evolution has given rise to a similar deflection of the epistoma in relation to varying degrees of increased separation of the antennae. I have, therefore, no hesitation in synonymizing *Pericyclocera* with *Megaselia*. The species in the former genus are, therefore, hereby transferred to *Megaselia*. In the case of *Megaselia floricola* (Borgmeier, 1966a) **Comb. nov**. this has the effect of causing a homonym with *Megaselia floricola* Borgmeier, 1967a. The latter species is therefore renamed (after the original collector) as follows:

Megaselia peterseni **nom. nov.**
= *Megaselia floricola* Borgmeier, 1967a: 220.

A further 'genus' is rejected because it cannot be clearly distinguished from *Megaselia*. Malloch (1914a) proposed the genus *Paraphiochaeta* for *Megaselia* species with a row of differentiated antero-dorsal hairs on the hind tibia, in addition to the postero-dorsals. This proposal was rejected (e.g. Schmitz, 1929g). Beyer (1958a) endeavoured to reinstate *Paraphiochaeta* as a subgenus. Again this was rejected (e.g. Borgmeier, 1961a, 1968b). However, Borgmeier (1961a) restored the genus for three Neotropical species with antero-dorsals on the hind tibia, a longer than average costa, and other small deviations from typical *Megaselia* species. However, every character cited is to be found in numerous *Megaselia* species in other parts of the world. There is no set of derived character states by which *Paraphiochaeta* can be characterized. I therefore, herewith, abolish the latter genus. The three species are assigned to *Megaselia*, two being returned to the genus and one being a new combination:

Megaselia exorbitans (Borgmeier, 1961a: 82.) **Comb. nov.**

Brown (1988) proposed a new genus *Kerophora*. While these flies form a distinct monophylum, it seems most probable that they are no more than a segregate within *Megaselia*. The initial description overlooked Dufour's crop mechanism in the females (see Disney, 1992b), thus giving rise to the erroneous view that they should be excluded from the Megaseliini. The three

Nearctic species are keyed by Brown (1988). Brown (1992b) has proposed the transfer of *Megaselia styloprocta* and *M. tristis* to this genus. I agree that the ovipositor of *M. styloprocta* resembles that of *Kerophora*. However, the male epandrium and anal tube are very different (cf. Figs 120–123 in Brown, 1988, and Fig. 368 in Disney, 1989a). I consider it as likely that the resemblance of the ovipositors is due to convergence. I therefore consider it premature to transfer these species to *Kerophora*. Indeed there are other species of Old World *Megaselia* which are far stronger candidates for being assigned to *Kerophora*. For example, the ovipositor of *M. aristica* is indistinguishable from that of *Kerophora* and the hypopygium of the male is far closer to that of *Kerophora* than that of *M. styloprocta*. I propose treating *Kerophora* as a subgenus within *Megaselia*. Unfortunately this action creates two new homonyms. Replacement names are as follows:

Megaselia brianbrowni **nom. nov.**
Kerophora ferruginea Brown, 1988: 309, nec *Aphiochaeta ferruginea* Brunetti, 1912a: 84.
Megaselia brokawi **nom. nov.**
= *Kerophora brunnea* Brown, 1988: 310, nec *Aphiochaeta brunnea* Schmitz 1920c: 136.

Today the priority in the genus *Megaselia* must be to facilitate the naming of specimens. It will be many years before a phylogenetic classification of the subgroups within *Megaselia* can be achieved. Until then we should aim at a stable nomenclature. Estimates suggest that only about 10% of the species have been described for the world. While some females are very distinctive, it is now clear that our knowledge of *Megaselia* taxonomy will be best advanced by only describing new species on the basis of both sexes or the males only. Descriptions based on females only are now an impediment. A prime requirement for all regions of the world is user-friendly, adequately-illustrated, keys to males. I review below the existing keys, and supplementary papers, required for each region of the world.

Afrotropical

Beyer (1965b) provides keys which, however, are seriously flawed by errors and omissions. Bridarolli's (1951) pioneering paper is still a useful companion (especially for the photographs of wings). Species in genera subsequently synonymized with *Megaselia* also need to be considered (Colyer and Elberg, 1969; Disney, 1986b, 1990d). A species assigned to '*Pericyclocera*' by Schmitz (1940a) runs to couplet 7 at the top of page 51 in Beyer's key. Additional species are covered by Borgmeier (1967b), Compton and Disney (1991) and Disney (1978d, 1982o, 1985b, 1989b, 1991a,b, 1991h).

Beyer (1965b) erected a new genus, '*Mollicauda*', for two Afrotropical species, one of which he had previously assigned to *Megaselia*. He distinguished the 'genus' from *Megaselia*, in his generic key, on the basis of the somewhat elongated, pear-shaped, third antennal segments in the males. Perhaps aware that this is a character of degree only, he included the 'genus' in the key to *Megaselia* species as well (loc. cit. p. 51). Here he distinguished it on the basis of the divergence of the antial bristles on the frons. Neither of these characters is sufficient to exclude numbers of undoubted *Megaselia* species. Indeed I can find no characters that will diagnose Beyer's genus. I accordingly, herewith, synonymise *Mollicauda* with *Megaselia*. The two species thus become:

Megaselia discrepans (Beyer). **Comb. nov.**
Megaselia subpyricornis Beyer. **Orig. desig.**

Australasian and Oriental

Borgmeier (1967a,e) provides keys (including a key to '*Plastophora*' species and a '*Pericyclocera*' – 1967e – which are now included in *Megaselia*). These keys need supplementing by Disney (1981k, 1982m, 1986b, 1987g, 1990d, 1991i, 1993b and 1994b), Disney *et al.* (1992), Gotô (1985a) and Johal and Disney (in press).

Nearctic

Borgmeier (1964b, 1966a) provides keys, including two species of '*Pericyclocera*' which are merely aberrant *Megaselia*. Species formerly in '*Plastophora*' are covered by Colyer and Elberg (1969). The subgenus *Kerophora* (see above) is keyed by Brown (1988). Important revisions and additions are given by W. H. Robinson (1977, 1978b, 1981). Further additions are covered by Robinson and Wisseman (1983), Disney (1981o), Brown (1990b) and Barnes (1991a).

Neotropical

A primary key is provided by Borgmeier (1962c). This needs augmenting with supplementary keys (Borgmeier, 1969a,b and 1971a, including a key to species formerly placed in '*Plastophora*' – on p. 65). Additions are covered by Disney (1982l, 1989h) and Wakeford and Disney (1994). The five species of '*Pericyclocera*' covered by Borgmeier (1967e, 1969b) are only aberrant *Megaselia* species. The three species of '*Paraphiochaeta*' will run to couplet 5 (page 301) of Borgmeier's (1962c) key. Borgmeier (1958, 1961a) covers the recognition of these species.

Palaearctic

The primary, but incomplete, monograph is the contribution of Schmitz to Lindner's *Die Fliegen der palaearktischen Region* (Schmitz, 1956b, 1957a,b; Schmitz and Beyer, 1965, 1974; Schmitz and Delage, 1974, 1981). Colyer and Elberg (1969) cover species formerly placed in '*Plastophora*'. The entire British fauna is covered by a key to males (Disney, 1989a) supplemented by Disney (1991j, 1993c). This key not only incorporates much revision of European species but also adds many species not covered by Schmitz's monograph. Additional Palaearctic species are covered by Lundbeck (1922),

Colyer (1962, 1969), Delage and Lauraire (1970), Naumov (1979), Gotô, (1985a), Disney (1988h, 1989e,i, 1990l, 1991i) and Čakar and Disney (1991).

It needs to be emphasized that undescribed species are likely to turn up in almost any collection of Phoridae. In many parts of the world the majority of the species encountered may prove to be undescribed.

Melaloncha Brues
Thirty-two Neotropical species are known. Borgmeier (1971a) provides a key, one species is more fully described by Borgmeier and Prado (1975) and Barnes (1991a) has recorded one species from the Nearctic Region.

Melittophora Brues
The female of the single Neotropical species is covered by Brues (1928a) with supplementary figures by Salt (1929). Brown (1993a) describes the male.

Menozziola Schmitz
Eight species have been described and a ninth reported. Beyer (1965b) describes two Afrotropical species, Borgmeier (1961a) describes two Neotropical species and Brown (1992b) transfers a third species to this genus, from *Stylusa* (Borgmeier and Prado, 1975). Disney (1990b) keys the three Palaearctic species. An undescribed species is reported from the Nearctic Region (Brown, 1988; Brown *et al.*, 1991). Whether all these really belong to a single genus remains to be demonstrated.

Mesopathusa Schmitz
Females of the three Afrotropical species are covered by Disney and Kistner (1989c). I briefly characterize the undescribed male sex below, by describing the first male for one of the known species.

Mesopathusa modesta Disney
Male: Frons brown, broader than high, and with three ocelli. Four supra-antennal bristles, the lower being weaker and closer together. A pair of pre-ocellar bristles level with the lower margin of anterior ocellus. A pair of ocellar bristles and four other bristles on vertex. Median furrow absent, but represented by a bare frontal vitta. With 3–4 bristles on each cheek and also on each jowl. Third antennal segment globose, pale brown, with subterminal arista. The latter pale brown, with the swollen base of segment 3 longer than segment 1, which is longer than 2. The hairs of the flagellum of segment 3 well developed (about 0.015 mm long). Eyes hairy. Palps somewhat broadened distally but narrowing to apex, being brown apically but paler basally. With two strong apical bristles, and scattered hairs behind. Proboscis relatively small, with brown labrum and paler, relatively simple labella.

Thorax brown, being darkest on top. Mesopleuron with a furrow and bare. With no differentiated humeral bristle. With three bristles on notopleuron (but only two on one side in one specimen). An anterior and a posterior supra-alar, and a pre-scutellar dorsocentral bristle each side. The scutellum with a posterior pair of long bristles and an anterior pair of short, fine, hair-like bristles.

Abdomen with brown tergites bearing short hairs, which are longer at hind margins. Tergite 6 longer than any of preceding tergites. Venter pale brownish grey and with small, obscure, hairs on segment 4–6 only. Hypopygium and anal tube brown and as Fig. 8.54(b,c).

Legs pale brown. All tibiae devoid of dorsal hair palisades and bristles, apart from short apical spurs. Front leg as Fig. 8.59(c).

Wing as Fig. 8.59d. Length 1.49–1.61 mm. Costal index 0.56–0.58. Costal ratios 0.66–0.71 : 1. Costal cilia 0.03–0.04 mm long. Thick veins pale brown, thin veins dark brown. No hair at base of vein 3 and no axillary bristles. Membrane distinctly greyish-brown tinged. Haltere with dark greyish brown knob.

Material examined: Several males, including pairs in copula, from Zimbabwe, Harare, July and August 1991 (coll. Meg S. Cumming), at mounds of the termites *Odontotermes latericius* and *O. transvaalensis*.

Metopina Macquart

More than 30 species of this cosmopolitan genus are known. Many undescribed species are also known. Confusion has been caused by naming species on the basis of descriptions of one sex only. Modern species descriptions should be based on both sexes or the females alone. Males alone should not be named. The discovery of polymorphism among the females of one species (Disney, 1988d) revealed that in another species one morph had been assigned to a separate genus!

A single Afrotropical species is redescribed by Disney and Kistner (1989c). Other undescribed species are known from this region. The two species known from Australia and New Zealand are covered by Disney (1994b). The two Nearctic species are covered by Borgmeier (1963d). The Neotropical species are keyed by Borgmeier (1969b), who also redescribes one species under the generic name '*Typhlophorina*'. Further species are added by Borgmeier and Prado (1975). Amplified descriptions of some species are given by Borgmeier (1969a) and Disney and Kistner (1989a). A distinctive species, the only one known with a forked vein 3, was omitted by Borgmeier (1969b). It is covered by Schmitz (1929g). The Palaearctic species are covered by Disney (1983a) supplemented by Disney (1979j). The Oriental species and tropical Australasian species are currently being revised, with the addition of several new species. The published species are covered by Schmitz (1927e), Hardy and Beyer (1964), Beyer (1966) and Disney (1988d, 1994b).

Microplatyphora Schmitz

The single Afrotropical species, known only in the female sex, is covered by Schmitz (1915d).

Microselia Schmitz

This genus may only be a clade within *Pseudacteon*. It is still poorly known and species recognitions are based on the female sex in the first instance. The single Nearctic species and single Neotropical species are covered by Disney (1982k), supplemented by Borgmeier (1969b). The three European species are covered by Disney (1988c). The four Afrotropical species are covered by

Disney (1983g, 1991a). I have an undescribed species from Papua New Guinea.

Misotermes Schmitz
A species from Java is covered by Schmitz (1938i) and a species from Thailand by Borgmeier (1967e), who gives a key to the two females.

Morahania Schmitz
The single Afrotropical species, known only in the female sex, is described by Schmitz (1958e).

Multinevra Disney
The male only of a single Afrotropical species has been described (Disney, 1979e). Several undescribed Afrotropical species are also known.

Myopiomyia Disney
A single species from Sulawesi, known only in the female sex, is covered by Disney (1987g).

Myrmomyia Silvestri
The three Neotropical species, known only as females, are only doubtfully distinct from *Puliciphora*. They are covered by Borgmeier (1958, 1960b) and Silvestri (1911).

Myrmosicarius Borgmeier
The single Nearctic species is covered by Borgmeier (1963d). The females of ten Brazilian species are keyed by Borgmeier (1931a) and another Brazilian species is added by Borgmeier and Prado (1975).

Necperissa Disney
A single Afrotropical species was described (as a *Perissa*) by Disney (1990j), before the abolition of the Alamirinae, but has been subsequently placed in a new genus (Disney, 1993a).

Neodohrniphora Malloch
This genus is in need of revision. Prado (1976) covers the currently recognized species. These are seven Neotropical species and one Nearctic species. The latter was placed in a separate genus (*Borgmeieria*) by Prado, but was returned to *Neodohrniphora* by Brown (1988), who describes its female. The male of this species is also covered by Borgmeier (1966a).

Neophora Borgmeier
The single Neotropical species is covered by Borgmeier (1958) with additional notes by Borgmeier and Prado (1975).

Obscuriphora Disney
The single European species is described by Disney (1986a) and included in the handbook on British Phoridae (Disney, 1989a).

Odontoxenia Schmitz
The single Oriental species, known only in the female sex, is described by Schmitz (1916a).

Oligophora Silvestri
The single Oriental species, known only in the female sex, is described by Silvestri (1947b).

Pachyneurella Brues
This Neotropical genus had long been synonymized with *Puliciphora*, when Borgmeier (1969a) reinstated it. However, it was only readily distinguished from the latter genus in the male sex, and on the basis of a single, seemingly somewhat trivial, character. With the taxonomy of *Puliciphora* being based primarily on the female sex the problem of recognition is compounded. I suspect that *Pachyneurella* is merely a Neotropical subgenus within the cosmopolitan genus *Puliciphora*.

The recognized species are covered by Borgmeier (1969a) and Borgmeier and Prado (1975). However, the genus *Puliciphora* should also be considered when trying to identify specimens, especially in the case of females.

Palpiclavina Silvestri
The two Oriental species are keyed by Disney and Kistner (1992).

Palpocrates Schmitz
This 'genus' probably represents two further species of *Triphleba* with enlarged and peculiar palps: there being several (subsequently discovered) species of Nearctic and Palaearctic *Triphleba* with enlarged and aberrant palps in the male sex. The discovery of the unknown females is likely to resolve this. The two species, both from New Zealand, are covered by Schmitz (1939a).

Paraphiura Beyer
This genus is known from a single male from Tasmania. It is described by Beyer (1966).

Parasyneura Malloch
A single species from Java, known only in the female sex, is described by Malloch (1912b) supplemented by Borgmeier (1962c).

Paratermitoxenia Beyer
The single Afrotropical species, known only in the female sex, is described by Beyer (1965c) and more fully illustrated by Kistner (1982).

Paurophora Silvestri
A single Brazilian species, known only in the female sex, is covered by Silvestri (1947a).

Pelidnophora Borgmeier
Two species are both known only from males from Brazil. They are covered by Borgmeier (1969b).

Penthaplus Schmitz
A single Afrotropical species is described by Schmitz (1958e).

Pericyclocerina Borgmeier
A single Neotropical species, known only in the female sex, is described by Borgmeier (1971a).

Perissa Borgmeier
This 'genus' has proved to be the 'missing' males of the Termitoxeniinae (Disney and Cumming, 1992). One Oriental species remains in the 'genus' by default, until it can be linked up with its termitoxeniine female. The Termitoxeniinae are currently being revised. In the interim '*Perissa*' species not yet re-assigned to their termitoxeniine genera are covered by Disney (1990j).

Perittophora Disney
The single Afrotropical species, known only in the female sex, is described by Disney (1990j).

Phalacrotophora Enderlein
About 50 species are known in this cosmopolitan genus. The New World species are keyed by Borgmeier (1971a), who also keys the Nearctic species separately (Borgmeier, 1963d). The Australasian, Oriental and Japanese species are keyed by Borgmeier (1967e). The Afrotropical species are covered by Beyer (1965b) supplemented by Schmitz (1932e). British species are covered by Disney (1983a), and the rest of European species by Delage and Lauraire (1974b) and Disney (1979k).

Pheidolomyia Schmitz
Three species are known from Brazil. The females are keyed by Borgmeier (1956), who also describes the male of one species.

Phlebothrix Borgmeier
The single Afrotropical species is covered by Schmitz (1929g), under its discarded generic name *Nossibea*.

Phora Latreille
More than 50 species are known, with most being north temperate in distribution. The Nearctic species are keyed by Borgmeier (1963d), including Holarctic species. Beyer (1965b) deals with an Afrotropical species and (1958a) with an Oriental species. The European species are covered by Disney (1983a) supplemented by Schmitz (1953a, 1955b) and Disney (1989a, 1989c, 1993c). The Asian species are covered by Schmitz (*loc. cit.*) supplemented by Gotô's (1986a) key to Japanese species, and additions by Zaitzev (1977) and Michailovskaya (1986).

Phymatopterella Brues
The five Neotropical species are keyed by Borgmeier (1971a) and a Nearctic species is described by Barnes (1990b).

Physoptera Borgmeier
There are 13 known species in this genus, two Nearctic species keyed by Borgmeier (1966a) and also, along with 11 Neotropical species, by Borgmeier and Prado (1975). Whether this is really a distinct genus, or merely a subgenus within *Megaselia*, remains to be demonstrated.

Placophorina Borgmeier
A single Neotropical species, known only in the female sex, is redescribed by Borgmeier (1958).

Plastophorides Beyer
The females of the three Afrotropical species are keyed by Beyer (1959f). The male of the type species is described by Disney (1991a), who also amplifies the description of the female. An Oriental species is described by Disney (1993b).

Platydipteron Borgmeier and Prado
The single Neotropical species, known only in the female sex, is covered by Borgmeier and Prado (1975).

Plectanocnema Schmitz
The single European species is covered by Schmitz (1956b) and Disney (1983a).

Plethysmochaeta Schmitz
This genus is currently being revised. There are supposedly eight Afrotropical species, but with several known in only one sex, one from Malaya and one from Java. The last two are poorly known, but are treated by Borgmeier (1967a). There is no key to the Afrotropical species. Reference should be made to Schmitz (1953d, 1958e) and Beyer (1965b) in the first instance.

Pleurophorina Borgmeier
This genus is known from a single male from Costa Rica (Borgmeier, 1969b).

Postoptica Disney
This single species is known from males only (Disney, 1987a), being recorded from Sulawesi and the Philippines.

Pradea Borgmeier
The two Neotropical species are keyed by Prado (1976).

Procliniella Borgmeier
The single Neotropical species, known only in the female sex, is covered by Borgmeier (1925c, 1926c).

Pseudacteon Coquillett
More than 40 species are known in this genus. The females of the Nearctic and Neotropical species are keyed by Borgmeier (1969b), which needs supplementing by Borgmeier and Prado (1975) and Disney (1982k, 1991f). The Palaearctic species are keyed by Schmitz (1938m), with the British species covered by Disney (1983a). Borgmeier (1967a) redescribes a species from Java and Beyer (1966) describes a species from the Solomon Islands.

Pseudohypocera Malloch
The single Neotropical species, *P. kerteszi*, has been redescribed several times under a number of synonyms. Malloch (1912b) provides a brief description in English. Borgmeier (1930b) provides the most detailed description.

Pseudotermitoxenia Shiraki
The single Oriental species, known only in the female sex, is too poorly described to be adequately distinguished from other Termitoxeniinae (Borgmeier, 1964a). A fresh description is required when new material is available (the type material being seemingly lost).

Psyllomyia Loew
The females of the four Afrotropical species are keyed by Schmitz (1951d). The only male known is described by Beyer (1965b).

Ptochomyia Silvestri
Three Afrotropical species are known. The genus is currently being revised. For the time being reference should be made to Disney (1993a).

Pulicimyia Borgmeier
The three Neotropical species, known only as females, are keyed by Borgmeier (1960b).

Puliciphora Dahl
About 70 species are known in this cosmopolitan genus, or more if the Neotropical *Pachyneurella* is merely a subgenus of *Puliciphora* (see above). The boundaries of the genus require clarification. The females of the Old World species are keyed by Disney (1988b), along with an updated key to the Australasian and Oriental species (Disney, 1990f) and additional information on Afrotropical species (Disney and Kistner, 1989c; Disney, 1991a). Nearctic species are keyed by Borgmeier (1963d), who also keys the Neotropical species (Borgmeier, 1960b). The latter needs supplementing by Borgmeier (1969b), Borgmeier and Prado (1975) and Disney and Kistner (1989a). The genera *Pachyneurella* and *Myrmomyia* should be taken into consideration as well. There is need for further revision and keys to males. Descriptions of new species should only be based on both sexes or females. Descriptions based on males only will merely add confusion.

Razorfemora Disney
A single Palaearctic species is known. The male is described by Disney (1990b). The undescribed female is described below (based on a specimen from the Zaragoza Region of Spain, collected in 1990 by J. Blasco-Zumeta).

Razorfemora nussbaumi Disney
Female: Frons similar to male but the bristles longer and stronger (including the outer pair on the vertex). No supra-antennal bristles, as in male. Proboscis, palps and antennae as in male.

Thorax as male, that is with a humeral, two notopleurals, a pre-alar, an intra-alar and a posterior dorsocentral bristle each side. Scutellum with four bristles as in male.

Abdomen with brown tergites and brownish grey venter. The latter with short hairs along median band on segments 2–6 and longer bristles postero-laterally on segments 3–6 and also on flanks of 5 and 6. Tergites 1–6 as Fig. 8.79(a). The terminalia form a heavily sclerotized dark brown capsule with a yellow, sharply-pointed, ovipositor (Fig. 8.79c). Internally without a Dufour's crop mechanism and with only two rectal papillae. Beneath tergites 4 and 5 lie a pair of glands joined together posteriorly, where they appear to discharge at rear of tergite 5 (Fig. 8.79a).

Legs as male except hind femur lacks the basal projection of the male and the last tarsal segments tend to be more elongate and a little tapered, especially on mid and hind legs (e.g. Fig. 8.79b). In the case of the front leg the whole tarsus is longer than in male.

Wing 1.73 mm long. Costal index 0.45–0.46. Costal ratios 2.09:0.64:1. Costal cilia 0.08 mm long. Otherwise it and haltere similar to male.

Rhabdomyia Borgmeier
The validity of this genus, known only in the female sex, has been called into question by Disney and Kistner (1989b), who described an Oriental species. Two species from Mexico were described by Brues (1925a) (under *Ecitomyia* and *Ecituncula*). The procurement of males should resolve the taxonomic problems posed by this genus.

Rhopica Schmitz
Only two Australasian species have been described. They are both characterized by Gotô (1985b). However, I have a further four Australasian and an Oriental species awaiting description, and I hope to produce a key in the near future.

Rhynchomicropteron Annandale
A dozen species are known from the Oriental Region. The females are keyed by Disney (1990c), supplemented by Disney (1991g, 1992d). Brown (1992a) has shown that the, subsequently described, genus '*Gymnoselia*' represents the hitherto missing males. At present males can only be named by association with their females, unless covered by the 1992d paper cited above or Schmitz (1929g). Eventually it is likely that two *Rhynchomicropteron* species will be found to be the females of the two described species of '*Gymnoselia*'.

Rhyncophoromyia Malloch
The eight Neotropical species and two Nearctic species are keyed by Prado (1976), and the two Nearctic species are also covered by Borgmeier (1963d).

Sphinctomyia Borgmeier
The single Neotropical species, known only in the female sex, is covered by Borgmeier (1954).

Spiniphora Malloch
Nineteen species have been described, from the Nearctic, Palaearctic and Oriental Regions. However, *S. bergenstammi* is cosmopolitan, probably through the agency of man. The Nearctic species are keyed by Borgmeier (1963d). The Palaearctic species are keyed by Schmitz (1941c) supplemented by Beyer (1958b), Borgmeier (1962c), Colyer (1969) and Disney (1983a, 1990l). The Oriental species are covered by Brues (1936a), Schmitz (1940e), Colyer (1955a) and Borgmeier (1962c).

Stenodicrania Borgmeier
A single species is known from Malaya (Borgmeier, 1967a).

Stenoneurellys Borgmeier
Only two species are known, both from Brazil. They are covered by Borgmeier (1924a, 1927d) and Borgmeier and Prado (1975).

Stenophorina Borgmeier
The single Nearctic species, known only in the female sex, is described by Borgmeier (1963d) and also figured by Peterson (1987).

Stethopathusa Schmitz
The single species from Sumatra, known only in the female sex, is described by Schmitz (1921b).

Stichillus Enderlein
Twenty-five species are known in this cosmopolitan genus. An Afrotropical species is covered by Beyer (1960b). Australasian species are covered by Beyer (1966). Nearctic species are keyed by Borgmeier (1963d). Neotropical species are keyed by Borgmeier (1959), supplemented by Borgmeier and Prado (1975). Oriental species are keyed by Beyer (1958a), supplemented by Kaneko *et al.* (1961). Palaearctic species are covered by Schmitz (1951c, 1953a), Beyer (1958a) and Kaneko *et al.* (1961). It should be noted that *S. matsumurai* Schmitz is a synonym of *S. japonicus* (Matsumura) and the *S. matsumurai* of Kaneko *et al.* is a misidentification of *S. brunneicornis* Beyer.

Styletta Borgmeier
The female of the single Neotropical species is described by Borgmeier (1961a) and the male by Brown *et al.* (1991). Other species assigned to this genus have been transferred to *Trucidophora*.

Synaptophora Brown
A single Afrotropical species, known only in the female sex, is covered by Brown (1992b).

Synclinusa Borgmeier
Two species known from Brazil are covered by Borgmeier (1971a) and Borgmeier and Prado (1975).

Syneura Brues
There are nine Neotropical species, one of which occurs in the Nearctic also. The species are keyed by Prado (1976).

Syneurina Borgmeier
The single Australian species, known only in the male sex, is covered by Borgmeier (1963b).

Syntermitoxenia Schmitz
This Afrotropical genus is currently undergoing revision. The females of three species are referred to by Borgmeier (1964a). Two species assigned to *Perissa*, including a male, are covered by Disney (1990j), but are herewith transferred to this genus:

Syntermitoxenia latiptera (Disney, 1990j:313). **Comb. nov.**
Syntermitoxenia lewisi (Disney, 1979g: 143). **Comb. nov.**

Syntermophora Seevers
The single Neotropical species, known only in the female sex, is described by Seevers (1941).

Tarsocrates Schmitz
A single species is known from New Zealand (Schmitz, 1939a).

Termitomyia Wasmann
The single Afrotropical species, known only in the female sex, is described by Wasmann (1900a) supplemented by Schmitz (1955f).

Termitophilomyia Schmitz
Three Afrotropical species are known. Borgmeier (1964a) refers to the females. The male, along with additional figures of the female, of one species is covered by Disney and Cumming (1992). The genus is currently undergoing revision.

Termitophorides Borgmeier
The two Neotropical species, known only in the female sex, are covered by Borgmeier (1967b) supplemented by Borgmeier (1923c, 1924a) and Silvestri (1947a).

Termitophorina Borgmeier
The single Afrotropical species, known only in the female sex, is covered by Borgmeier (1967b).

Termitosagma Reichensperger
Two Afrotropical species, known only as females, have been described. Borgmeier (1959) redescribed the 'two' species, but his distinctions have proved unreliable. Fresh material (collected by David Kistner) indicates that *T. ephippium* Schmitz (1951a: 77) is a synonym of *T. henningsi* (Reichensperger 1931: 303). **Syn. nov.**

Termitoscrofa Schmitz
Two Afrotropical species, known only in the female sex, are covered by Schmitz (1939d) and Delachambre (1966).

Termitosphaera Wasmann
The two Oriental species, known only as females, are covered by Borgmeier (1964a).

Termitostroma Reichensperger
Four Afrotropical species, known only as females, have been described. A key is provided by Delachambre (1965). *Alamira peckorum*, covered by Disney (1990j), is almost certainly the flying stage of one of these species. The genus is currently being revised.

Termitoxenia Wasmann
A single Afrotropical species, known only from females, is included in this genus (Borgmeier, 1964a), which is currently being revised. However, with the abolition of the Alamirinae, a species of *Perissa* (known in both sexes) is transferred to this genus herewith:

Termitoxenia oligoseta (Disney, 1990j: 316). **Comb. nov.**

Thalloptera Borgmeier and Schmitz
Nine Neotropical species, known only in the female sex, are keyed by Borgmeier (1936), supplemented by Borgmeier (1971a).

Thaumatoxena Breddin and Börner
The eight Afrotropical species are keyed by Disney and Kistner (1992).

Tranopeltoxenos Brues
The single Neotropical species, known only in the female sex, is described by Brues (1923a).

Travassophora Borgmeier
A single species from Brazil is known only in the male sex (Borgmeier, 1938a).

Trichocerina Borgmeier and Prado
This genus is known from a single male from Brazil (Borgmeier and Prado, 1975).

Trineurocephala Schmitz
The three Neotropical species are keyed by Borgmeier and Prado (1975).

Triphleba Rondani
About 90 species (including those formerly placed in *Citrago*) are known in this genus, which is recorded from all regions except the Afrotropical. However, the genus is primarily Holarctic. The Nearctic species are keyed by Borgmeier (1963d). The Palaearctic species are keyed by Schmitz (1943) supplemented by Schmitz (1949c, 1955b), Delage and Lauraire (1970), Disney (1983a, 1987j, 1994c), Gotô and Takeno (1983b) and Michailovskaya (1986). Oriental species are covered by Beyer (1958a) and Australasian species by Schmitz (1939a).

Trophithauma Schmitz
This genus in the male sex 'cannot be distinguished from *Megaselia* by any remarkable character' (Borgmeier, 1963d). The females are indeed distinctive, but so are numerous *Megaselia* females. It must remain doubtful that this is a valid genus, rather than some *Megaselia* species with similarly striking females. There are five reported species. The Nearctic species is covered by Borgmeier (1963d). A Neotropical and Oriental species are covered by Schmitz (1929g) (with *T. splendidum* being a synonym of *T. portentum*) and Gotô (1984c) covers two Japanese species.

Trophodeinus Borgmeier
This genus is known from the Nearctic and Neotropical Regions. The males have long been known under the name *Bactropalpus*, until correctly associated with the females by Brown (1986). The males of seven of the nine named species are keyed by Borgmeier (1963c), with a synonym amendment by Brown (1986). The three known females are keyed by Borgmeier (1963d), but two of these are only known as females. Their males are probably described under another name.

Tropophleba Schmitz
Schmitz (1929i) assigned two Neotropical species, known only in the female sex, to separate subgenera because of considerable differences. The latter are clearly more significant than the resemblances and so I have raised *Dicranodeina* to the generic rank (see above).

Trucidophora Brown
Brown *et al.* (1991) erected a new genus for a Nearctic and a Palaearctic species of *Styletta*. The former is described by Brown (1988) and the latter by Disney (1990b).

Tubicera Schmitz
There is supposedly one Neotropical and one Palaearctic species in this genus, but the former almost certainly belongs to a different genus (Disney, 1990b). It is covered by Borgmeier (1928c). The European species is covered by Schmitz (1920c).

Tubiceroides Borgmeier
The two Neotropical species, known only in the female sex, are keyed by Borgmeier and Prado (1975).

Veruanus Schmitz
This genus is of doubtful validity, and now only includes the original European species (Disney, 1990b). The latter is covered by Schmitz (1929g).

Wandolleckia Cook
The single Afrotropical species is covered by Baer (1953, which figures the male anal tube as the penis in error). The female is also figured by Schmitz (1916b) (under a synonym).

Woodiphora Schmitz
Forty species are known in this cosmopolitan genus. The Afrotropical species are keyed by Disney (1985d), supplemented by Disney (1991h). The Australasian and Oriental species are keyed by Disney (1989f). The single European species is covered by Disney (1983a). The Nearctic and Neotropical species are keyed by Borgmeier (1967e).

Woodiphorides Beyer
A single species, known only in the male sex, is described from New Guinea (Beyer, 1961).

Xanionotum Brues
Thirteen species are known from the Nearctic and Neotropical Regions. The Nearctic species are keyed by Borgmeier (1963d) and the females of the 12 species known in this sex are keyed by Borgmeier (1968a). The Brazilian species known only in the male sex is described (as *Metopina photophila*, the female being a different species) by Borgmeier (1959).

Zikania Borgmeier
The two Neotropical species, known only in the female sex, are covered by Borgmeier (1925a, 1936).

8.4 FOSSIL PHORIDAE

The fossil Phoridae that have been studied are preserved in amber. They are sometimes found in amber beads. Fossils in amber are reviewed by Poinar (1992). Work on fossil scuttle flies is patchy and lacks any critical evaluation. The described species are catalogued by Borgmeier (1968b) and updated by Spahr (1985), the major addition being by Hong (1981). Further additions are made by Disney (1987n), Grimaldi (1989) and Grimaldi *et al.* (1989). Some of the fossil Sciadoceridae (McAlpine and Martin, 1966; Brown and Pike, 1990) have been treated as Phoridae by some authors. Whether these interesting flies should be included in the Phoridae is still in dispute (see Disney, 1985a, 1991m). Whereas the modern Sciadoceridae and Phoridae are readily distinguished, the placement of some of the fossil genera is not so easy. Only when a consensus is achieved with regard to the groundplan of the Phoridae (8.5) can these disputes be resolved.

The genus *Eophora* (Handschin, 1944) is known only from puparia, and is therefore essentially unrecognizable in our present state of knowledge. In addition no type species was designated, so that the name has been declared invalid by Brown (1990c). However, if at any time a specimen were to be designated the holotype and be given a specific epithet, the name could be reinstated.

Grimaldi (1989) reported the genus '*Metopina*' from the late Cretaceous and lower Miocene. However, these fossils clearly have the arista dorsal in position. This will immediately distinguish them from all modern *Metopina* (Borgmeier, 1963d). Furthermore Wada (1991) has indicated that an apical arista is the ancestral (plesiomorphic) state within the Aschiza, so that the dorsal position of the arista in Grimaldi's fossils can be regarded as a diagnostic feature. Therefore it seems wiser to assign them to a separate genus, than leave standing the less probable hypothesis that the genus *Metopina* has been in existence for at least 80 million years. I propose, therefore, to assign Grimaldi's fossils to a **new genus** *Pseudometopina*. I accordingly rename the Cretaceous species

Pseudometopina goeleti (Grimaldi, 1989: 68) **comb. nov.** (the type species of the genus *Pseudometopina*)

Only further studies can evaluate the relationship between *Metopina* and *Pseudometopina*. The similarities could be largely due to convergence. (*Note:* Grimaldi referred to the 'ptilinal suture', in his description of these interesting fossils. This, however, is an error. It is clear, from his excellent photographs of the modern *Metopina oligoneura*, that he is referring to the incipient transverse ridge, which is the precursor of the post-ptilinal ridge – see Disney, 1988g, 1991k.)

KEY TO EXTINCT GENERA OF PHORIDAE

Most of the Tertiary fossils belong to genera still found today. I give below a key to genera known only as fossils. This key is designed *only* to be used for specimens that will *not* run out to any known genus of modern Phoridae in the keys above.

1. Veins M1 and M2 present as separate branches of vein M1+2, even though they may not be joined at the actual forking (as in Fig. 8.96b) ... Sciadoceridae
– Vein M1+2 (vein 4, the first thin vein) not forked (as in Fig. 8.96c) ... Phoridae 2
2. Vein 3 (Rs) with hairs along the length of dorsal face. Hind tibia with longitudinal hair palisades but no isolated bristles above apical quarter ... 3
– Vein 3 bare beyond any hair at base. Hind tibia with or without isolated bristles above apical quarter, but without any longitudinal hair palisades ... 4
3. Scutellum with 4 bristles ... *Protoplatyphora* Brues
– Scutellum with 8 bristles on margin ... *Protophorites* Brues
4. With no isolated bristles above apical quarter of hind tibia. Vein 6 with

sudden bend near middle, opposite and opposed to curve of vein 5 (as in Fig. 8.56a). (Arista dorsal in position.) *Pseudometopina* Disney
- Hind tibia with isolated bristles above apical quarter. Vein 6 without such a bend in the middle 5

5. Frontal bristles include a transverse row of six bristles between the four upturned supra-antennal bristles and the four bristles of the median row (the medio-laterals plus pre-ocellars) *Electrophora* Brues
- Frons with only eight bristles between the supra-antennals and posterior ocelli 6

6. Supra-antennal bristles downturned. Third antennal segment globose, with pre-apical arista *Hypoceridites* Brues
- Supra-antennal bristles upturned. Third antennal segment greatly elongated, with dorsal arista well before apex *Rhoptrocera* Hong

8.5 THE CURRENT DEBATE ON PHORID CLASSIFICATION

The affinities of the Phoridae have been a matter of dispute in the past, but the current consensus places them in the section of the Aschiza which includes the Lonchopteridae plus Phoroidea but excludes the Syrphoidea (1.3). The Empidoidea is regarded as the sister group of the Cyclorrhapha (i.e. the Aschiza plus Schizophora). Which of these two groups should include the Opetiidae is unresolved (Disney, 1987d). Recently reviewed evidence, from the study of the thoracic musculature (Ulrich, 1991) and of the larval mandibles (Sinclair, 1992), tends to support the current consensus. The latter view, summarized by McAlpine (1989), needs to be considered along with more specific data relating to the inferred groundplan of the Phoridae (e.g. Disney, 1986c,d, 1988g, 1989f, 1991m; Brown, 1992b). Rival interpretations of these data form the basis of the current debate relating to the supra-generic classification of the Phoridae. There is currently only a partial consensus.

The unresolved disputes mainly relate to the inferred groundplan of the Phoridae. In particular the paper by Brown (1992b) offers views differing considerably from my own. His paper includes many interesting new data, which he has made the basis for some proposals for radical changes to the current supra-generic classification. While his new data are greatly to be welcomed, I regard most of his proposals as being premature. He himself remarks of his hypotheses that 'the large number of character state reversals and convergences is distressing'. His proposals, however, are worthy of consideration. One can never prove a phylogenetic hypothesis. They serve to direct attention to interesting data and gaps in knowledge. New data may undermine or support hypotheses. In the case of Brown's proposals, I would caution against acceptance of his radical revisions of the classification of phorid genera until a consensus is reached. My disagreements are both general and specific.

My general disagreement is with regard to the application of the outgroup comparison method for the resolution of the polarities of postulated transformation series of homologous character states. This issue has been discussed elsewhere (Disney, 1993e). The current consensus with regard to the Phoroidea (e.g. McAlpine and Martin, 1966, Brown, 1992b) places the Sciadoceridae as the sister group of the Phoridae, the Ironomyiidae as the sister group of the Sciadoceridae, and the Platypezidae as the sister group of the Ironomyiidae. Oldroyd (1964) remarks that Bezzi called *Sciadocera rufomaculata* (Fig. 8.95) 'the most wonderful fly in the world' 'because it combines features of Empids, Dolichopodids, Phorids, and Platypezids, at one of the most obscure points in the evolutionary plan of the flies'. It is precisely because it is such a mosaic of ancestral (plesiomorphic) and derived (apomorphic) character states that *Sciadocera* is such an intriguing insect. Which are the derived states and which the ancestral?

The basic problem, in relation to two postulated homologous states, A and B, which both occur in the Phoridae, is whether A evolved from B or B from A. The outgroup method examines the sister group for the presence of A and B. If A is restricted to the Phoridae but B is not, it is argued that B is the groundplan state for the Phoridae and A is the derived state. In applying this method Brown tends to compare the Phoridae with the Sciadoceridae, then the latter with the Ironomyiidae, and then this family with the Platypezidae, and finally the latter with the Empidoidea. By contrast I tend to use the Sciadoceridae plus the Ironomyiidae plus the Platypezidae (and, if necessary, the Empidoidea and other Brachycera as well) as the outgroup for the purposes of comparison. This difference can give rise to very different interpretations. For example, tergite 8 of the female abdomen is present in some Phoridae and absent in others. It is absent in both the Sciadoceridae and Ironomyiidae, but present in Platypezidae and Empidoidea. Brown concludes that the groundplan of the Phoridae has no tergite 8. He then postulates that it evolved afresh (possibly as an 'underlying synapomorphy' or 'atavism' which re-emerged after being dormant for millions of years) in many Phoridae, as a 'new', derived, state. By contrast I conclude that tergite 8 is part of the groundplan of the Phoridae, and has been independently lost in Ironomyiidae, Sciadoceridae and some Phoridae.

I have defended my approach (Disney, 1993e) with reference to the prevalence of mosaic evolution (without which all phylogenetic reconstruction would be impossible – as emphasized by Ax, 1987) and by pointing to the long separation of the Sciadoceridae and Phoridae. They have been separate for at least 80 million years (McAlpine and Martin, 1966; Grimaldi, 1989). The re-emergence of a lost character is, of course, a possibility. However, I maintain that it should never be assumed and this assumption then be employed in the construction of a cladogram. The latter should be produced without the aid of the disputed transformation series, which is then subsequently mapped onto the cladogram to test the hypotheses of disappearance followed by re-emergence of one of the character states of the series.

Some of Brown's specific proposals I consider to be insufficiently justified. An example concerns the hypandrial bristles (Fig. 8.14a). Brown postulates that these are a novelty, and then proposes the transfer of *Triphleba* from the Phorinae to the Metopininae. It is true that *Sciadocera* lacks such bristles. However, several hypandrial bristles are the norm in Ironomyiidae and

Platypezidae. An alternative homology is that hypandrial bristles in Phoridae represent the last remnants of the gonopods (Fig. 8.15). Either interpretation leads me to the conclusion that hypandrial bristles are a groundplan feature of the Phoridae (Disney, 1994c). There are other examples of unresolved polarities of postulated transformation series. For example, a two-segmented palp (e.g. Fig. 8.41b, c) was proposed as a groundplan feature of the Phoridae (Disney, 1988g). Brown (1992b) counter-proposes that a one-segmentd palp is the groundplan state, because it is so in *Sciadocera*. However, evidence from the subfamily Thaumatoxeninae (cf. Figs 8.66b and 8.89b) indicates that a two-segmented palp is a groundplan feature of this subfamily at least (Disney and Kistner, 1992).

Apart from polarity problems, there are differences of opinion regarding the homology, or otherwise, of critical character states. For example, Brown (1992b) proposes that the dorso-lateral sclerotized patches on the seventh abdominal segment of *Dohrniphora* females (Fig. 8.40c) and those of *Aenigmatias* females (Fig. 8.21b, c) are a synapomorphy (a shared, uniquely-derived, character state). I have counter proposed (Disney, 1993c) that in *Dohrniphora* these patches are the lateral portions of tergite 7, and that the essential novelty is the loss of the sclerotisation of the median part of the tergite. By contrast in *Aenigmatias* I consider tergite 7 to be absent and the postero-lateral sclerotized patches to be a novel development, associated with the musculature operating the undisputed novelty of the two series of hooks at the rear of the segment. In short, I reject the postulated homology between the patches in the two genera. Likewise, the homologies of the thin veins of the wing are in dispute (Disney, 1991c, 1991m; Brown, 1992b). The homologies I believe to be best supported by the available evidence are shown in Fig. 8.96. The principal difference from Brown's preference is that he regards vein 4 (the first thin vein) as being M1 and vein 5 as M2. This difference of interpretation affects the proposed family assignments of certain Cretaceous genera. According to my interpretation the mutant shown in Fig. 1.1b represents a throw-back mutation, restoring the lost outer cross-vein of the phorid ancestors. This cross-vein is retained in many other families of Diptera (e.g. Fig. 8.96a).

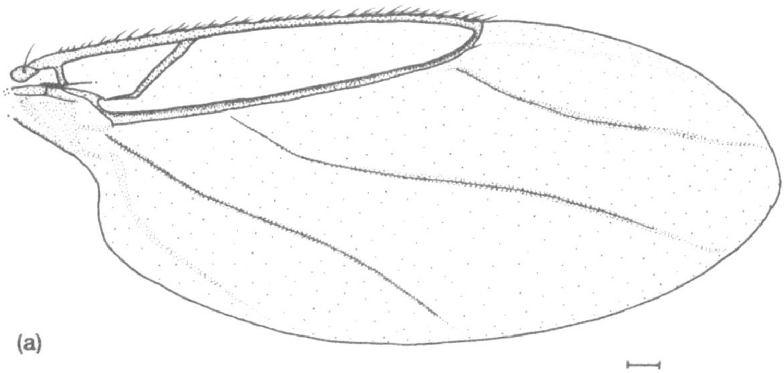

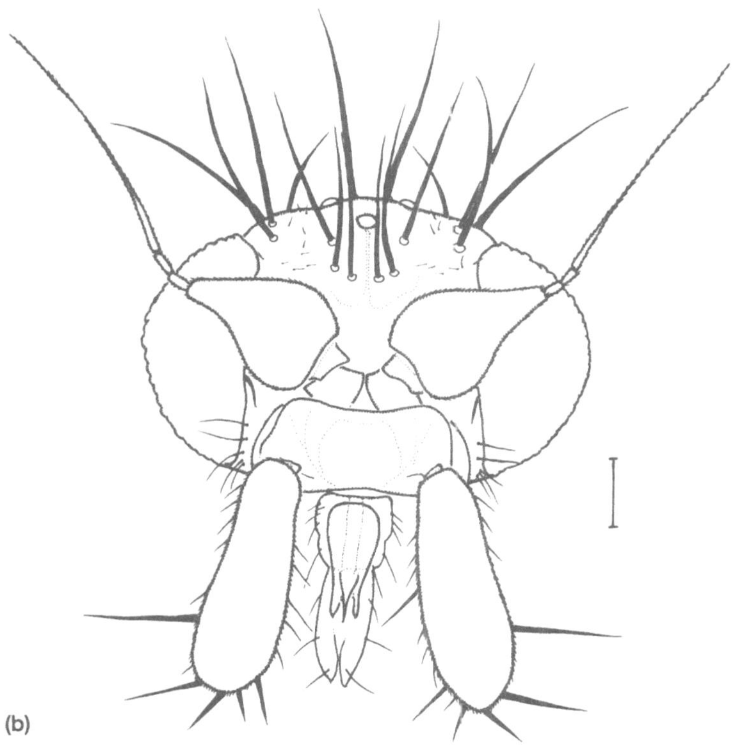

Figure 8.16 *Adelopteromyia*, male (a) Right wing (b) Frontal view of head

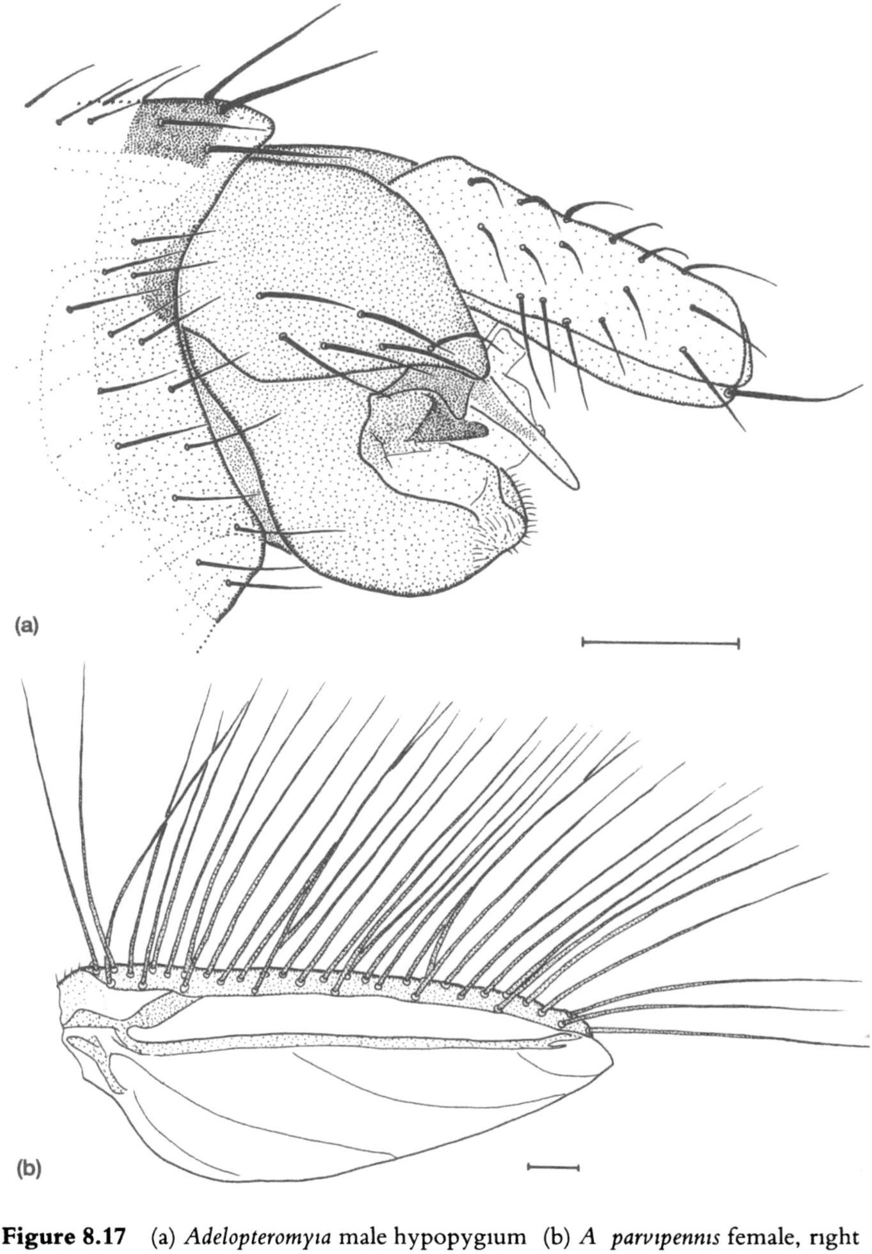

Figure 8.17 (a) *Adelopteromyia* male hypopygium (b) *A parvipennis* female, right wing

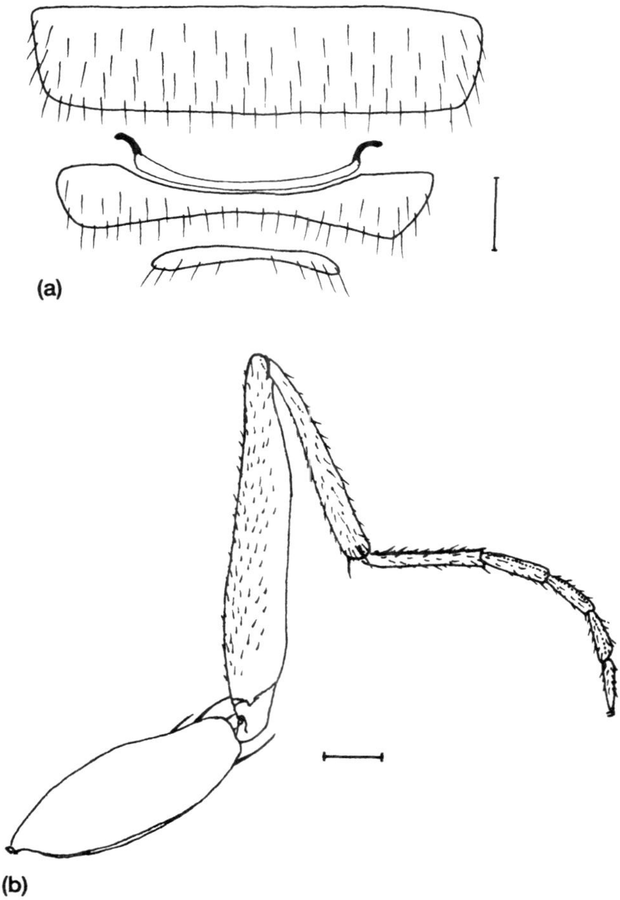

Figure 8.18 (a) *Adenophora pulchella* female, abdominal tergites 4–6. (b) *Auxanommatidia medleri* female, left front leg.

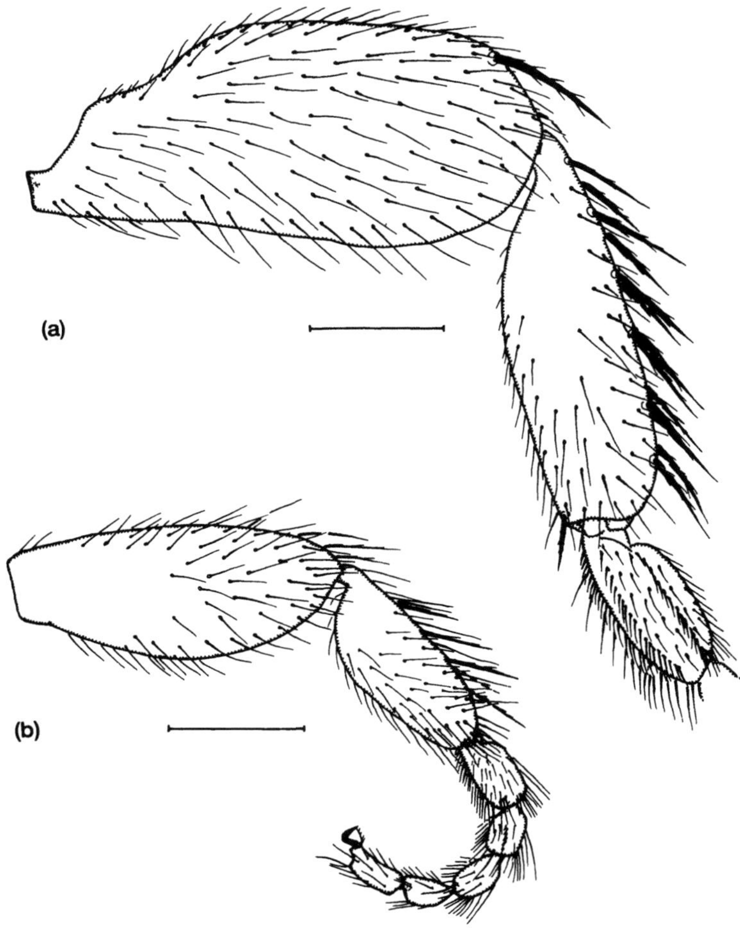

Figure 8.19 *Aenictacantha setifera* female, anterior faces of legs. (a) Hind femur, tibia and metatarsus. (b) Front leg (omitting coxa and trochanter).

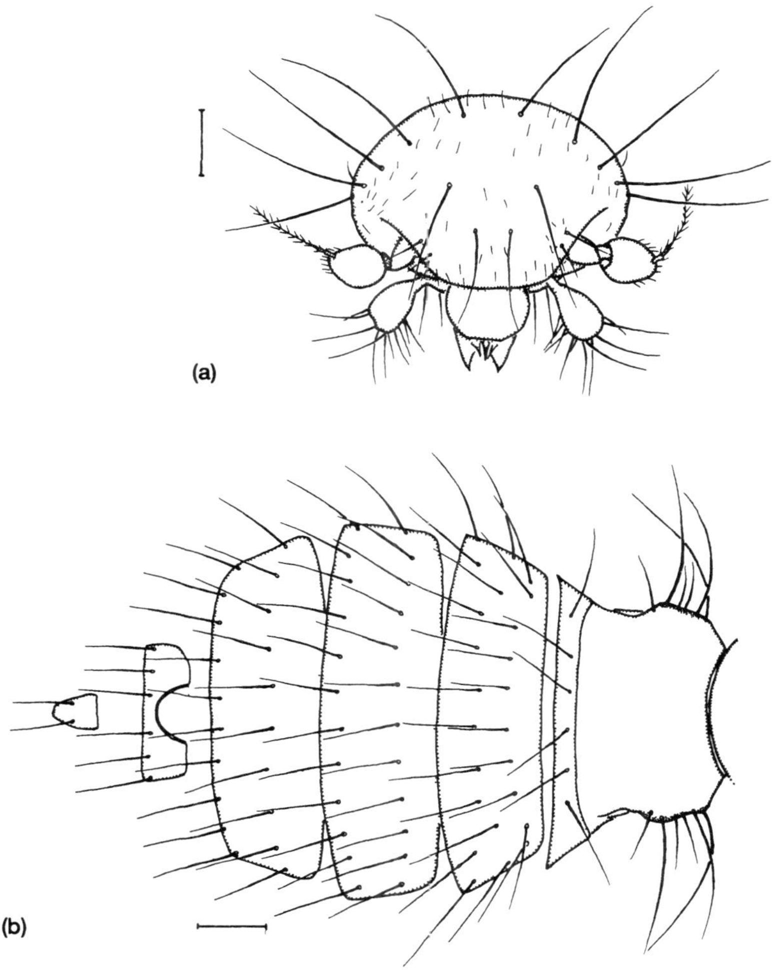

Figure 8.20 *Aenictacantha setifera* female. (a) Frontal view of head. (b) Dorsal view of thorax and abdominal tergites.

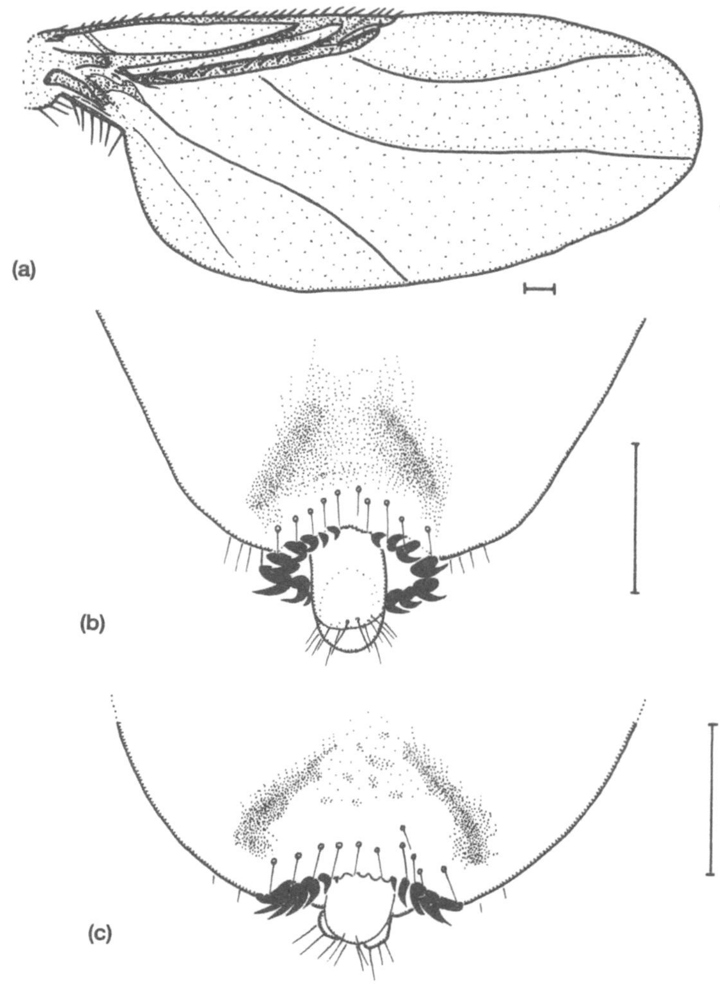

Figure 8.21 *Aenigmatias* (a) *A lubbocki* male, right wing (b), (c) Terminal abdominal segments from above in *A lubbocki* (b) and *A franzi* (c)

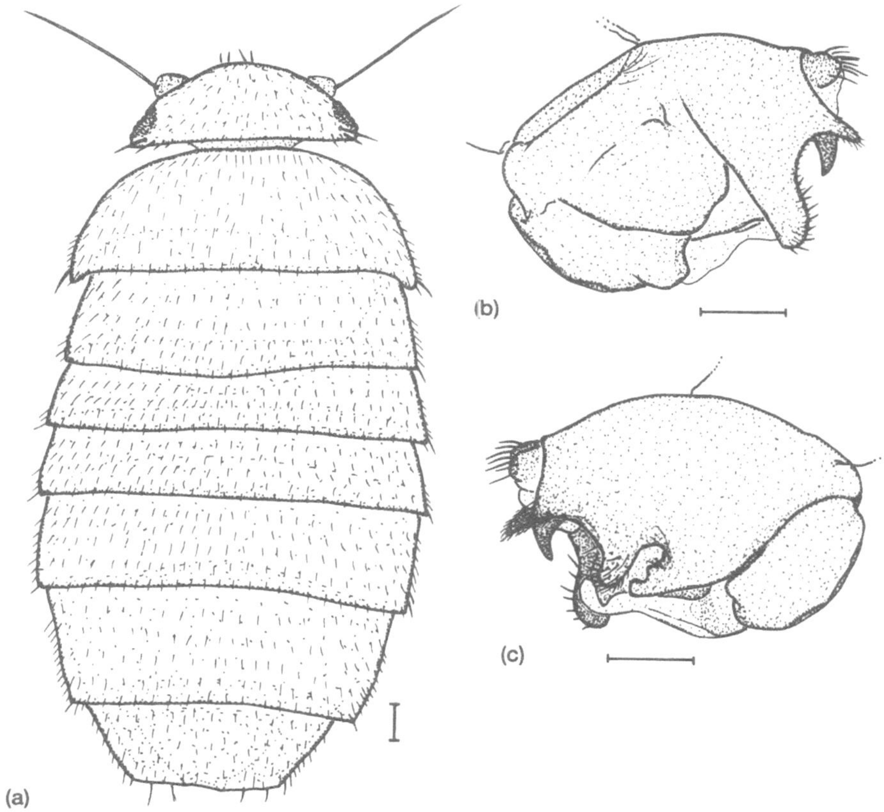

Figure 8.22 *Aenigmatias lubbocki* (a) Female from above (terminal segments withdrawn) (b), (c) left and right faces of male hypopygium

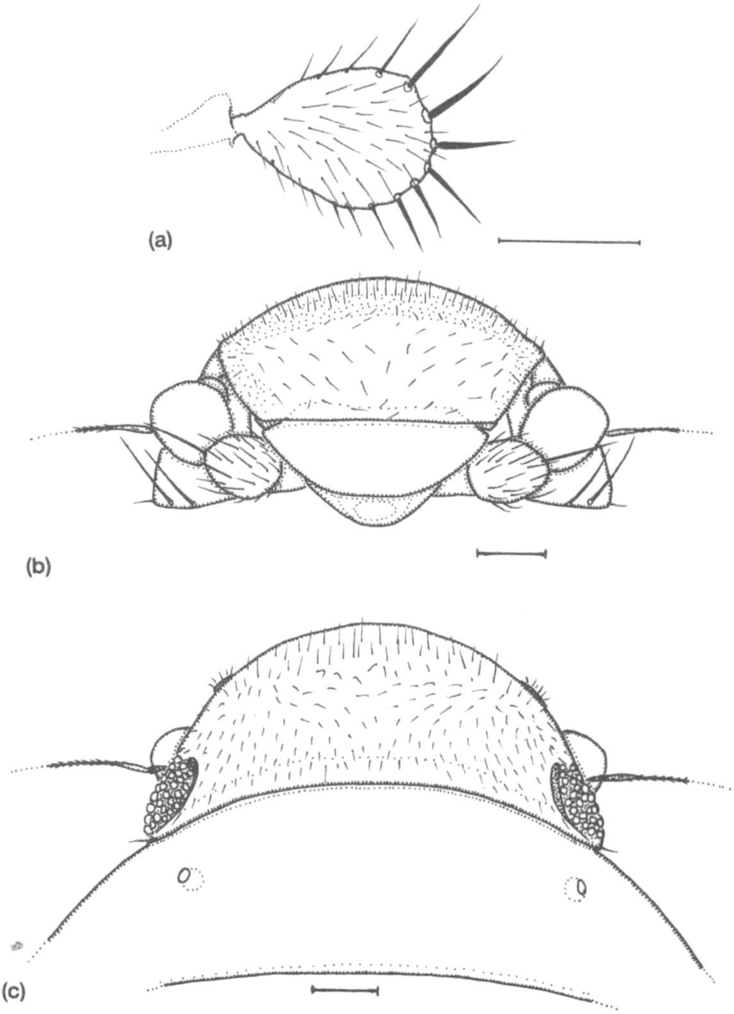

Figure 8.23 *Aenigmatopoeus kohli* female, right palp from below. (b), (c) *Aenigmatistes latifrons* female, head from below (b) and head and thorax from above (c).

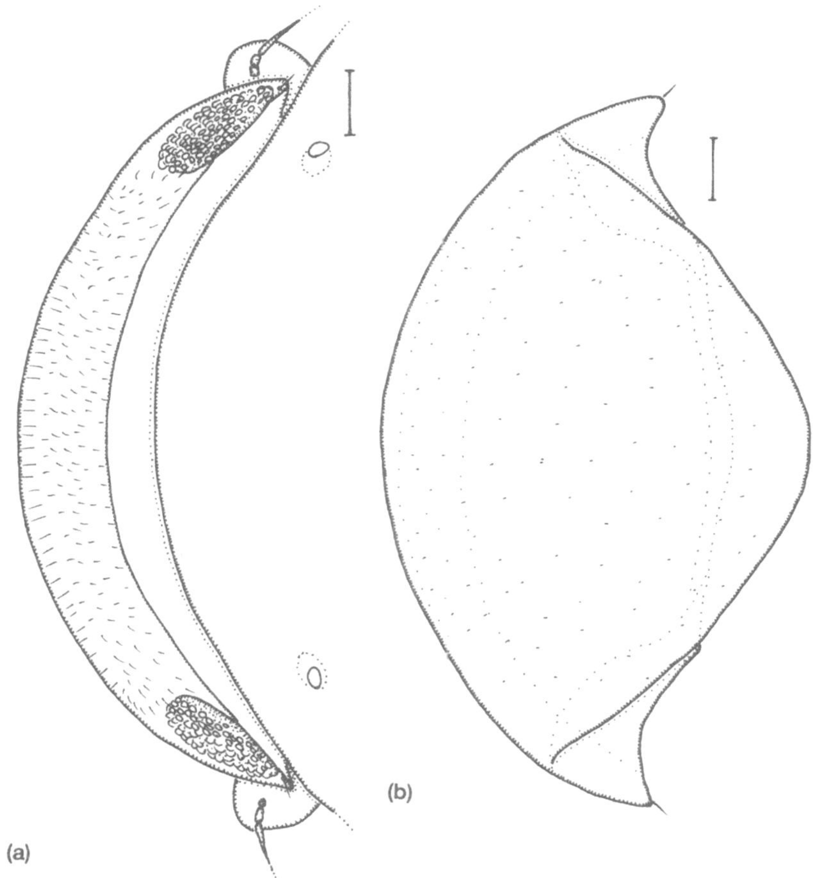

Figure 8.24 *Aenigmatistes herero* female, head facing left, viewed from above (a) and below (b)

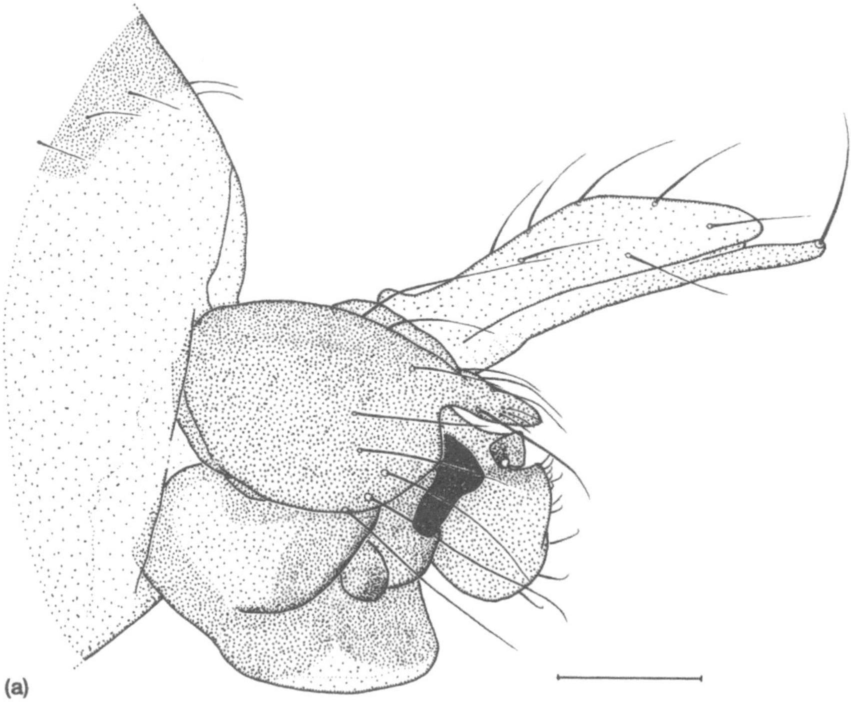

(a)

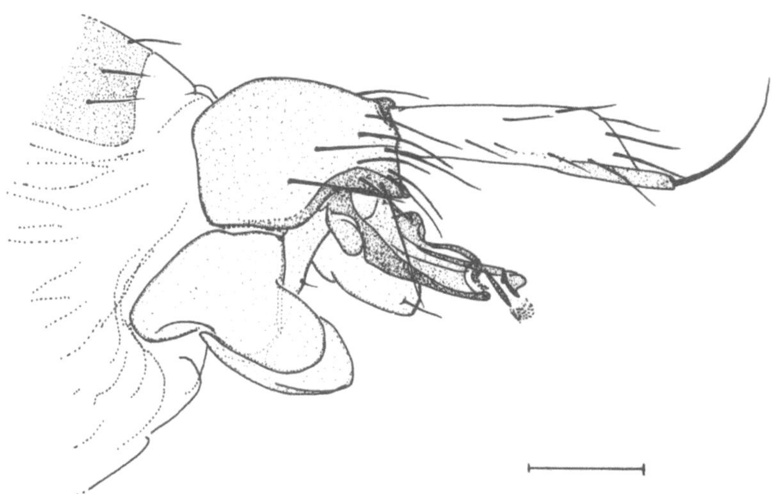

(b)

Figure 8.25 *Apocephalus* male hypopygia, left faces (a) *A borealis* (b) *A laceyi*

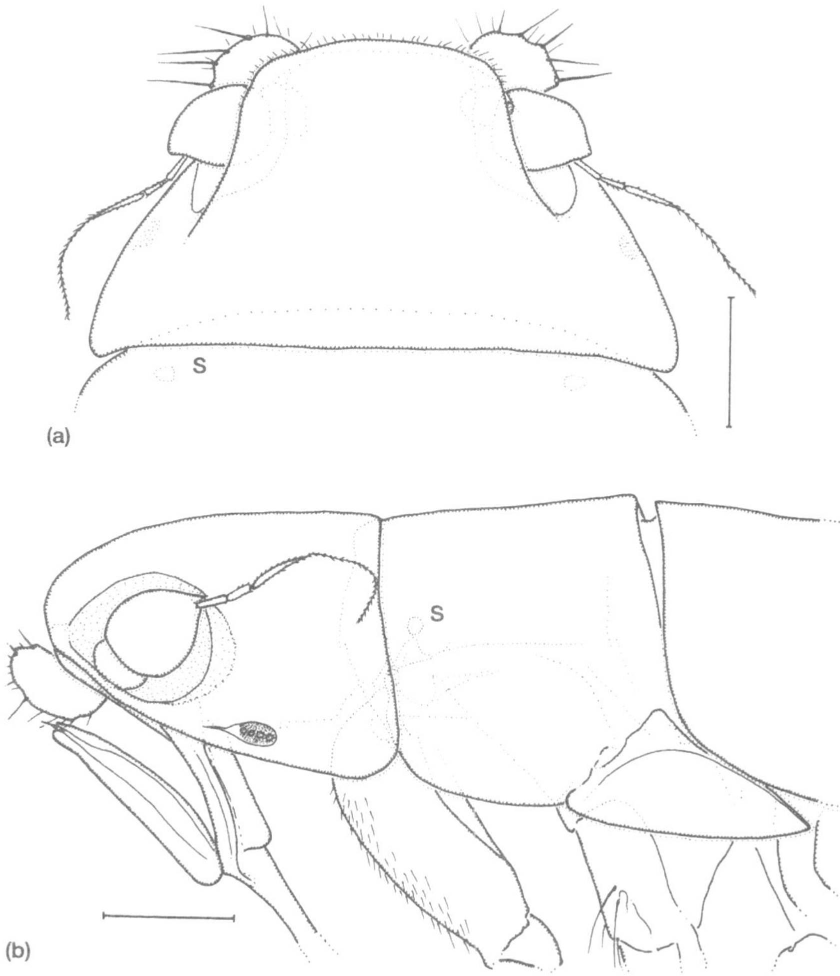

Figure 8.26 *Assmutherium rostratum* female (a) Head and front of thorax from above (b) Head, thorax and front of abdomen from side S = position of anterior thoracic spiracle on side, beneath overhang of thoracic notum

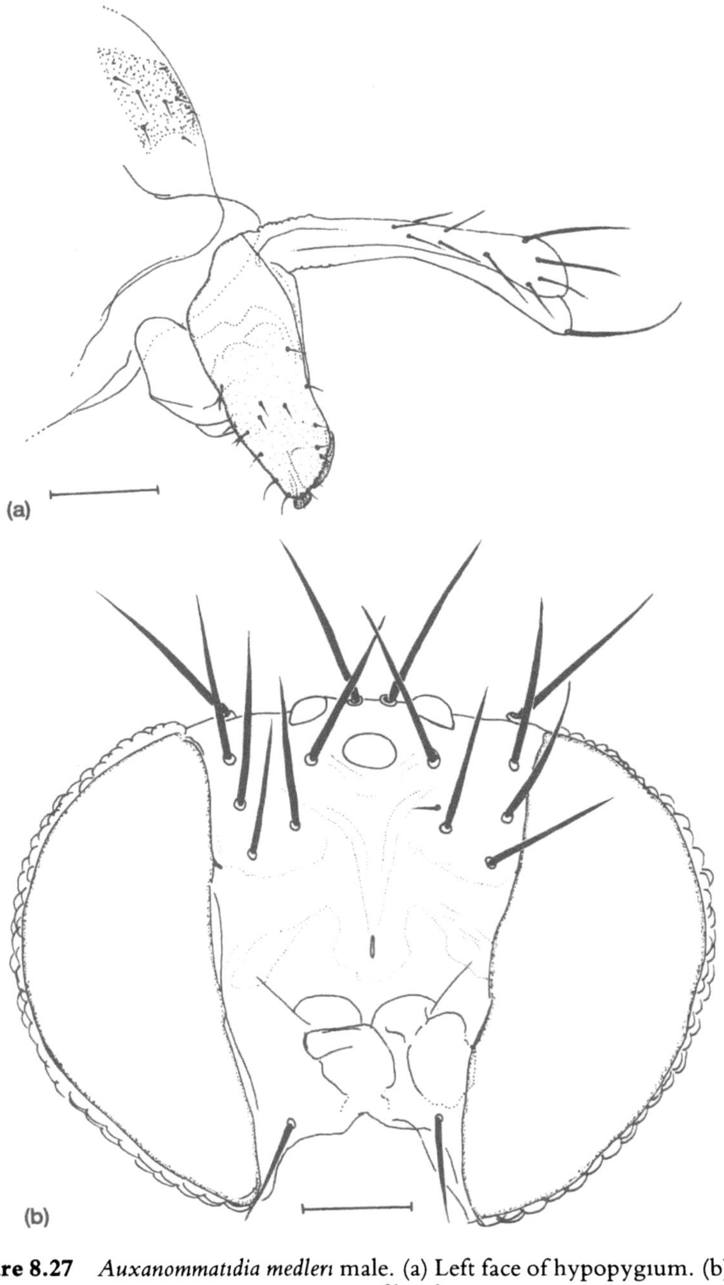

Figure 8.27 *Auxanommatidia medleri* male. (a) Left face of hypopygium. (b) Frontal view of head.

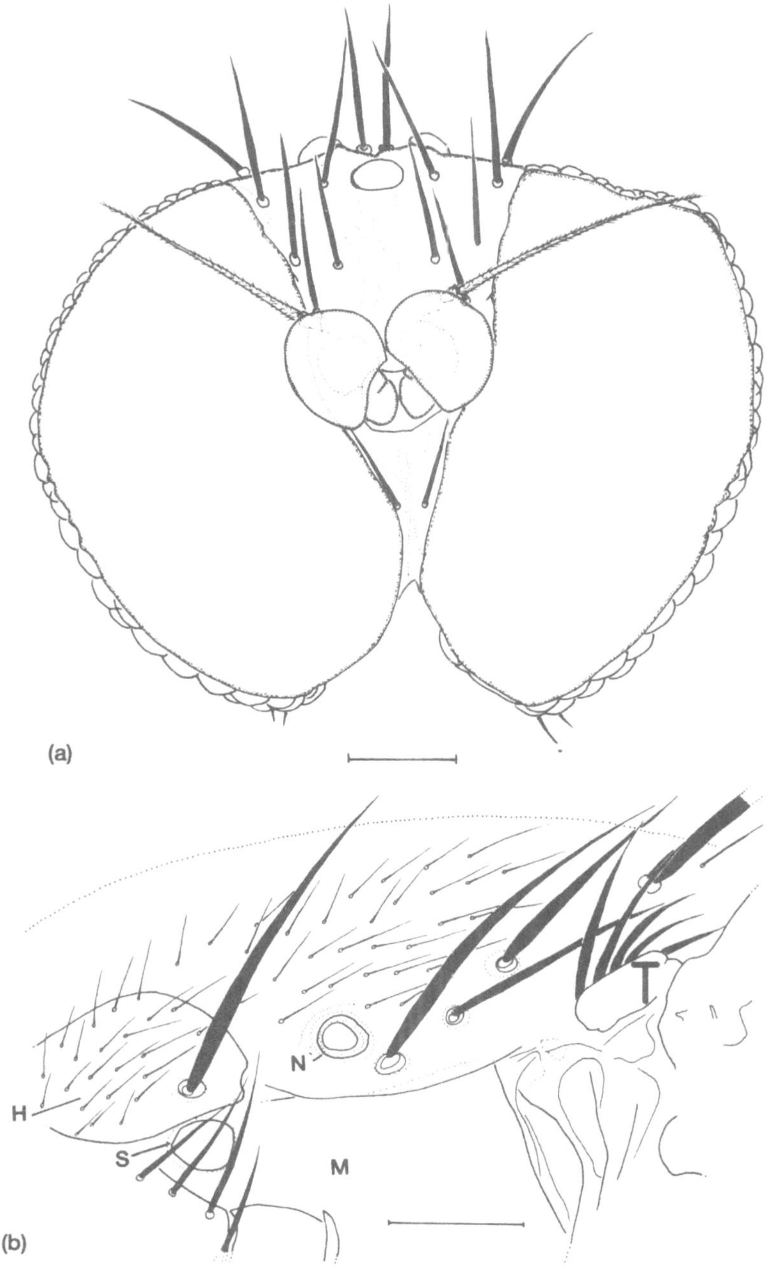

Figure 8.28 (a) *Auxanommatidia medleri* female, frontal view of head. (b) *Beckerina umbrimargo*, left humerus and notopleuron of male (H = humerus, M = mesopleuron, N = opening of notopleural gland, S = spiracle, T = tegula).

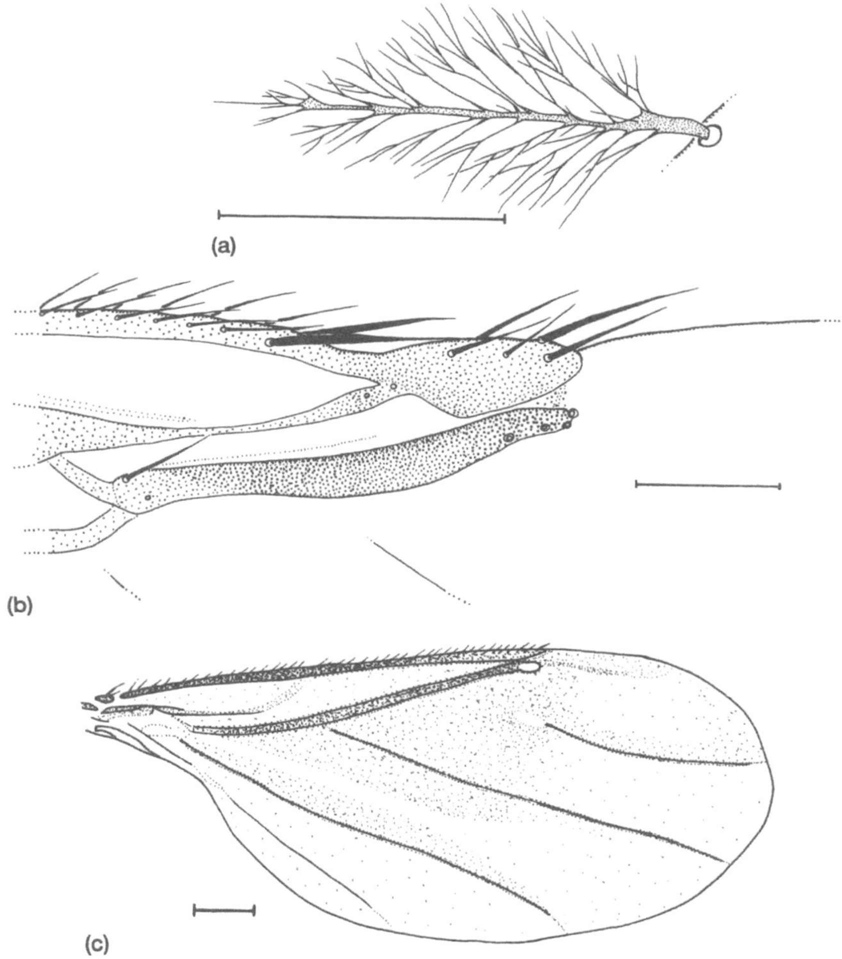

Figure 8.29 (a), (b) *Beyermyia spinulicosta:* (a) axillary bristle; (b) tips of costa and vein plus all of vein 3. (c) *Chonocephalus heymonsi*, right wing.

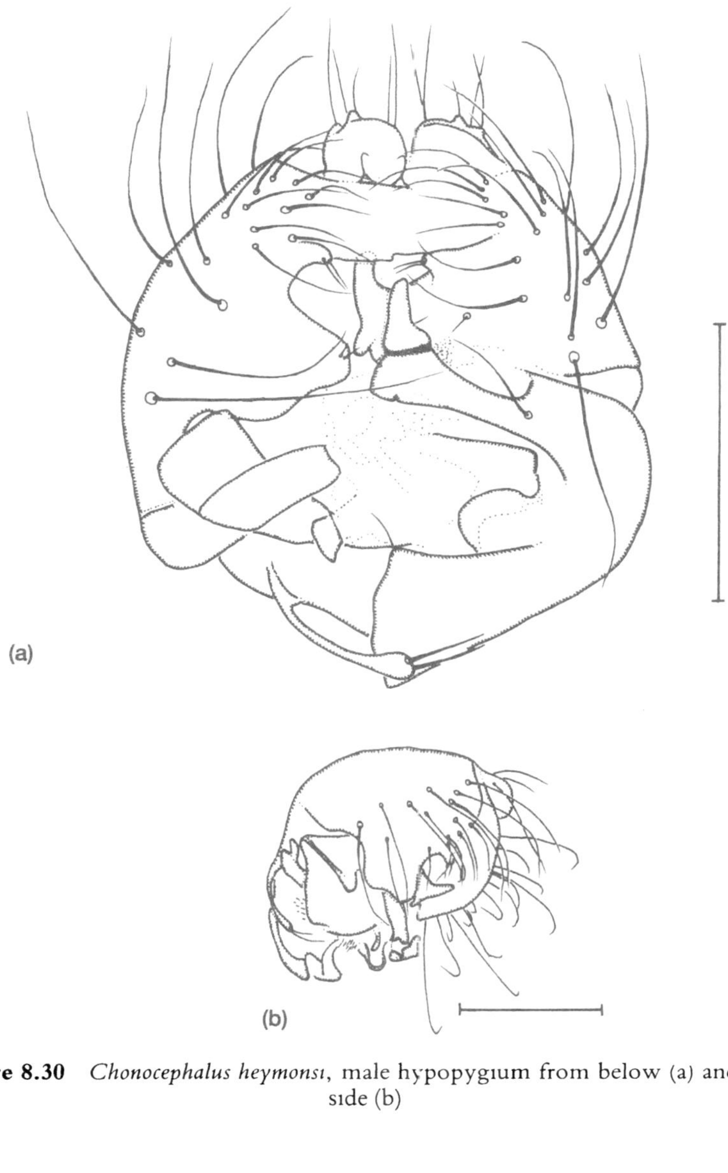

Figure 8.30 *Chonocephalus heymonsi*, male hypopygium from below (a) and from side (b)

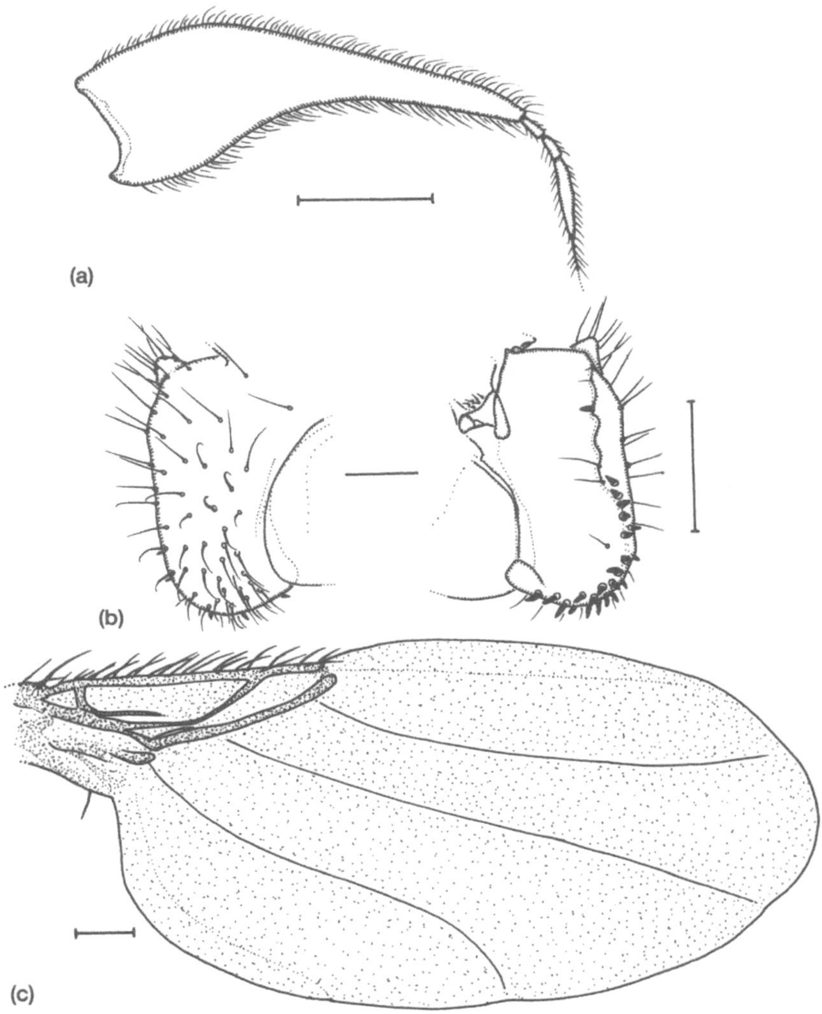

Figure 8.31 *Conicera* males (a) *C malae*, third antennal segment and arista (b) *C crassicosta*, outer (left) and inner face (right) of right clasper (surstylus) of hypopygium (c) *C schnittmanni*, right wing

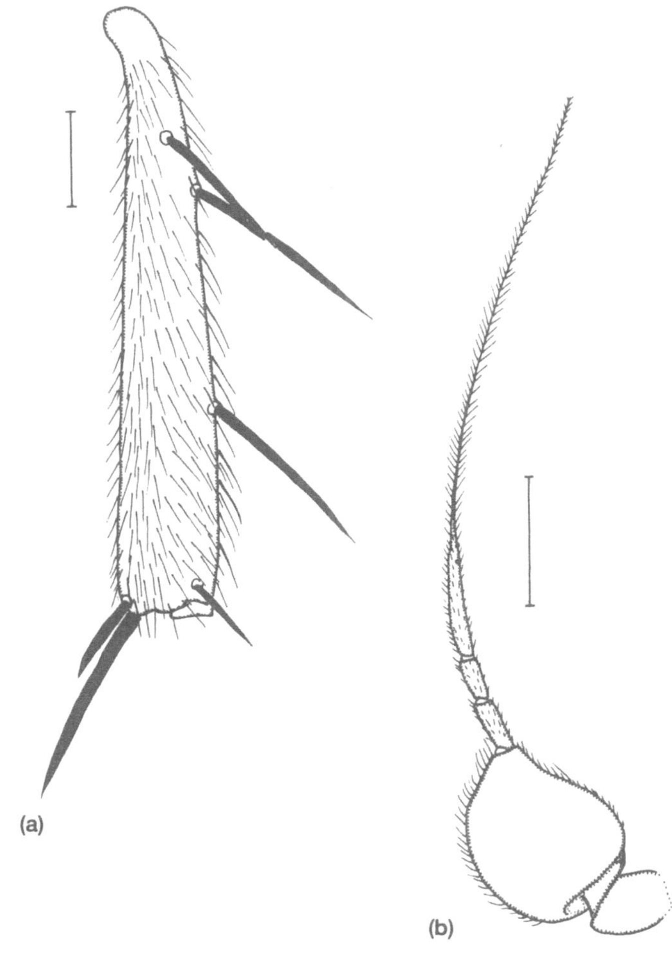

Figure 8.32 *Conicera crassicosta* male. (a) Anterior face of hind tibia. (b) Antenna.

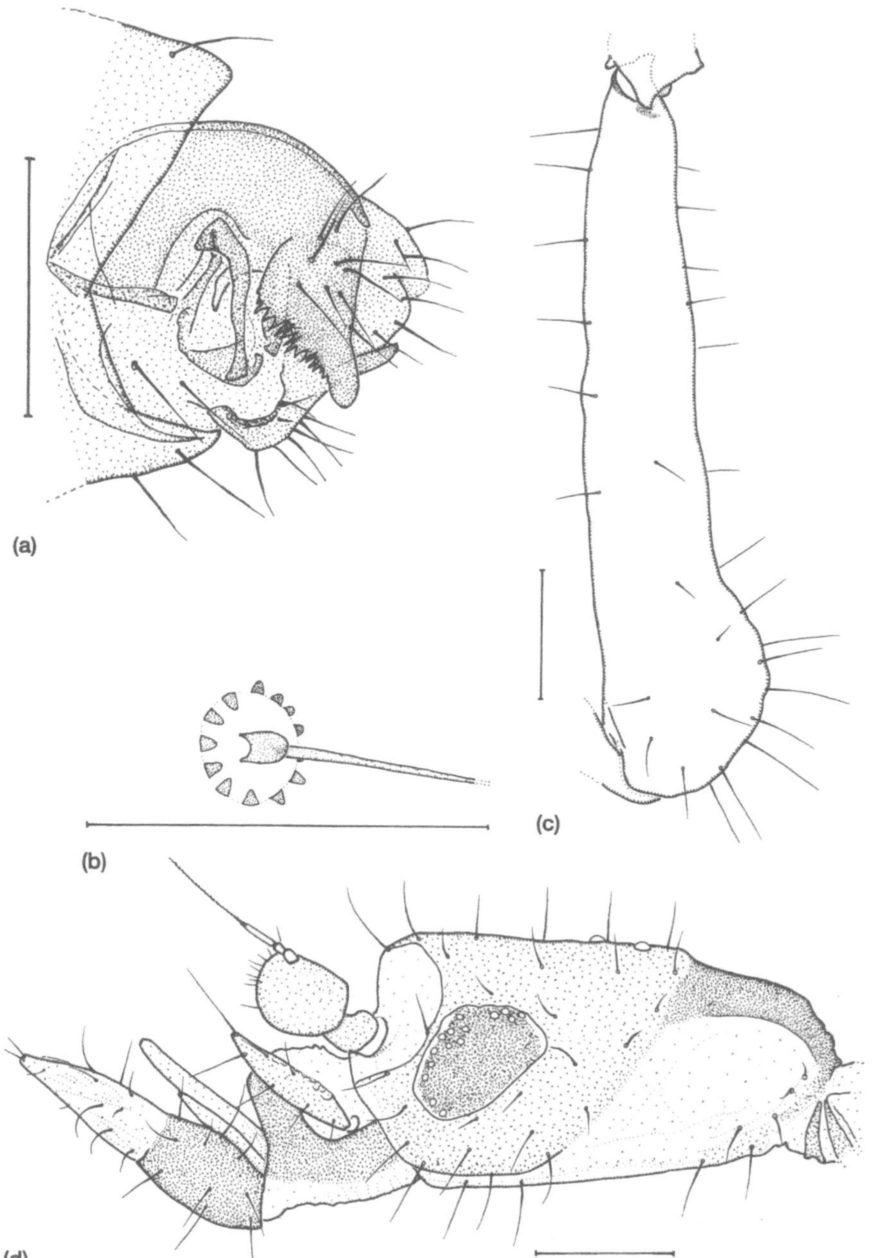

Figure 8.33 (a) *Perissa pakistanensis* male, left face of hypopygium (b)–(d) *Clitelloxenia peradeniyae* (b) base of bristle on abdomen, (c) posterior face of hind femur, (d) left face of head

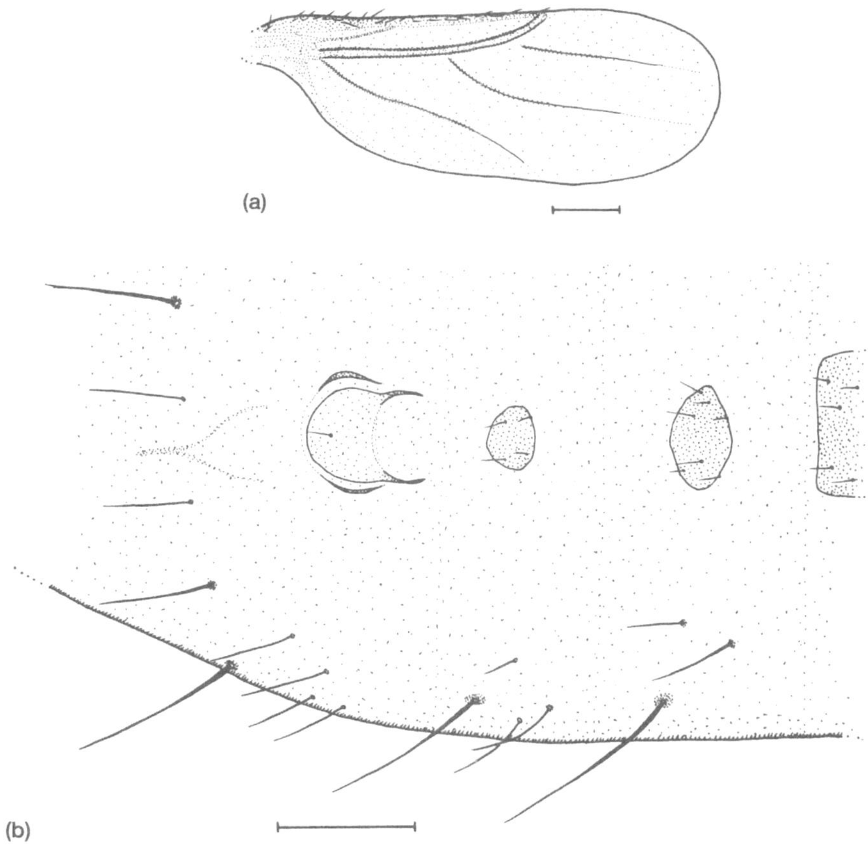

Figure 8.34 *Crasilla setifera* female. (a) Right wing. (b) Abdomen from above, from rear of tergite 2 to segment 6.

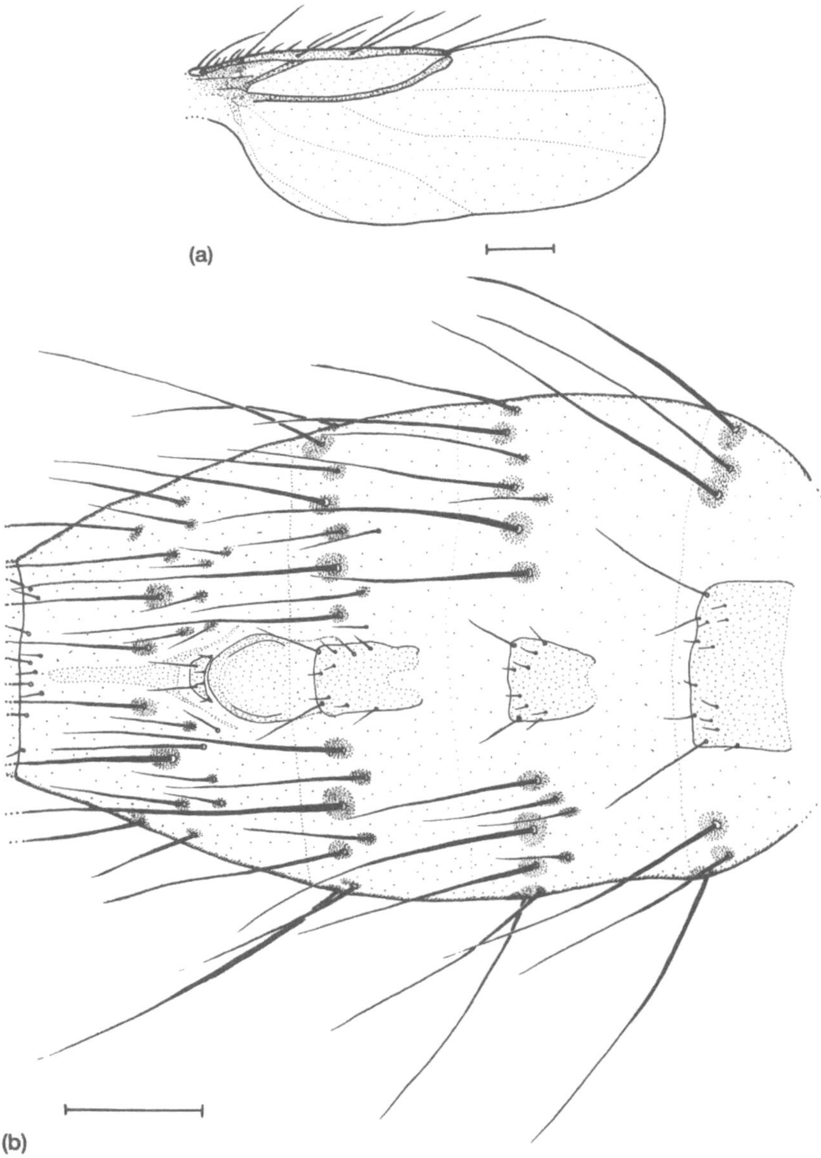

Figure 8.35 *Crasilla longiseta* female. (a) Right wing. (b) Abdominal segments 2–6 from above.

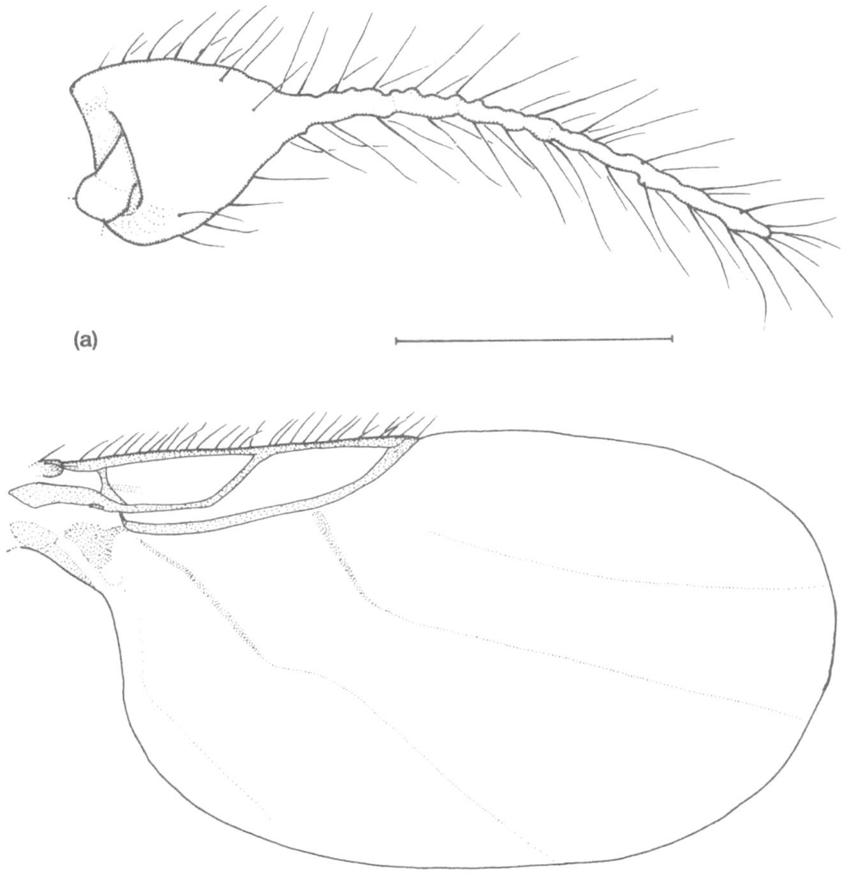

(a)

(b)

Figure 8.36 *Dahliphora sigmoides* male (a) Third antennal segment (b) Right wing

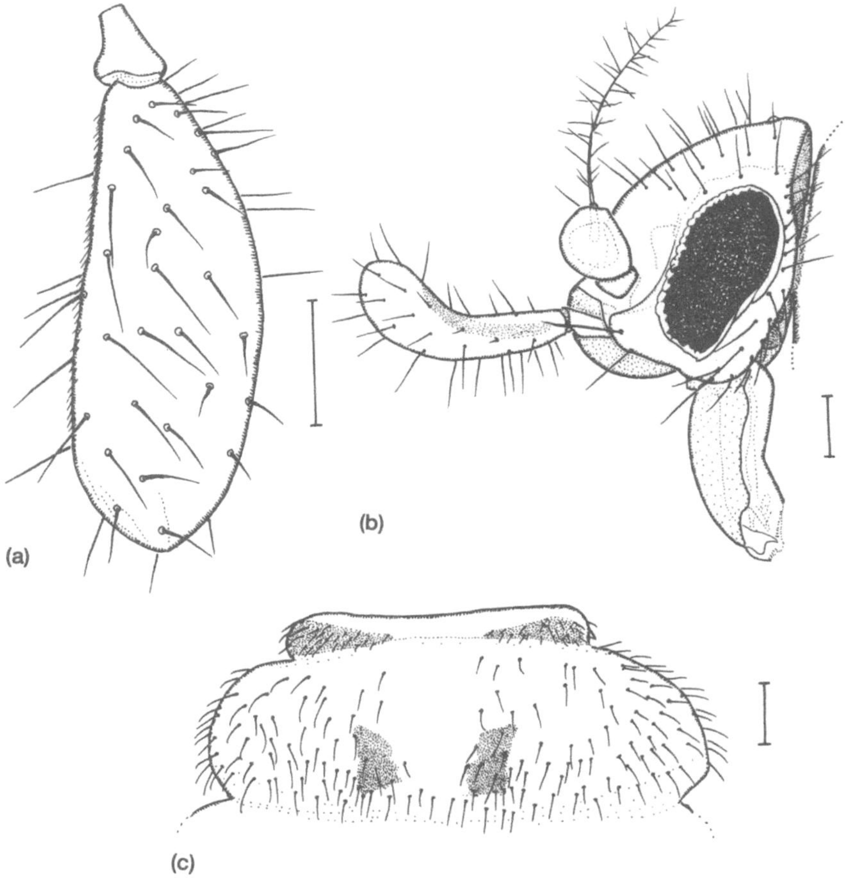

Figure 8.37 *Dicranopteron setipennis* female (a) Right palp, inner face (b) Head from left side (c) Abdominal segments 1 and 2 from above

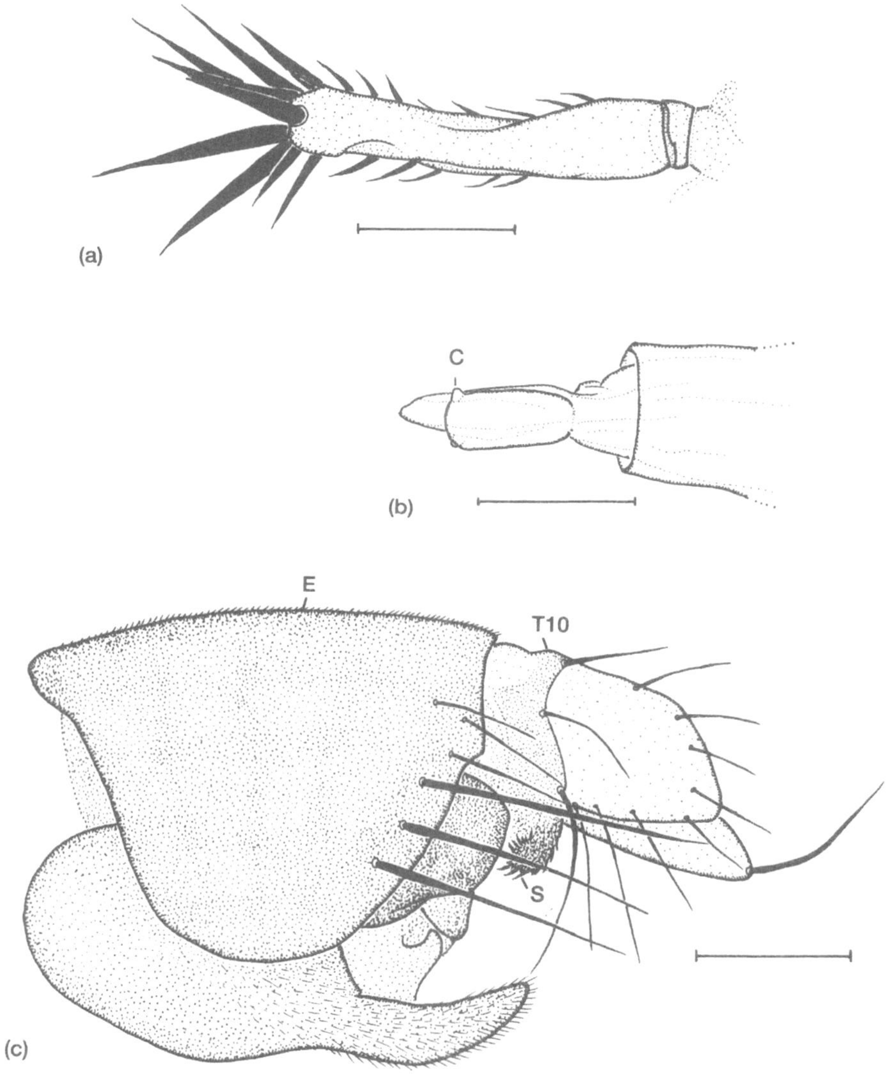

Figure 8.38 *Diocophora appretiata* (a) Left palp of female from above (b) Ovipositor from above (c = vestigial cercus) (c) Male hypopygium, left face (E = epandrium, T10 = tergite 10, S = surstylus)

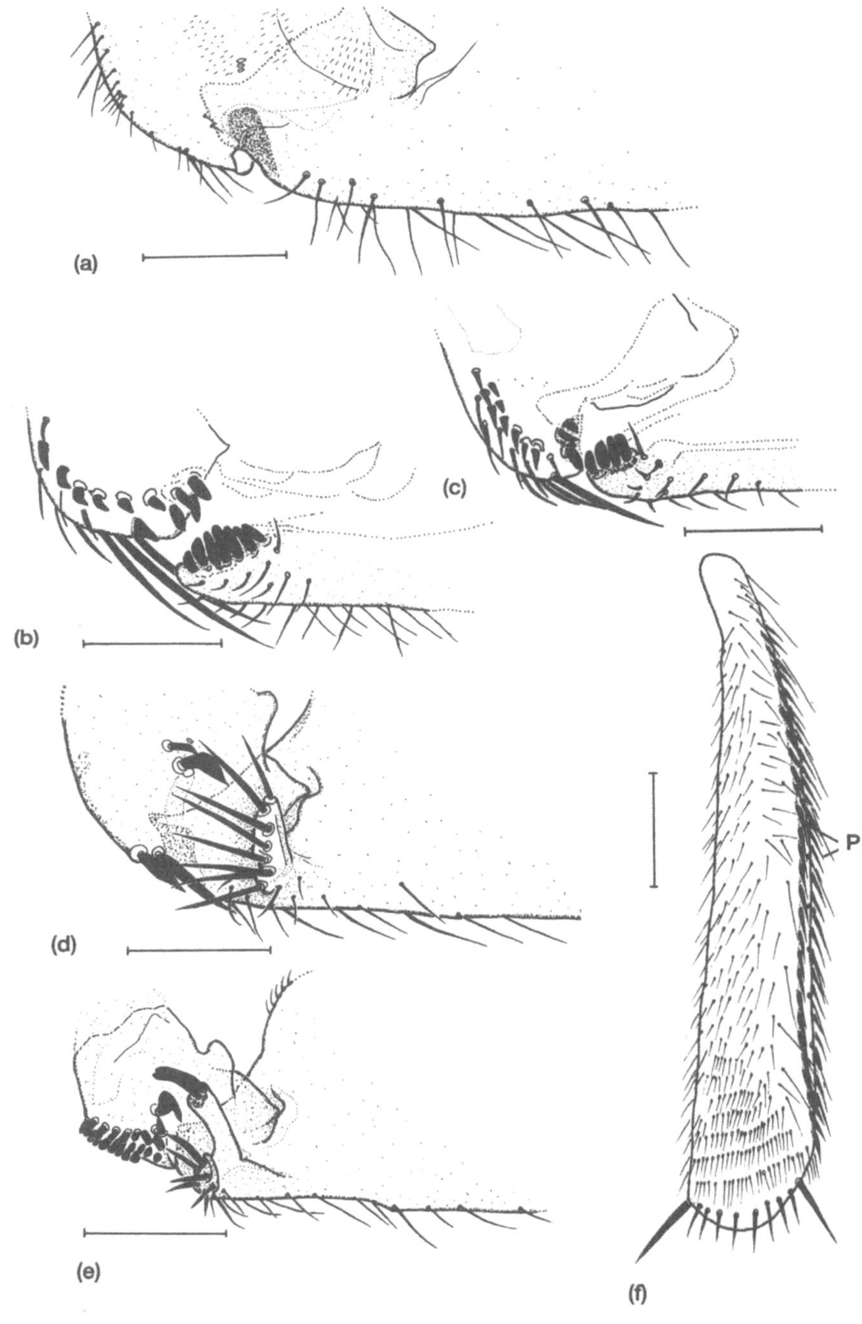

Figure 8.39 *Diplonevra* males, details of legs. (a)–(e) Posterior faces of bases of hind femora and trochanters: (a) *D. abbreviata*; (b), (c) *D. florea*; (d) *D. nitidula*; (e) *D. funebris*. (f) *D. watsoni*, posterior face of hind tibia (P = hair palisades).

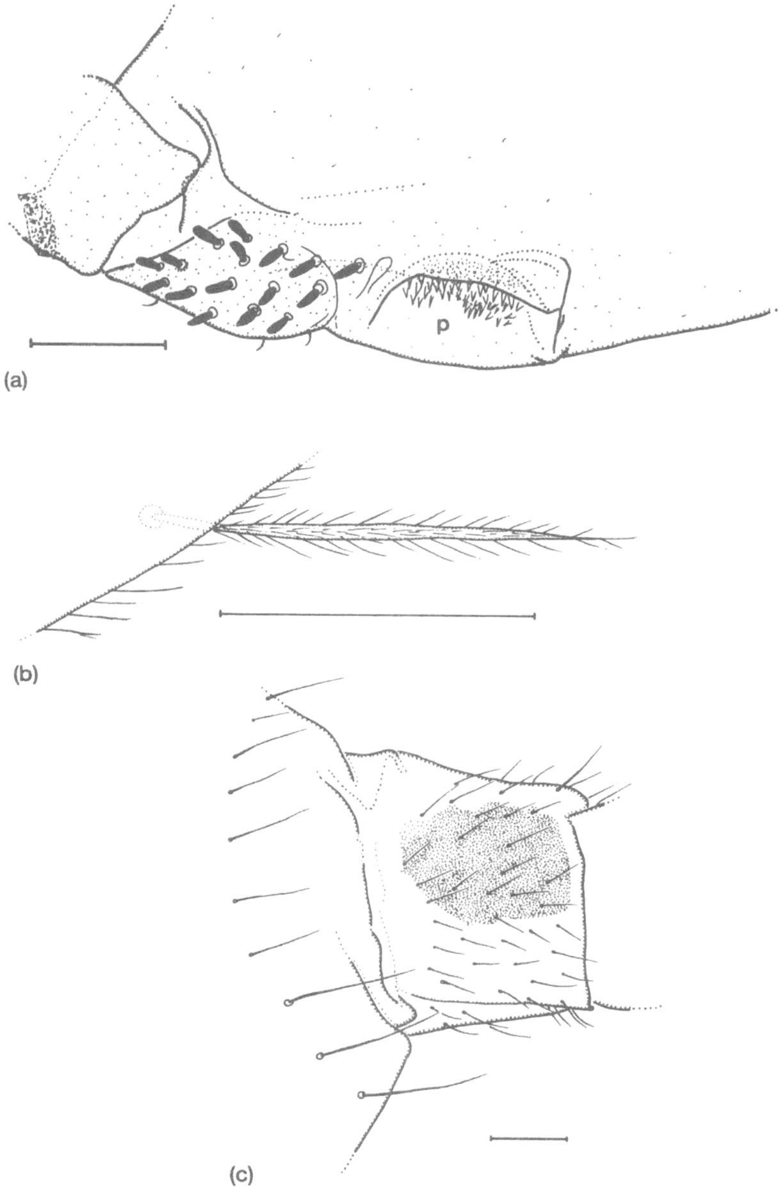

Figure 8.40 *Dohrniphora*. (a) *D. trudiae* male, posterior face of base of hind femur (P = sensory pit). (b) *D. kistneri* female, axillary bristle from above. (c) *D. cornuta* female, rear of abdominal segment 6 and segment 7, from side.

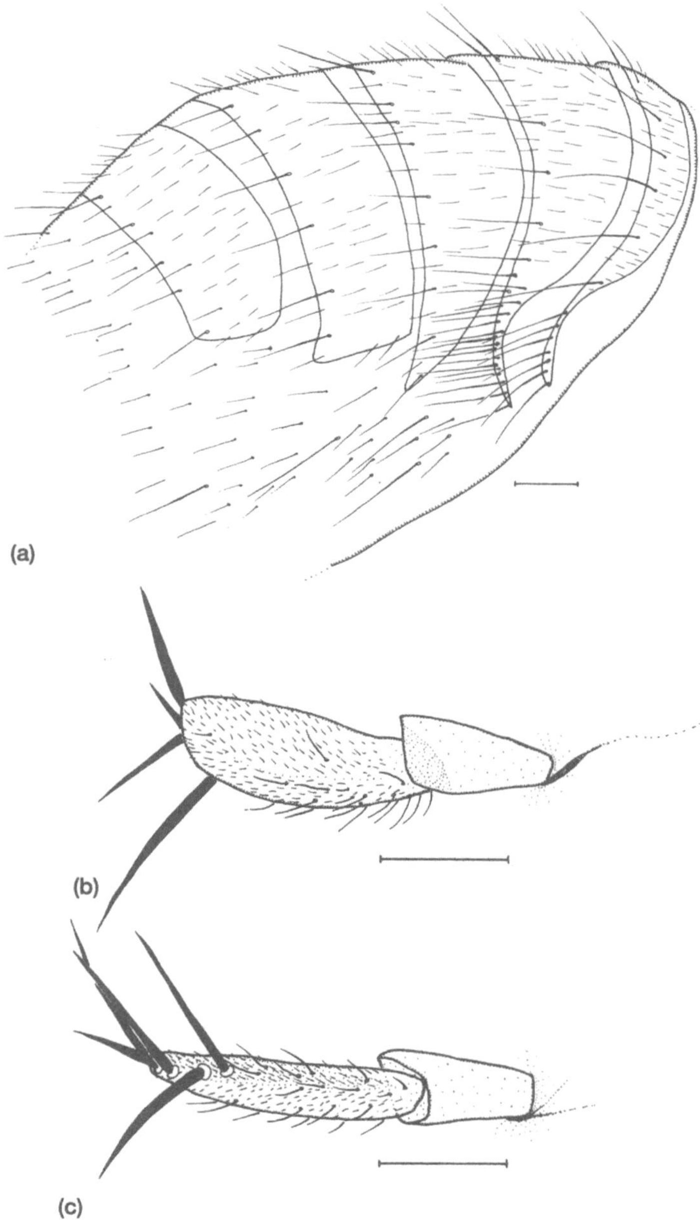

Figure 8.41 (a) *Dundophora fimbriiterga* female abdominal tergites 1–5 from right side. (b), (c) *Dohrniphora diminuens* female palp, inner face (b) and from below (c).

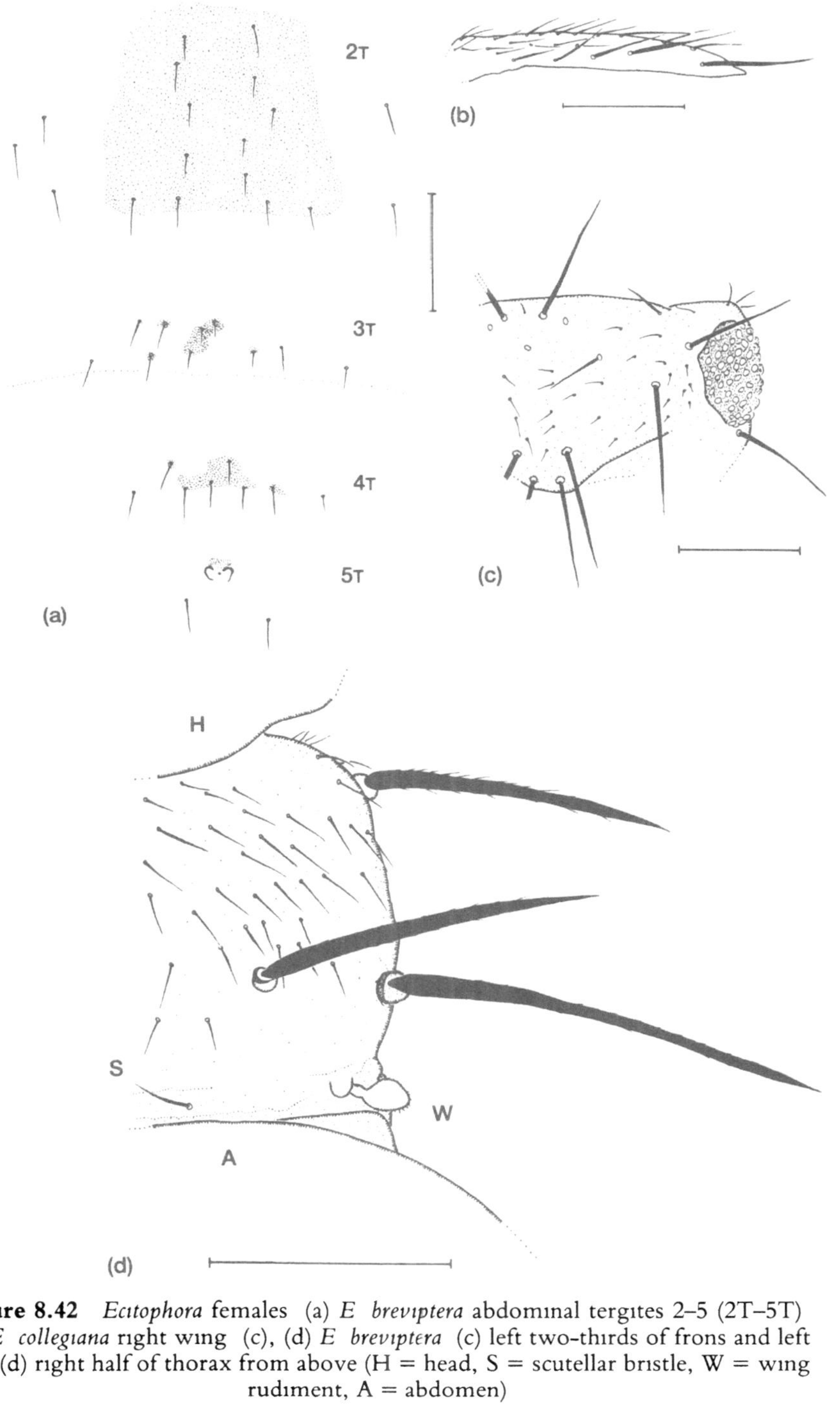

Figure 8.42 *Ecitophora* females (a) *E. breviptera* abdominal tergites 2–5 (2T–5T) (b) *E. collegiana* right wing (c), (d) *E. breviptera* (c) left two-thirds of frons and left eye, (d) right half of thorax from above (H = head, S = scutellar bristle, W = wing rudiment, A = abdomen)

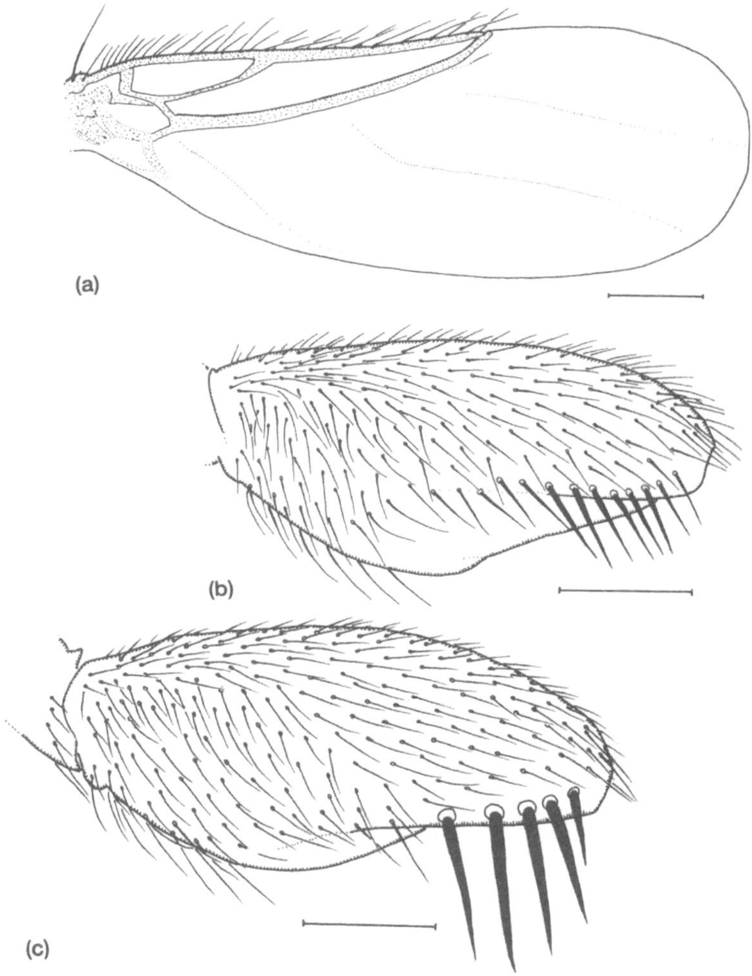

Figure 8.43 (a) *Ecitoptera humeralis* female, right wing (b), (c) *Epicnemis* females, anterior faces of mid femora (b) *Epicnemis* sp (probably *E testis*), (c) *E flavidula*

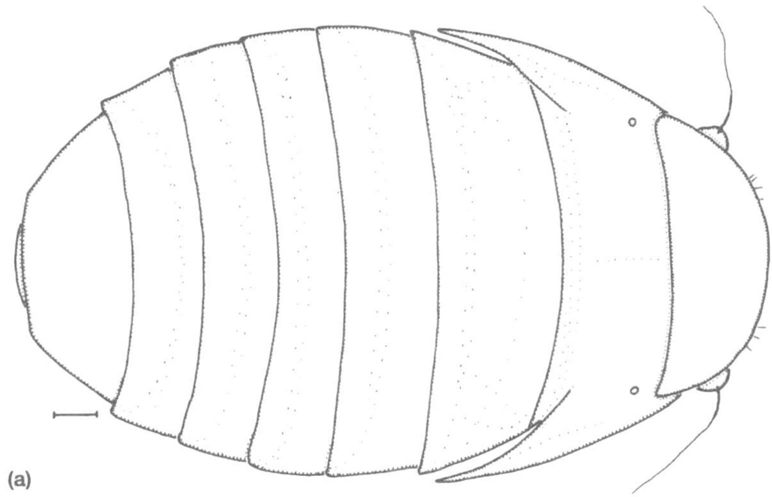

(a)

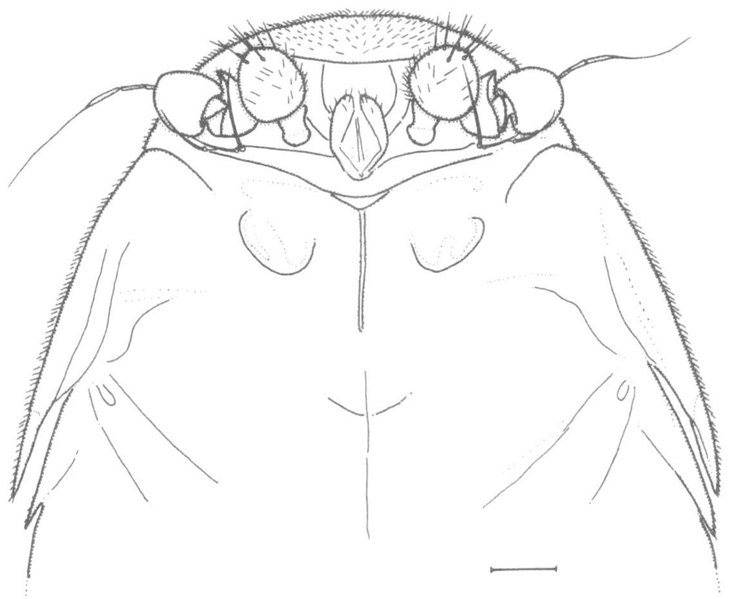

(b)

Figure 8.44 *Epicnemis flavidula* female. (a) From above. (b) Head and thorax from below (legs omitted).

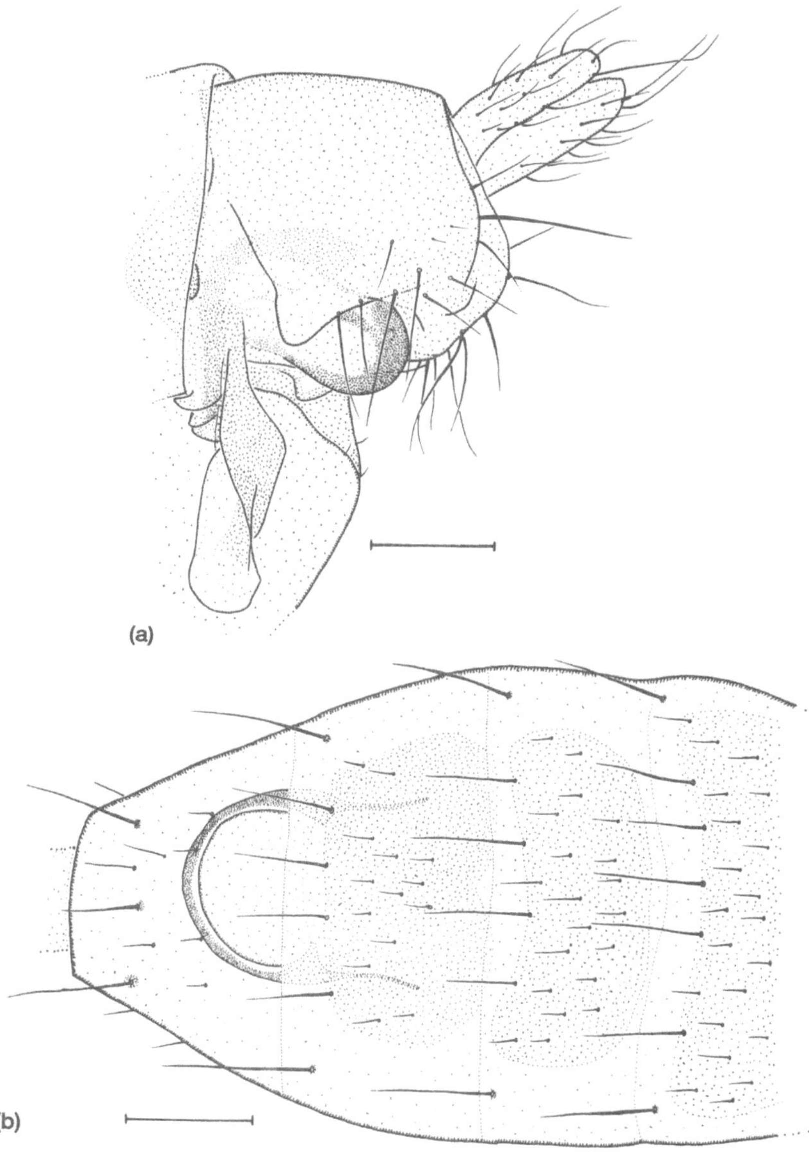

Figure 8.45 (a) *Epicnemis testis*, left face of male hypopygium (b) *Franssenia malaysiae* female, abdominal tergites 2–5

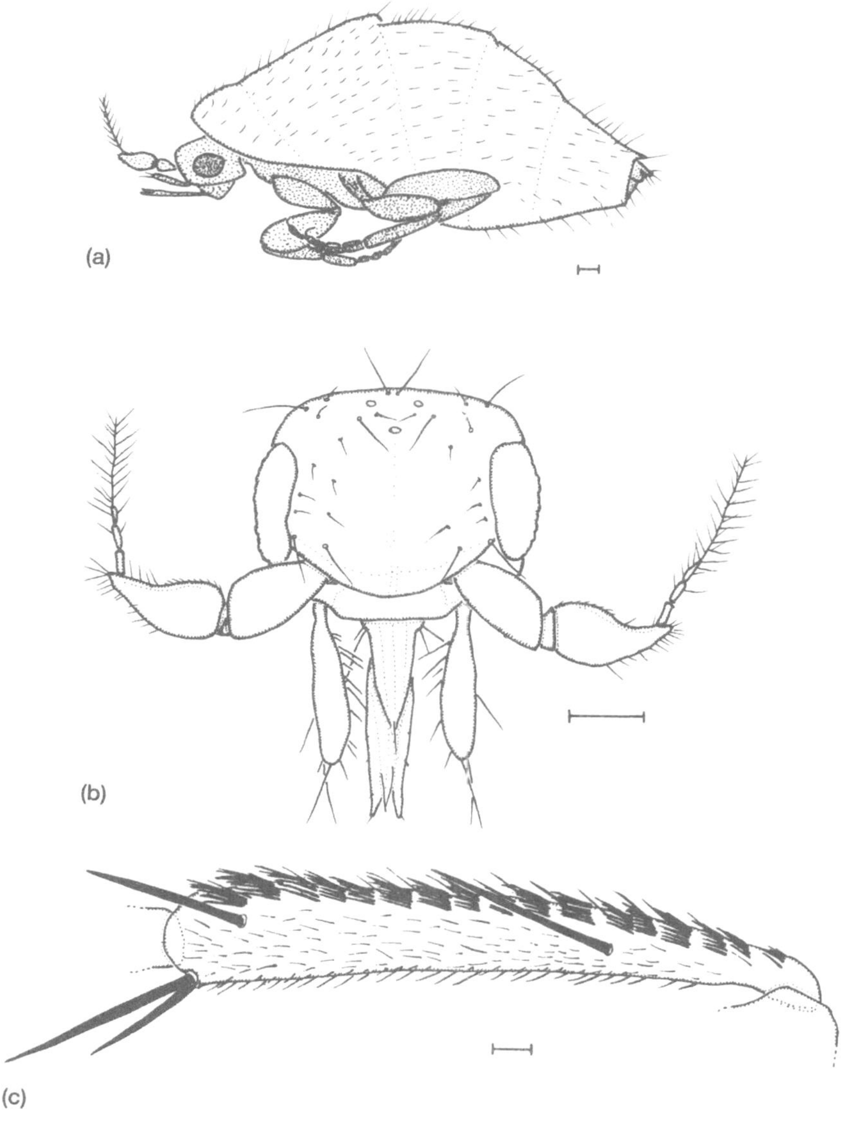

Figure 8.46 (a), (b) *Eutermiphora abdominalis* female (a) side view of whole fly (b) frontal view of head (c) *Hypocera mordellaria*, anterior face of hind tibia

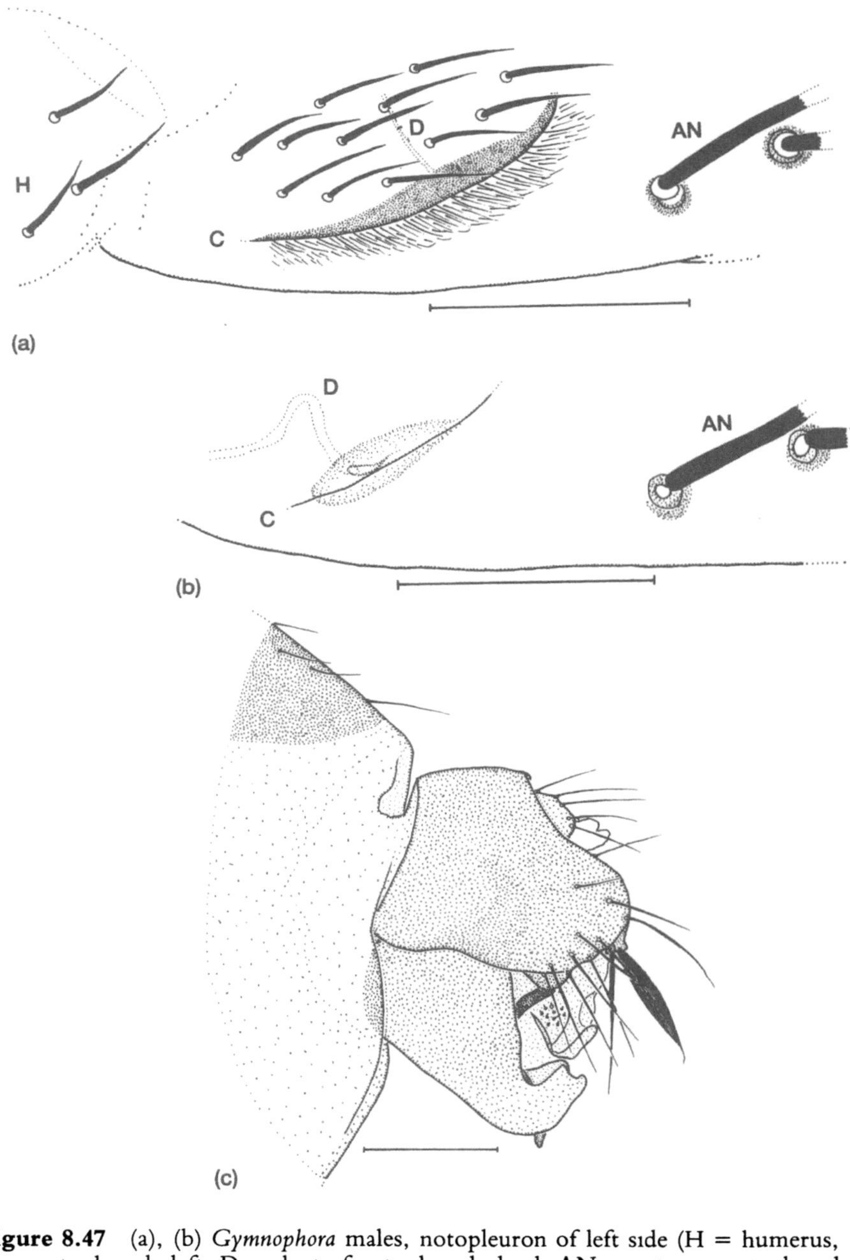

Figure 8.47 (a), (b) *Gymnophora* males, notopleuron of left side (H = humerus, C = notopleural cleft, D = duct of notopleural gland, AN = anterior notopleural bristle) (a) *G quartomollis*, (b) *G arcuata* (c) *Gymnoptera longicostalis* male hypopygium, left face

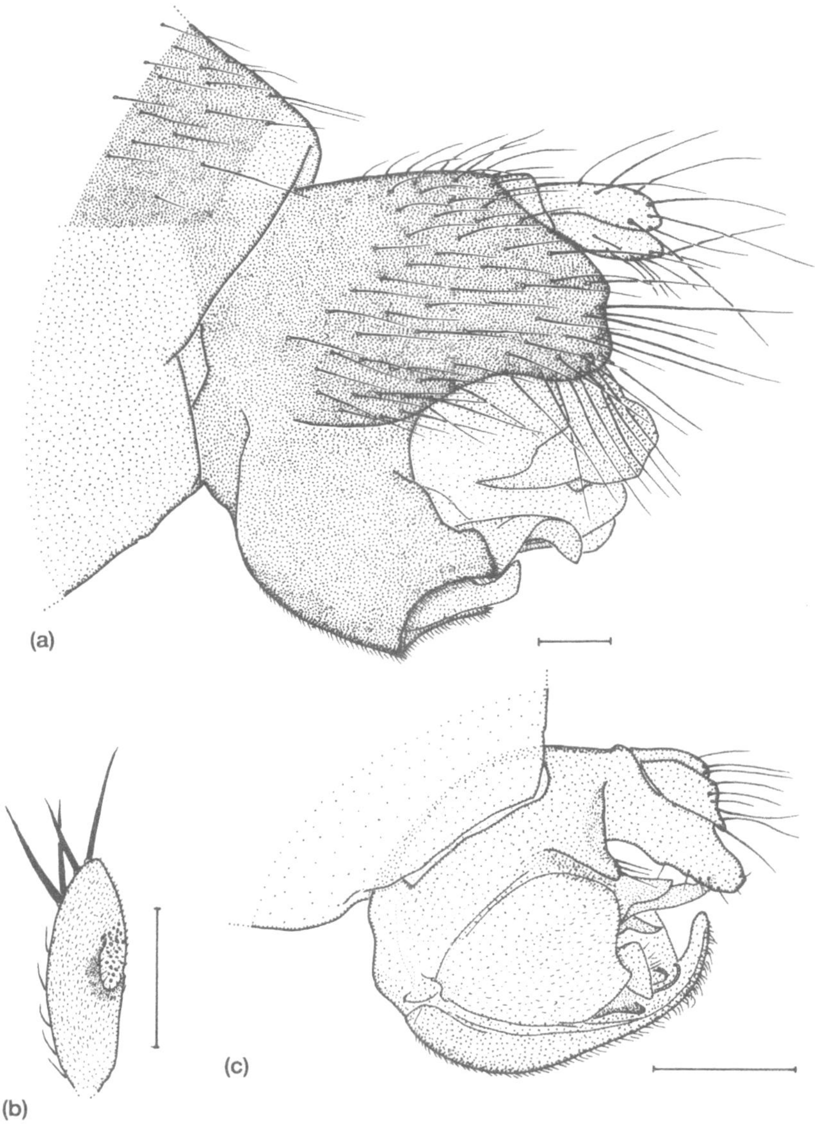

Figure 8.48 (a) *Hypocera mordellaria* male, left face of hypopygium (b), (c) *Hypocerides nearcticus* male (b) left palp, (c) left face of hypopygium

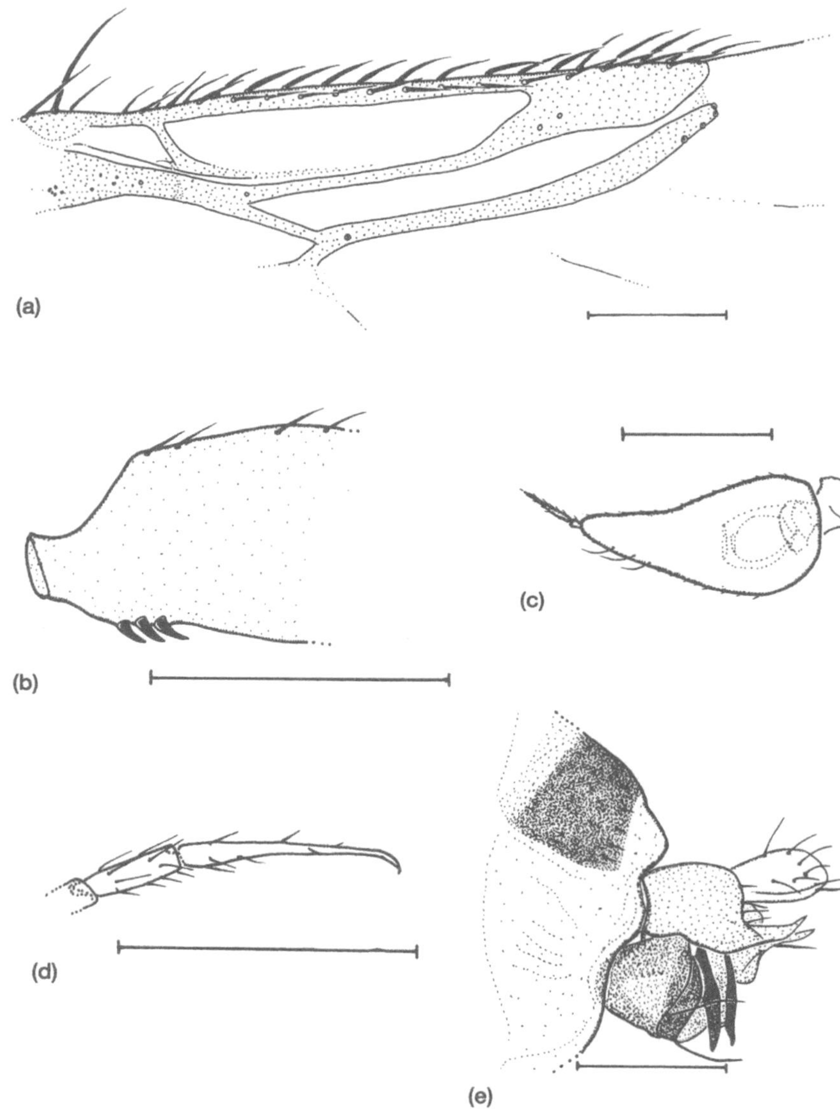

Figure 8.49 (a) *Hypocerides nearcticus* male, basal region of front half of wing (b)–(e) *Iridophora clarki* (b) base of hind femur of male, (c) male antenna, (d) tarsal segments 4 and 5 of front leg of female, (e) left face of male hypopygium

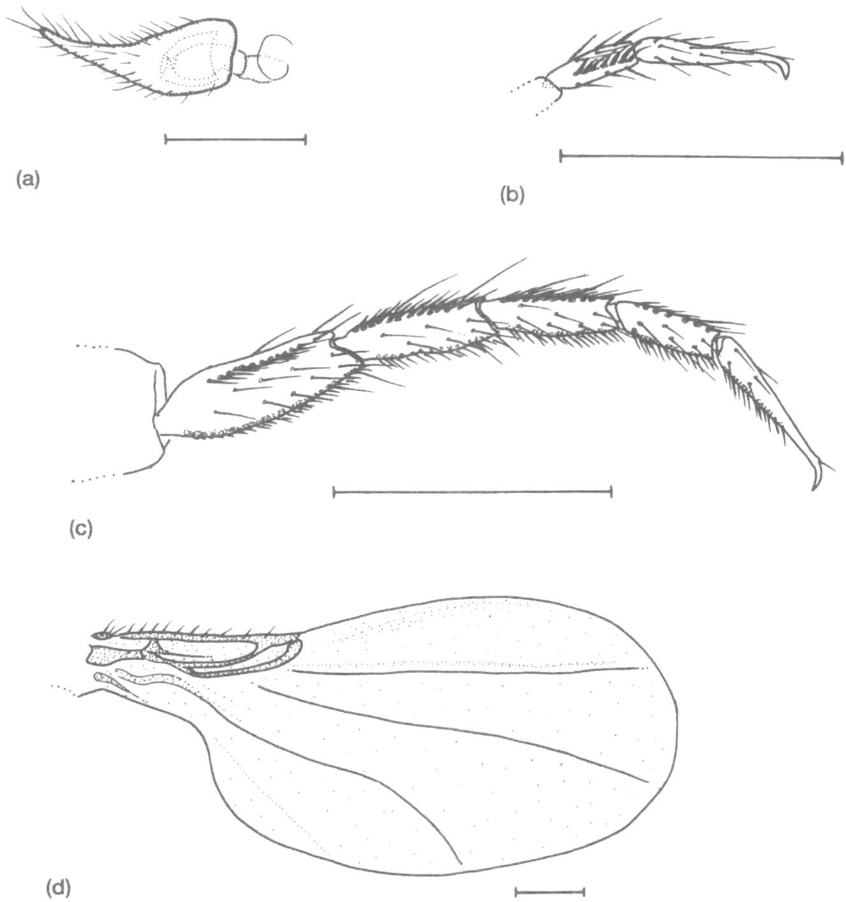

Figure 8.50 *Iridophora clarki* female (a) Antenna (b) Tarsal segments 4 and 5 of middle leg (c) Tarsus of hind leg (d) Right wing

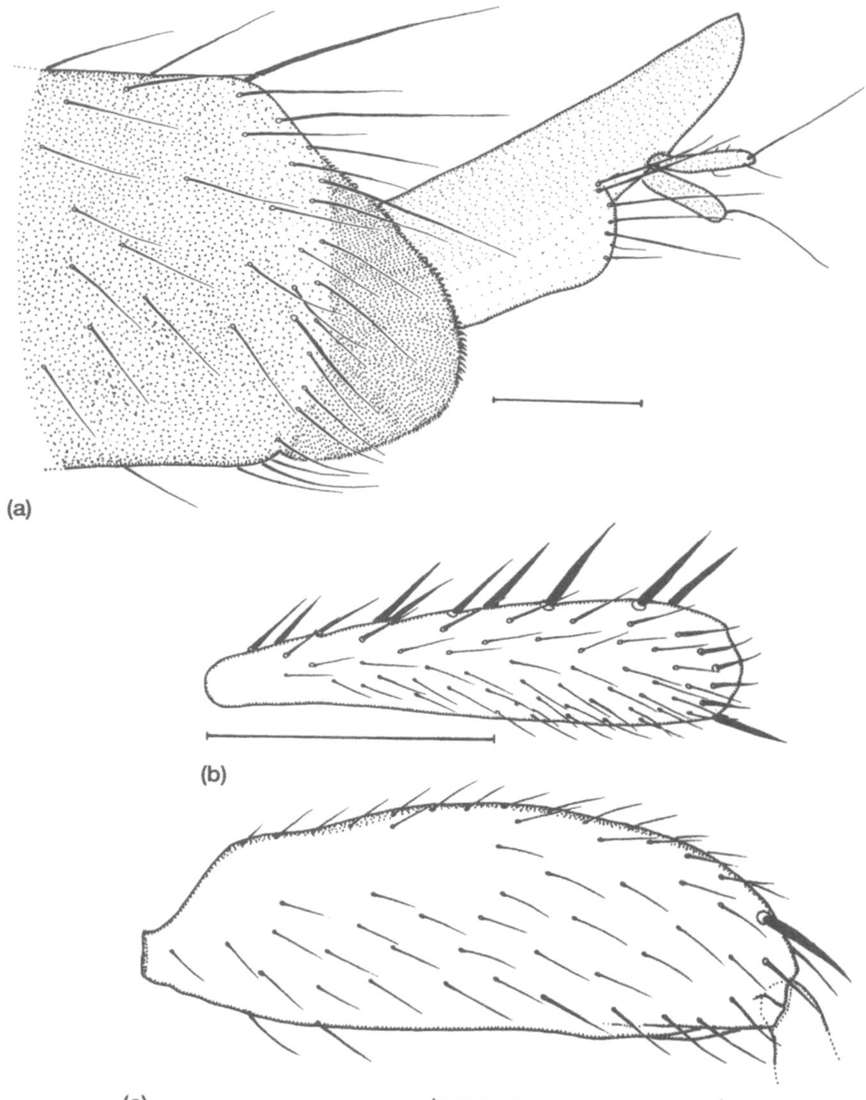

Figure 8.51 (a) *Megaselia brokawi* ovipositor, left face (b), (c) *Maculiphora aenictophila* parts of hind leg of female (b) posterior face of tibia, (c) anterior face of femur

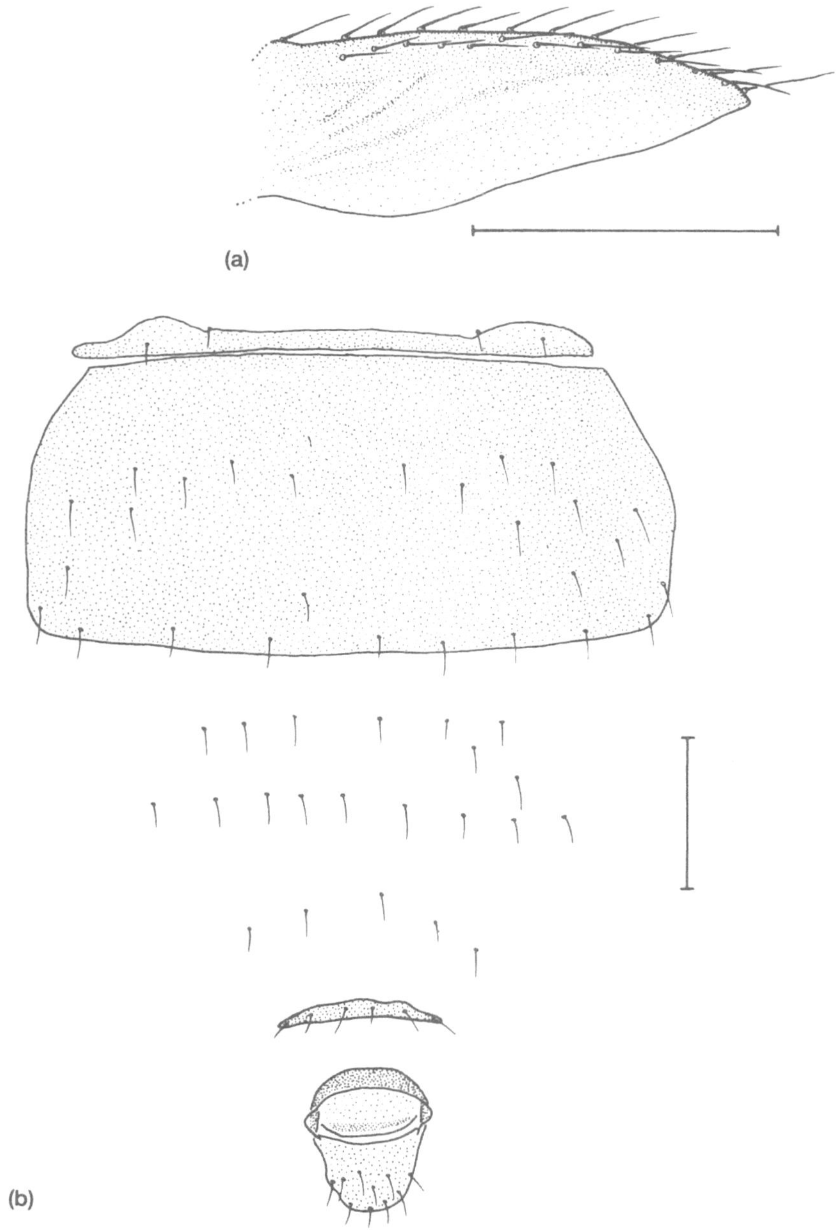

Figure 8.52 *Lepidophoromyia labida* female (a) Right wing (b) Abdominal tergites 1–5

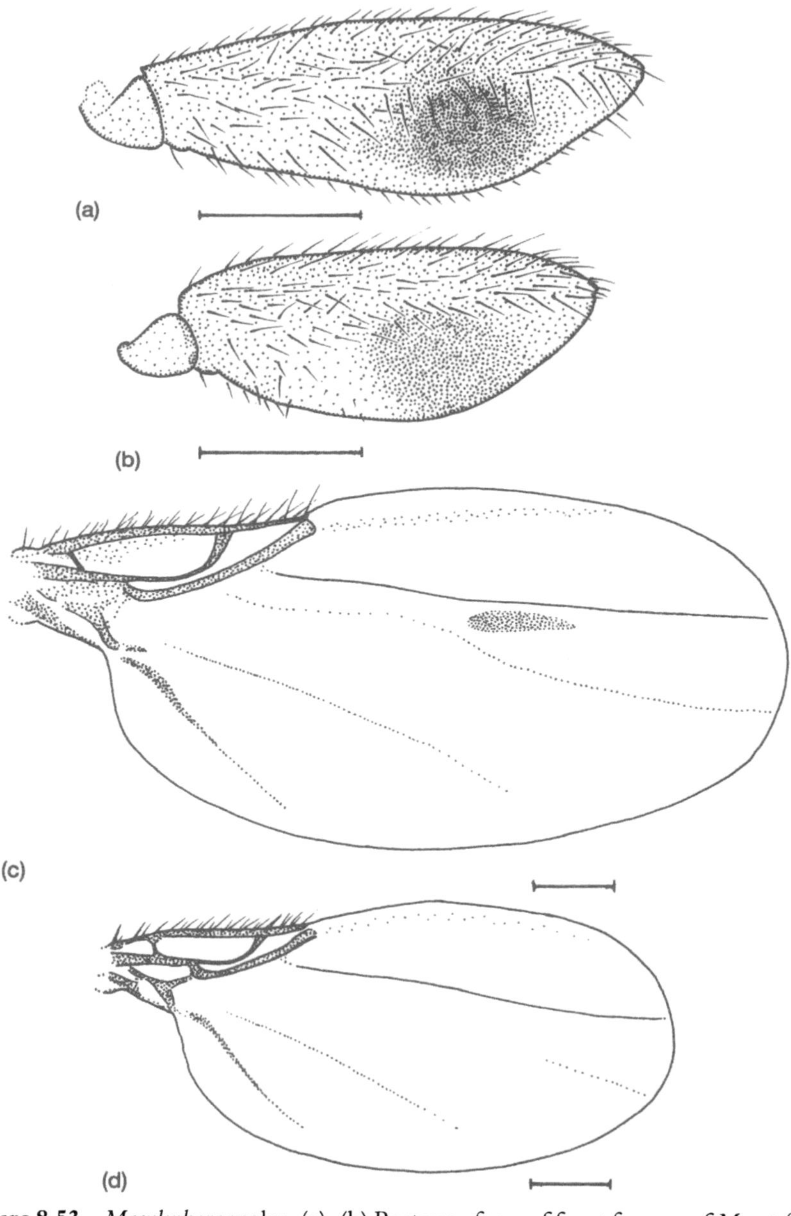

Figure 8.53 *Maculıphora* males (a), (b) Posterıor faces of front femora of *M setıfera* (a) and *M aenıctophıla* (b) (c), (d) Rıght wıngs of *M setıfera* (c) and *M aenıctophıla* (d)

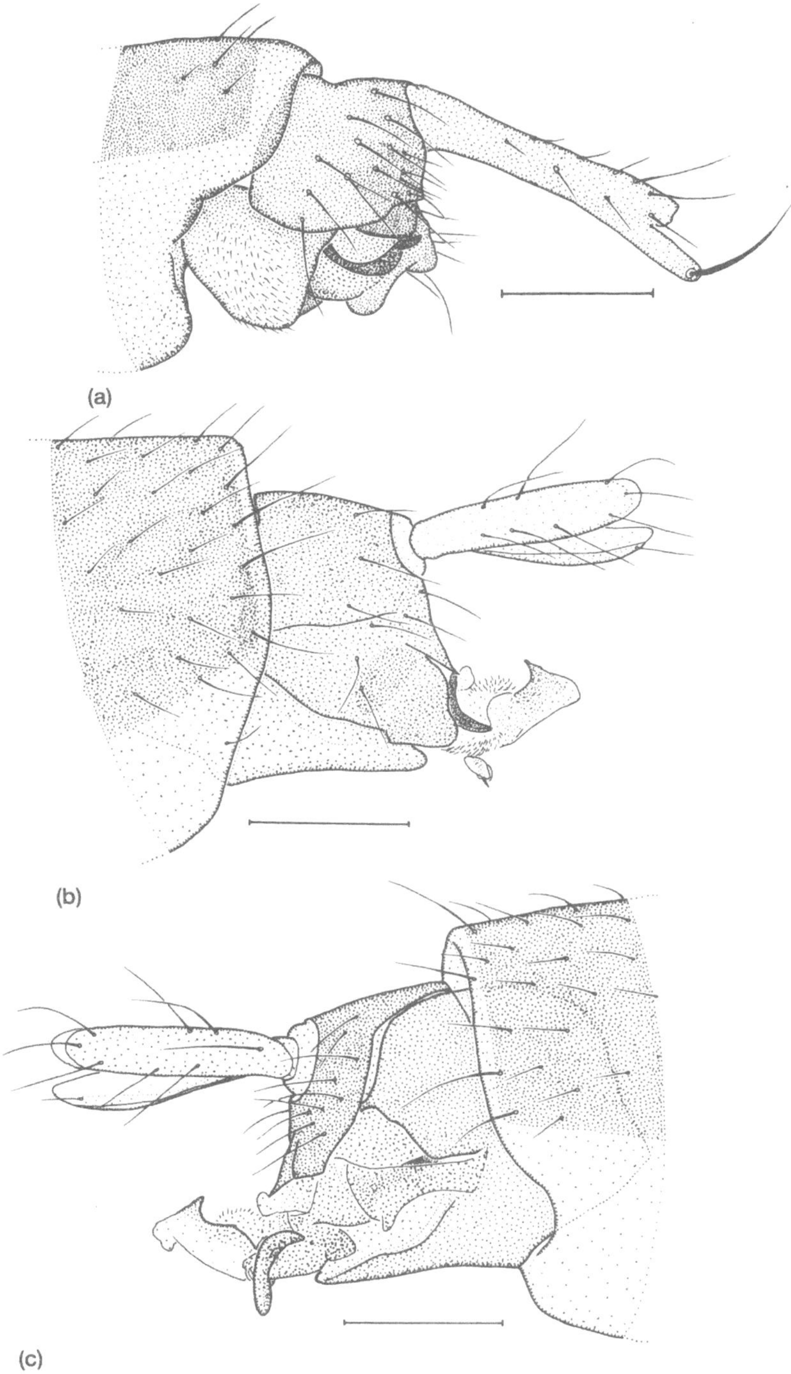

Figure 8.54 (a) *Menozziola obscuripes* male, left face of hypopygium. (b), (c) *Mesopathusa modesta* male, hypopygium from left (b) and right (c) sides.

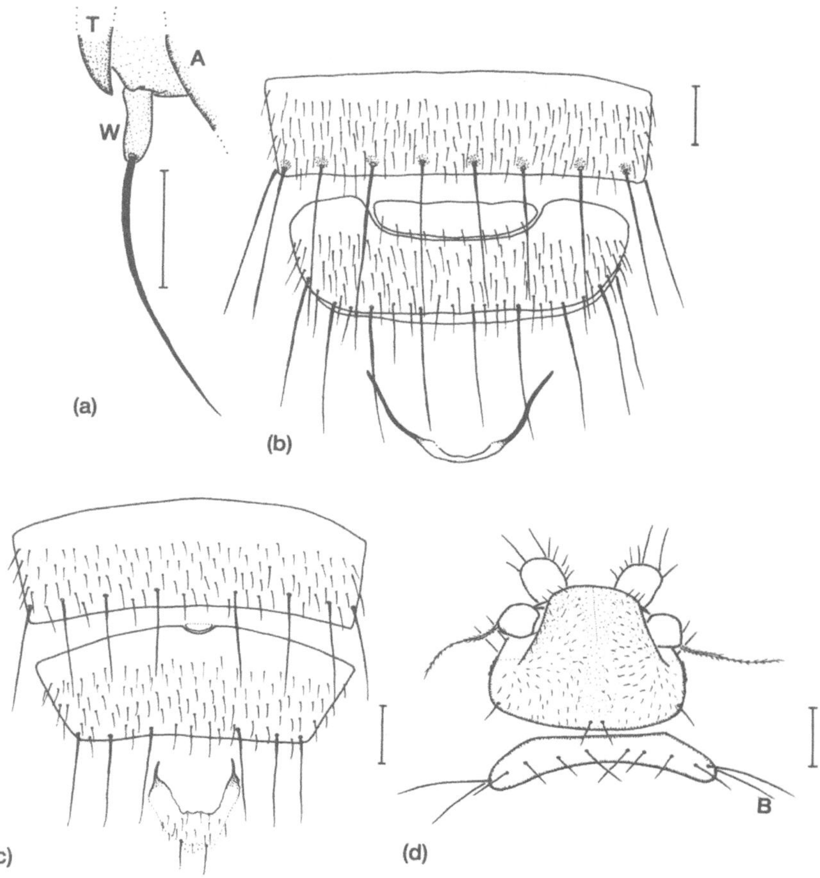

Figure 8.55 *Mesopathusa* females. (a) *M. cloveri* wing rudiment from above (T = thoracic notum, A = abdomen, W = wing rudiment). (b) *M. cloveri* abdominal tergites 4–6. (c) *M. modesta* abdominal tergites 4–6. (d) *M. modesta* head and thorax from above.

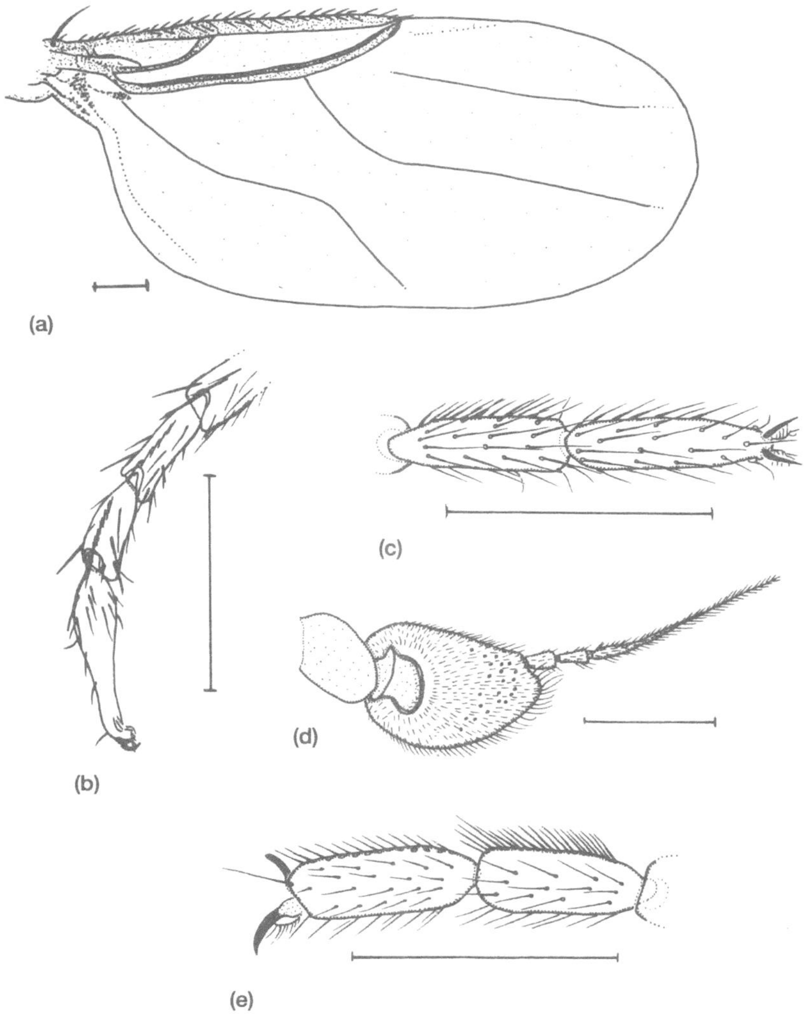

Figure 8.56 (a) *Metopina crassinervis* female, right wing. (b)–(e) *Microselia*: (b) *M. beaveri* female, segments 3–5 of front tarsus. (c)–(e) *M. bingana*: (c) female segments 4 and 5 of front tarsus; (d) female left antenna; (e) male segments 4 and 5 of front tarsus.

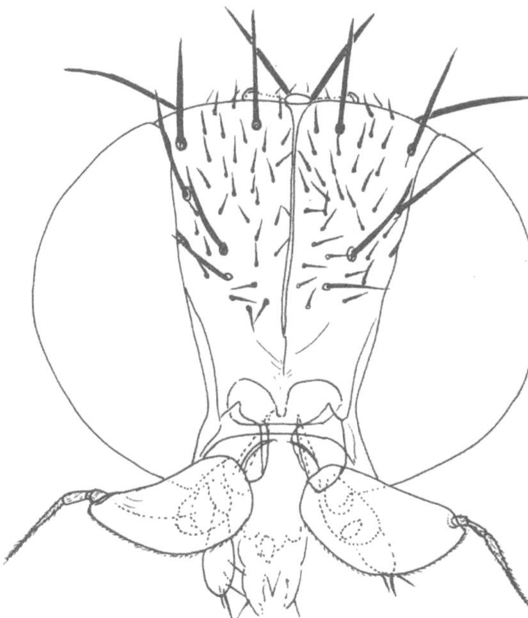
(a)

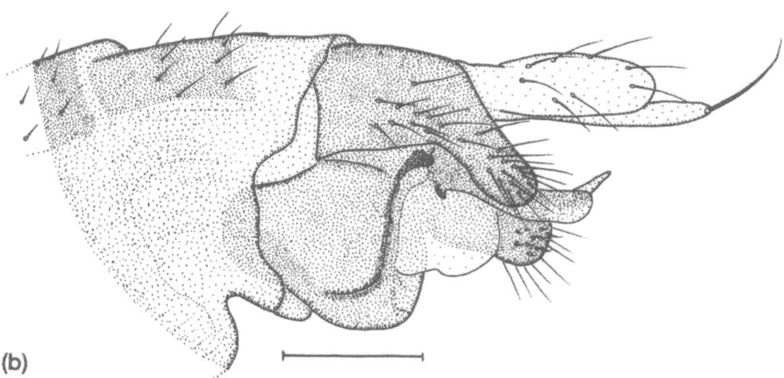
(b)

Figure 8.57 *Microselia*. (a) *M. beaveri* female, frontal view of head. (b) *M. bingana* male, left face of hypopygium.

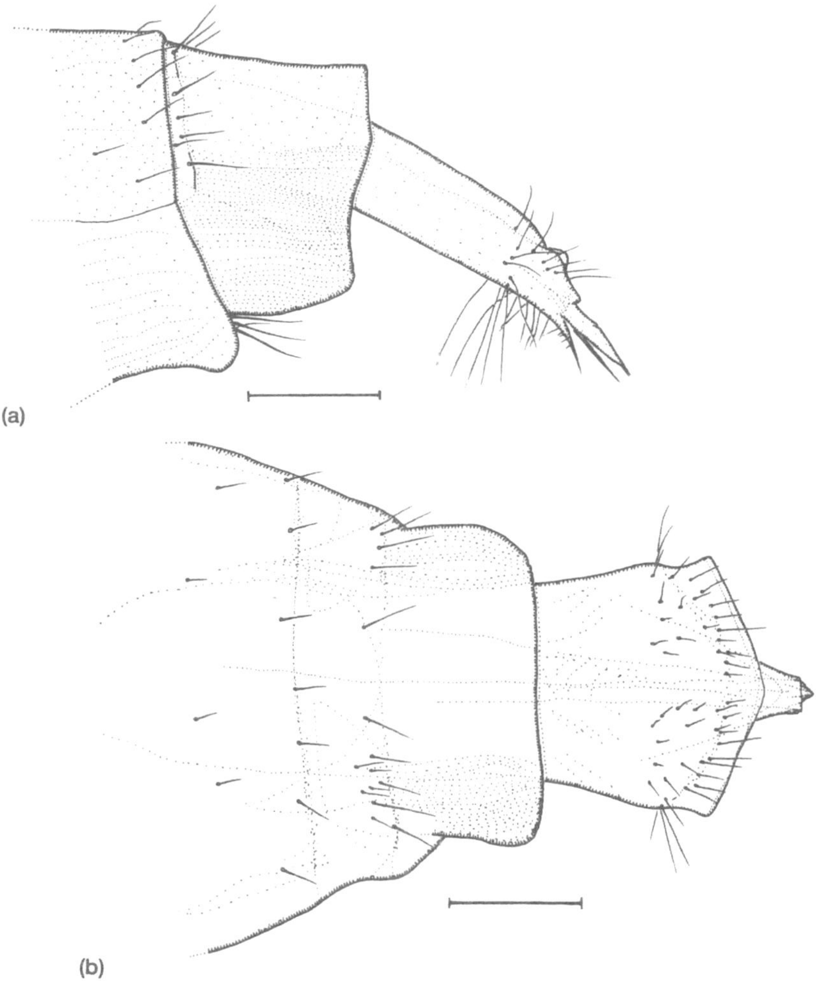

Figure 8.58 *Microselia bingana* female, ovipositor from left side (a) and above (b)

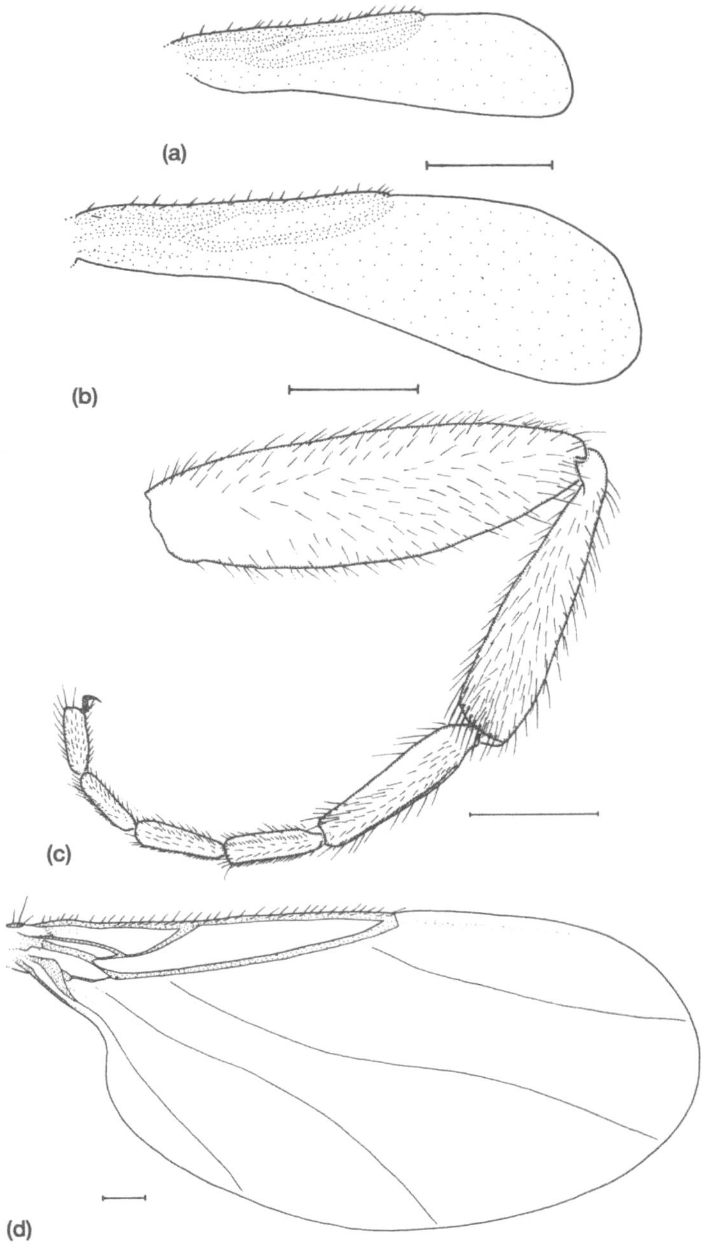

Figure 8.59 (a), (b) *Metopina ciceri* male, right wings of two different specimens. (c), (d) *Mesopathusa modesta* male: (c) front leg, anterior face; (d) right wing.

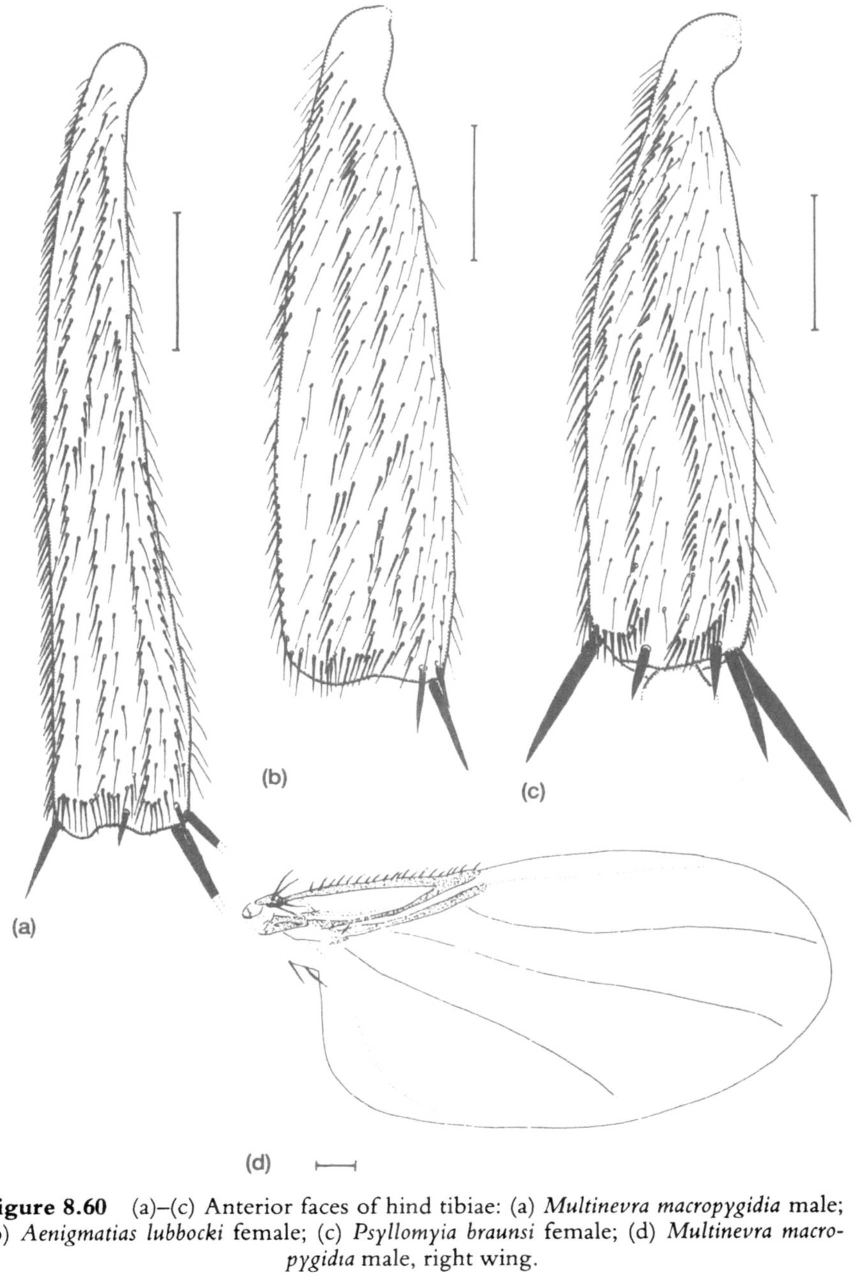

Figure 8.60 (a)–(c) Anterior faces of hind tibiae: (a) *Multinevra macropygidia* male; (b) *Aenigmatias lubbocki* female; (c) *Psyllomyia braunsi* female; (d) *Multinevra macropygidia* male, right wing.

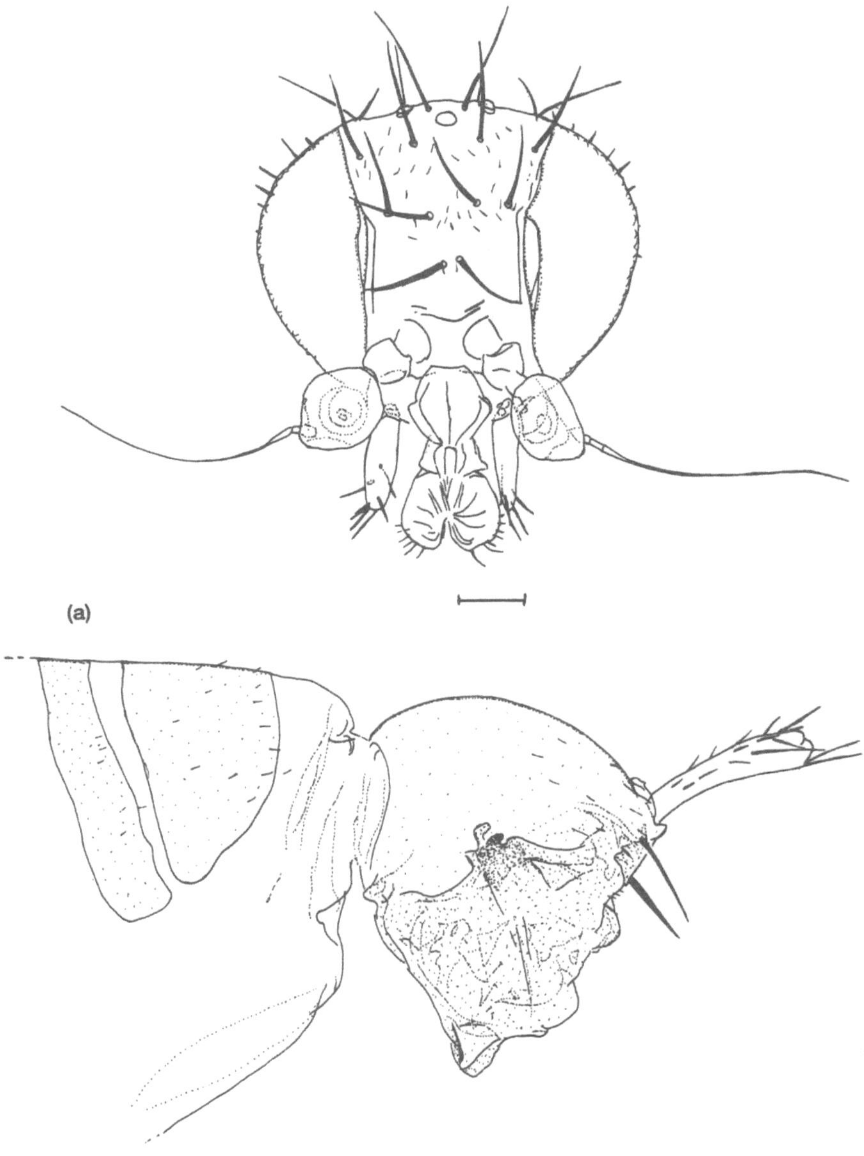

Figure 8.61 *Multinevra macropygidia* male (a) Frontal view of head (b) Left face of hypopygium

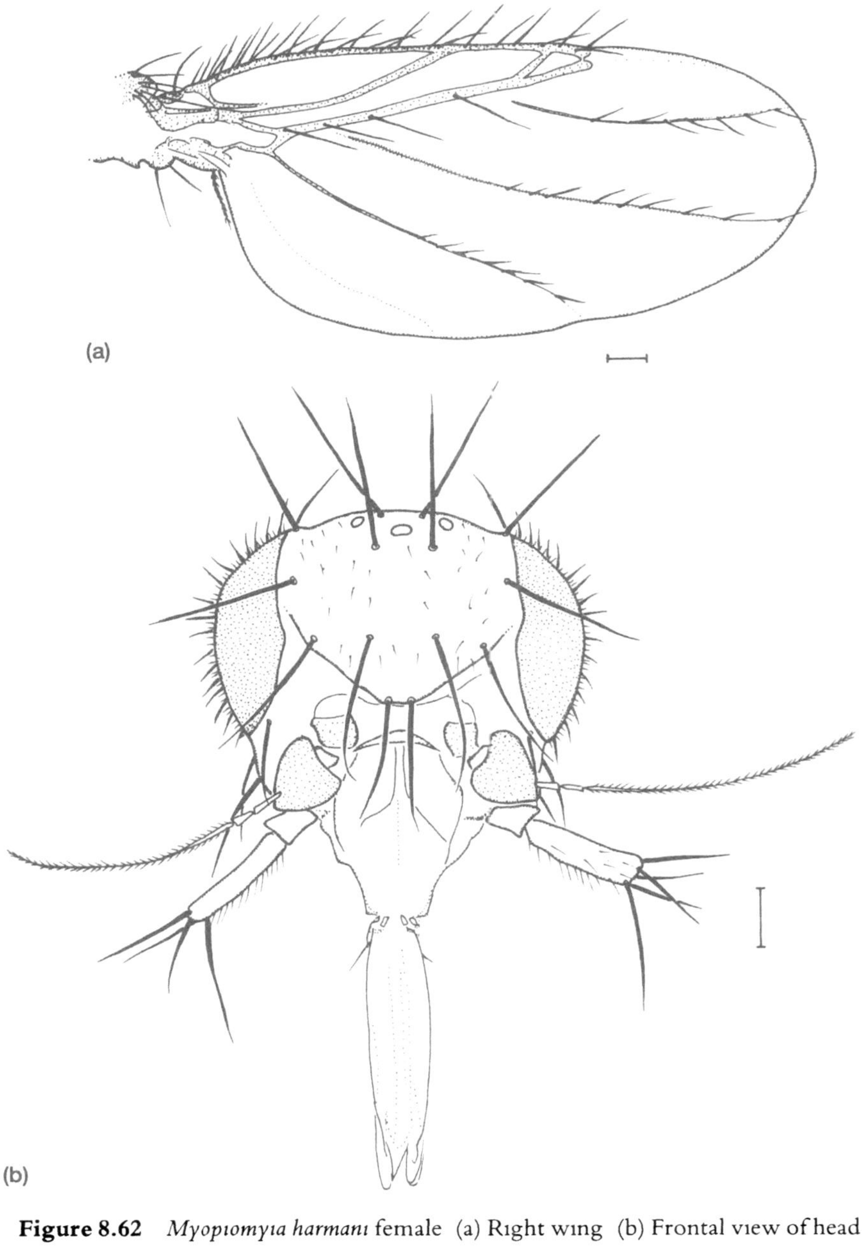

Figure 8.62 *Myopiomyia harmani* female (a) Right wing (b) Frontal view of head

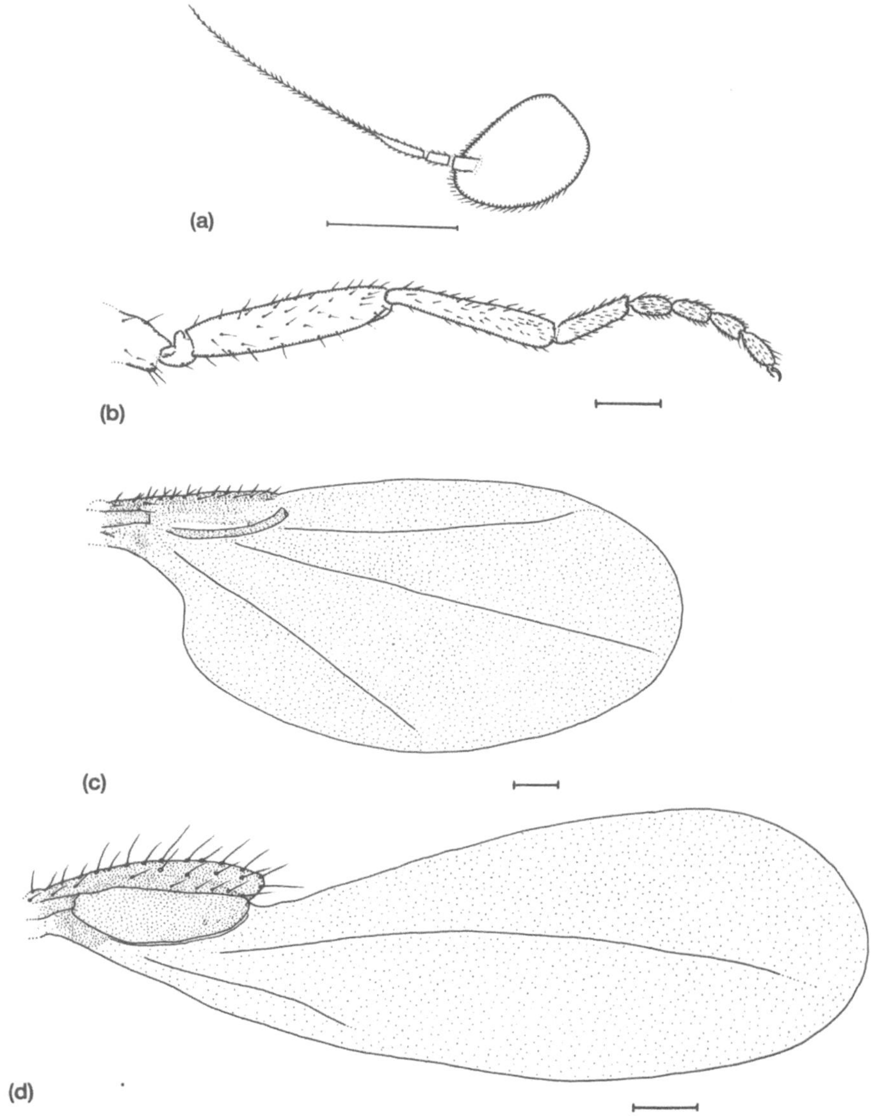

Figure 8.63 *Necperissa tinglei.* (a) Left antenna of male. (b) Front leg of male. (c) Right wing of male. (d) Right wing of female.

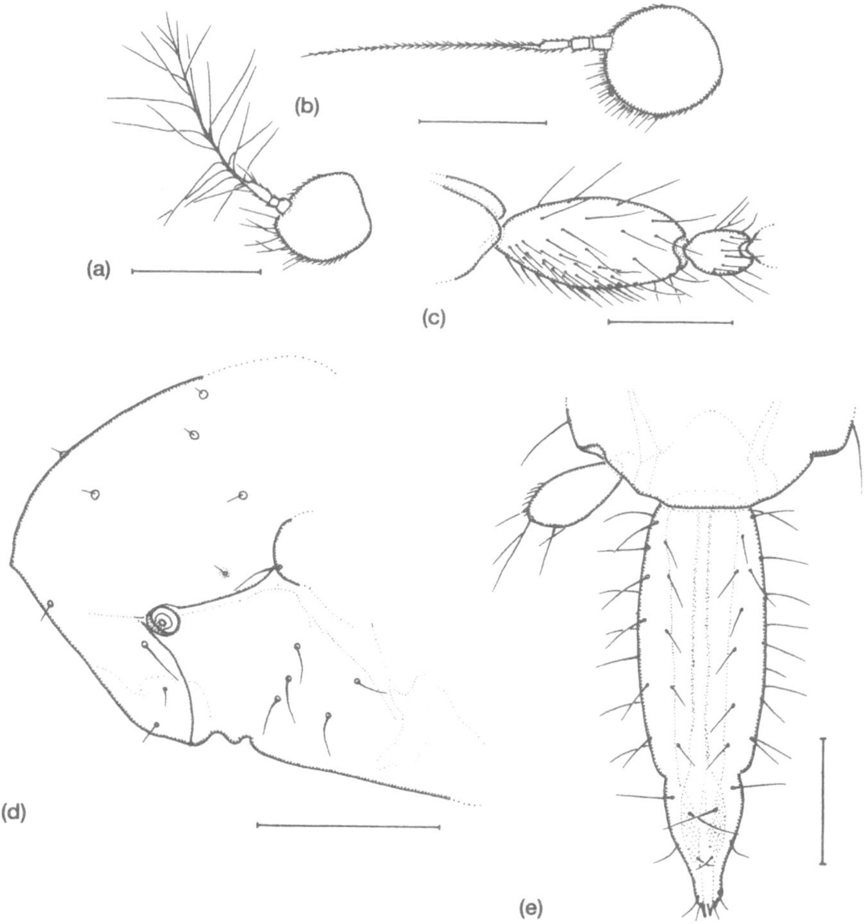

Figure 8.64 Females of *Necperissa* and *Ptochomyia* (a) *P cummingae* left antenna (b) *N tinglei* left antenna (c) *P cummingae* tarsal segments 1 and 2 of front leg (d) *P afra*, left face of front part of thorax (e) *N tinglei* proboscis and right palp

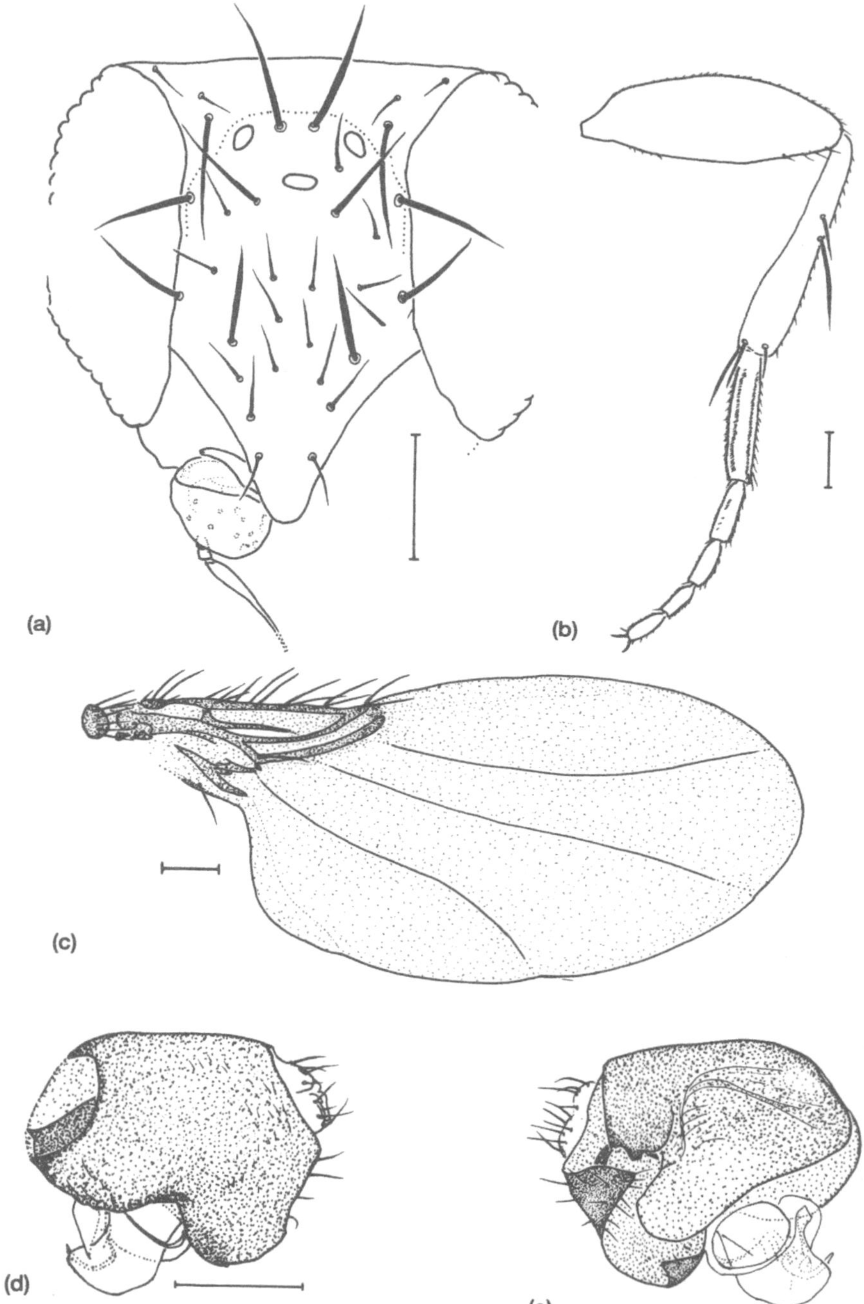

Figure 8.65 *Obscuriphora sheppardi* (a) Female frons (b) Male, anterior face of hind leg (c) Male, right wing (d), (e) Left and right faces of hypopygium

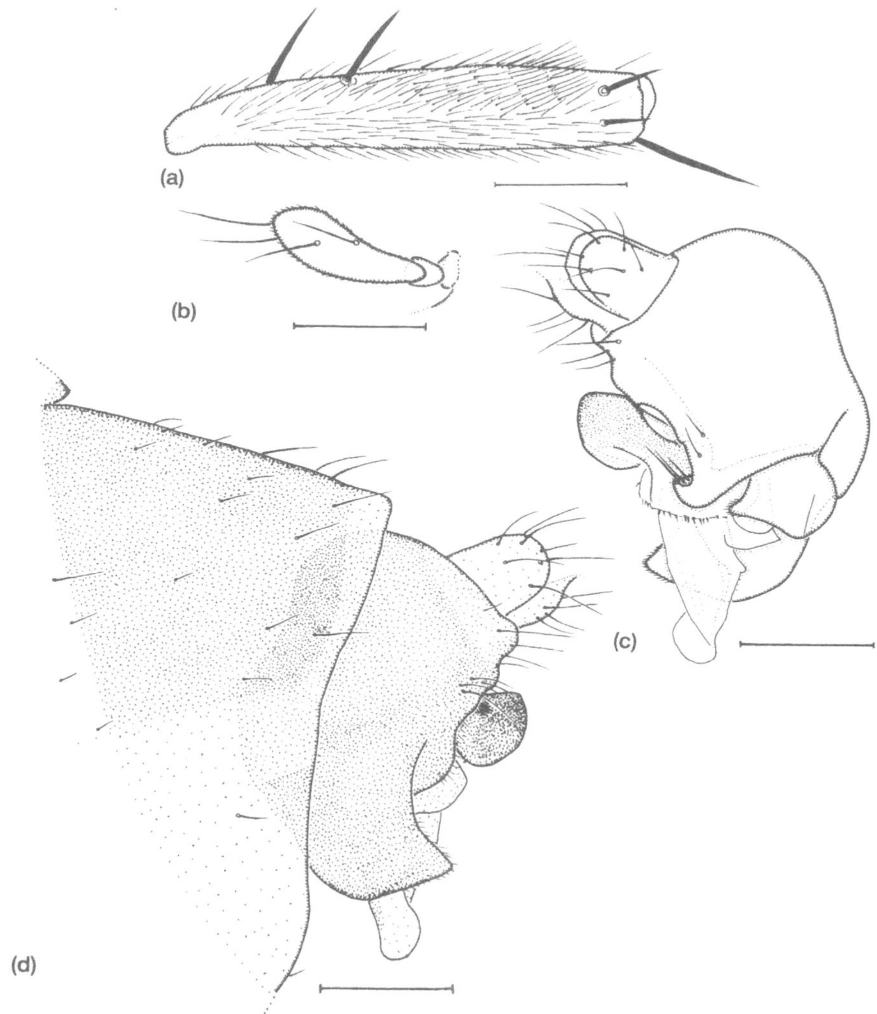

Figure 8.66 *Palpiclavina kistneri* male (a) Anterior face of mid tibia (b) Right palp (c), (d) Right and left faces of hypopygium

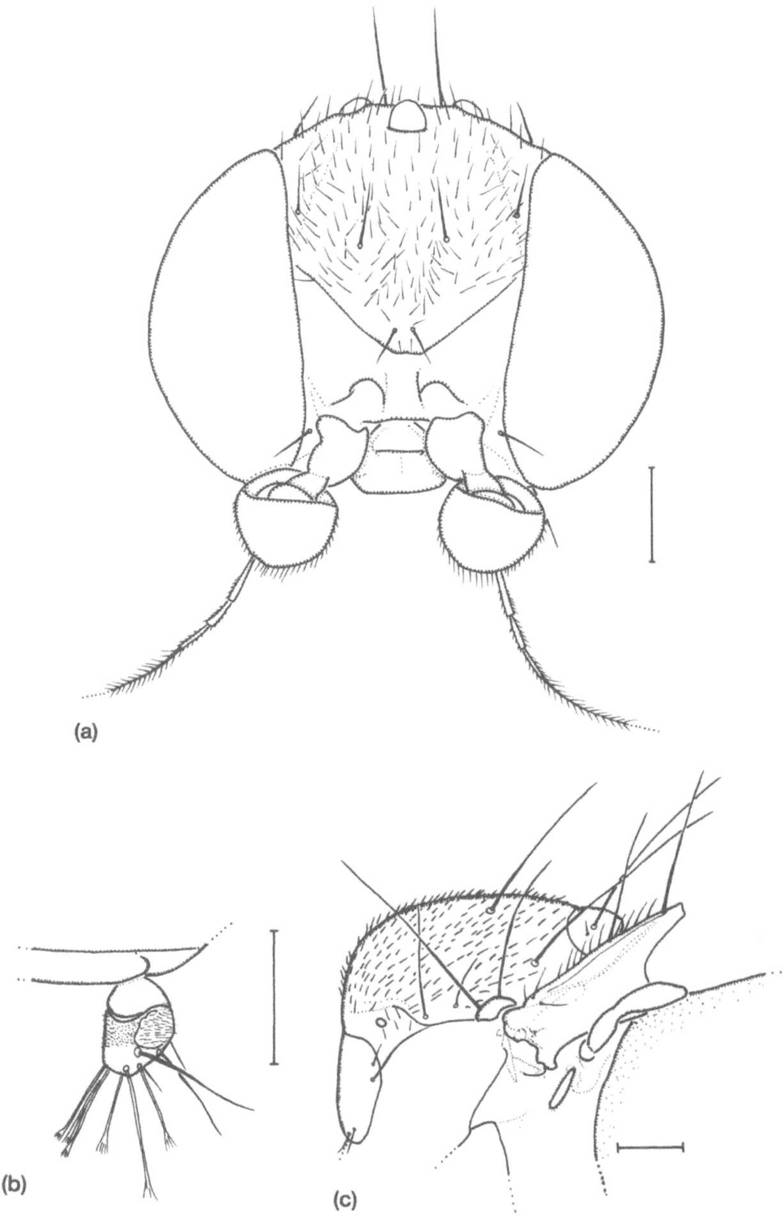

Figure 8.67 *Palpiclavina kistneri* (a) Male, frontal view of head (b) Female, left palp (c) Female, left face of thorax

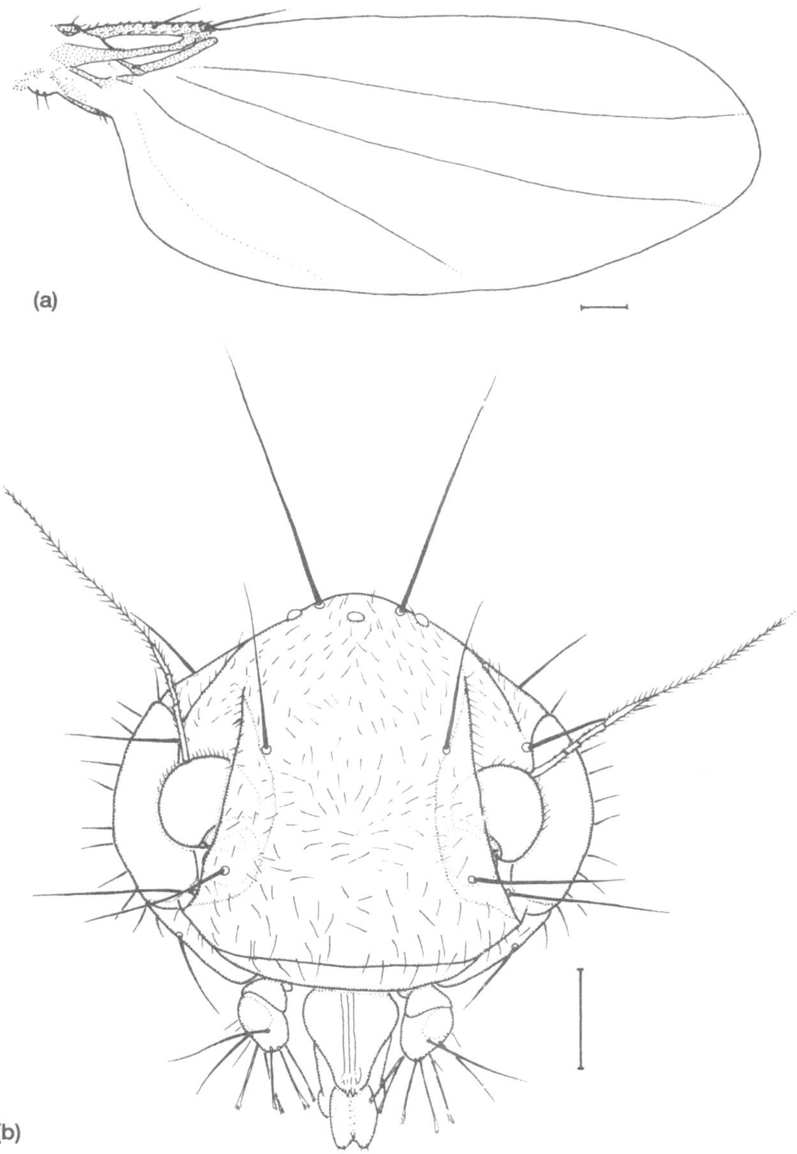

Figure 8.68 *Palpicalvina* females (a) *P tonkinensis*, right wing (b) *P kistneri* frontal view of head

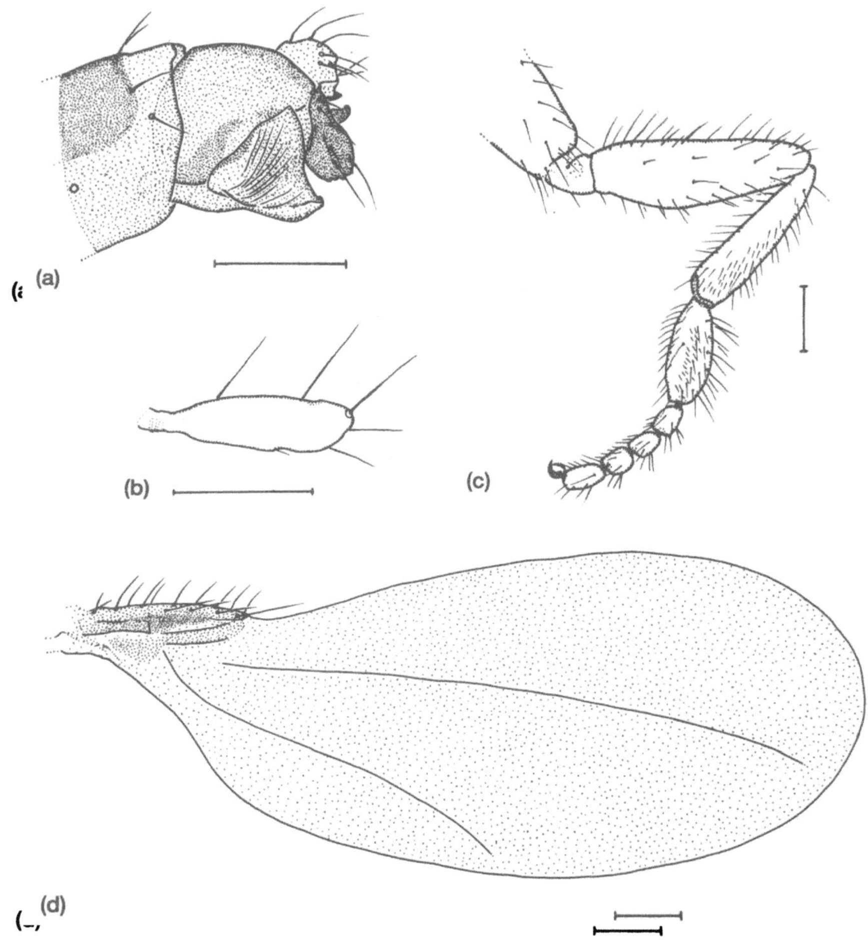

Figure 8.69 (a) *Necperissa tinglei* male hypopygium. (b)–(d) *Perittophora couturieri* female: (b) left palp from above, (c) front leg, (d) rıght wıng.

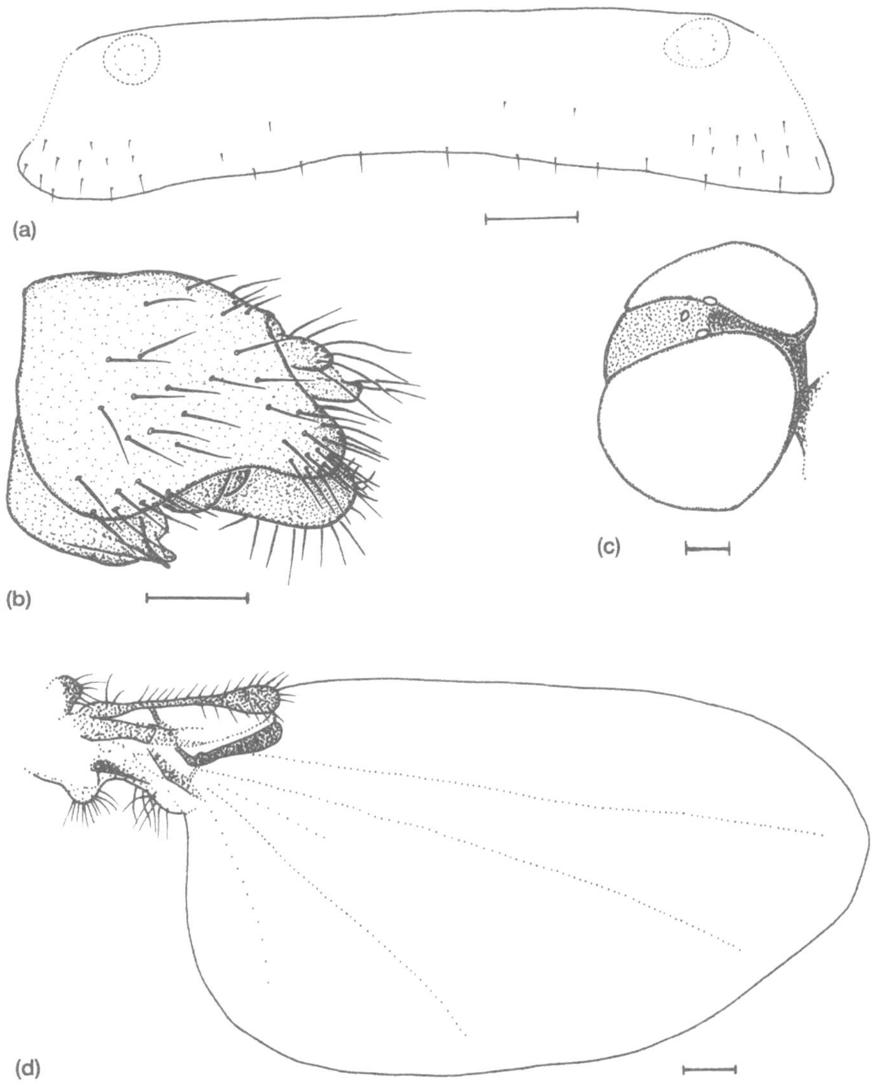

Figure 8.70 (a) *Phalacrotophora berolinensis* female, abdominal tergite 3. (b) *Plectanocnema nudipes* male hypopygium, left face. (c), (d) *Postoptica platypezoidea* male: (c) dorso-lateral view of head, (d) right wing.

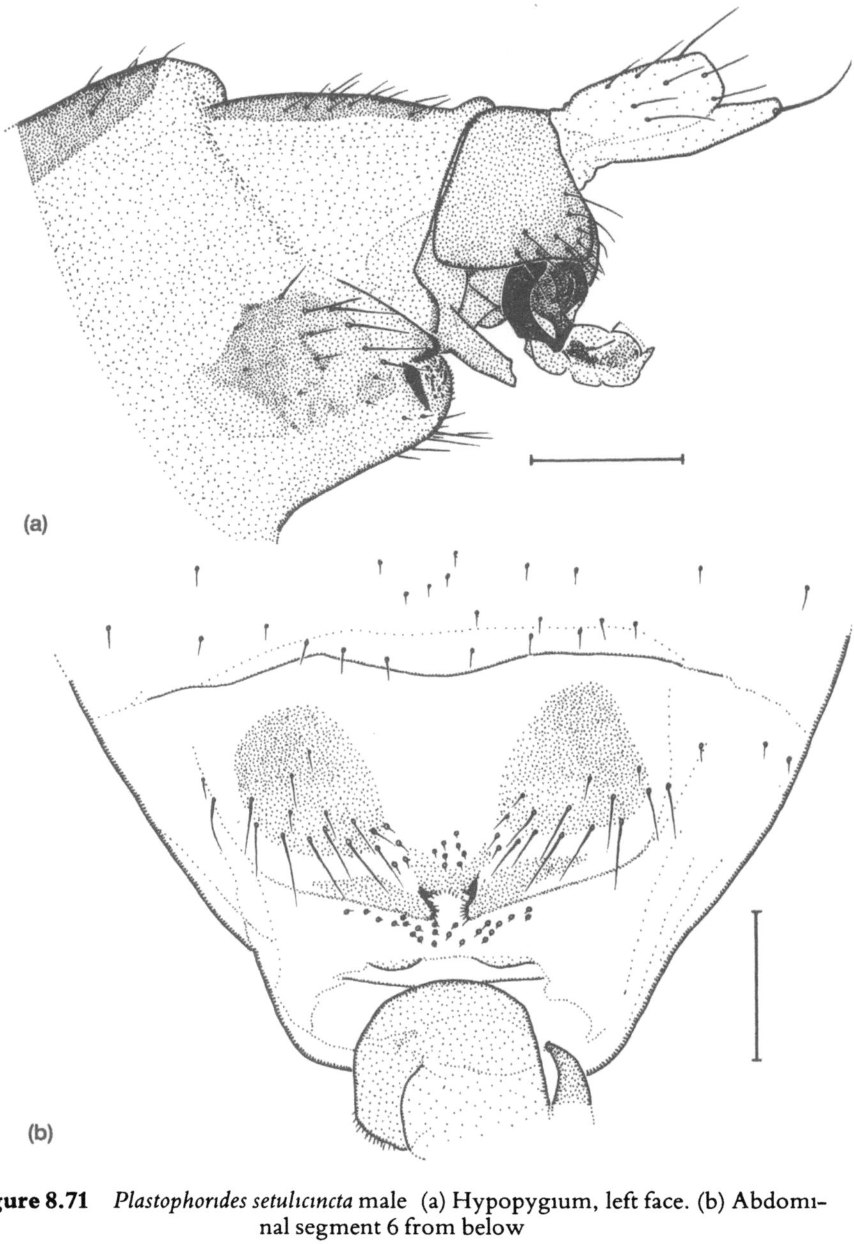

Figure 8.71 *Plastophorides setulicincta* male (a) Hypopygium, left face. (b) Abdominal segment 6 from below

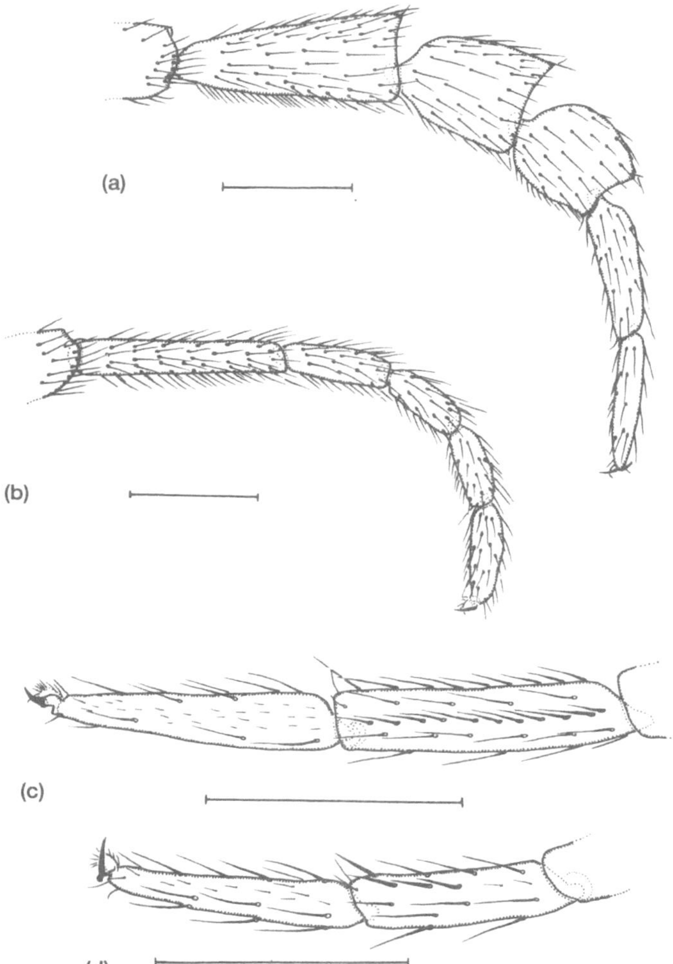

Figure 8.72 *Plastophorides setulicincta* tarsal segments (a) Female front leg (b) Male front leg (c) Female hind leg, segments 4 and 5 (d) Female middle leg, segments 4 and 5

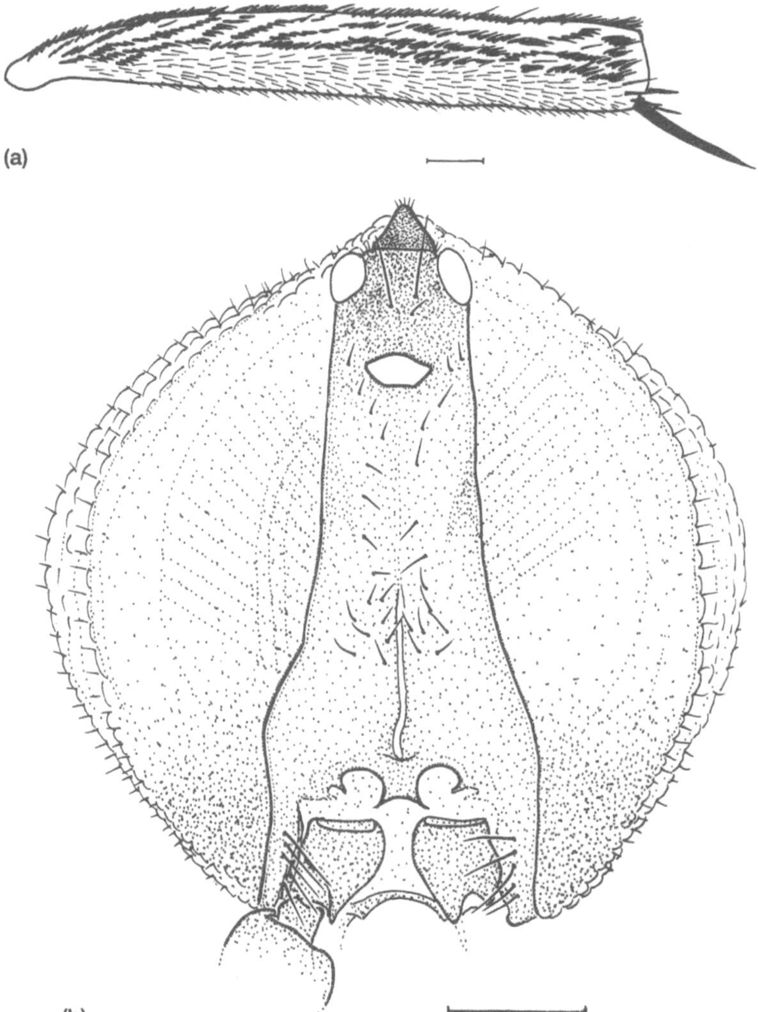

(a)

(b)

Figure 8.73 (a) *Plectanocnema nudipes* male, anterior face of hind tibia. (b)*Postoptica platypezoidea* male, frontal view of head.

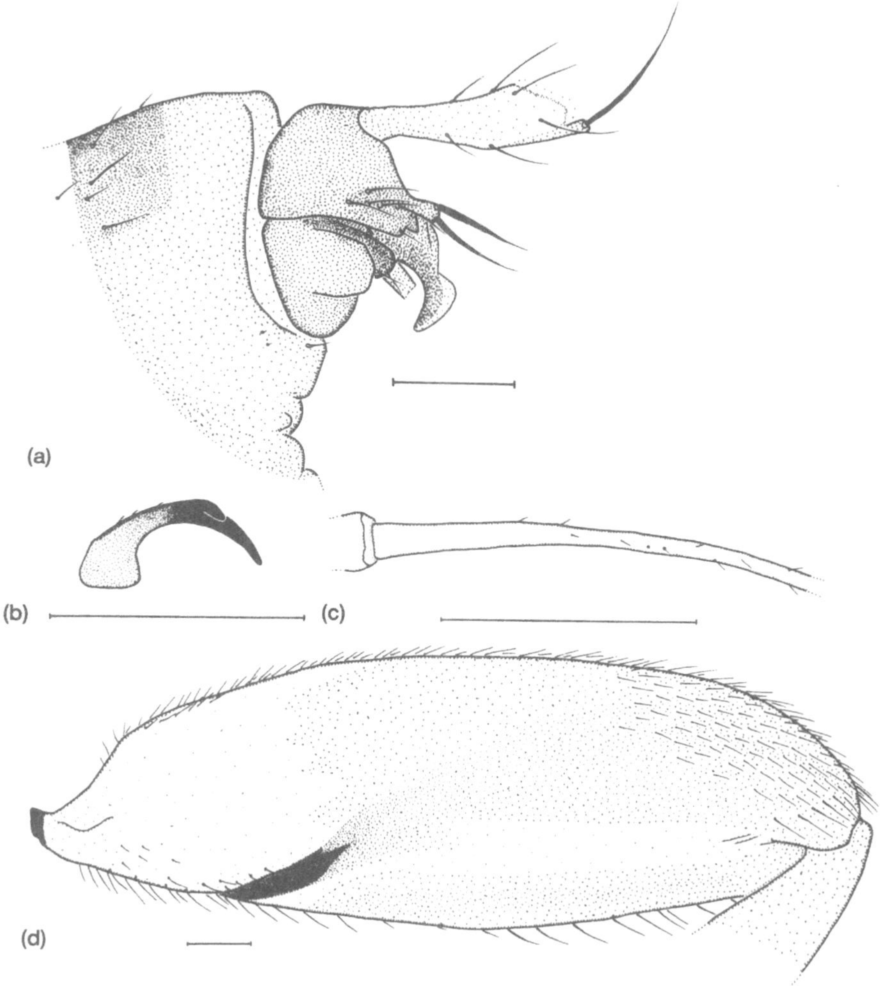

Figure 8.74 (a) *Pseudacteon spatulatus* male, left face of hypopygium (b)–(d) *Pseudohypocera kerteszi* (b) hind tarsal claw of male, (c) base of third segment of female arista, (d) posterior face of hind femur of male

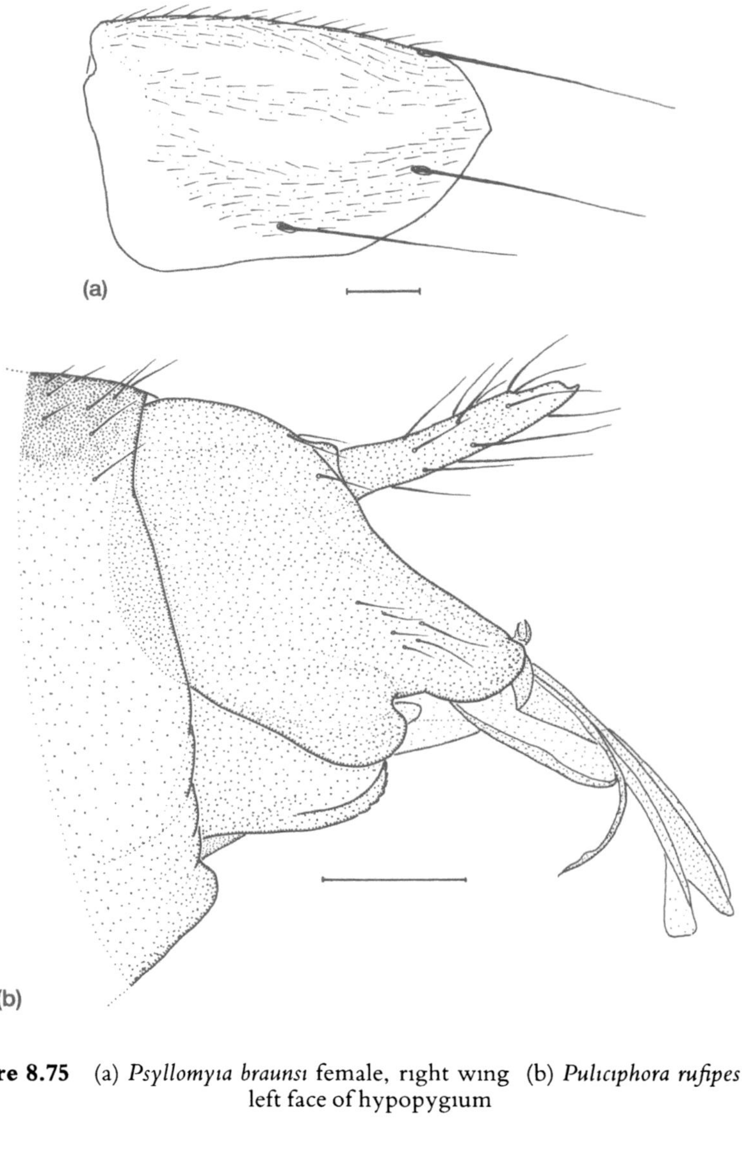

Figure 8.75 (a) *Psyllomyia braunsi* female, right wing (b) *Puliciphora rufipes* male, left face of hypopygium

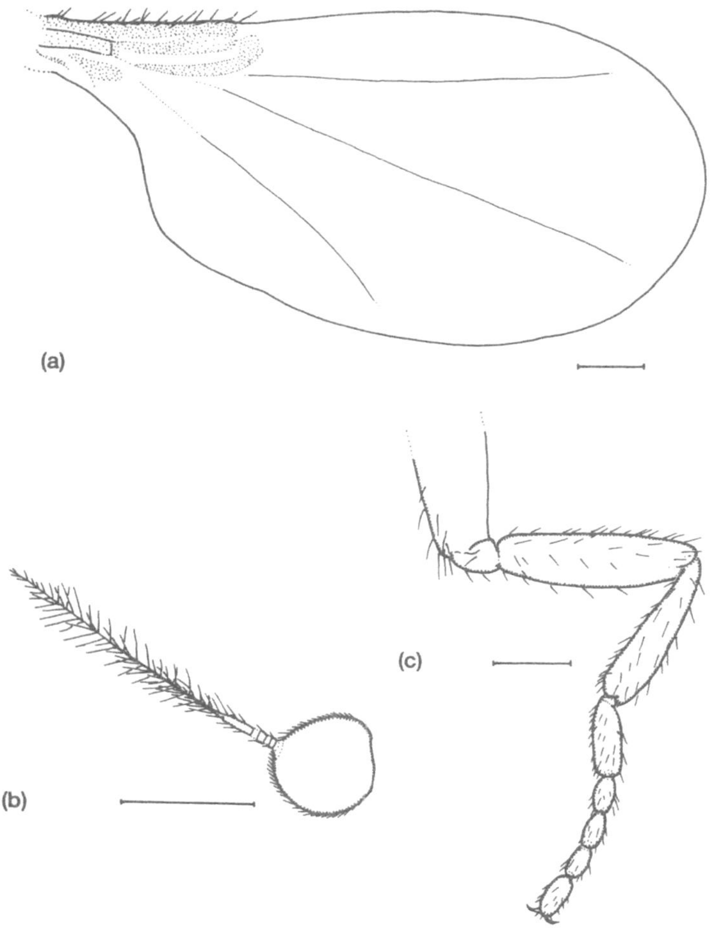

Figure 8.76 *Ptochomyia cummingae* male. (a) Right wing. (b) Left antenna. (c) Right front leg.

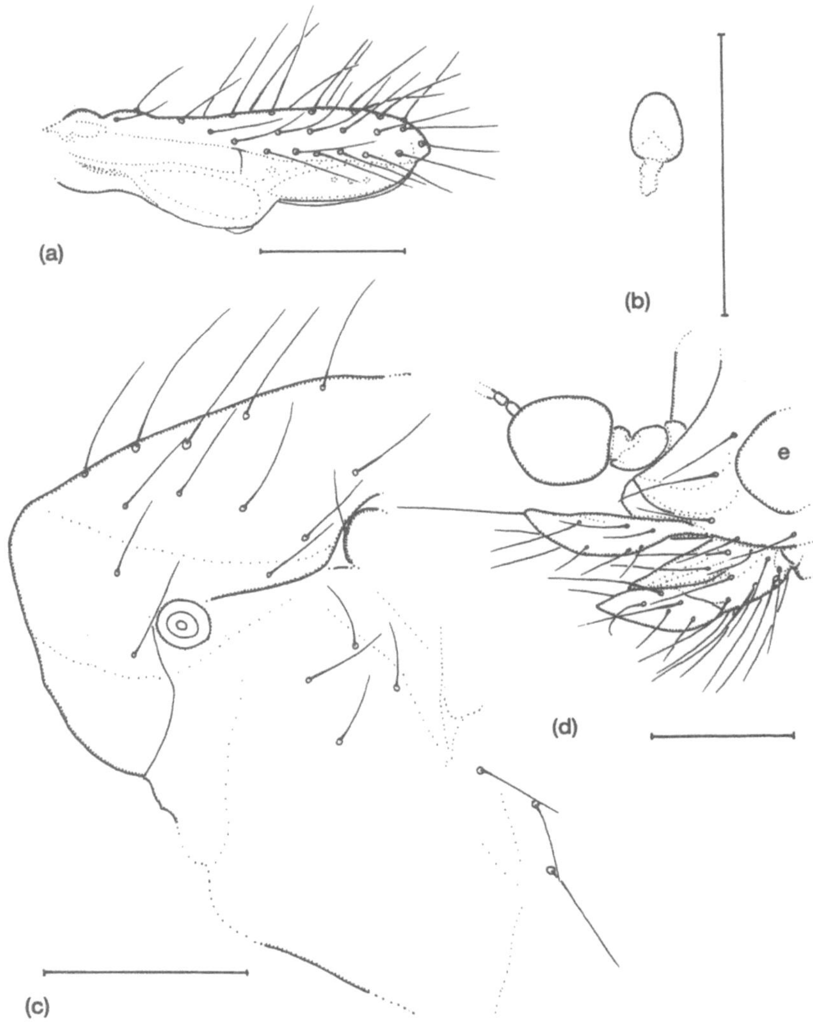

Figure 8.77 *Ptochomyia cummingae* female (a) Right wing stump (b) Haltere (c) Front end of left face of thorax (d) Proboscis and antenna, left face (e = eye)

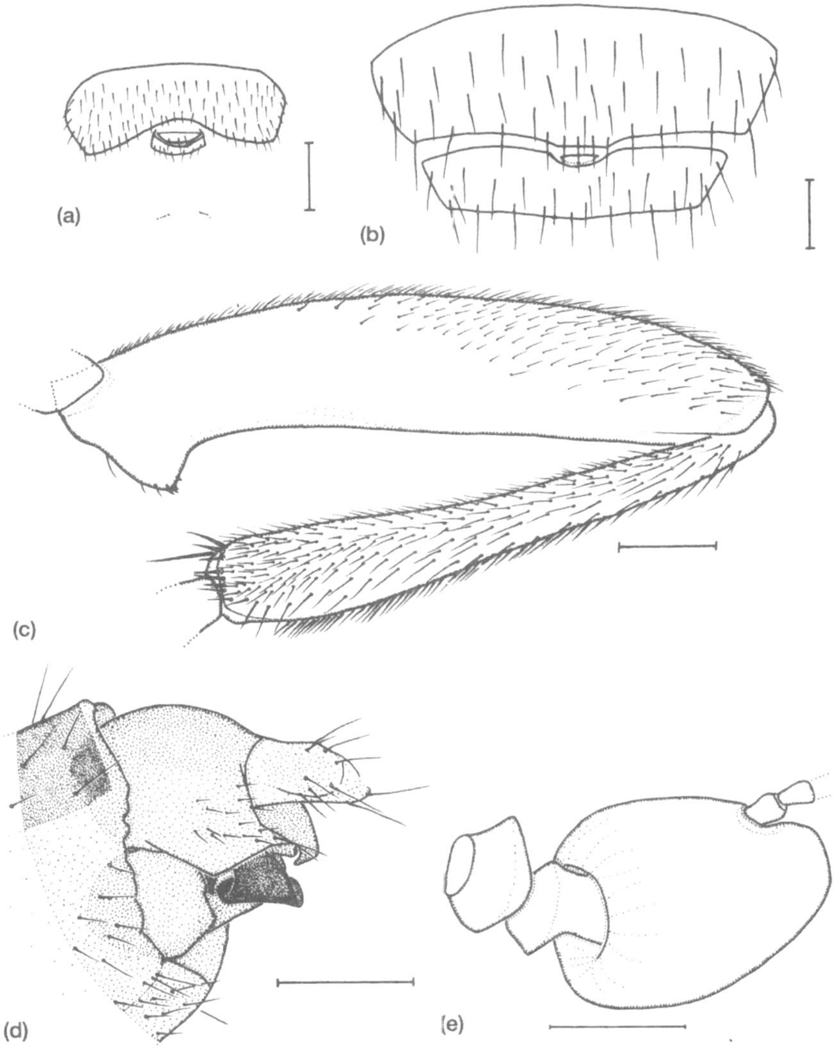

Figure 8.78 (a), (b) *Puliciphora* females, abdominal tergites 4 and 5: (a) *P. knighti*, (b) *P. rosei*. (c)–(e) *Razorfemora nussbaumi* male: (c) posterior face of hind femur and tibia; (d) left face of hypopygium; (e) left antenna.

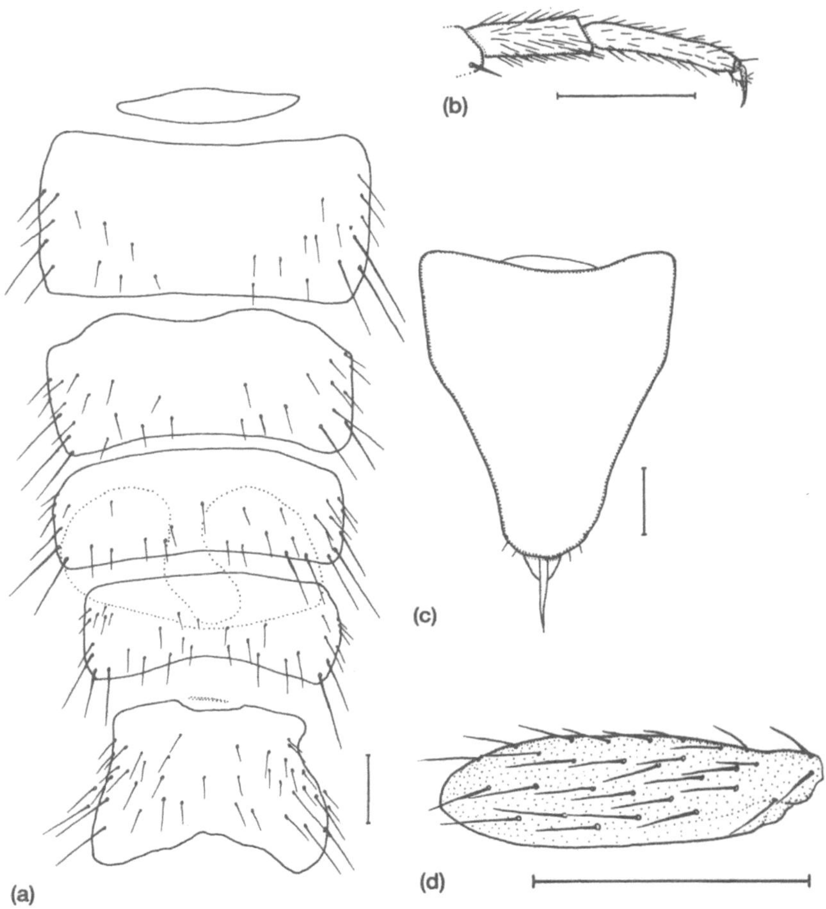

Figure 8.79 (a)–(c) *Razorfemora nussbaumi* female (a) abdominal tergites 1–6, (b) anterior faces of mid-tarsal segments 4 and 5, (c) ovipositor from above, (d) *Rhabdomyia kistnerorum* female, left wing

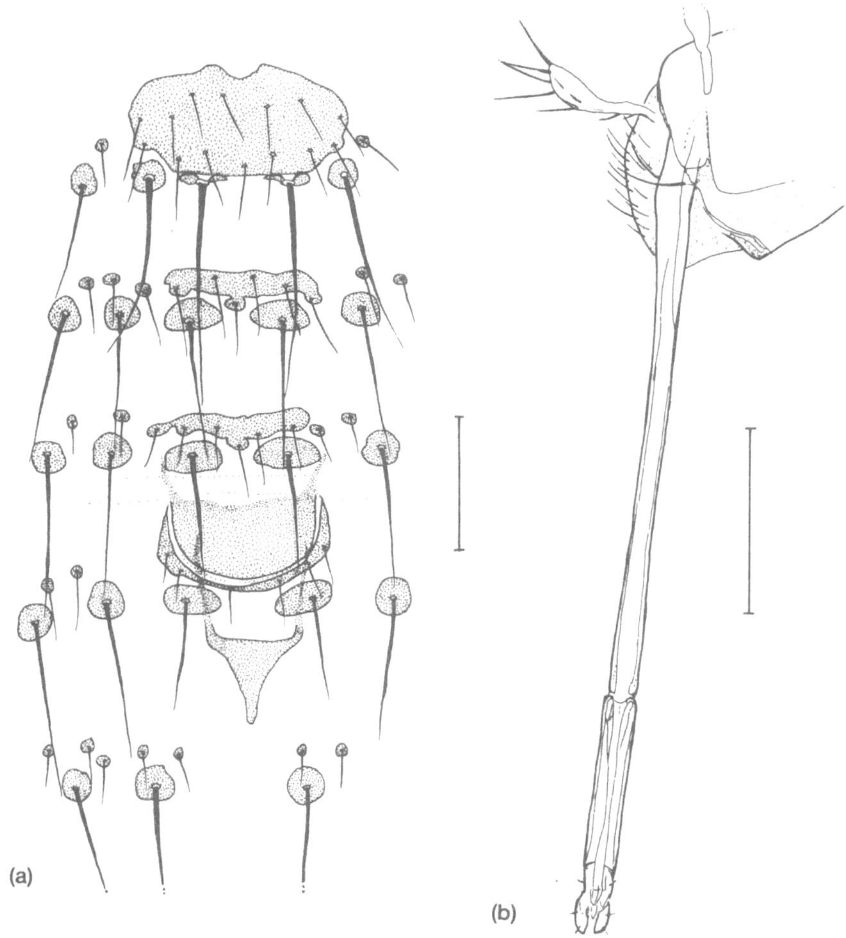

Figure 8.80 (a) *Rhabdomyia kistnerorum* female, abdominal tergites 2–6 (b) *Rhynchomicropteron beaveri* female, proboscis

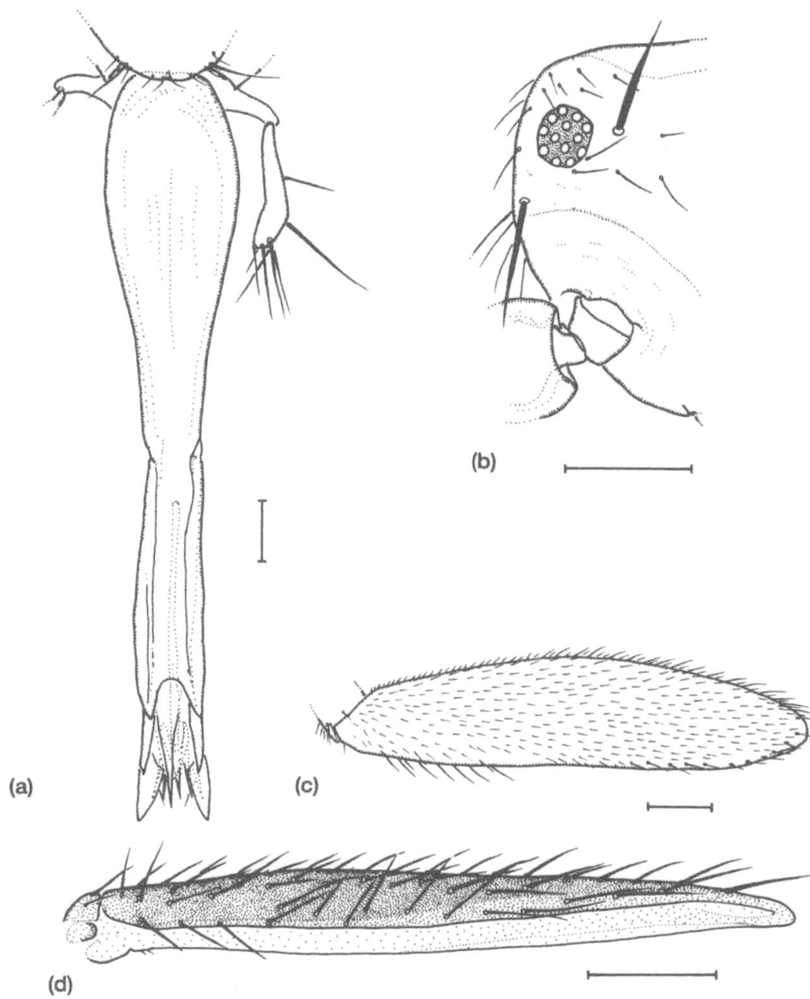

Figure 8.81 *Rhychomicropteron necaphidiforme* female. (a) Proboscis. (b) Right side of frons. (c) Anterior face of hind femur. (d) Right wing.

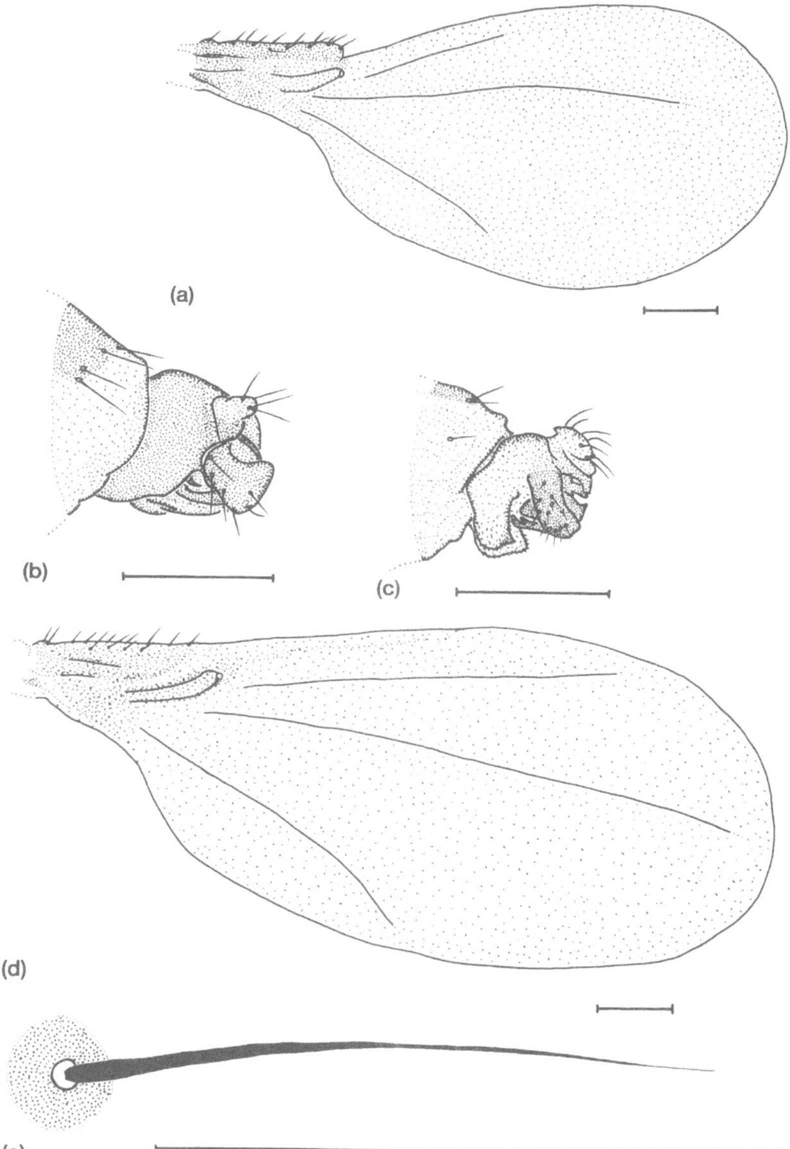

Figure 8.82 *Syntermitoxenia* (a)–(d) Males (a) *S latiptera*, right wing, (b) *S lewisi*, left face of hypopygium, (c) *S latiptera*, left face of hypopygium, (d) *S lewisi*, right wing, (e) *S apiogaster*, abdominal bristle of female

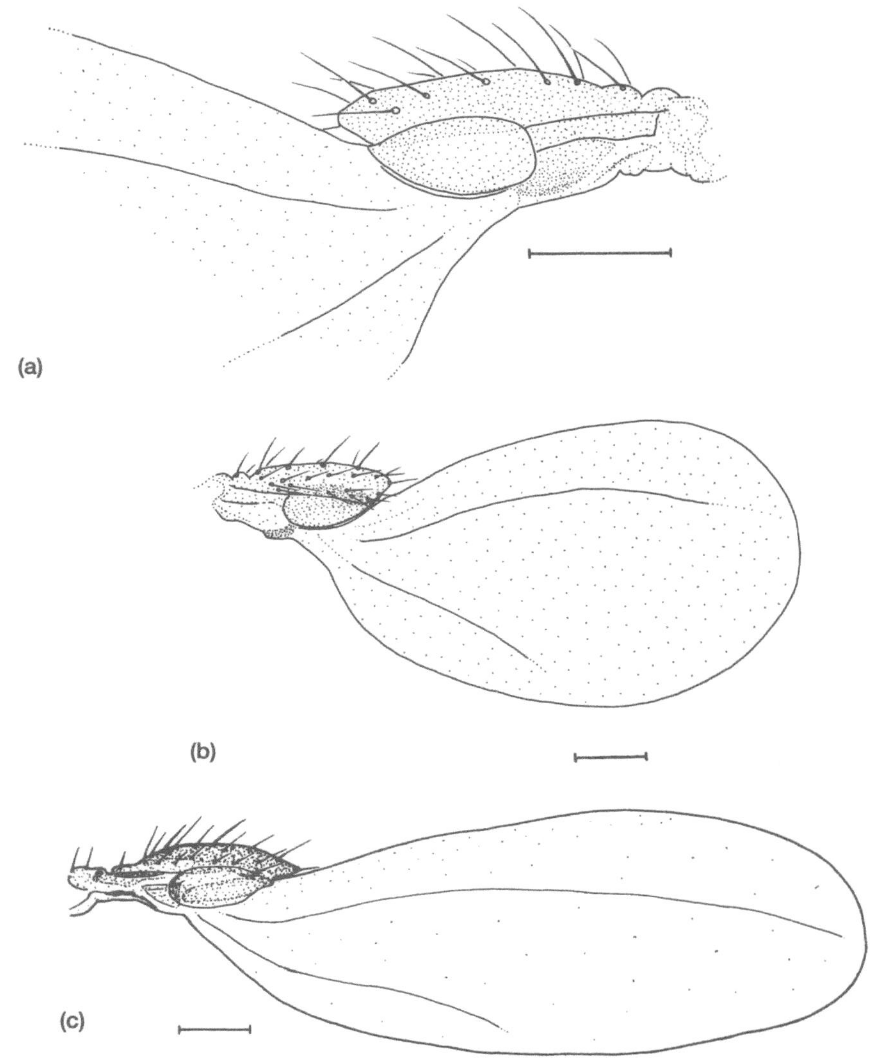

Figure 8.83 *Syntermitoxenia* females, right wings. (a) *S. latiptera*, base of wing from below. (b) *S. latiptera*. (c) *S. lewisi*.

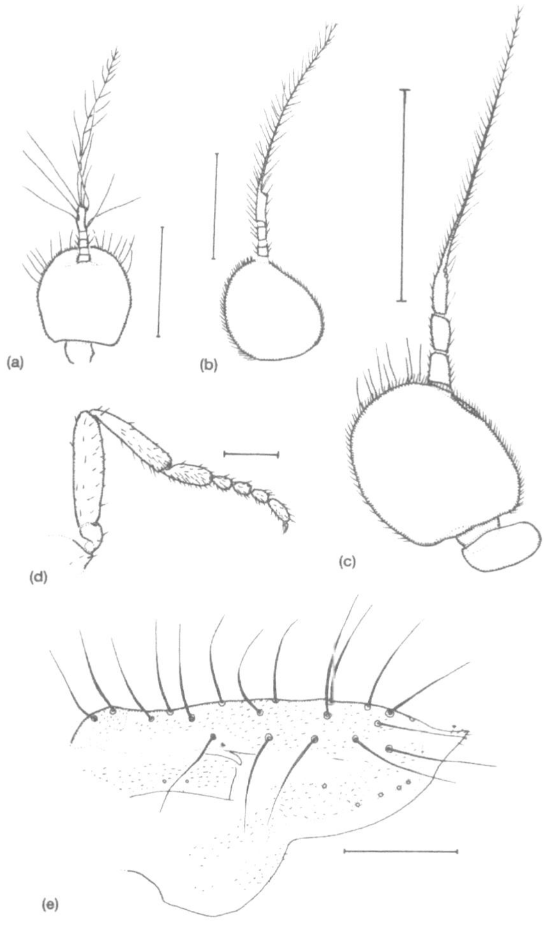

Figure 8.84 (a), (b) *Termitophilomyia braunsi* antennae: (a) female, (b) male. (c) *Termitoxenia oligoseta* male, left antenna (at a larger scale than (a) and (b)). (d) *T oligoseta* male, front leg. (e) *Termitosagma henningsi* female, right wing stump

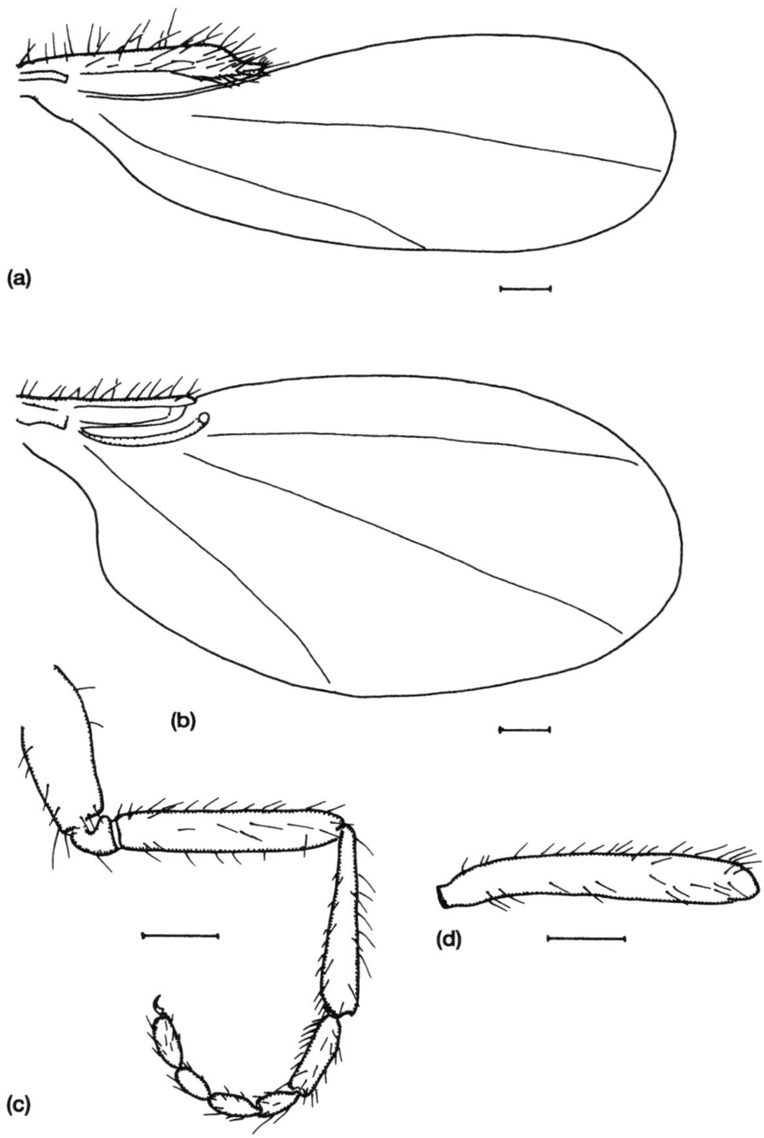

Figure 8.85 *Termitophilomyia braunsi* (a) Female, right wing (b) Male, right wing (c) Male, right front leg (d) Male, anterior face of hind femur

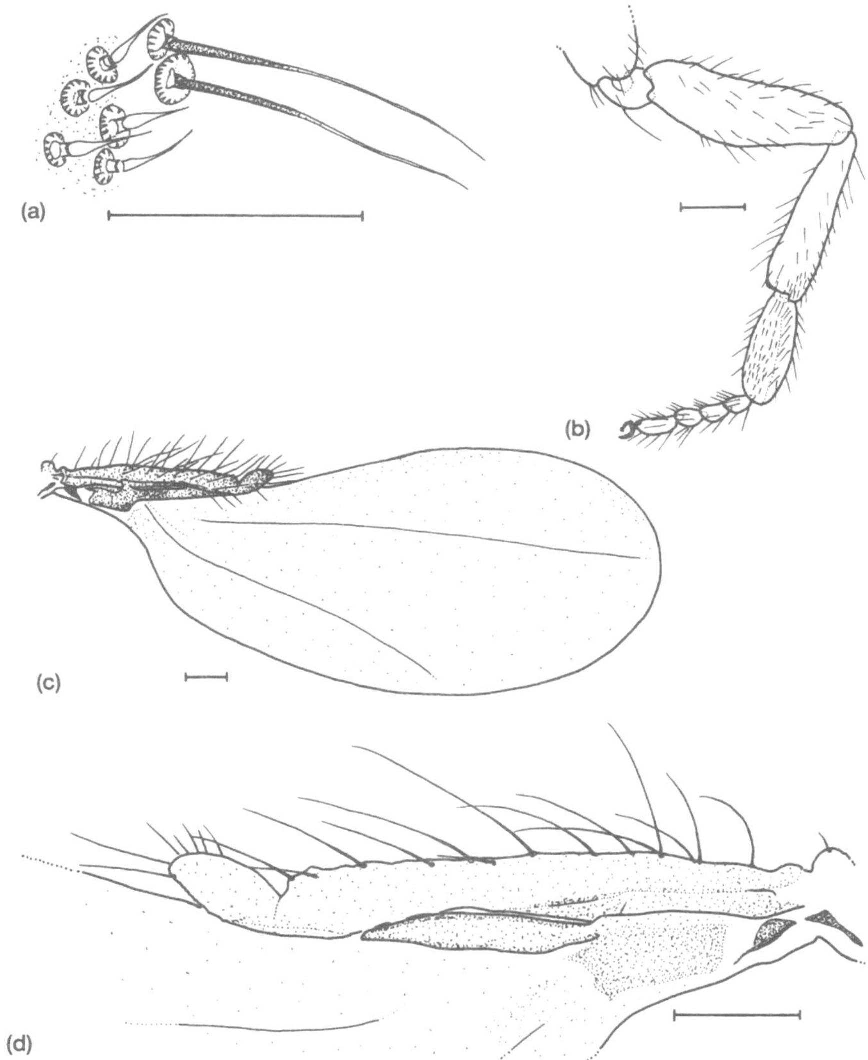

Figure 8.86 *Termitostroma* sp , female (a) Bristles at base of abdominal segment 2 (b) Front leg (c) Right wing (d) Base of right wing from below

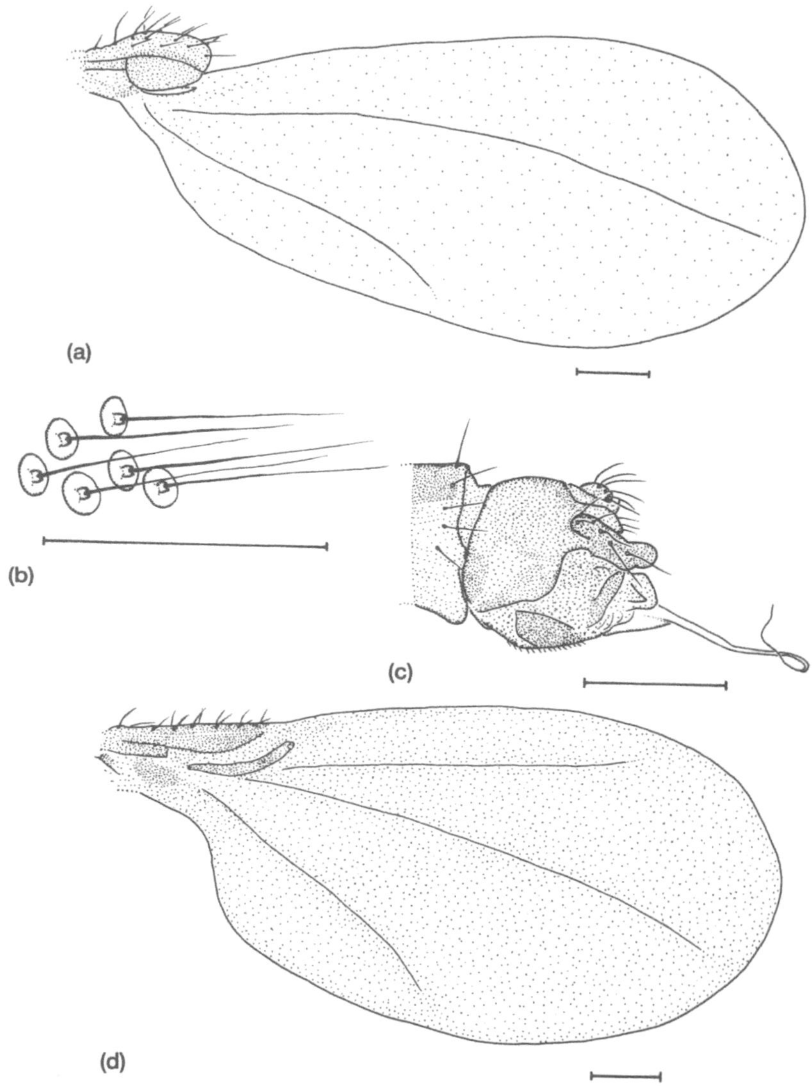

Figure 8.87 *Termitoxenia oligoseta* (a) Female, right wing (b) Female, bristles at side of abdominal segment 3 (c) Male, left face of hypopygium (d) Male, right wing

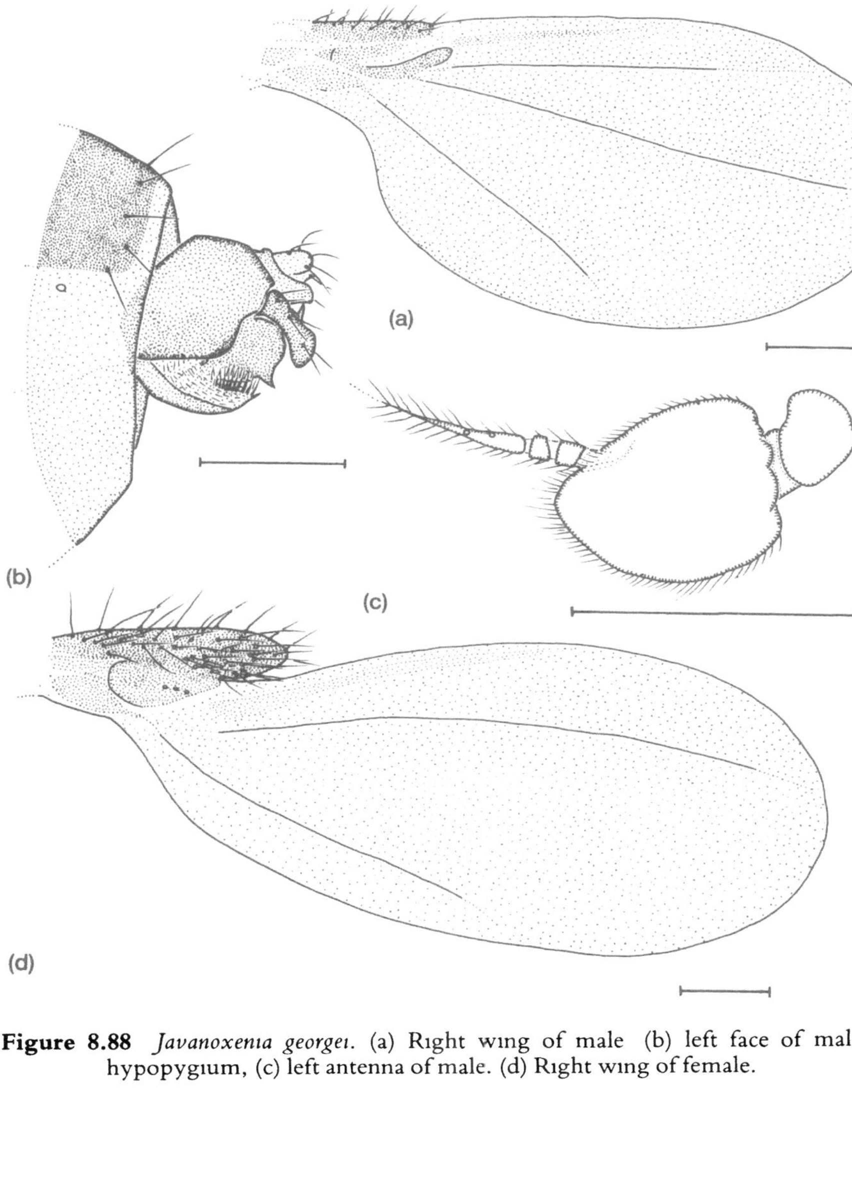

Figure 8.88 *Javanoxenia georgei*. (a) Right wing of male (b) left face of male hypopygium, (c) left antenna of male. (d) Right wing of female.

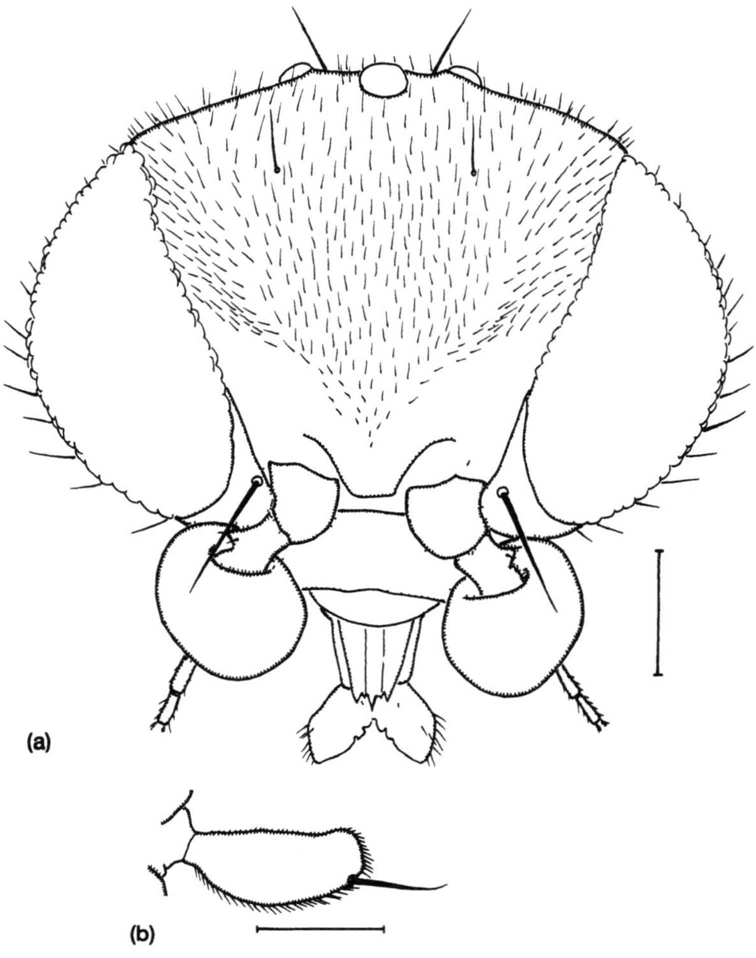

Figure 8.89 *Thaumatoxena andreinii* male (a) Frontal view of head (b) Right palp

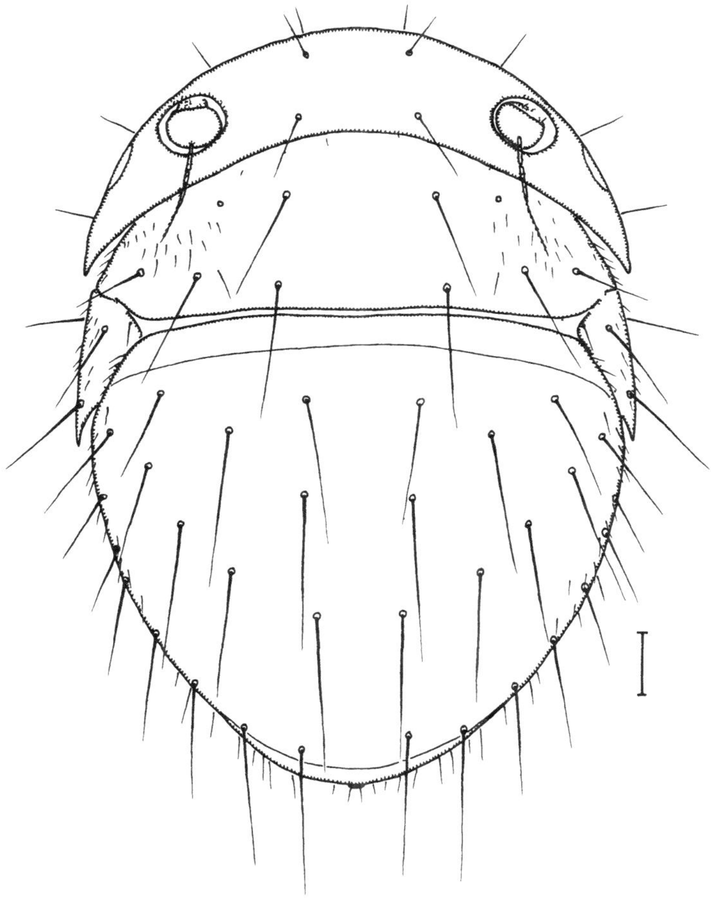

Figure 8.90 *Thaumatoxena andreinii* female, from above

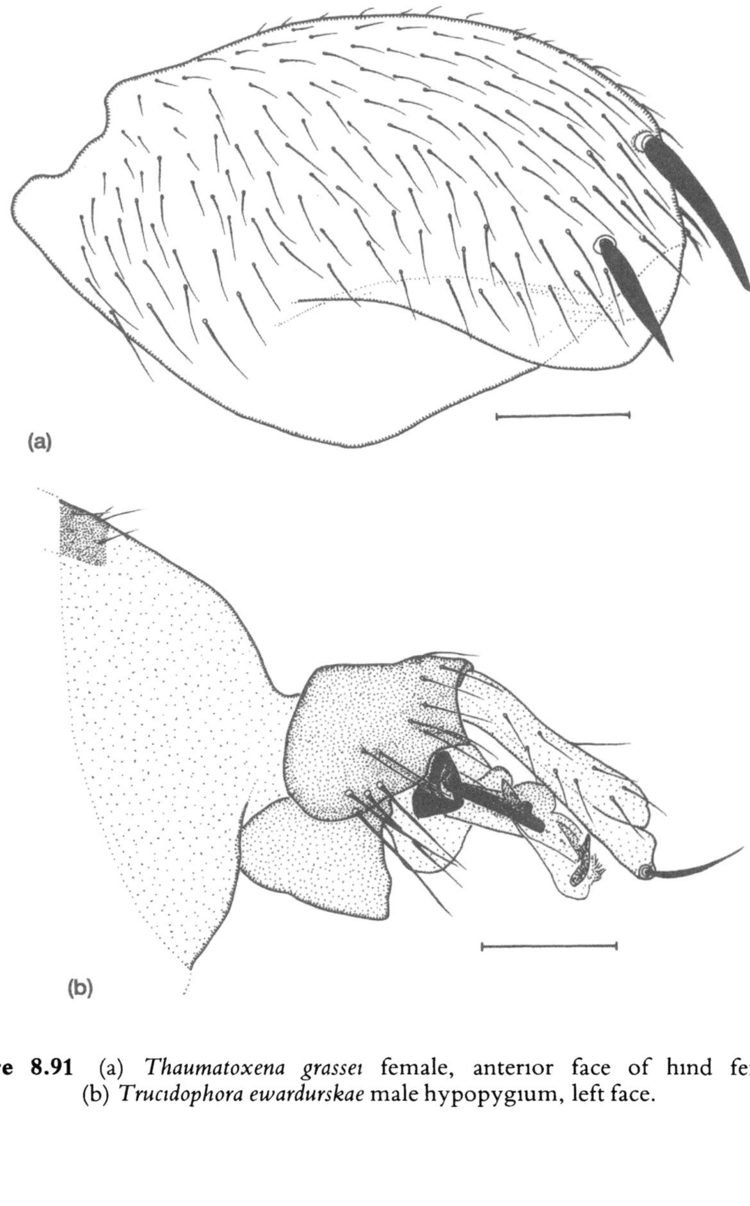

Figure 8.91 (a) *Thaumatoxena grassei* female, anterior face of hind femur. (b) *Trucidophora ewardurskae* male hypopygium, left face.

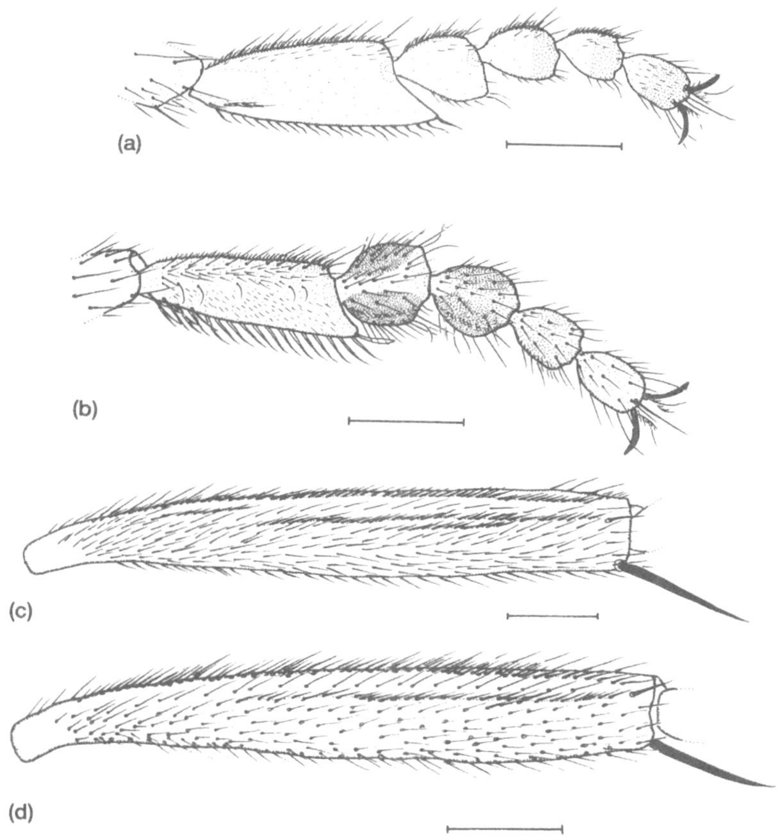

Figure 8.92 *Woodiphora*. (a) *W. distans* male, front tarsus. (b) *W. dormanae* male front tarsus. (c) *W. multinevra* female, anterior face of hind tibia. (d) *W. parvula* female, hind tibia.

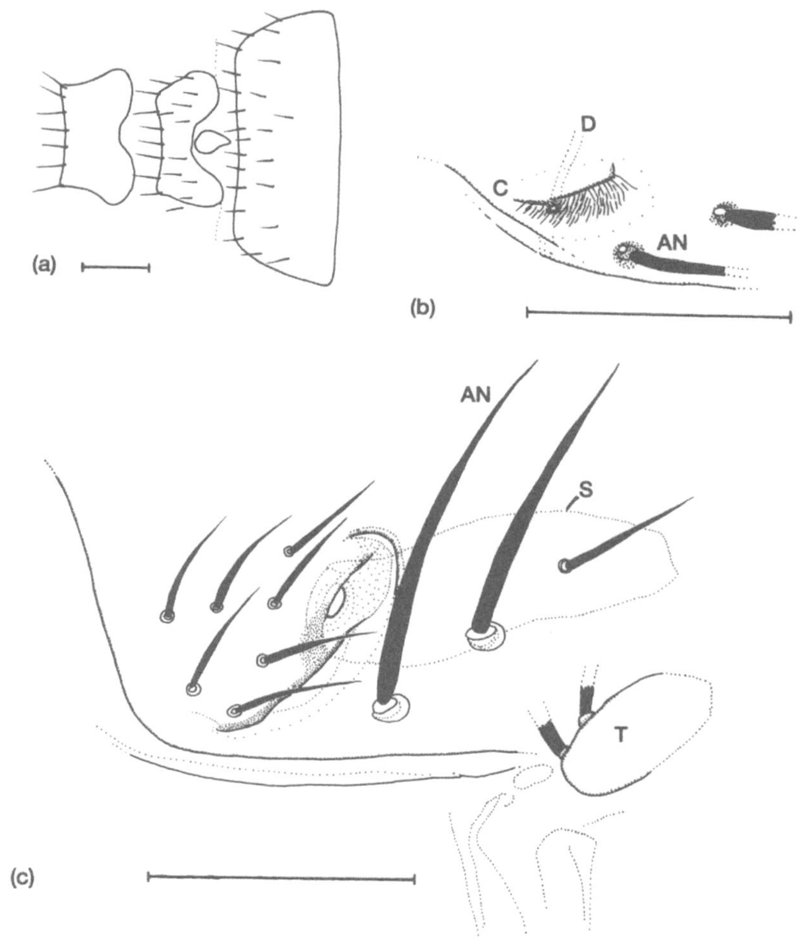

Figure 8.93 *Woodıphora* (a) *W pallıdınervıs* female, abdomınal tergıtes 5–7 (b) *W santoshı* male, left notopleuron (c) *W afrotropıca* male, left notopleuron (C = notopleural cleft, D = duct from notopleuron gland, AN = anterıor notopleural brıstle, S = sac-lıke notopleuron gland plus duct, T = tegula at base of wıng)

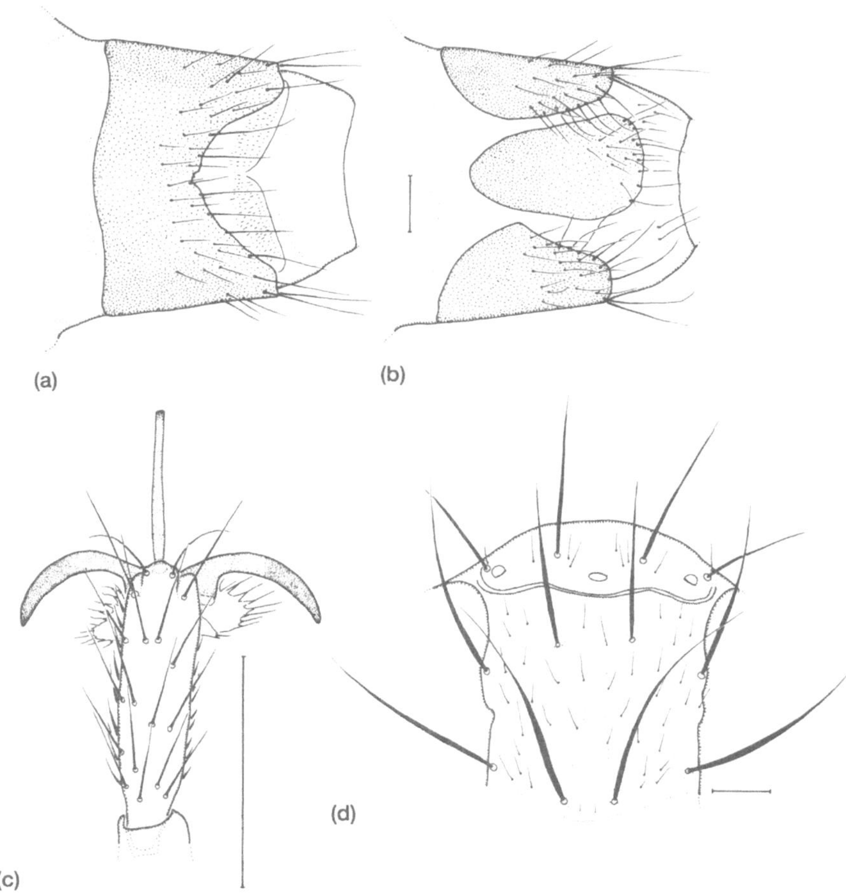

Figure 8.94 (a)–(c) *Spiniphora bergenstammi* female (a)–(b) dorsal and ventral views of abdominal segment 7, (c) last tarsal segment of hind leg (d) *Stichillus* sp (undescribed) male, frons

Figure 8.95 *Sciadocera rufomaculata* male. (After Oldroyd, 1964.)

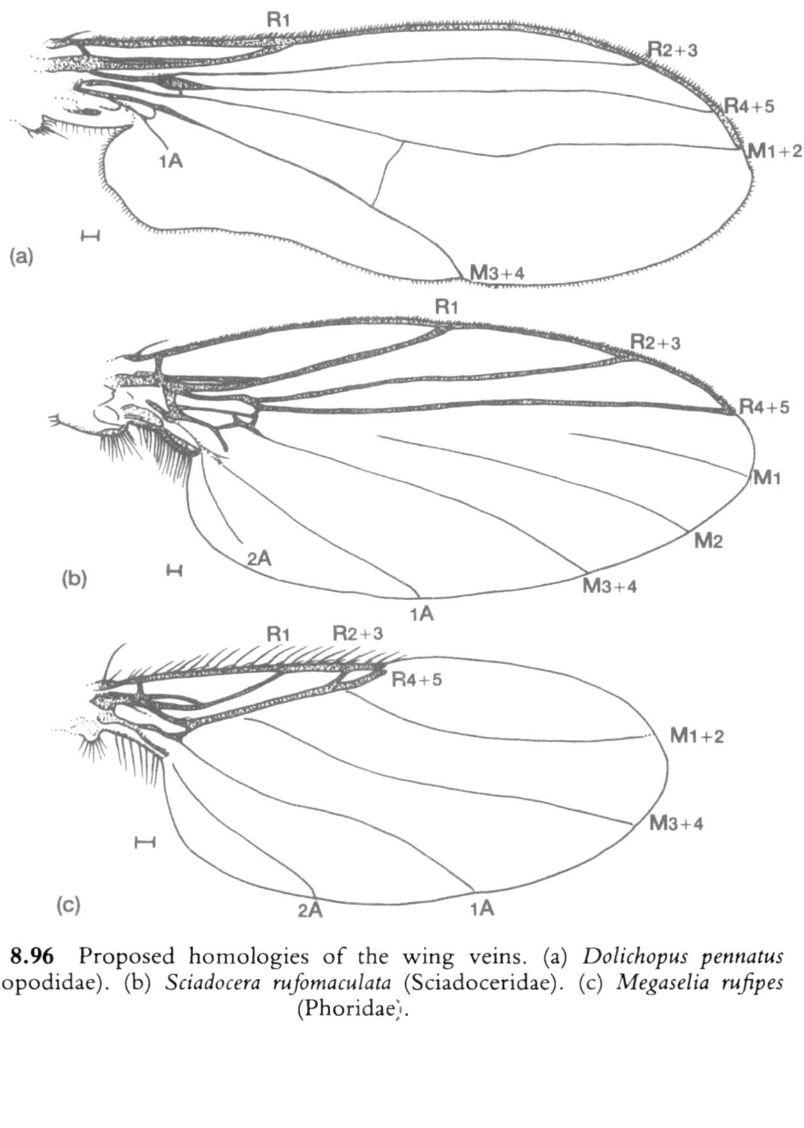

Figure 8.96 Proposed homologies of the wing veins. (a) *Dolichopus pennatus* (Dolichopodidae). (b) *Sciadocera rufomaculata* (Sciadoceridae). (c) *Megaselia rufipes* (Phoridae).

9

Methods

All scientific work is easier with appropriate instrumental aids; some kinds of scientific work absolutely depend on particular instrumental aids. But science itself is an attitude, a method of thinking, not an accumulation of gadgets (Bates, 1950).

9.1 INTRODUCTION

This chapter does not aim to provide a comprehensive coverage of methods for the study of scuttle flies. Numerous books on ecological methods are now available. In any research programme it is problems that are the primary focus of attention. Methods employed by others for other studies may or may not be suitable. Particular problems frequently require novel methods or adaptations of those used before in other studies. My aim, in the notes below, is to draw attention to some methods that have been employed in the study of Phoridae. Hopefully these will at least provide a point for departure. A section on the methods used to control pest species is also included.

9.2 COLLECTING AND SAMPLING

Any collecting technique for insects is likely to turn up Phoridae. Fine-mesh insect nets, especially sweep nets, are effective. A vacuum aspirator has been employed with success (e.g. Brown and Marshall, 1984). I have done much collecting with a simple pooter (mouth-operated aspirator). This allows the securing of individuals observed behaving in an interesting manner. For example, I have used a pooter to collect flies visiting flowers (6.2), leaf-searching (6.3), parasitoids running up to their hosts (e.g. 6.5, 3.5) or those at windows. The most convenient design, first developed for collecting Dixidae (Disney, 1975) but now widely used in field work (e.g. Chalmers and Parker, 1989), is shown in Fig. 9.1. It is made from two lengths of polythene tubing, the mouthpiece section being of a smaller diameter so that one end will fit tightly into the capture section after first being covered with a small square of muslin. Captured phorids are blown straight into a tube of 70% alcohol, unless there is a reason to keep the fly alive. When not in use the pooter's mouthpiece can be inserted into the open end of the capture section,

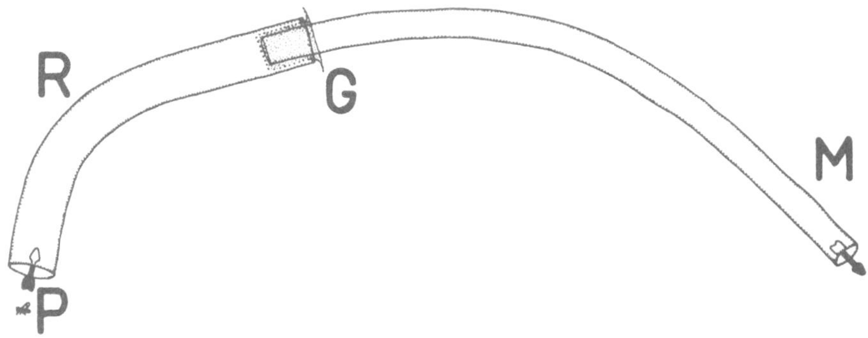

Figure 9.1 Pooter made from two lengths of clear flexible plastic (polythene) tubing of different diameters, separated by a square of cotton gauze. P = phorid fly about to be aspirated into receptacle section (R); G = cotton gauze; M = mouthpiece section. Arrows indicate direction of air flow during aspiration. The flow is reversed to deliver the fly into tube of alcohol.

and the pooter can then be easily carried in the pocket without getting dirt in it. The pooter is cleaned by dismantling it and pushing a piece of tissue through each section with a straight length of thick wire (such as commonly used for coathangers). With repeated use condensation in the capture section may become a problem. I have found it convenient to use a succession of pooters when engaged in prolonged fieldwork and to save the cleaning of them all until afterwards.

When collecting phorids off carrion, dung or other unhygienic situations, it is advisable to use a blow pooter. A robust design is described by Whitfield (1987).

For quantitative studies replicated traps are useful. Some of these are reviewed below.

9.2.1 EMERGENCE TRAPS

There are now many designs of emergence trap for soil dwelling insects, most of which have proved useful for sampling Phoridae. Szadziewska (1977) constructed one of the simplest traps using inverted washing up bowls. An exit hole in the bottom had a fitted tube leading into a collecting pot. Cheng and Hogue (1974) describe an emergence trap for use over intertidal muds in mangrove swamps, in which they obtained the first records of an intertidal phorid (see Chapter 7).

Büchs (1988) describes an emergence trap for bark-breeding insects, successful with Phoridae, which is attached to a branch or the trunk of a tree.

9.2.2 FLIGHT TRAPS

Comparisons of different types of trap have rarely been attempted. Some data indicate that white water traps tend to be effective in catching numbers

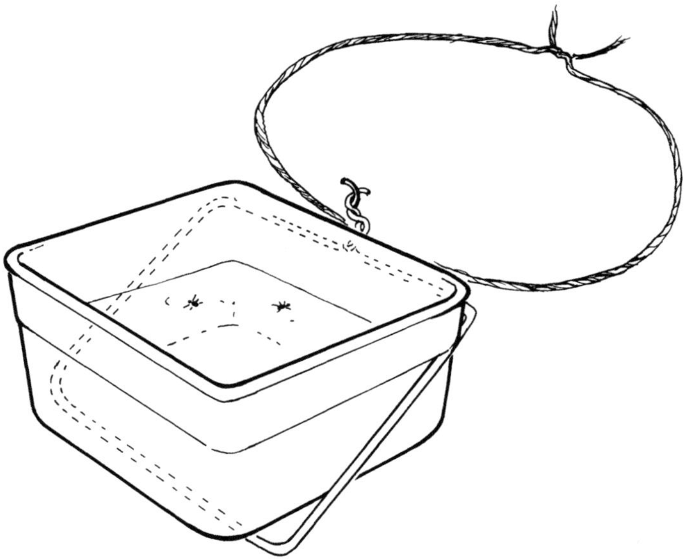

Figure 9.2 A method of attachıng a water trap to a tree trunk.

of Phoridae with minimum effort. Malaise traps are also effective but it takes time to sort the Phoridae from the rest of the catch. They give a good return for the effort but exhibit a high variance between adjacent traps. For this reason they are more useful as a collecting device than as a sampling device. Yellow water traps are less effective than white water traps, and pitfall traps are the least effective (Disney *et al.*, 1982). Each sort of trap, however, tends to select certain species. In experimental comparisons, *Triphleba citreiformis* was readily caught in white water traps, was not caught in yellow traps and only rarely in a Malaise trap. In contrast *Megaselia tumida* was mainly caught in yellow traps, more rarely in white traps and not in the Malaise trap. Some workers have favoured yellow traps (e.g. Vincent, 1975; Maruszewska, 1976). Durska (1981) sampled Phoridae in the crowns of trees using yellow water traps, as employed by Moericke (1951) for other insects. Kirk (1984) has pointed out that white traps can be divided into white and 'black' in ultra violet light. It is the truly white traps that are the most effective for Phoridae. If in doubt a 'white' bowl should be viewed in ultra violet light.

A water trap can be set on the ground, raised so that its rim is just above the top of the herb layer, be attached to a tree trunk (Figs 9.2, 9.3) (Disney, 1987o) or otherwise set above the ground.

Nelson (1965) used 10 × 10 cm sheets of celluloid coated with banding gum, some sheets being mounted on floating wooden rafts on a stream. He removed the flies, which included Phoridae, by washing the sheets in petrol. Sticky traps, using polybutenes spread on the surface of cellulose acetate sheets or polythene bags, have been employed for trapping *Megaselia halterata* (Hussey and Gurney, 1963; Binns *et al.*, 1979, Binns, 1980). Hussey (1965) placed his traps 0.6 m above ground level.

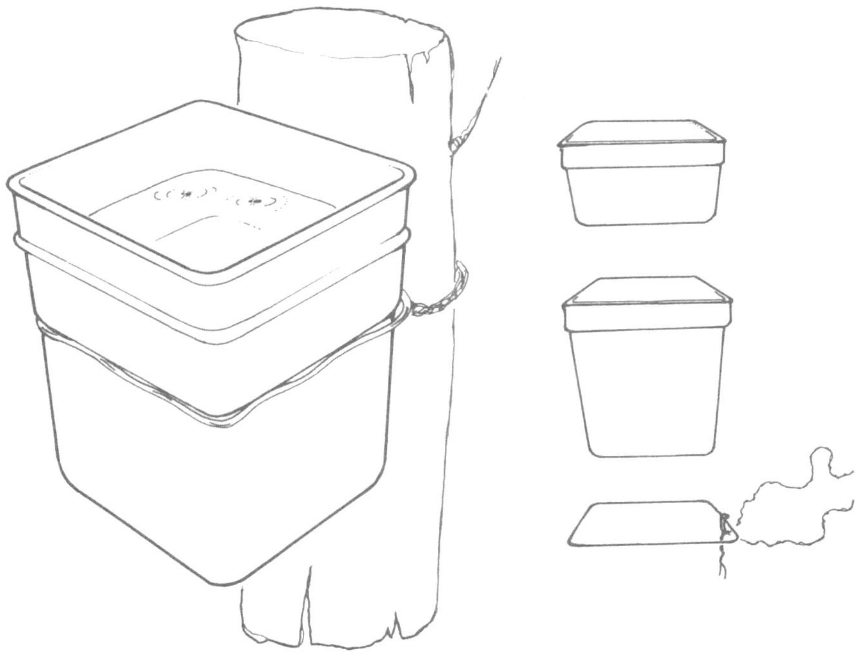

Figure 9.3 Another method of attaching a water trap to a tree trunk.

9.2.3 LIGHT TRAPS

Light traps have been employed for sampling Phoridae (e.g. Baumann, 1976b, 1977a; Brown and Marshall, 1984). The indications are that these are more effective in the tropics. Hardy and Beyer (1964) recorded large numbers of *Chonocephalus* attracted to lights and in light traps.

9.2.4 PITFALL TRAPS

Pitfall traps have been used by a number of workers (e.g. Baumann, 1977b; Disney *et al.*, 1981). Brown and Marshall (1984) baited their traps with decaying fungus (*Agaricus bisporus*). However, pitfall traps are highly selective. Out of more than 7000 specimens of Phoridae collected in sets of pitfall traps, at 50 sites in Northern England, more than 60% belonged to one species, *Megaselia longicostalis*. The mean number of species was 10.4 (range 1–18) per site (Disney *et al.*, 1981). By contrast two small, white water traps set in a similar site in the same region produced 21 species, and *M. longicostalis* represented only 1.1% of the specimens caught (Disney, 1979d). However, in mopane woodland in Zimbabwe Phoridae were second only to Cecidomyiidae among Diptera caught in pitfalls. While the returns for Diptera were poor, the Phoridae were well represented by males and flying-stage females of the Termitoxeniinae (Tingle *et al.*, 1992).

Barber (1931) employed a bait of rotting minced meat and cheese in a vial held in the mouth of a pitfall trap, so that insects fell in before reaching the bait. When employed in caves this type of baited pitfall caught large numbers of Phoridae.

9.2.5 OTHER TRAPS AND BAITS

In studies of Diptera in vole burrows Baumann (1977b) employed funnel traps and Hackman (1963) employed a range of baits. Phoridae were trapped at rotting apples, rotting fungi, rodent droppings, cheese, meat and carrion.

Feener (1987) employed tuna fish baits to attract fire ants and he observed their recruitment trails for the presence of parasitoid phorids hovering above the ants.

Beavers *et al.* (1972) used sticky traps plus a chemical lure for trapping Diptera. They reported that 2,4-hexadienyl butyrate is effective in attracting Phoridae. Traps baited with 10% 1-phenylethanol have proved to be a strong attractant for some *Megaselia* species, especially *M. pleuralis* (Kamm *et al.*, 1987).

Young *et al.* (1987b) compared sticky traps and water-filled McPhail traps baited with steam-distilled floral oils of various Sterculiaceae and controls. The baited traps attracted *Dohrniphora* and *Megaselia* species.

9.2.6 EXTRACTING LARVAE FROM SOIL

So far the only method developed for extracting phorid larvae from soil samples, which has proved at least partly successful, is a combined sieving and flotation procedure described by Heynen (1990).

9.3 REARING

Rearing larvae collected in the field, or from eggs laid by a gravid female, is a way of obtaining novel biological data or valuable series for taxonomic studies. For taxonomic purposes, the isolation of specimens in individual rearing tubes is recommended. Likewise the isolation of suspect host organisms in individual tubes is a useful way of establishing the hosts of parasitoid species. Rettenmeyer and Akre (1968) used rearing tubes one-third filled with plaster of paris plus charcoal, kept moist with distilled water, and capped with nylon netting.

A major problem is to avoid contamination of rearing tubes by roving females of polyphagous saprophage species, who can lay eggs through the holes in even quite fine gauze. Indeed Blickle (1964) reported a gravid female *Megaselia* (possibly *M. scalaris*) entering a vial (with horsemeat) through a double layer of nylon netting. Likewise Garris (1983) reported the invasion of laboratory colonies of ticks by *Megaselia scalaris*, noting that the presence of a double layer of organdie cloth did not prevent infestation. Furthermore Roberg (1915) reported this species escaping from his containers through eight layers of stretched gauze (7.11.2). The capping of rearing tubes with tall narrow polythene bags is to be recommended (e.g. Rocha *et al.*, 1984) or the placing of the individual tubes in a larger closed box (e.g. Disney, 1979f). The latter procedure allows the maintenance of high humidity in the box without the need for too much moisture in the individual tubes. An excess of water in the tubes tends to encourage fungal contamination.

The pre-adult Termitoxeniinae most easily collected in the field are the long-lived eggs. They are very liable to become mouldy if not regularly groomed by worker termites. Franssen (1933b) handled them with a fine brush, placing them on fresh potato slices daily. However, mould was more readily kept at bay by placing them between layers of moist filter paper along with termite workers to groom them. Fresh termite workers, however, need to be provided at regular intervals.

9.4 CULTURING IN THE LABORATORY

The saprophagous species are the phorids most easily cultured in the laboratory. Austin and Pitcher (1936) described a simple culturing procedure and emphasized the importance of maintaining a high humidity. Prawirodisastro and Benjamin (1979) give details of a successful larval diet and general culturing procedure for *Megaselia scalaris.* Benner and Ostermeyer (1980) describe a yeast, corn meal, sorghum syrup, *Drosophila* agar medium used for culturing this species. Barnes (1990a) describes a procedure for culturing *Dohrniphora cornuta.* Blickle (1964) reared a *Megaselia* species (possibly *M. scalaris*) using fresh horsemeat in 250 cm^3 flasks and with paper-towel strips for pupariating. The strips with the pupae were then transferred to another flask, to which fresh meat was added when the flies emerged. In this way the phorids were cultured for several generations. *Megaselia abdita* has been readily cultured on the protein-rich flakes sold as food for aquarium fish (Klaus Sander, *in litt.*).

Richardson and Hesling (1977a) should be consulted for successful procedures used to culture the mushroom pest *Megaselia halterata.*

9.5 MARKING FLIES FOR POPULATION STUDIES

Megaselia halterata have been successfully marked by feeding them with sugar solution containing a fluorescent dye (Hussey and Gurney, 1964b).

9.6 CONTROL OF PEST SPECIES

Most work on the control of phorids has been directed at the mushroom pests.

Megaselia nigra used to be a significant pest of cultivated mushrooms in England, until it was discovered that it will oviposit only in the light. The simple expedient of blacking out the doors and windows of mushroom-growing houses has largely controlled this species (Binns, 1978). Fletcher *et al.* (1989) review control methods for this species.

Megaselia halterata has been subjected to a range of control techniques. Apart from proper management procedures, the majority of mushroom growers have tended to employ chemical pesticides when control measures have been called for. These have been reviewed by a number of authors (e.g. Hussey, 1965; Spencer *et al.*, 1977; van Zaayen, 1983; Cantelo and McDaniel, 1978; Binns *et al.*, 1979; White, 1979, 1983). Non-technical and down to earth summaries of control procedures are given by van de Geijn (1983) and Fletcher *et al.* (1989). Also MAFF Leaflet 583 provides a summary of mushroom pests and their control (see Gratwick, 1992). The biological control of this species looks increasingly promising, using Nematodes (e.g. Rinker and Snetsinger, 1984). *Bacillus thuringiensis* has also been tried (Hussey and Gurney, 1964b).

Sandhu and Bhattal (1989) review the chemical control of *Megaselia agarici*. In addition Sandhu and Arora (1990) have successfully controlled this species using nylon nets (mesh sizes 35 or 40) to screen houses against the entry of flies. They also evaluate different insecticides and modes of application.

The control of *Megaselia rufipes*, causing a nuisance by emerging in numbers from sewage filters, has been achieved by slowing the distributor arms for 1.5 h a day. This produced a scouring action sufficient to upset the phorids (Painter, 1980). The same author detailed chemical controls found to be effective.

9.7 SLIDE MOUNTING PHORIDAE

For most Phoridae confident identification will require the slide-mounting of specimens. In the past phorids have been mounted on micropins. Some workers still use this method. It has, however, been responsible for much misidentification of specimens, not only with small flies like *Metopina* but also with larger flies in difficult genera like *Megaselia*. Others freeze dry the flies and then attach them to the side of a pin with a spot of gum (e.g. Brown, 1992b). This is a considerable improvement, but there is no doubt that far more detail can be seen on slide-mounted specimens. A further consideration is that for any quantitative ecological studies it is far easier to be able to store samples straight away in 70% alcohol, and to subsequently slide mount specimens as necessary, than to have to pin-mount all the specimens at the time of collection. With several thousand specimens such a task becomes irksome. An alternative to preservation in alcohol is to freeze the specimens, between layers of tissue, until they are subsequently mounted on pins or slides.

For slide-mounting I have found the most satisfactory mounting medium to be Berlese Fluid. This can be purchased from some commercial suppliers. One should specify the formula when ordering from one of these suppliers, as some unsatisfactory variants have been appearing on the market. I use the following formula, which has been found to be the best, in terms of its clearing properties and durability without discolouration (Lewis, 1973; Disney, 1983a):

Best quality gum arabic ('picked lumps') 12 g
Chloral hydrate crystals 20 g
Glacial acetic acid 5 ml
50% w/w glucose syrup 5 ml
Distilled water 30–40 ml

To make up the mixture the gum arabic is enclosed in a cotton bag and immersed in the water until dissolved. Impurities are retained by the cotton bag. The other ingredients are then added to the solution in the order given. A drop of the mixture is tested by allowing it to dry on a slide. If it does not dry to an amber-like texture, but is somewhat granular and whitish opaque, then the mixture contains too much glucose syrup for the particular batch of gum arabic used. Dilute the mixture with the same quantity of another mixture made up without the glucose syrup and test another drop. If necessary repeat this process until the tested drop dries to an amber-like texture.

Upton (1993) has reviewed the gum chloral slide-mounting media and concluded that their durability is suspect. An alternative medium is Euparal. However, some internal structures, such as the Dufour's crop mechanism of many female phorids, are better observed in specimens mounted in Berlese Fluid. It is perhaps best, therefore, to mount most specimens in Euparal but, when several duplicates are available, to mount some in Berlese Fluid. When using Euparal, it will be found that the drop in which the dissection of the fly is carried out tends to become tacky before the operation is complete. This can be overcome by using Euparal diluted with Euparal Essence for this initial drop of mountant, the detached parts then being transferred to drops of plain Euparal in readiness for placement of the coverslips (see below).

To mount a specimen it is first removed from the alcohol. The fluid is removed by briefly placing it against a piece of tissue paper. This specimen is then placed in a drop of Berlese Fluid or Euparal on a slide. Under the dissecting microscope the fly is dissected as far as is necessary.

I have adopted the following conventions when making a standard slide mount. Using a pair of fine (watchmaker's) forceps and a mounted needle the right wing, right haltere, and all the legs (severed as near the base of the coxae as possible) are detached. With a finer mounted needle the head is then removed. The parts are then displayed under separate coverslips as in Fig. 9.4, with the top (front) of the head being uppermost. In the case of some females I also detach the abdomen and display its dorsal face under a separate coverslip. In the case of *Triphleba* females it is the ventral face that is displayed uppermost and likewise with the males of *Metopina*, whose ventral face of the abdomen is taxonomically important. In *Dohrniphora* males the hind legs need detaching with care as the hind coxal lobes are important taxonomically. The coverslips used are 10 mm diameter, but the haltere is covered with a 6 mm diameter coverslip or half a 10 mm one. One label is the

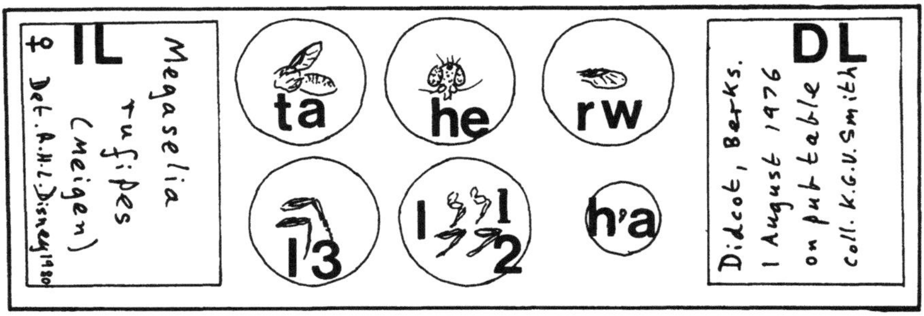

Figure 9.4 The standard layout for a fully dissected slide-mounted phorid. IL = identification label; ta = thorax and abdomen; he = head; rw = right wing; l3 = hind legs; DL = data label; l1 and 2 = front and middle legs; ha = right haltere.

data label and the other is the identification label. Procedures for remounting old pinned specimens on slides are given by Disney (1983a).

For permanent storage slide mounts in Berlese Fluid are dried on a hotplate and then the coverslips ringed with a modern tough nail varnish. It is recommended that slide mounts are stored horizontally.

References

Note For authors who have published more than one paper in a single year these are distinguished by lower-case letters after the date However, these may not form a complete sequence, as not all an author's papers published in a year may have been cited in the text In order to facilitate cross-reference to other published bibliographies, especially that of Borgmeier (1968b), I have refrained from re-lettering these incomplete listings

Akre RD, Rettenmeyer CW (1968) Trail-following by guests of army ants (Hymenoptera Formicidae Ecitonini) *Journal of the Kansas Entomological Society* **41,** 165–74

Aldrich JM (1892) New species of *Phora* *Canadian Entomologist* **24,** 142–6

Alford DV (1975) *Bumblebees* London Davis-Poynter

Altmuller R (1976) Okoenergetische Untersuchungen an Dipterenpopulationen im Buchenwald (*Luzulo-Fagetum*) *Verhandlungen der Gesellschaft für Okologie (Gottingen)* **5,** 133–8

Altmuller R (1979) Untersuchungen uber den Energieumsatz von Dipterenpopulationen im Buchenwald (*Luzulo-Fagetum*) *Pedobiologia* **19,** 245–78

Amerling C (1862) Noch etwas uber die Schmarotzer der Bienen *Lotos* **12,** 239–40

Annandale N (1912) Description of a micropterous fly of the family Phoridae associated with ants *Spolia Zeylanica, Colombo* **8,** 85–9

Araujo RL (1977) *Catálogo dos Isoptera do Nuovo Mundo* Rio de Janeiro Academia Brasileira de Ciências

Ardo P (1953) Likflugan, *Conicera tibialis,* i Sverige (Dipt , Phoridae) *Opuscula Entomologica* **18,** 33–6

Armstrong EA (1953) Nidicoles and parasites of the wren *Irish Naturalists' Journal* **1,** 57–64

Ashraf M, Berryman AA (1969) Biology of *Scolytus ventralis* (Coleoptera Scolytidae) attacking *Abies grandis* in Northern Idaho *Melanderia* **2,** 1–23

Assmuss E (1860) Muthmassliche Entstehungsweise der Faulbrut *Bienen-zeitung, Eichstaedt* **16,** 11–12

Assmuss E (1865) *Die Parasiten der Honigbiene und die durch dieselben bedingten Krankheiten dieses Insects* Berlin

Assmuth J (1913) *Termitoxenia assmuthi* Wasm , Anatomisch- histologische Untersuchung *Nova Acta Academiae Caesarea Leopoldino-Carolinae Germanicum Naturae Curiosorum* **98,** 191–316

Aston A (1957) The Diptera of Suffolk Family XXVI Phoridae Hump-backed flies *Transactions of the Suffolk Naturalists' Society* **10,** 223–7

Austen EE (1910) Some Dipterous insects which cause myiasis in man *Transactions of the Society of Tropical Medicine and Hygiene* **3,** 215–42

REFERENCES

Austin MD, Pitcher RS (1936) A laboratory method for rearing *Sciara* and phorid flies *Entomologist's Monthly Magazine* **72,** 12–15

Auten M (1925) Insects associated with spider nests *Annals of the Entomological Society of America* **18,** 240–50

Autuori M (1928) *Syneura infraposita* Borgm -Schmitz, um novo parasita de *Icerya purchasi* Mask *Archivos do Instituto biologico Sao Paulo* **1,** 193–200

Ax P (1987) *The Phylogenetic System* Chichester Wiley

Azevedo e Silva F, Serrão Nogueira CD (1965) Notas sobre a entomofauna florestal Portuguesa-Lisbon *Archivos do Museu Bocage* **1,** 57–66

Bachli G (1969) Beitrig zur Kenntnis der Uberwinterungsstatten von Insekten im Wald *Vierteljahrsschrift der Naturforschenden Gesellschaft in Zurich* **114,** 455–60

Bachli G (1970) Der Waldboden als Entwicklungssubstrat von Dipteren *Vierteljahrsschrift der Naturforschenden Gesellschaft in Zurich* **115,** 255–9

Baer JG (1953) Notes de faunistique eburneene III Contribution a l'etude morphologique et biologique de *Wandolleckia achatinae* Cook, Phoride (Diptera) commensal d'Achatines de la forêt tropicale *Acta Tropica, Basel* **10,** 73–9

Bahrmann R (1984) Die Zweiflugler der Kraut- und Strauchschicht des Leutratals bei Jena (Thuringen) – ein okofaunistischer Vergleich *Zoologische Jahrbucher Abteilung fur Systematik* **111,** 175–217

Bahrmann R (1987) Untersuchungen der Dipterenfauna in natur- und industrienahen Rasenbiotopen Thuringens (DDR) mittels Bodenfallen (Diptera Brachycera) *Deutsche entomologische Zeitschrift* **34,** 85–106

Baker EW, Wharton GW (1952) *An Introduction to Acarology* New York Macmillan

Baker GH (1985) Parasites of the millipede *Ommatoiulus moreletii* (Lucas) (Diplopoda Iulidae) in Portugal, and their potential as biological control agents in Australia *Australian Journal of Zoology* **33,** 23–32

Baker R, Parton AH, Rao, VB, Rao VJ (1982) The isolation, identification and synthesis of 3,6–dimethylheptan-2, 4–dione, a pheromone of the mushroom fly, *Megaselia halterata* (Diptera Phoridae) *Tetrahedron Letters* **23(30),** 3103–4

Balduf WV (1928) Notes on the habits of *Aphiochaeta aletiae* *Ohio Journal of Science* **28,** 237–45

Banks N (1911) A curious habit of one of our phorid flies *Proceedings of the Entomological Society of Washington* **13,** 212–14

Barber HS (1931) Traps for cave-inhabiting insects *Journal of the Elisha Mitchell Scientific Society* **1931,** 259–66

Barnes JK (1990a) Life history of *Dohrniphora cornuta* (Bigot) (Diptera Phoridae), a filth-inhabiting humpbacked fly *Journal of the New York Entomological Society* **98,** 474–83

Barnes JK (1990b) First record of genus *Phymatopterella* in the Nearctic Region and description of *P ovatimacula*, a new humpbacked fly from Florida (Diptera Phoridae) *Florida Entomologist* **73,** 644–9

Barnes JK (1990c) Biology and immature stages of *Sciomyza varia* (Diptera Sciomyzidae), a specialised parasitoid of snails *Annals of the Entomological Society of America* **83,** 925–38

Barnes JK (1991a) Additions to the phorid fauna (Diptera Phoridae) of North America north of Mexico *Florida Entomologist* **74,** 305–10

Barnes JK (1991b) Taxonomic status of the Neartic species of *Auxanommatidia* Borgmeier (Diptera Phoridae) *Proceedings of the Entomological Society of Washington* **93**, 930–2

Bates M (1950) *The Nature of Natural History* New York Scribner (Reprinted 1990, Princeton University Press, New Jersey)

Batra SWT (1965) Organisms associated with *Lasioglossum zephyrum* (Hymenoptera Halictidae) *Journal of the Kansas Entomological Society* **38,** 367–89

REFERENCES

Batra SWT, Schuster JC (1977) Nests of *Centris, Melissodes*, and *Colletes* in Guatemala (Hymenoptera Apoidea) *Biotropica* **9,** 135–8

Baumann E (1976a) Literaturnachtrag zu Borgmeier's 'Catalogue of the Phoridae of the world' (1968) und 'Supplement' (1971) (Diptera Phoridae) *Studia Entomologica, Petropolis* **19,** 375–407

Baumann E (1976b) Rennfliegen aus den Rheinauenwalden des Naturschutzgebietes, Hordter Rheinaue i Phorinae (Diptera, Phoridae) *Mitteilungen der Pollichia des Pfälzischen Vereins für Naturkunde und Naturschutz* **64,** 188–93

Baumann E (1977a) Buckelfliegen aus Lichtfangen im Naturpark Hoher Vogelsberg (Diptera Phoridae) *Entomologische Zeitschrift* **87,** 27–33

Baumann E (1977b) Untersuchungen uber die Dipterenfauna subterraner Gangsysteme und Nester von Wuhlmausen (*Microtus, Chlethrionomys*) auf Wiesen der montanen Region im Naturpark Hoher Vogelsberg *Zoologisches Jahrbucher, Abteilung für Systematik* **104,** 368–414

Baumann E (1978a) Rennfliegen (Diptera Phoridae) als Blutenbesucher Kritische Sichtung der Literatur *Flora, Jena* **167,** 301–14

Baumann E (1978b) Phorinae (Phoridae/Diptera Rennfliegen) als Blutenbesucher auf Doldengewachsen (Umbelliferae) *Plant Systematics and Evolution* **130,** 43–52

Baumann E (1979) Rennfliegen aus den Auenwalden des Natursschutzgebietes 'Hordter Rheinaue' II Die Gattung *Gymnophora* mit Anmerkungen zur Systematik und Biologie (Diptera Phoridae) *Mitteilungen der Pollichia des Pfälzischen Vereins für Naturkunde und Naturschutz* **64,** 184–93

Beaver RA (1972) Ecological studies on Diptera breeding in dead snails 1 Biology of the species found in *Cepaea nemoralis* (L) *Entomologist* **105,** 41–52

Beaver RA (1979) Biological studies of the fauna of pitcher plants (*Nepenthes*) in West Malaysia *Annales de la Societé Entomologique de France* **15,** 3–17

Beaver RA (1983) The communities living in *Nepenthes* pitcher plants fauna and food webs In Frank, J H and Lounibos, L P (eds), *Phytotelmata Terrestrial Plants as Hosts for Aquatic Insect Communities* Medford, USA Plexus Publishing Inc , 129–59

Beaver RA (1987) Biological studies of non-muscoid flies (Diptera) bred from mollusc carrion in Southeast Asia *Japanese Journal of Sanitary Zoology* **38,** 187–95

Beavers JB, McGovern TP, Beroza M, Sutton RA (1972) Synthetic attractants for some Diptera species *Journal of Economic Entomology* **65,** 1740–1

Becker T (1901) Die Phoriden *Abhandlungen der zoologisch-botanischen Gesellschaft in Wien* **1,** 1–100

Benner DB (1985) Oocyte development and fecundity in *Megaselia scalaris* (Phoridae Diptera) *International Journal of Entomology* **27,** 280–8

Benner DB (1991) The mechanism of sperm transfer in *Megaselia scalaris* (Diptera Phoridae) *Giornale Italiano di Entomologia* **5,** 351–8

Benner DB, Curtis SK (1988) Internal reproductive organs of the female humpbacked fly *Megaselia scalaris* Loew (Diptera Phoridae) *International Journal of Insect Morphology and Embryology* **17,** 197–205

Benner DB, Ostermeyer EC (1980) Some observations on the life history of the fly *Megaselia scalaris* Loew (Phoridae) with special reference to the eclosion pattern *Journal of the Tennessee Academy of Sciences* **55,** 103–5

Bequaert JC (1926) Medical Report of the Hamilton Rice Seventh Expedition to the Amazon, in Conjunction with the Department of Tropical Medicine of Harvard University XVIII A Dipterous parasite of a snail from Brazil, with an account of the arthropod enemies of mollusks *Contributions from the Harvard Institute for Tropical Biology and Medicine* **4,** 292–303

Berland L (1933) Sur le parasitisme des Phorides (Diptères) *Bulletin de la Société Zoologique de France* **57,** 529–30 (1932)

Berry PA, Parker HL (1949) Investigations on a South American *Epilachna* sp and the importation of its parasite *Lydinolydella metallica* Tns into the United States (Coleoptera Coccinellidae, Diptera Larvaevoridae) *Proceedings of the Entomological Society of Washington* **51,** 93–103

Beyer E (1958a) Die ersten Phoriden von Burma (Dipt Phor) *Commentationes Biologicae, Helsingfors* **18,** 3–72

Beyer EM (1958b) Drei neue Phoriden aus Japan und Finnland *Notulae Entomologicae* **38,** 104–8

Beyer E (1959a) Neue Phoriden aus Tanganyika *Stuttgarter Beitrage zur Naturkunde aus dem Staatlichen Museum fur Naturkunde in Stuttgart* **21,** 1–11

Beyer E (1959d) Neue Phoridengattungen und -Arten aus Angola (Phoridae, Diptera) *Publicacoes Culturais da Companhia de Diamantes de Angola (Museu do Dundo Estudos Diversos XVII)* **45,** 53–76

Beyer E (1959f) Zur naheren Kenntnis einiger als *Aphiochaeta* beschriebenen palaotropischen Phoriden (Diptera) *Brotéria* **28,** 105–8

Beyer EM (1959g) Diptera (Cyclorrhapha) Phoridae) *South African Animal Life* **6,** 376–89

Beyer EM (1959h) Gattung *Pericyclocera* Schmitz in Ostasien *Entomologische Zeitschrift, Stuttgart* **69,** 167–9

Beyer EM (1960a) XXX Diptera Phoridae Mission zoologique de l'I R S A C en Afrique orientale (1957) *Annales du Musée Royal du Congo Belge, Tervuren, Sciences Zoologiques* **81,** 412–32

Beyer EM (1960b) Neue und wenig bekannte Phoriden aus Suedafruika *Journal of the Entomological Society of Southern Africa* **23,** 390–402

Beyer EM (1960e) Wenig bekannte Phoriden aus Ost- und Sued-afrika *Stuttgarter Beitrage zur Naturkunde aus dem Staatlichen Museum fur Naturkunde in Stuttgart* **35,** 1–2

Beyer EM (1961) Eine neue Phoridengattung aus Neu-Guinea *Nova Guinea, Zoology* **11,** 215–17

Beyer EM (1963) Die erste *Auxanommatidia* aus Nordamerika *Entomologische Zeitschrift Frankfurt a M* **73,** 141–3

Beyer EM (1965b) Phoridae (Diptera Brachycera) *Exploration du Parc National Albert, Mission G F De Witte (1933–1935)* **99,** 1–211

Beyer EM (1965c) Ueber neue und wenig bekannte termitophile Phoriden aus Suedafrika *Journal of the Entomological Society of Southern Africa* **27,** 181–7

Beyer EM (1966) Neue und wenig bekannte Phoriden, zumeist aus dem Bishop Museum, Honolulu *Pacific Insects* **8,** 165–217

Beyer EM (1967) Diptera Phoridae *Insects of Micronesia* **13,** 329–60

Beyer E, Schmitz H (1957) Eine neue *Metopina* (Phoridae, Diptera) aus dem Rheinland *Decheniana* **109,** 213–17

Biegel W (1953) Zur Biologie und Okologie sozialer Waspen *Sitzungsberichte der physikalisch-medizinischen Sozietat zu Erlangen* **76,** 115–53

Biery TL, Clegern RW, Hart WW (1978) Two cases of phorid (Diptera Phoridae) myiasis in Texas *Journal of Medical Entomology* **15,** 122–3

Bigot JMF (1890) New species of Indian Diptera *Indian Museum Notes* **1,** 191–5

Bin F (1972) Ritrovamento Dell'Ospite E Decrizione del Maschio di *Basalys helicicola* (Kieff) (Hymenoptera, Proctotrupoidea, Diapriidae) *Entomologica, Piacenza* **8,** 35–44

Binns ES (1978) The Worthing phorid – a 25-year history *The Mushroom Journal* **66,** 1–7

Binns ES (1979) *Scutacarus bacultarsus* Mahunka (Acarina Scutacaridae) phoretic on the mushroom phorid fly *Megaselia halterata* (Wood) *Acarologia* **21,** 91–107

Binns ES (1980) Mating behaviour, fecundity and feeding in the mushroom phorid, *Megaselia halterata* (Wood) (Dipt) *Entomologist's Monthly Magazine* **116,** 45–57

Binns ES (1981) Physogastry in *Scutacarus baculitarsus* Mahunka (Acarina, Scutacaridae) *Entomologist's Monthly Magazine* **116,** 157–9 (1980)

Binns ES, Gurney B, Wyatt IJ, White PF (1979) Populations of the phorid fly *Megaselia halterata* on an experimental mushroom unit over four years *Annals of Applied Biology* **92,** 159–71

Blackman MW, Stage HH (1924) On the succession of insects living in the bark and wood of dying, dead and decaying hickory *Technical Publication No 17 of the New York State College for Forestry* **24(22),** 3–35

Blair KG (1922) Notes on the life history of *Rhizophagus parallelocollis* Gyll (Col) *Entomologist's Monthly Magazine* **58,** 80–3

Blanchard EE (1937) Dipteros Argentinos nuevos o poco conocidos *Revista de la Sociedad Entomologica Argentina* **9,** 35–58

Blickle RL (1964) Horsemeat as food for rearing Diptera (Phoridae and Tabanidae) *Florida Entomologist* **47,** 29–30

Bohart GE (1947) The phorid flies of Guam *Proceedings of the United States National Museum* **96,** 397–416

Bohart GE, Gressitt JL (1951) Filth-inhabiting flies of Guam *Bulletin of the Bernice P Bishop Museum* **204,** 1–152

Bolton B (1990) Army ants reassessed the phylogeny and classification of the doryline section (Hymenoptera, Formicidae) *Journal of Natural History* **24,** 1339–64

Bon M (1987) *The Mushrooms and Toadstools of Britain and North-western Europe* London Hodder & Stoughton

Boness M (1953) Die Fauna ser Wiesen unter besonderer Berucksichtigung der Maha (ein Beitrag zur Agrarokologie) *Zeitschrift für Morphologie und Okologie der Tiere* **42,** 225–77

Boness M (1958) Biocoenotische untersuchungen uber die Tierwelt von Klee- und Luzernefeldern *Zeitschrift für Morphologie und Okologie der Tiere* **47,** 309–73

Boness M (1975) Arthropoden in Hochwassergenist von Flussen *Bonner zoologisches Beitrage* **26,** 383–401

Bonnamour S (1921) Note sur deux Diptères parasites nouveaux de la Pieride de chou *Bulletin de la Societe Entomologique de France* **26,** 217–19

Borgmeier T (1922a) Zur Lebenweise von *Pseudacteon Borgmeieri* Schmitz (in litt) (Diptera – Phoridae) *Deutscher Verein für Wissenschaft und Kunst in São Paulo* **2,** 238–248 (1921)

Borgmeier T (1923a) Uma nova especie termitophila de *Dohrniphora* Dahl (Diptera – Phoridae) *Revista do Museu Paulista, São Paulo* **13,** 1215–24

Borgmeier T (1923c) Neue Phoridengattungen aus Brasilien *Zeitschrift deutscher Verein für Wissenschaft und Kunst in São Paulo* **3,** 155–81

Borgmeier T (1923d) Contribuição para o conhecimento dos Phorideos do Brasil (Phoridae – Diptera) *Archivos do Museu Nacional, Rio de Janeiro* **24,** 323–46

Borgmeier T (1923e) Novos Phorideos brasileiros *Boletim do Museu Nacional do Rio de Janeiro* **1,** 51–9

Borgmeier T (1924a) Novos generos e especies de Phorideos do Brasil *Boletim do Museu Nacional do Rio de Janeiro* **1,** 167–202

Borgmeier T (1924c) Um novo genero de Phorideos do Parana (Dipt) *Boletim do Museu Nacional do Rio de Janeiro* **1,** 283–8

Borgmeier T (1925a) Eine neue myrmecophile Phoridengattung aus Brasilien *Entomologische Mitteilungen* **14,** 237–9

Borgmeier T (1925b) Eine neue myrmecophile Phoride aus Brasilien *Societas entomologica, Stuttgart* **40,** 45

Borgmeier T (1925c) Novos subsidios para o conhecimento da familia Phoridae *Archivos do Museu Nacional, Rio de Janeiro* **25,** 85–281

Borgmeier T (1926a) Phorideos myrmecophilos da Argentina (Dipt) *Boletim do Museu Nacional do Rio de Janeiro* **2,** 1–31

Borgmeier T (1926c) Phorideos novos on pouco conhecidos do Brasil *Boletim do Museu Nacional do Rio de Janeiro* **2,** 39–52

Borgmeier T (1927a) Eine neue ecitophile Phoridgattung aus Brasilien (Dipt) *Entomologische Mitteilungen* **16,** 411–14

Borgmeier T (1927b) Dois generos novos de Phorideos *Boletim do Museu Nacional do Rio de Janeiro* **3,** 31–6

Borgmeier T (1927c) Eine neue *Pheidolomyia*-Arten aus Brasilien *Zoologischer Anzeiger* **74,** 160–4

Borgmeier T (1927d) Zwei neue myrmecophile Phoriden aus Brasilien (Dipt) *Eos, Madrid* **3,** 505–11

Borgmeier T (1928a) Zur Kenntnis der myrmecophilen Phoridengattung *Commoptera* Brues *Wiener entomologische Zeitung* **45,** 1–7

Borgmeier T (1928b) Nota prévia sobre alguns Phorideos que parasitam formigas cortadeiras dos generos *Atta* e *Acromymex* *Boletim Biologico Laboratorio de Parasitologia, São Paulo* **14,** 119–26

Borgmeier T (1928c) Investigações sobre Phorideos myrmecophilos (Diptera, Phoridae) *Archivos do Instituto Biologico de Defesa Agricola e Animal, São Paulo* **1,** 159–92

Borgmeier T (1929) Ueber attophle Phoriden *Zoologischer Anzeiger* **80,** 493–517

Borgmeier T (1930a) Eine neue myrmecophile *Apterophora*-Art *Zoologischer Anzeiger* **89,** 57–62

Borgmeier T (1930b) Zur Morphologie und Biologie von *Pseudohypocera nigrofascipes* Borgmeier and Schmitz *Zoologischer Anzeiger* **90,** 92–104

Borgmeier T (1931a) Sobre alguns Phorideos que parasitam a saúva e outras formigas cortadeiras (Diptera – Phoridae) *Archivos do Instituto Biologico de Defesa Agricola e Animal, São Paulo* **4,** 209–28

Borgmeier T (1931b) Especies arachnophiles e outras especies dos genero *Megaselia* Rondani (1856) e *Pericyclocera* Schmitz (1928) (Dipt , Phoridae) *Revista de Entomologia, São Paulo* **1,** 142–50

Borgmeier T (1932a) Sobre alguns Phorideos neotropicaes do Museu de Hamburgo (Dipt Phoridae) *Revista de Entomologia, São Paulo* **2,** 43–52

Borgmeier T (1932b) Revisão do genero ecitophilo *Xanionotum* Brues, com descripção de duas especies novas (Dipt Phoridae) *Revista de Entomologia, São Paulo* **2,** 369–80

Borgmeier T (1933b) Um novo genero myrmecophilo de Phorideos da Argentina (Dipt) *Revista de Entomologia, Rio de Janeiro* **3,** 45–8

Borgmeier T (1934a) Sobre alguns Forideos da Guiana Inglezu, colecionados pela expedição da Universidade de Oxford (1929) e conservados no Museu Britanico (Dipt Phoridae) *Archivos do Instituto de Biologia Vegetal, Rio de Janeiro* **1,** 27–34

Borgmeier T (1934b) O genero *Melaloncha* Brues (Dipt Phoridae) *Revista de Entomologia, São Paulo* **4,** 167–89

Borgmeier T (1935a) Um novo genero de Phorideos de Costa Rica *Revista de Entomologia, Rio de Janeiro* **5,** 136–8

Borgmeier T (1935d) Eine augenlose termitophile Phoride aus Brasilien nebst Beschreibung einer neuen *Pheidolomyia* (Dipt Phoridae) *Revista de Entomologia, Rio de Janeiro* **5,** 488–95

Borgmeier T (1935g) Sobre o cyclo evolutivo de *Chonocephalus* Wandolleck e uma nova espécie de *Melaloncha* Brues, endoparasita de abelhas *Archivos do Instituto de Biologia Vegetal, Rio de Janeiro* **2,** 255–65

Borgmeier T (1936) Alguns Phorideos myrmecophilos de Costa Rica e do Brasil (Dipt Phoridae) *Revista de Entomologia, Rio de Janeiro* **6,** 23–37

Borgmeier T (1937) Uma nova especie de *Apocephalus* (Dipt Phoridae), endoparasita do *Chauliognathus fallax* Germ (Col Cantharidae) *Revista de Entomologia, Rio de Janeiro* **7,** 207–16

Borgmeier T (1938a) Um novo genero de Phorideos de Santa Catharina (Dip Phoridae) In *Livro Jubilar Prof Travassos* Rio de Janeiro, **3,** 87–9

Borgmeier T (1938b) Algunas Phorideos myrmecophilos de Costa Rica e do Brasil e quatro especies novas de *Melaloncha* Brues (Dipt Phoridae) *Revista de Entomologia, Rio de Janeiro* **9,** 39–53

Borgmeier T (1938c) Phorideos parasitas da formiga argentina (Dipt Phoridae) *Revista de Entomologia, Rio de Janeiro* **9,** 261–2

Borgmeier T (1950a) Revision der Gattung *Coniceromyia* Borgmeier, nebst Beschreibung von sechs neuen Arten (Dipt Phoridae) *Revista de Entomologia, Rio de Janeiro* **21,** 281–99

Borgmeier T (1950b) Die myrmecophile Phoridengattung *Chelidophora* Borgmeier, nebst Beschreibung einer neuen Art aus Brasilien (Dipt Phoridae) *Revista de Entomologia, Rio de Janeiro* **21,** 301–4

Borgmeier T (1954) Eine neue myrmekophile Dipterengattung aus Brasilien *Zoologischer Anzeiger* **153,** 292–8

Borgmeier T (1956) Zur Kenntnis der myrmecophilen Phoridengattung '*Pheidolomyia*' Schmitz (Diptera, Phoridae) *Revista Brasileira de Biologia* **16,** 415–20

Borgmeier T (1957) Basic questions of systematics *Systematic Zoology* **6,** 53–69

Borgmeier T (1958) Neue Beitraege zur Kenntnis der neotropischen Phoriden (Diptera, Phoridae) *Studia Entomologica, Petropolis* **1,** 305–406

Borgmeier T (1959) Neue und wenig bekannte Phoriden aus der neotropischen Region, nebst einigen Arten aus dem Belgischen Congo-Gebiet (Diptera, Phoridae) *Studia Entomologica, Petropolis* **2,** 129–208

Borgmeier T (1960a) Revision der ecitophilen Phoridengattung *Ecitoptera* Borgmeier & Schmitz, mit Beschreibung von sechs neuen Arten (Diptera, Phoridae) *Anais da Academia Brasileira de Ciências* **32,** 125–41

Borgmeier T (1960b) Gefluegelte und ungefluegelte Phoriden aus der neotropischen Region, nebst Beschreibung von sieben neuen Gattungen (Diptera, Phoridae) *Studia Entomologica, Petropolis* **3,** 257–374

Borgmeier T (1961a) Weitere Beitraege zur Kenntnis der neotropischen Phoriden, nebst Beschreibung einiger *Dohrniphora*-Arten aus der indo-australischen Region (Diptera, Phoridae) *Studia Entomologica, Petropolis* **4,** 1–112

Borgmeier T (1962c) Versuch einer Uebersicht ueber die neotropischen *Megaselia*-Arten, sowie neue oder wenig bekannte Phoriden verschiedener Gattungen (Dipt Phoridae) *Studia Entomologica, Petropolis* **5,** 289–488

Borgmeier T (1962d) Cinco espécies novas do gênero *Pseudacteon* Coquillett *Arquivos do Museu Nacional* **52,** 27–30

Borgmeier T (1963b) New or little known phorid flies from Australia (Dipt Phoridae) *Anais da Academia Brasileira de Ciências, Rio de Janeiro* **35,** 109–24

Borgmeier T (1963c) A new genus of phorid flies, with the description of seven new species (Diptera, Phoridae) *Revista Brasileira de Biologia, Rio de Janeiro* **23,** 165–70

Borgmeier T (1963d) Revision of the North American phorid flies Part I The Phorinae, Aenigmatiinae and Metopininae, except *Megaselia* (Diptera, Phoridae) *Studia Entomologica, Petropolis* **6,** 1–256

Borgmeier T (1963e) New or little known *Coniceromyia*, and some other Neotropical and Paleotropical Phoridae (Diptera) *Studia Entomologica, Petropolis* **6,** 449–80

Borgmeier T (1964a) A generic revision of the Termitoxeniinae, with the description of a new species from Burma *Studia Entomologica, Petropolis* **7,** 73–95

Borgmeier T (1964b) Revision of the North American Phorid flies Part II The species of the genus *Megaselia*, subgenus *Aphiochaeta* (Diptera, Phoridae) *Studia Entomologica, Petropolis* **7,** 257–416

Borgmeier T (1966a) Revision of the North American Phorid flies Part III The species of the genus *Megaselia*, subgenus *Megaselia* (Diptera, Phoridae) *Studia Entomologica, Petropolis* **8,** 1–160 (1965)

Borgmeier T (1967a) Studies on Indo-Australian Phorid flies, based mainly on material of the Museum of Comparative Zoology and the United States National Museum (Diptera, Phoridae) *Studia Entomologica, Petropolis* **9,** 129–328 (1966)

Borgmeier T (1967b) Ueber einige afrikanische, bei Termiten gefundene Phoriden, nebst Bemerkungen ueber die Gattung *Termitophorides* (Diptera, Phoridae) *Studia Entomologica, Petropolis* **9,** 417–28 (1966)

Borgmeier T (1967c) A new genus and species of phorid fly from Canada (Diptera Phoridae) *Canadian Entomologist* **90,** 221–3

Borgmeier T (1967e) Studies on Indo-Australian phorid flies, based mainly on material of the Museum of Comparative Zoology and the United States National Museum Part II *Studia Entomologica, Petropolis* **10,** 81–276

Borgmeier T (1967f) New or little known phorid flies, with description of two new genera *Anais da Academia Brasileira de Ciências, Rio de Janeiro* **39,** 195–210

Borgmeier T (1967h) Miscellaneous studies on phorid flies II *Revista Brasileira de Biologia, Rio de Janeiro* **27,** 395–405

Borgmeier T (1968a) Miscellaneous studies on phorid flies III *Revista Brasileira de Biologia, Rio de Janeiro* **28,** 167–76

Borgmeier T (1968b) A catalogue of the Phoridae of the world (Diptera, Phoridae), *Studia Entomologica, Petropolis* **11,** 1–367

Borgmeier T (1969a) Bredin-Archbold-Smithsonian Biological Survey of Dominica The Phoridae of Dominica (Diptera) *Smithsonian Contributions Zoology* **23,** 1–69

Borgmeier T (1969b) New or little-known Phorid flies, mainly of the Neotropical Region (Diptera, Phoridae) *Studia Entomologica, Petropolis* **12,** 33–132

Borgmeier T (1970a) On a small collection of phorid flies (Dipt) taken with termites (Isopt) *Studia Entomologica, Petropolis* **13,** 253–6

Borgmeier T (1971a) Further studies on Phorid flies, mainly of the Neotropical Region (Diptera, Phoridae) *Studia Entomologica, Petropolis* **14,** 1–172

Borgmeier T (1971b) Supplement to 'A Catalogue of the Phoridae of the World (Diptera, Phoridae)' *Studia Entomologica, Petropolis* **14,** 177–224

Borgmeier T, Prado AP do (1975) New or little-known Neotropical Phorid flies, with description of eight new genera (Diptera, Phoridae) *Studia Entomologica, Petropolis* **18,** 3–90

Borgmeier T, Schmitz H (1923a) Beitrag zur Kenntnis der Phoriden Brasilens *Deutscher Verein für Wissenschaft und Kunst in São Paulo* **3,** 127–48 (1922)

Brasse D (1975) Die Arthropodenfauna von Getreidefeldern auf verschiedenen Boden im Braunschweiger Raum *Pedobiologia* **15,** 405–14

Brauns A (1950) Zur Kenntnis der Schadinsekten an Champignonkulturen *Nachrichtsblatt für den deutschen Pflanzenschutzdienst, Braunschweig* **2,** 153–6

Breddin G, Borner C (1904) Ueber *Thaumatoxena wasmanni*, den Vertreter einer neuen Unterordnung der Rhynchoten *Sitzungsberichte der Gesellschaft naturforschender Freunde zu Berlin* **1904,** 84–93

Bridarolli A (1937b) Los Termitoxenidos y los estadios de superiodo larval *Estudios, Buenos Aires* **56,** 121–38

Bridarolli A (1940a) Morfologia externa de la *Clitelloxenia hemicyclia* Schmitz en sus tres primeros estados *International Congress of Entomology (1935)* **6,** 163–71

Bridarolli A (1951) Diptera Phoridae de Kivu (Gen *Megaselia*) del Museo de Congo Belga (Tervuren) *Annales du Musee Royale du Congo Belge, Zoologie* **7,** 1–67

Brischke G (1868) Kleinere Beobachtungen uber Insecten *Schriften der naturforschenden Gesellschaft in Danzig* **2,** 1–5

Bristowe WS (1925) Notes on the habits of insects and spiders in Brazil *Transactions of the Entomological Society of London* **1924,** 475–86

Bristowe WS (1928) Some notes on the biology of hunting wasps *Entomologist's Monthly Magazine* **64,** 7–11

Bristowe WS (1941) *The Comity of Spiders* Volume II London Ray Society

Britten H (1942) *Aphiochaeta rufipes* Mg feeding on dead hive-bees *North Western Naturalist* **17,** 251

Brooks DR, McLennan DA (1991) *Phylogeny, Ecology, and Behavior, a research program in comparative biology* Chicago University of Chicago Press

Brown BV (1986) Striking sexual dimorphism and its taxonomic consequences among new world phorid flies (Diptera) *Proceedings of the Entomological Society of Washington* **88,** 787–8

Brown BV (1987a) Revision of the *Gymnophora* (Diptera Phoridae) of the Holarctic Region classification, reconstructed phylogeny and geographic history *Systematic Entomology* **12,** 271–304

Brown BV (1987b) Classification, reconstructed phylogeny and geographic history of the Neotropical phorid flies of the genus *Gymnophora* (Diptera Phoridae) *Journal of Natural History* **21,** 1477–524

Brown BV (1988) Additions to the Phorid chapter in the 'Manual of Nearctic Diptera, Volume 2' (Diptera Phoridae) *Canadian Entomologist* **120,** 307–22

Brown BV (1989a) Testing phylogenetic hypotheses new species of *Gymnophora* Macquart (Diptera Phoridae) from Japan and Nepal, with an updated world classification *Canadian Journal of Zoology* **67,** 2543–56

Brown BV (1990a) *Triphleba vitrinervis* (Malloch) an unrecognised species of *Crinophleba* Borgmeier (Diptera Phoridae) *Proceedings of the Entomological Society of Washington* **92,** 271–3

Brown BV (1990b) New Nearctic Region records of Palaearctic *Megaselia* species (Diptera Phoridae) *Quaestiones Entomologicae* **26,** 701–2

Brown BV (1990c) Update of World Catalogue of the Phoridae (Borgmeier, 1968), plus supplement (Borgmeier, 1971) *Fly Times* **4,** 12–18

Brown BV (1992a) Life history, immature stages and undescribed male of *Rhynchomicropteron* (Diptera Phoridae) *Journal of Natural History* **26,** 407–16

Brown BV (1992b) Generic revision of Phoridae of the Nearctic Region and phylogenetic classification of Phoridae, Sciadoceridae and Ironomyiidae (Diptera) Phoridea *Memoirs of the Entomological Society of Canada* **164,** 1–144

Brown BV (1993a) Convergent adaptations in Phoridae (Diptera) living in the nests of social insects a review of the New World Aenigmatiinae *Memoirs of the Entomological Society of Canada* **165,** 115–37

Brown BV (1993b) Taxonomy and preliminary phylogeny of the parasitic genus *Apocephalus* subgenus *Mesophora* (Diptera Phoridae) *Systematic Entomology* **18,** 191–230

Brown BV, Feener DH (1991a) Life history parameters and description of the larva of *Apocephalus paraponerae* (Diptera Phoridae), a parasitoid of the giant tropical ant *Paraponera clavata* (Hymenoptera Formicidae) *Journal of Natural History* **25,** 221–31

Brown BV, Feener DH (1991b) Behavior and host location cues of *Apocephalus paraponerae* (Diptera Phoridae), a parasitoid of the giant tropical ant, *Paraponera clavata* (F) (Hymenoptera Formicidae) *Biotropica* **23,** 182–7

Brown BV, Feener DH (1993) Life history and immature stages of *Rhyncophoromyia maculineura*, an ant-parasitizing phorid fly (Diptera Phoridae) from Peru *Journal of Natural History* **27,** 429–34

Brown BV, Francoeur A, Gibson RL (1991) Review of the genus *Styletta* (Diptera Phoridae), with description of a new genus *Entomologica Scandinavica* **22,** 241–50

Brown BV, Marshall SA (1984) Phorid flies (Diptera Phoridae) associated with mushrooms in Southern Ontario *Proceedings of the Entomological Society of Ontario* **115,** 77–80

Brown BV, Pike EM (1990) Three new fossil phorid flies (Diptera Phoridae) from Canadian late Cretaceous amber *Canadian Journal of Sciences* **27,** 845–8

Brown ES (1945) *Trupheoneura sublugubris* (Wood) (Dipt , Phoridae) in wasps' nest *Entomologist's Monthly Magazine* **81,** 119

Bruch C (1928) Estudios mirmecologicos *Anales del Museo Nacional de Buenos Aires* **34,** 341–60

Brues CT (1901) Two new myrmecophilous genera of aberrant Phoridae from Texas *American Naturalist* **35,** 337–56

Brues CT (1902a) Notes on the larvae of some Texan Diptera *Psyche* **9,** 351–4

Brues CT (1902b) New and little-known guests of the Texan legionary ants *American Naturalist* **36,** 365–78

Brues CT (1903b) A monograph of North American Phoridae *Transactions of the American Entomological Society* **29,** 331–404

Brues CT (1906a) Diptera, Fam Phoridae *Genera Insectorum* **44,** 1–21

Brues CT (1906b) Two new species of Phoridae *Bulletin of the Wisconsin Natural History Society* **4,** 100–102

Brues CT (1907c) A remarkable new phorid from Cape Colony *Entomological News* **18,** 391–2

Brues CT (1908c) Some new North American Phoridae *Journal of the New York Entomological Society* **16,** 199–201

Brues CT (1909b) Some further remarks on the affinities of the Phoridae, with descriptions of two North American species *Bulletin of the Wisconsin Natural History Society* **7,** 103–8

Brues CT (1915b) Some new Phoridae from Java *Journal of the New York Entomological Society* **23,** 184–92

Brues CT (1915c) Some flies of the family Phoridae obtained by the expedition, with notes on a species possibly associated with external myiasis in man *Appendix II Report of the first expedition to South America, 1913* Cambridge, USA Harvard University Press

Brues CT (1915d) A synonymic catalogue of the dipterous family Phoridae *Bulletin of the Wisconsin Natural History Society* **12,** 85–152 (1914)

Brues CT (1919c) The occurrence of wingless Phoridae on the Fiji Islands *Psyche* **26,** 49

Brues CT (1923a) A new genus of myrmecophilous Phoridae, with notes on some related forms *Psyche* **30,** 18–22

Brues CT (1923c) Two myrmecophilous Phoridae from British Guiana *Zoologica, Scientific Contributions of the New York Zoological Society* **3,** 435–40

Brues CT (1924b) Notes on Phoridae from South Africa *Psyche* **31,** 87–91

Brues CT (1924c) Another 'snow' phorid (Diptera) *Psyche* **31,** 92

Brues CT (1925a) Some myrmecophilous Phoridae from the Neotropical Region *Psyche* **32,** 303–12

Brues CT (1928a) Some Colombian Phoridae from the nests of stingless bees *Psyche* **35,** 134–7

Brues CT (1928c) Some Cuban Phoridae which visit the flowers of *Aristolochia elegans* *Psyche* **35,** 160–1

REFERENCES

Brues CT (1930) *Aenictomyia*, a new myrmecophilous phorid from the Philippines *Psyche* **37,** 163–6

Brues CT (1932a) Phoridae associated with ants and termites in Trinidad *Psyche* **39,** 134–8

Brues CT (1932b) Notes on some tropical Phoridae *Psyche* **39,** 139–44

Brues CT (1936a) Philippine Phoridae from Mount Apo region in Mindanao *Proceedings of the American Academy of Arts and Sciences, Boston* **70,** 365–466

Brues CT (1936b) A new ecitophilous North American phorid fly *Bulletin of the Brooklyn Entomological Society* **31,** 68–70

Brunetti E (1912a) New Oriental Diptera *Record of the Indian Museum, Calcutta* **7,** 83–6

Brunetti E (1912b) New Oriental Diptera *Record of the Indian Museum, Calcutta* **7,** 445–513

Bruns TD (1984) Insect mycophagy in the Boletales fungivore diversity and the mushroom habitat In Wheeler Q, Blackwell M (eds), *Fungus-Insect Relationships, Perspectives in Ecology and Evolution* New York Columbia University Press

Buchs W (1988) *Stamm- und Rinden-zoozonosen verschiedener Baumarten des Hartholzauenwaldes und uhr Indikatorwert für die Fruherkennung von Baumschaden* Bonn Rheinische Friederich-Wilhelms-Universitat (Ph D Thesis)

Buczacki S (1989) *Fungi of Britain and Europe* London Collins

Bugnion E (1913) *Termitoxenia,* etude anatomo-histologique *Annales de la Societe Entomologique Belgique* **57,** 23–44

Buhr H (1965) *Bestimmungstabellen der Gallen (Zoo- und Phytocecidien) an Pflanzen Mittel- und Nordeuropas Band II* Jena Gustav Fischer Verlag

Burges RJ (1979) A rare fly and its parasitic behaviour towards an ant (Diptera Phoridae, Hymenoptera Formicidae) *Florida Entomologist,* **62,** 413–14

Burla H, Bachli G (1968) Beitrag zur Kenntnis der schweizerischen Dipteren, insbesondere *Drosophila*-Arten, die sich in Fruchtkorpen *Vierteljahresschrift der naturforschenden Gesellschaft in Zurich* **113,** 311–36

Butani DK (1957) A tachinid fly parasite of *Chilo zonellus* Swinhoe *Indian Journal of Entomology* **19,** 62–3

Buttiker W (1969) Parasiten und Nidicolen der Uferschwalbe (*Riparia riparia* (L)) in der Schweiz *Mitteilungen der schweizerischen entomologischen Gesellschaft* **42,** 205–20

Buxton PA (1961) British Diptera associated with fungi III Flies of all families reared from about 150 species of fungi *Entomologist's Monthly Magazine* **96,** 61–94

Čakar Lj, Disney RHL (1991) A leaf-mining scuttle fly (Diptera Phoridae) as a pest of chickpeas (Leguminosae) in Yugoslavia *Bulletin of Entomological Research* **81,** 21–4

Cantelo WW, McDaniel JS (1978) Mushroom flies controlled by incorporating diazinon *Journal of Economic Entomology* **71,** 670–3

Canzanelli A (1941) La fauna dei funghi freschi II Contributo La ditterofauna fungicola *Commentationes, Pontificia Academia Scientiarum, Roma* **5,** 211–82

Carareto CMA, Mourao CA (1988) Dimorfismo sexual do tamanho de pupas em *Megaselia scalaris* *Ciência e Cultura, São Paulo* **40,** 995–7

Carpenter GDH (1937) A note on some parasites of *Zygaena* *Journal of the Society for British Entomology* **1,** 176–8

Carpenter JM (1982) The phylogenetic relationships and natural classification of the Vespoidea *Systematic Entomology* **7,** 11–38

Carpenter TL, Chastain DO (1992) Facultative myiasis by *Megaselia* sp (Diptera Phoridae) in Texas a case report *Journal of Medical Entomology* **29,** 561–3

Carr JW (1924) The Diptera pollinating the flowers of *Aristolochia sipho* *Entomologist's Monthly Magazine* **60,** 258

Carter JB (1977) Survival of *Tipula paludosa* (Dipt , Tipulidae) larvae after infection with larvae of *Megaselia paludosa* (Dipt , Phoridae) *Entomologist's Monthly Magazine* **112,** 245 (1976)

Chalmers N, Parker P (1989) *The OU Project Guide fieldwork and statistics for ecological projects* Second edition London Field Studies Council Occasional Publication No 9

Chandler P (1966) Two new rearing records of Phoridae (Diptera) *Entomologist's Record and Journal of Variation* **78,** 185

Chandler P (1973) Some Diptera and other insects associated with decaying elms (*Ulmus procera* Salisbury) at Bromley, Kent, with some additional observations on these and related species *Entomologist's Gazette* **24,** 329–46

Chandler P (1978) Associations with plants, fungi In Stubbs A, Chandler P (eds), *A Dipterist's Handbook* London The Amateur Entomologist **15,** Chapter 6 199–211

Chang ST (1991) Recent trends in world production of cultivated edible mushrooms *The Mushroom Journal* **503,** 15–18

Chaud-Netto J (1980) Biological studies on *Pseudohypocera kerteszi* (Phoridae, Diptera) *Experientia* **36,** 61–2

Chazeau J, Bouye E, Bonnet de Larbogne L (1989) Lutte biologique contre le psylle *Heteropsylla cubana*, ravageur du faux-mimosa *Leucaena leucocephala* en Nouvelle Calédonie *Centre ORSTOM de Noumea, Conventions Sciences de la vie, Zoologie appliquee* **1989,** 1–82

Cheng L, Hogue CL (1974) New distribution and habitat records of biting midges and mangrove flies from the coasts of southern Baja California, Mexico (Diptera Ceratopogonidae, Culicidae, Chironomidae, and Phoridae) *Entomological News* **85,** 211–18

Cherrett JM, Powell RJ, Stradling DJ (1989) The mutualism between leaf-cutting ants and their fungus In Wilding *et al* (eds), 1989 (see below) Chapter 4 93–120

Chevalier L (1925) *Phora pygmaea* Zett Diptère mangeur de mouches *Bulletin de la Société de Sciences de Seine-et-Oise* **6,** 93–6

Clapham AR, Tutin TG, Moore DM (1987) *The Flora of the British Isles* Third Edition Cambridge Cambridge University Press

Clements AN (1963) *The Physiology of Mosquitoes* London Pergamon Press

Clift AD (1979) The identity, economic importance and control of insect pests of mushrooms in New South Wales, Australia *Mushroom Science* **10,** 367–83

Clift AD, Larsson SF (1987) Phoretic dispersal of *Brennandania lambi* (Kcrzal) (Acari Tarsonemida Pygmephoridae) by mushroom flies (Diptera Sciaridae and Phoridae) in New South Wales, Australia *Experimental and Applied Acarology* **3,** 11–20

Coaton WGH, Sheasby JL (1972) Preliminary report on a survey of the termites (Isoptera) of South West Africa *Cimbebasia Memoir* **2,** 1–130

Coggins RE (1970) *Megaselia paludosa* (Wood) (Dipt , Phoridae) parasitising *Tipula paludosa* Mg (Dipt , Tipulidae) *Entomologist's Monthly Magazine* **106,** 108

Cole LC (1946) A study of the cryptozoa of an Illinois woodland *Ecological Monographs* **16,** 51–86

Coles EC (1890) Entomology notes 1 – *Trycolyga bombycis*, the Bengal silkworm fly *Indian Museum Notes* **1,** 83–8

Collart A (1933) Diptères élevées des nids de guêpes *Bulletin du Musee Royal d'Histoire Naturelle de Belgique* **9,** 1–8

Colless DH, McAlpine DK (1991) Diptera (flies) In CSIRO *The Insects of Australia* 2nd Edition Melbourne Melbourne University Press, Chapter 39 717–86

Collin JE (1939) On various new or little-known British Diptera, including species bred from the nests of birds and mammals *Entomologist's Monthly Magazine* **75,** 134–54

Collins CW, Baker WL (1934) Exploring the upper air for windborne gipsy moth larvae *Journal of Economic Entomology* **27,** 320–7

Colyer CN (1950) Notes on the breeding of *Diploneura pilosella* Schmitz and *Megaselia rufipes* Mg (Dipt , Phoridae) and on the puparium of the former *Entomologist's Monthly Magazine* **86,** 320–2

Colyer CN (1952a) Notes on the life-histories of the British species of *Phalacrotophora* Enderlein (Dipt , Phoridae) *Entomologist's Monthly Magazine* **88,** 135–9

Colyer CN (1952b) Notes on *Triphleba minuta* F (Dipt , Phoridae) and its immature stages *Entomologist's Monthly Magazine* **88,** 226–9

Colyer CN (1954a) The 'coffin' fly, *Conicera tibialis* Schmitz (Dipt , Phoridae) *Journal of the Society for British Entomology* **4,** 203–6

Colyer CN (1954b) A new species of *Megaselia* (Dipt , Phoridae) from Britain, notes on fungicolous Phoridae *Entomologist's Monthly Magazine* **89,** 108–12

Colyer CN (1954c) 'Swarming' of Phoridae (Diptera) *Journal of the Society for British Entomology* **5,** 22–7

Colyer CN (1954e) More about the 'coffin' fly, *Conicera tibialis* Schmitz (Diptera, Phoridae) *Entomologist* **87,** 129–32

Colyer CN (1954f) Further emergences of *Conicera tibialis*, the 'coffin fly' (Diptera, Phoridae) *Entomologist* **87,** 234

Colyer CN (1954g) Further notes on the life-histories of the British species of *Phalacrotophora* Enderlein (Dipt , Phoridae) *Entomologist's Monthly Magazine* **90,** 208–10

Colyer CN (1955a) A new species of *Spiniphora* (Dipt , Phoridae) from Ceylon, notes on *Spiniphora genitalis* Schmitz *Entomologist's Monthly Magazine* **91,** 48–50

Colyer CN (1956a) A new species of *Conicera* (Diptera Phoridae) from Hawaii, with observations on the genera *Parafannia* Bohart and *Gymnoptera* Lioy *Proceedings of the Hawaiian Entomological Society* **16,** 232–40

Colyer CN (1958) A new species of *Aenigmatopoeus*, and the hitherto unknown male in the genus (Diptera, Phoridae) *Studia Entomologica, Petropolis* **1,** 247–58

Colyer CN (1961a) A new species of *Phalacrotophora* (Diptera, Phoridae) from India *Broteria* **30,** 93–7

Colyer CN (1962) Ten new species of Phoridae from Yugoslavia (Diptera) *Studia Entomologica, Petropolis* **5,** 257–76

Colyer CN (1966) Two new species of Phoridae (Diptera) from the Kuril Islands, Sakhalin District, USSR *Eesti NSV Teaduste Akadeemia Toimetised (Bioloogia)* **15,** 72–6

Colyer CN (1969) Some Phoridae (Diptera) from Southern Spain and Majorca, with descriptions of two new species *Entomologiske Meddelelser* **37,** 9–26

Colyer CN (unpublished) Notebooks [Held by Diptera Section, Department of Entomology, Natural History Museum, London]

Colyer CN, Elberg K (1969) New data on Phoridae (Diptera) *Eesti NSV Teaduste Akadeemia Toimetised (Bioloogia)* **18,** 154–69

Compton SG, Disney RHL (1991) New species of *Megaselia* (Diptera Phoridae) whose larvae live in fig syconia (Urticales Moraceae), and adults prey on fig wasps (Hymenoptera Agaonidae) *Journal of Natural History* **25,** 203–19

Cook OF (1897) A new wingless fly from Liberia *Science* **6,** 886

Coquillett DW (1895) Synopsis of the Dipterous genus *Phora* *Canadian Entomologist* **27,** 103–7

Coquillett DW (1907) A new phorid genus with horny ovipositor *Canadian Entomologist* **39,** 207–8

Corbet SA (1985) Insect chemosensory responses a chemical legacy hypothesis *Ecological Entomology* **10,** 143–53

Cowie RH (1992) Evolution and extinction of Partulidae, endemic Pacific island land snails *Philosophical Transactions of the Royal Society, Series B Biological Sciences* **335,** 167–91

Creighton JT (1936) Report on cotton leaf-worm investigations in Florida, including temperature and hibernation studies *Journal of Economic Entomology* **29,** 88–94

Crosby TK, Watt JC, Kistemaker AC, Nelson PE (1986) Entomological identification of the origin of imported cannabis *Journal of the Forensic Science Society* **26,** 35–44

Cross MJ, Jacobs L (1968) Some observations on the biology of *Verticillium malthousei* *Mushroom Science* **7,** 239–44

Cross WH, McGovern WL (1969) New parasites, *Zatropis perdubius* and *Megaselia aletiae*, on the boll weevil, *Anthonomus grandis* *Annals of the Entomological Society of America* **62,** 674

Crosskey RW (ed) (1980) *Catalogue of the Diptera of the Afrotropical Region* London British Museum (Natural History)

Crowder AA, Pearson MC, Grubb PJ, Langlois PH (1990) Biological flora of the British Isles *Drosera* L *Journal of Ecology* **78,** 233–67

Curtis J (1833) *British Entomology*, Volume 10 London

Curtis SK, Benner DB, Cowden RR (1987) Ultrastructure of the larval salivary glands of *Megaselia scalaris* Loew (Diptera, Phoridae) *Journal of Morphology* **191,** 265–88

Curtis SK, Benner DB, Musil G (1989) Ultrastructure of the spermatozoon of *Megaselia scalaris* Loew (Diptera Brachycera Cyclorrhapha Phoridea Phoridae) *Journal of Morphology* **200,** 47–61

Cuthbertson A (1936) Biological notes on some Diptera of Southern Rhodesia *Occasional Papers of the Rhodesian Museum* **5,** 46–63

Dabrowska-Prot E, Karg J (1974) Preliminary analysis of Diptera of rye and potato fields *Bulletin de l'Academie Polonaise des Sciences, Serie des Sciences Biologiques* **22,** 385–91

Daggett PM, Dollahon N, Janovy J (1972) *Herpetomonas megaseliae* sp n (Protozoa Trypanosomatidae) from *Megaselia scalaris* (Loew, 1866) Schmitz, 1929 (Diptera Phoridae) *Journal of Parasitology* **58,** 946–9

Decou-Burghele A (1961) Sur la biologie de *Megaselia melanocephala* von Roser, phoride parasite des cocons de *Meta menardi* Latr *Annales du Laboratoire Souterrain, Hans-sur-Lesse* **11,** 16–22

Delachambre J (1965) Recherches sur les Termitoxeniidae (Diptera) I Description d'une novelle espèce africaine et de sal arve *Insectes Sociaux* **12,** 273–83

Delachambre J (1966) Recherches sur les Termitoxeniidae (Diptera) II Descriptions de deux novelle espèce africaines *Insectes Sociaux* **13,** 105–16

Delage A, Lauraire M-C (1970) Trois nouvelles espèces de Phoridae (Diptères, Brachycères, Cyclorrhaphes), trouvées à Peyresq (Alpes de Provence) *Bulletin des Recherches Agronomiques de Gembloux* **5,** 450–8

Delage A, Lauraire M-C (1971c) A propos de deux Aenigmatiinae (Dipt Phoridae) nouvelles pour la faune de France *Entomologiste* **27,** 159–62

Delage A, Lauraire M-C (1974b) Mise au point sur le genre *Phalacrotophora* (Dipt Phoridae) et description de l'ovipositor d'une femelle parasite *Annales de Parasitologie Humaine et Comparee* **49,** 495–500

Deleporte S (1986) Biologie et écologie du Diptère Sciaridae *Bradysia confinis* (Winn , Frey) d'une litière de feuilles (Bretagne intérieure) *Revue d'Ecologie et de Biologie du Sol* **23,** 39–76

Della Santa E (1993) *Camponotus cruentatus* Latreille, 1802 (Hymenoptera Formicidae), hte du phorid myrmcophile *Microselia rivierae* Schmitz, 1934 (Diptera Phoridae) *Mitteilungen der schweizerischen entomologischen Gesellschaft* **66,** 135–136

Delucchi V (1953) *Aphidecta obliterata* L (Coleoptera, Coccinellidae) als Rauber von *Dreyfusia (Adelges) piceae* Ratz *Pflanzenschutzberichte* **11,** 73–83

Dely-Draskovits A, Babos M (1976) Fenologische Zusammenhange zurischen Fliegen und Hutpilzen *Folia Entomologica Hungarica* **29,** 23–38

Dessart P (1993) Histoire des Termitoxeniides ou les errements de la science *Les Naturlaistes Belges* **74,** 61–75

Dias Macieira OJ, Chaud-Netto J, Zanon AM (1983) Oviposition rate and relative viability of descendants from couples of *Megaselia scalaris* (Diptera Phoridae) reared in different experimental conditions *Revista Brasileira de Biologia* **43,** 223–228

Disney RHL (1969) The timing of adult eclosion in blackflies (Dipt , Simuliidae) in West Cameroon *Bulletin of Entomological Research* **59,** 485–503

Disney RHL (1970) A note on the uptake of sugar by some blood-sucking flies *Proceedings of the Royal Entomological Society of London (A)* **45,** 51–54

Disney RHL (1973) Flies reared from pupae found in shells of garden snails *Entomologist's Monthly Magazine* **108,** 87 (1972)

Disney RHL (1975) A key to the larvae, pupae and adults of the British Dixidae (Diptera) the meniscus midges *Freshwater Biological Association Scientific Publication* **31,** 1–78

Disney RHL (1976a) A survey of blackfly populations (Dipt , Simuliidae) in West Cameroon *Entomologist's Monthly Magazine* **111,** 211–227 (1975)

Disney RHL (1976b) A further case of a Nematoceran fly (Diptera Sciaridae) parasitized by a species of scuttle fly (Diptera Phoridae) *Entomologist's Gazette* **27,** 91–98

Disney RHL (1977a) Scuttle flies (Diptera, Phoridae) from Inishbofin, Co Galway *Irish Naturalists' Journal* **19,** 57–61

Disney RHL (1977b) A further case of a scuttle fly (Dipt , Phoridae) whose larvae attack slug eggs *Entomologist's Monthly Magazine* **112,** 174 (1976)

Disney RHL (1978a) Some scuttle flies (Dipt Phoridae) from North Wales *Nature in Wales* **16,** 25–31

Disney RHL (1978d) A new species of afrotropical *Megaselia* (Diptera Phoridae), with a re-evaluation of the genus *Plastophora* *Zeitschrift für angewandte Zoologie* **65,** 313–319

Disney RHL (1979a) A planidium-type larva in an adult scuttle fly (Dipt , Phoridae) *Entomologist's Monthly Magazine* **113,** 199–200 (1977)

Disney RHL (1979d) Some scuttle flies (Diptera, Phoridae) from Tow Hill Nature Reserve, North Yorkshire *Naturalist, Hull* **104,** 39–42

Disney RHL (1979e) *Multinevra macropygidia* n gen , n sp , a curious afrotropical scuttle fly (Diptera Phoridae) *Entomologica Scandinavica* **10,** 79–80

Disney RHL (1979f) Natural history notes on some British Phoridae (Diptera) with comments on a changing picture *Entomologist's Gazette* **30,** 141–150

Disney RHL (1979g) *Couturiera kensmithi* n gen n sp and *C lewisi* n sp of the afrotropical Alamirinae (Diptera Phoridae) *Entomologica Scandinavica* **10,** 142–146

Disney RHL (1979j) The British *Metopina* (Diptera Phoridae) with description of a new species *Zoological Journal of the Linnean Society* **67,** 97–113

Disney RHL (1979k) A new species of *Phalacrotophora* (Dipt Phoridae) from France *Annales de Parasitologie Humaine et Comparee* **54,** 533–536

Disney RHL (1980c) Records of flower visiting by scuttle flies (Diptera Phoridae) in the British Isles *Naturalist, Hull* **105,** 45–50

Dısney RHL (1980d) Varıatıon ın *Megaselıa pulıcarıa* (Fall) (Dıpt , Phorıdae) wıth the recognıtıon of new synonymıes *Entomologıst's Monthly Magazıne*, **115,** 97–103

Dısney RHL (1980f) Some soıl-ınhabıtıng scuttle flıes (Dıpt , Phorıdae) *Entomologıst's Monthly Magazıne* **115,** 231–232

Dısney RHL (1980g) A scuttle fly (Dıpt , Phorıdae) that parasıtıses larvae of *Trıchosıa* Wınnertz (Dıpt , Scıarıdae) *Entomologıst's Monthly Magazıne* **116,** 5–6

Dısney RHL (1980h) A new specıes of *Chonocephalus* (Dıptera Phorıdae) from Panama *Entomologıca Scandınavıca* **11,** 363–365

Dısney RHL (1980ı) A new specıes of *Apocephalus* (Dıptera Phorıdae) that parasıtıses *Atta texana* (Hymenoptera Formıcıdae) ın Texas *Zeıtschrıft für angewandte Zoologıe* **67,** 47–50

Dısney RHL (1980k) *Chaetopleurophora bohemannı* (Becker)(Dıptera Phorıdae) added to the Brıtısh Lıst *Entomologıst's Gazette* **31,** 245

Dısney RHL (1981d) *Apocephalus laceyı* n sp (Dıptera Phorıdae) attackıng *Camponotus femoratum* (F) (Hymenoptera Formıcıdae) ın Brazıl *Entomologıca Scandınavıca* **12,** 31–34

Dısney RHL (1981h) A new, and first Afrotropıcal, specıes of *Auxanommatıdıa* (Dıptera Phorıdae) *Zeıtschrıft für angewandt Zoologıe* **67,** 323–330

Dısney RHL (1981k) Four new specıes of *Megaselıa* (Dıptera Phorıdae) from Srı Lanka *Zeıtschrıft für angewandte Zoolologıe* **67,** 389–398 (1980)

Dısney RHL (1981o) A Palaearctıc scuttle fly recorded from Texas (Dıpt , Phorıdae) *Entomologıst's Monthly Magazıne* **117,** 158

Dısney RHL (1982e) Notes on the genus *Corıdophora* Schmıtz (Dıpt , Phorıdae) wıth a re-evaluatıon of the genus *Edaphomyıa* Borgmeıer *Entomologıst's Monthly Magazıne* **118,** 9–10

Dısney RHL (1982h) A scuttle fly (Dıptera Phorıdae) that appears to be a parasıtoıd of a snaıl (Stylommatophora Zonıtıdae) and ıs ıtself parasıtızed by a Braconıd (Hymenoptera) *Entomologıst's Record and Journal of Varıatıon* **94,** 151–154

Dısney RHL (1982k) Three new specıes of scuttle fly (Dıptera Phorıdae) that parasıtıze ants (Hymenoptera Formıcıdae) ın North Amerıca *Journal of Zoology, London* **197,** 473–481

Dısney RHL (1982l) A curıous new specıes of *Megaselıa* from Brazıl (Dıptera Phorıdae) *Zeıtschrıft für angewandte Zoologıe* **68,**415–418 (1981)

Dısney RHL (1982m) *Megaselıa argıopephaga* n sp (Dıptera Phorıdae), an Orıental scuttlefly whose larvae feed on spıder eggs *Entomologıca Scandınavıca* **13,** 321–324

Dısney RHL (1982n) A new specıes of *Megaselıa* (Dıptera Phorıdae) that breeds ın pıtchers of *Nepenthes* ın Srı Lanka *Ceylon Journal of Scıence (Bıologıcal Scıences)* **14,** 89–101 (1981)

Dısney RHL (1982o) Two new specıes of Afrotropıcal *Megaselıa* (Dıptera Phorıdae) *Entomologıst's Monthly Magazıne* **118,** 189–193

Dısney RHL (1983a) Scuttle flıes Dıptera, Phorıdae (except *Megaselıa*) *Handbooks for the Identıficatıon of Brıtısh Insects* **10(6),** 1–81

Dısney RHL (1983b) A useful new character ın the gıant genus *Megaselıa* (Dıptera Phorıdae), wıth two new specıes from Brıtaın *Zeıtschrıft für angewandte Zoologıe* **70,** 225–234

Dısney RHL (1983d) A synopsıs of the taxonomıst's tasks, wıth partıcular attentıon to phylogenetıc cladısm *Fıeld Studıes* **5(5),** 841–865

Dısney RHL (1983g) Two new specıes of Afrotropıcal *Mıcroselıa* (Dıptera Phorıdae) *Entomologıca Scandınavıca* **14,** 407–410

Dısney RHL (1983h) Four new specıes of *Dohrnıphora* (Dıptera Phorıdae) from Panama *Entomologıca Scandınavıca* **14,** 452–456

REFERENCES

Disney RHL (1985a) Re-interpretation of the wing veins in the Phoridae (Diptera) *Entomologist's Monthly Magazine* **121,** 55–58

Disney R L (1985b) Replacement names for three species of *Megaselia* (Dipt , Phoridae) *Entomologist's Monthly Magazine* **121,** 97

Disney RHL (1985c) Records of scuttle flies (Diptera Phoridae) from Flatford Mill, including two species new to Britain *Transactions of the Suffolk Naturalist's Society* **21,** 13–17

Disney RHL (1985d) Afrotropical *Woodiphora* with description of four new species (Diptera, Phoridae) *Polskie Pismo Entomologiczne* **55,** 271–281

Disney RHL (1985h) Watching scuttle flies on Project Wallace *Antenna* **9,** 139–141

Disney RHL (1986a) A new genus of scuttle-fly (Diptera Phoridae) from England *Zoolological Journal of the Linnean Society* **87,** 85–89

Disney RHL (1986b) A new genus and three new species of Phoridae (Diptera) parasitizing ants (Hymenoptera) in Sulawesi *Journal of Natural History* **20,** 777–787

Disney RHL (1986c) Morphological and other observations on *Chonocephalus* (Phoridae) and phylogenetic implications for the Cyclorrhapha (Diptera) *Journal of Zoology, London (A)* **210,** 77–87

Disney RHL (1986d) Two remarkable new species of scuttle-fly (Diptera Phoridae) that parasitize termites (Isoptera) in Sulawesi *Systematic Entomology* **11,** 413–422

Disney RHL (1986e) Two new species of scuttle fly (Diptera, Phoridae) from Malham Tarn, North Yorkshire *Naturalist, Hull* **111,** 113–121

Disney, RHL (1986g) Assessments using invertebrates posing the problem In Usher MB (ed), *Wildlife Conservation Evaluation* London Chapman & Hall, Chapter 12 271–293

Disney RHL (1986h) Inventory surveys of insect faunas discussion of a particular attempt *Antenna* **10,** 112–116

Disney RHL (1987a) A most remarkable new genus of scuttle fly (Diptera Phoridae) from Sulawesi *Systematic Entomology* **12,** 29–32

Disney RHL (1987b) Observations on a peculiar mechanism in the crop of some Phoridae (Diptera) and its taxonomic value *Journal of Natural History* **21,** 275–280

Disney RHL (1987d) Evidence that *Opetia nigra* Meigen (Dipt , Opetiidae) is not a Cyclorrhaphan *Entomologist's Monthly Magazine* **123,** 47

Disney RHL (1987f) A preliminary survey of the scuttle flies of Hayley Wood, with descriptions of three new species *Proceedings and Transactions of the British Entomological and Natural History Society* **20,** 27–34

Disney RHL (1987g) A new genus and two new species of Phoridae (Dipt) from nests of ants (Hym , Formicidae) in Sulawesi *Entomologist's Monthly Magazine* **123,** 157–161

Disney RHL (1987j) A new species of *Triphleba* from Spain and two new synonyms in this genus (Dipt , Phoridae) *Entomologist's Monthly Magazine* **123,** 191–194

Disney RHL (1987k) The undescribed male and holarctic status of *Megaselia abdita* Schmitz (Diptera Phoridae) *Entomologica Scandinavica* **18,** 263–264

Disney RHL (1987n) Four species of scuttle fly (Diptera Phoridae) from Dominican amber *Pan-Pacific Entomologist* **63,** 377–380

Disney RHL (1987o) The use of rapid sample surveys of insect fauna In Luff ML (ed), *The Use of Invertebrates in Site Assessment for Conservation* Newcastle-upon-Tyne The University of Newcastle-upon-Tyne Agriculture Environment Research Group, 19–28

Disney RHL (1988b) Biology and taxonomy of Old World *Puliciphora* (Diptera Phoridae) with descriptions of nine new species *Oriental Insects* **22,** 267–286

Disney RHL (1988c) A new species of *Microselia* (Dipt Phoridae) from France *Annales de Parasitologie Humaine et Comparee* **63,** 85–88

Disney RHL (1988d) A remarkable new species of scuttle fly (Diptera Phoridae) whose larvae infest chickpea root nodules in India *Journal of Natural History* **22,** 611–616

Disney RHL (1988e) Scuttle flies in Hayley Wood *Nature in Cambridgeshire* **30,** 21–22

Disney RHL (1988f) New synonyms in the *Megaselia dahli* complex (Diptera Phoridae) and two species new to the British List *Entomologist's Record and Journal of Variation* **100,** 169–174

Disney RHL (1988g) An interesting new species of *Megaselia* from Sulawesi, the ground plan of the Phoridae (Diptera) and phylogenetic implications for the Cyclorrhapha *Systematic Entomology* **13,** 433–441

Disney RHL (1988h) The Palaearctic species resembling *Megaselia pygmaea* (Diptera, Phoridae), including two new species *Annales Entomologici Fennici* **54,** 153–161

Disney RHL (1988i) The form of articulation between the pedicel and first flagellar segment of the antenna in flies (Diptera) *Entomologist* **107,** 99–103

Disney RHL (1989a) Scuttle Flies – Diptera Phoridae Genus *Megaselia* *Handbooks for the Identification of British Insects* **10(8),** 1–155

Disney RHL (1989b) Two new species of *Megaselia* (Dipt , Phoridae) reared from termite-mound fungus in Nigeria *Entomologist's Monthly Magazine* **125,** 45–48

Disney RHL (1989c) A new species of *Phora* (Dipt , Phoridae) from France *Entomologist's Monthly Magazine* **125,** 81–82

Disney RHL (1989d) Re-appraisal of the Australian genus *Eutermiphora* Lea (Diptera Phoridae) *Entomologica Scandinavica* **19,** 435–436

Disney RHL (1989e) A new species of *Megaselia* (Diptera, Phoridae) from France and Poland *Polskie Pismo Entomologiczne* **59,** 249–252

Disney RHL (1989f) A key to Australasian and Oriental *Woodiphora* (Diptera Phoridae), affinities of the genus and descriptions of new species *Journal of Natural History* **23,** 1137–1175

Disney RHL (1989g) The 'missing' males of the Alamirinae/ Termitoxeniinae complex and the supposed hermaphroditism in Termitoxeniinae (Diptera Phoridae) *Sociobiology* **16,** 109–120

Disney RHL (1989h) Two new species of *Megaselia* (Dipt , Phoridae) from the Falkland Islands (Malvinas), and a replacement name in the genus *Entomologist's Monthly Magazine* **125,** 183–186

Disney RHL (1989i) Six new species of *Megaselia* (Diptera, Phoridae) reared from fungi in Japan *Acta Entomologica Bohemoslovaca* **86,** 368–380

Disney RHL (1989j) Antidote to taxonomic instability *Nature* **342,** 488

Disney RHL (1990a) Some myths and the reality of scuttle fly biology *Antenna* **14,** 64–67

Disney RHL (1990b) Problem genera close to *Megaselia* (Diptera Phoridae) *British Journal of Entomology and Natural History* **3,** 11–22

Disney RHL (1990c) Two new species and a revised key to *Rhynchomicropteron* (Diptera, Phoridae) *Entomologica Fennica* **1,** 21–24

Disney RHL (1990d) A striking new species of *Megaselia* (Diptera, Phoridae) from Sulawesi, with re-evaluation of related genera *Entomologica Fennica* **1,** 25–31

Disney RHL (1990e) A key to *Diplonevra* males of the Australasian and Oriental Regions, including two new species (Diptera, Phoridae) *Entomologica Fennica* **1,** 33–39

Disney RHL (1990f) Revised key to *Puliciphora* (Diptera Phoridae) of the Australasian and Oriental Regions, with third report of a larval Parasitica (Hymenoptera) in an adult fly *Entomologica Scandinavica* **21,** 211–218

Disney RHL (1990g) Key to *Dohrniphora* males (Diptera Phoridae) of the Australasian and Oriental Regions with descriptions of new species *Zoological Journal of the Linnean Society* **99,** 339–387

REFERENCES

Disney RHL (1990j) Revision of the Alamirinae (Diptera Phoridae) *Systematic Entomology* **15,** 305–320

Disney RHL (1990k) A revised key to Australasian and Oriental *Conicera* (Diptera Phoridae), with three new species *Entomologica Scandinavica* **21,** 339–344

Disney RHL (1990l) Three new species of scuttle fly (Diptera, Phoridae) from the Canary Islands *Vieraea* **18,** 135–141

Disney RHL (1991a) Scuttle flies from Zimbabwe (Diptera, Phoridae) with the description of five new species *Journal of African Zoology* **105,** 27–48

Disney RHL (1991b) The Ascension Island scuttle fly (Diptera Phoridae) *Entomologist* **110,** 82–93

Disney RHL (1991c) Support for the revised interpretation of the wing veins of Phoridae (Diptera) *Entomologist's Monthly Magazine* **127,** 73–76

Disney RHL (1991d) Scuttle flies (Diptera Phoridae) as parasites of earthworms (Oligochaeta Lumbricidae) *British Journal of Entomology and Natural History* **4,** 11–13

Disney RHL (1991e) Family Phoridae In Soós A, Papp L (eds), *Catalogue of Palaearctic Diptera* Volume 7 (*Dolichopodidae – Platypezidae*) Budapest Akademiai Kiado, 143–204

Disney RHL (1991f) The fire-ant parasitoids of the *Pseudacteon spatulatus* complex (Diptera, Phoridae, Hymenoptera, Formicidae) *Sociobiology* **18,** 283–298, 332

Disney RHL (1991g) A new species of *Rhynchomicropteron* (Dipt ,Phoridae) from Nepal *Entomologist's Monthly Magazine* **127,** 119–121

Disney RHL (1991h) Scuttle flies of the Cape Verde Islands (Diptera, Phoridae) *Journal of African Zoolology* **105,** 205–241

Disney RHL (1991i) The aquatic Phoridae (Diptera) *Entomologica Scandinavica* **22,** 171–191

Disney RHL (1991j) Two rare scuttle flies (Diptera, Phoridae) from Yorkshire, including one new for Britain *Naturalist, Hull* **116,** 93–94

Disney RHL (1991k) *Aenigmatistes* (Diptera Phoridae), Aschiza with a ptilinum! *Bonner zoologisches Beitrage* **42,** 353–368

Disney RHL (1991l) A new species of *Conicera* (Diptera, Phoridae) from Germany *Zoologischer Anzeiger* **227,** 240–244

Disney RHL (1991m) Convergent and parallel evolution and the supra-generic classification of the Phoridae (Diptera) *Giornale Italiano di Entomologia* **5,** 263–287

Disney RHL (1992a) The 'missing' males of the Thaumatoxeninae (Diptera Phoridae) *Systematic Entomology* **17,** 55–58

Disney RHL (1992b) A further case of a scuttle fly (Dipt , Phoridae) parasitising a scale insect (Hem -Hom , Eriococcidae) *Entomologist's Monthly Magazine* **128,** 39–40

Disney RHL (1992c) 'Flightless' flies and fanciful 'facts' *Antenna* **16,** 150–158

Disney RHL (1992d) A further new species of *Rhynchomicropteron* (Dipt , Phoridae) from Malaysia *Entomologist's Monthly Magazine* **128,** 219–224

Disney RHL (1993a) A new species of Termitoxeniinae (Diptera Phoridae) from Zimbabwe and erection of a new genus for *Perissa tinglei* Disney *Entomologist's Monthly Magazine* **129,** 55–60

Disney RHL (1993b) New species of aquatic Phoridae (Diptera) from Malaysia *Aquatic Insects* **15,** 149–158

Disney RHL (1993c) Notes on European Phoridae (Diptera) *British Journal of Entomology and Natural History* **6,** 107–118

Disney RHL (1993d) The identities of phorid species infesting cultivated mushrooms *Mushroom Journal* **525,** 24

Disney RHL (1993e) Mosaic evolution and outgroup comparisons *Journal of Natural History* **27,** 1219–21

Disney RHL (1994a) From field studies to taxonomy *Field Studies* **8(2),** 197–216

Disney RHL (1994b) A new species and new records of Phoridae (Diptera) from New Zealand *Giornale Italiano di Entomologia* **6,** 119–124 (1992)

Disney RHL (1994c) A new species of *Triphleba* (Diptera Phoridae) from France and the subfamily assignment of this genus *Giornale Italiano di Entomologia* **6,** 135–139 (1992)

Disney RHL, Chazeau J (1990) The recognition and biology of *Phalacrotophora quadrimaculata* (Diptera Phoridae) parasitising *Olla v-nigrum* (Coleoptera Coccinellidae) used in attempts to control the *Leucaena* psyllid *Annales de Parasitologie Humaine et Comparee* **65,** 98–100

Disney RHL, Coulson JC, Butterfield J (1981) A survey of the scuttle flies (Diptera Phoridae) of upland habitats in Northern England *Naturalist, Hull* **106,** 53–66

Disney RHL, Cumming MS (1992) Abolition of the Alamirinae and ultimate rejection of Wasmann's theory of hermaphroditism in Termitoxeniinae (Diptera, Phoridae) *Bonner zoologisches Beitrage* **43,** 145–154

Disney RHL, Erzinçlioğlu YZ, Henshaw DJdeC, Howse D, Unwin DM, Withers P, Woods, A (1982) Collecting methods and the adequacy of attempted fauna surveys, with reference to the Diptera *Field Studies* **5(4),** 607–621

Disney RHL, Evans RE (1978) Two species of Phoridae (Dipt) reared from fungi *Entomologist's Monthly Magazine* **113,** 165 (1977)

Disney RHL, Evans RE (1979) Further records of Phoridae (Dipt) reared from fungi *Entomologist's Monthly Magazine* **114,** 166 (1978)

Disney RHL, Evans RE (1980) Phoridae (Diptera) whose larvae feed on eggs of spiders (Araneida) *Entomologist's Monthly Magazine* **115,** 21–22 (1979)

Disney RHL, Evans RE (1982) Records of Phoridae (Diptera) reared from fungi *Entomologist's Record and Journal of Variation* **94,** 104–105

Disney RHL, Evans RE (1988) New host records for fungus-breeding Phoridae (Diptera) *Entomologist's Record and Journal of Variation* **100,** 208–210

Disney RHL, Evans RE (1990) Some new records of fungus-breeding Phoridae (Diptera) *Dipterist's Digest* **7,** 20–23

Disney RHL, Evans RE (1994) Further new records of fungus-breeding Phoridae (Diptera) *Dipterist's Digest* (in press)

Disney RHL, Gunn LC (1980) Some scuttle flies (Dipt , Phoridae) from emergence traps over raspberry stools in Scotland *Entomologist's Monthly Magazine* **115,** 23–25 (1979)

Disney RHL, Henderson IF, Perry JN, Clements RO (1981) Phoridae (Diptera) from English pasture soils *Pedobiologia* **22,** 366–378

Disney RHL, Khochare NN, Godase SK (1992) A new parasitoid (Diptera Phoridae) of the mango shoot borer *Chlumetia transversa* (Lepidoptera Noctuidae) in India *Bulletin of Entomological Research* **82,** 191–195

Disney RHL, Kistner DH (1988) Phoridae collected from termite and ant colonies in Sulawesi (Diptera, Isoptera, Termitidae, Hymenoptera, Formicidae) *Sociobiology* **14,** 361–369

Disney RHL, Kistner DH (1989a) Neotropical Phoridae from termite colonies, including two new species (Diptera, Isoptera Termitidae) *Sociobiology* **15,** 79–84

Disney RHL, Kistner DH (1989b) Six new species of Oriental Phoridae from termite nests (Diptera, Isoptera, Termitidae and Rhinotermitidae) *Sociobiology* **15,** 325–348

Disney RHL, Kistner DH (1989c) Phoridae from termite colonies in Africa, including two new species (Diptera, Isoptera, Termitidae) *Sociobiology* **16,** 75–90

Disney RHL, Kistner DH (1989d) Neotropical Phoridae from army ant colonies, including two new species (Diptera, Hymenoptera, Formicidae) *Sociobiology* **16,** 149–174

Disney RHL, Kistner DH (1990a) Revision of the termite-parasitizing genus *Dicranopteron* (Diptera Phoridae) *Zoological Journal of the Linnean Society* **98,** 111–122

Disney RHL, Kistner DH (1990b) The genus *Adelopteromyia*, which associates with Neotropical army ants (Diptera, Phoridae, Hymenoptera, Formicidae) *Sociobiology* **16,** 259–263

Disney RHL, Kistner DH (1991) A new genus of Australasian/Oriental Phoridae associated with driver ants (Diptera, Hymenoptera, Formicidae) *Sociobiology* **18,** 269–281

Disney RHL, Kistner DH (1992) Revision of the termitophilous Thaumatoxeninae (Diptera Phoridae) *Journal of Natural History* **26,** 953–991

Disney RHL, Kurahashi H (1978) A case of urogenital myiasis caused by a species of *Megaselia* (Diptera Phoridae) *Journal of Medical Entomology* **14,** 717

Disney RHL, Majerus MEN, Walpole MJ (in press) Phoridae (Diptera) parasitising Coccinellidae (Coleoptera) *Entomologist*

Disney RHL, Nussbaum Y (1990) Phoridae (Diptera) from Israel *Israel Journal of Entomology* **24,** 107–114

Disney RHL, Schroth M (1989) Observations on *Megaselia persecutrix* Schmitz (Diptera Phoridae) and the significance of ommatidial size-differentiation *Entomologist's Monthly Magazine* **125,** 169–174

Disney RHL, Shaw MR (in press) The ant host (Hym , Formicidae) of a *Microselia* (Dipt , Phoridae) from France *Entomologist's Monthly Magazine*

Disney RHL, Szadziewski R (1979) Some scuttle flies (Diptera, Phoridae) from Poland *Polskie Pismo Entomolgiczne* **49,** 535–538

Donisthorpe HStJK (1913a) *Aenigmatias blattoides*, Meinert, captured in Scotland *Entomologist's Record and Journal of Variation* **25,** 277–278

Donisthorpe HStJK (1927) *The Guests of British Ants Their Habits and Life Histories* London Routledge

Donisthorpe H (1945) Habits of *Pseudacteon formicarum* (Dipt , Phoridae) *Entomologist's Monthly Magazine* **81,** 215

Dowding VM (1967) The function and ecological significance of the pharyngeal ridges occurring in the larvae of some cyclorrhaphous Diptera *Parasitology* **57,** 371–388

Downes JA, Dahlem, GA (1987) Keys to the evolution of Diptera role of Homoptera *Environment and Ecology* **16,** 847–854

Drabble E, Drabble H (1927) Some flowers and their Dipteran visitors *New Phytologist* **26,** 115–123

du Buysson H (1917) *Phora fasciata* Fall [Dipt] parasite par *Homalotylus eitelweinii* Pratz [Hym] *Miscellanea Entomologica* **25,** 66–67

Dunger W (1989) The return of soil fauna to coal mined areas in the German Democratic Republic In Majer JD (ed), *Animals in Primary Succession, the role of fauna in reclaimed lands* Cambridge Cambridge University Press, Chapter 13 307–337

Dunlop GA (1930) Some notes and correspondence on a 'mixed' nest of *Vespa North Western Naturalist* **5,** 87–90

Durska E (1981) Phoridae (Diptera) of Warsaw *Memorabilia Zoologica, Warszawa* **35,** 47–56

Edgar WD (1970) Prey of the wolf spider *Lycosa lugubris* (Walck) *Entomologist's Monthly Magazine* **106,** 71–73

Edmunds LR (1952) Some notes on the habits and parasites of native wood-roaches in Ohio (Orthoptera Blattidae) *Entomological News* **63,** 141–145

Edwards EE (1929) A survey of the insect fauna and other invertebrate fauna of permanent and arable land of certain soil types at Aberystwyth *Annals of Applied Biology* **16,** 299–323

Edwards FW, Schmitz H (1939) New data concerning *Thaumatoxena* Breddin & Borner

(Diptera Phoridae) *Proceedings of the Royal Entomological Society of London (B)* **8,** 79–87

Edwards JS, Banko PC (1976) Arthropod fallout and nutrient transport a quantitative study of Alaskan snowpatches *Arctic and Alpine Research* **8,** 237–245

Eggleton P, Gaston KJ (1990) 'Parasitoid' species and assemblages convenient definitions or misleading compromises? *Oikos* **59,** 417–421

Eibl-Eibesfeldt I, Eibl-Eibesfeldt E (1967) Das Parasitenabwehren der Minima-Arbeiterinnen der Blattschneider -Ameise (*Atta cephalotes*) *Zeitschrift für Tierpsychologie* **24,**278–281

Eisenschmidt H (1958) Ein starkes Auftreten der Buckelfliege *Megaselia rufipes* Mg als Parasit des Kartoffelkafers (*Leptinotarsa decemlineata* Say) im Jahre 1956 *Zeitschrift für angewandte Zoologie* **1958,** 11–20

Eisfelder I (1956) Die haufigsten Pilzbewohner (Fliegen als Pilzverzehrer) *Zeitschrift für Pilzkunde* **22,** 108–117

El-Miniawi SF, Moustafa MA (1965) On the biology of *Megaselia scalaris* (Loew) (Diptera Phoridae) *Bulletin Societé Entomologique d'Egypte* **49,** 81–91

Elton CS (1966) *The Pattern of Animal Communities* London Methuen

Emerson AE (1928) Termites of the Belgian Congo and the Cameroon *Bulletin of the American Museum of Natural History* **57,** 401–574

Enderlein G (1912b) Neue Gattungen und Arten aussereuropaescher Phoriden *Stettiner entomologische Zeitung* **73,** 46–52

Enderlein G (1924) Zur Klassifikation der Phoriden und uber vernichtende Kritik *Entomologische Mitteilungen aus dem zoologischen Staatsinstitut und Museum, Hamburg* **13,** 270–281

Enderlein G (1929) Entomologica Canaria IV *Wiener entomologische Zeitung* **46,** 95–109

Engel EO (1916) Bemerkungen zur Synonymie und Biologie der Phoriden *Wiener entomologische Zeitung* **35,** 57–58

Ennik F (1973) *Apocephalus borealis* Brues parasitic upon *Vespula* spp (Diptera Phoridae, Hymenoptera Vespidae) *Pan-Pacific Entomologist* **49,** 403–404

Ernst E (1964) Bermukungen zu den Soldaten von *Acanthotermes acanthothorax* Sjost (Isoptera) *Revue Suisse de Zoologie* **71,** 569–576

Erzinçlioğlu YZ (1983a) Exhumed homonyms *Antenna* **7,** 156

Erzinçlioğlu YZ (1983b) The application of entomology to forensic medicine *Medicine Science and the Law* **23,** 57–63

Erzinçlioğlu YZ (1985a) Immature stages of British *Calliphora* and *Cynomyia*, with a re-evaluation of the taxonomic characters of larval Calliphoridae (Diptera) *Journal of Natural History* **19,** 69–96

Erzinçlioğlu YZ (1985b) The entomological investigation of a concealed corpse *Medicine, Science and the Law* **25,** 228–230

Erzinçlioğlu YZ (1989) Entomology, zoology and forensic science the need for expansion *Forensic Science International* **43,** 209–213

Evans H (1989) Mycopathogens of insects of epigeal and aerial habitats In Wilding *et al* , 1989 Chapter **9,** 203–238

Evans RE (1969) Parasites on spiders and their eggs *Proceedings of the Birmingham Natural History Society* **21,** 156–168

Eyles AC (1965) Observations on some parasites of two *Wiseana* species (Lep Hepialidae) *New Zealand Journal of Agricultural Research* **8,** 951–958

Fabre JH (1918) *The Wonders of Instinct* London T Fisher Unwin Ltd

Fabricius JC (1794) *Entomologica systematica emendata et aucta Hafniae* **4,** 1–472

Fage L (1933) A propos du parasitisme des Phorides *Bulletin de la Société Zoologique de France* **58,** 90–92

Fain A, Greenwood MT (1991) Notes on a small collection of mites (Acari), phoretic on Diptera, mainly Phoridae, from the British Isles *Bulletin de l'Institut Royal des Sciences Naturelles de Belgique, Entomologie* **61,** 193–197

Fairbanks LD, Burch GE (1970) Whole-body turnover of chloride in adults of *Drosophila* (Drosophilidae) and *Megaselia* (Phoridae) (Dipt) *Annals of the Entomological Society of America* **63,** 1628–1631

Falcoz L (1912a) La recherche des arthropodes dans les terriers *Feuille des Jeunes Naturalistes* **42,** 178–180

Falcoz L (1912b) Contribution a la faune des terriers de Mammifères *Compte Rendu Hebdomadaire des Seances de l'Academie des Sciences, Paris* **154,** 1380–1383

Feener DH (1981) Competition between ant species outcome controlled by parasitic flies *Science* **214,** 815–817

Feener DH (1987) Size-selective oviposition in *Pseudacteon crawfordi* (Diptera Phoridae), a parasite of fire ants *Annals of the Entomological Society of America* **80,** 148–151

Feener DH (1988) Effects of parasites on foraging and defense behavior of a termitophagous ant, *Pheidole titanis* Wheeler (Hymenoptera Formicidae) *Behavioral Ecology and Sociobiology* **22,** 421–427

Feener DH, Brown BV (1992) Reduced foraging of *Solenopsis geminata* (Hymenoptera Formicidae) in the presence of parasitic *Pseudacteon* spp (Diptera Phoridae) *Annals of the Entomological Society of America* **85,** 80–84

Feener DH, Brown BV (1993) Ovipositon behavior of ant-parasitizing fly, *Neodohrniphora anivinervis* (Diptera Phoridae) and defense behaviour by its leaf-cutting ant host *Atta cephalotis* (Hymenoptera Formicidae) *Journal of Insect Behavior* **6,** 675–688

Feener DH, Moss KAG (1990) Defense against parasites by hitchhikers in leaf-cutting ants a quantitative assessment *Behavioral Ecology and Sociobiology* **26,** 17–29

Felt EP (1935) The occurrence of insects at some height in the air *Circular of the New York State Museum* **17,** 1–70

Ferrar P (1987) *A Guide to the Breeding Habits and Immature Stages of Diptera Cyclorrhapha Entomonograph* Volume 8 Leiden/Copenhagen EJ Brill/ Scandinavian Science Press

Fischer M (1974) *Aspilota*-Arten, gezogen aus Phoriden (Hymenoptera, Braconidae, Alysiinae) *Bolletino del Laboratorio di Entomologia Agraria 'Filippo Silvestri' di Portici* **31,** 253–267

Fitton MG, Shaw MR, Austin AD (1987) Hymenoptera associated with spiders in Europe *Zoological Journal of the Linnean Society* **90,** 65–93

Fletcher JT, White PF, Gaze RH (1989) *Mushrooms Pest and Disease Control* 2nd Edition Hampshire Intercept Ltd

Ford RLE (1979) Dipterous parasites of ladybirds *Proceedings of the Isle of Wight Natural History and Archaeological Society* **7,** 47 (1976)

Foresti L, Pereira da Silva V (1970) Nota sôbre a ocorrêcia de um forideo parasito de formigas sauvas (Diptera, Phoridae) *Ciência e Cultura, São Paulo* **22** (suppl) 287

Fox WH (1887) Note on a new parasite of *Camponotus pennsylvanicus* *Proceedings of the Entomological Society of Washington* **1,** 100

Frank JH, Lounibos LP (eds) (1983) *Phytotelmata terrestrial plants as hosts for aquatic insect communities* New Jersey Plexus Publishing Inc

Franssen CJH (1933a) *Termitoxenia's* *Natuurhistorisch Maandblad* **22,** 90–92

Franssen CJH (1933b) Biologische Untersuchungen an *Termitoxenia hemicyclia* Schmitz, *Termitoxenia punctiventris* Schmitz und *Odontoxenia brevirostris* Schmitz *Biologisches Zentralblatt* **53,** 337–358

Franssen CJH (1936) Aanteekeningen over de ontwikkelingscyclus der Termitoxeniidae (Dipt) *Entomologische Mededeelingen van Nederlandsch-Indie* **2,** 62–65

Franz H (1949) Erster Nachtrag zur Landtierwelt der mitteleren Hohen Tauern

Sitzungsberichte der Osterreichischen Akademie der Wissenschaften **158,** 1–77

Froese A (1992a) *Vergleichende Untersuchungen zur Biologie und Okologie der Dipteren auf integriert und konventionell bewirtschafteten Feldern* Tubingen Eberhard-Karls-Universitat (Ph D Thesis)

Froese A (1992b) Zur Morphologie und Okologie von *Triphleba nudipalpis* (Diptera, Phoridae), einer nekrophagen Buckelfliegen (Diptera Phoridae) *Entomologische Zeitschrift mit insektenborse* **102,** 21–30

Froese A (1992c) Zur Morphologie und Okologie von *Metopina oligoneura* Mik (Diptera, Phoridae) *Zoologisches Jahrbacher, Abteilung für Systematik* **119,** 383–395

Fuller ME, Lee DJ (1938) A new termitophilous phorid *Proceedings of the Linnaean Society of New South Wales* **63,** 75–80

Furukawa E, Kaneko K (1981) Studies of phorid flies (Phoridae, Diptera) in Japan Part IV Scanning electron microscopic observations of eggs of two *Megaselia Japanese Journal of Sanitary Zoology* **32,** 78–81

Garris GI (1983) *Megaselia scalaris* (Diptera Phoridae) infesting laboratory tick colonies *Journal of Medical Entomology* **20,** 688

Gauld I, Bolton B (1988) *The Hymenoptera* Oxford and London Oxford University Press and British Museum (Natural History)

Geigy R, Hecker H, Keiser F (1964) Parasitische Dipterenlarvenin Soldaten von *Bellicositermes bellicosus* (Smeath) (Isoptera) *Acta Tropica* **21,** 280–286

Gemesi O, Disney RHL (1991) A further case of parasitisation of larval Bibionidae by a scuttle fly (Diptera Phoridae) *Entomologist's Gazette* **42,** 67–69

Gimmerthal BA (1848) Beobachtete Metamorphose einer neuen Fliegenspecies und einer bereits bekannten Blattwespe *Arbeiten des Naturforscher-Vereins zu Riga* **1,** 329–330

Girard JE, Snetsinger R, Hendry LB (1972) Light, temperature and geotactic responses of a mushroom-infesting phorid fly *Megaselia bovista* (Diptera Phoridae) *Melsheimer Entomological Series* **10,** 1–7

Glick PA (1939) The distribution of insects, spiders and mites in the air, using traps mounted on a small aeroplane *US Department of Agriculture Technical Bulletin* **673,** 1–150

Goantsa K (1966) The parasites of the bituberculate soft scale (*Palaeolecanium bituberculatum* Tarq) and the spherical apple soft scale (*Eulecanium mali* Schr) in the Moldavian SSR *Trudy Moldavskogo Nauchno-issledovatel'skogo Instituta Sadovodstva Vinogradarstva i Vinodeliya* **13,** 59–69

Goidanich A (1928a) Contributi alli conoscenza dell'entomofauna della Canapa 1 Prospetto generale *Bollettino del Laboratorio di Entomologia del R Istituto Superiore Agrario di Bologna* **1,** 37–64

Goidanich A (1928b) Su di un dittero parassita endofago di formiche *Fiume, Revista Semestrale della Societe di Studi Fiumani* **6,** 244–256

Gol'berg AM (1979) Observations on the fungus *Entomophthora culicis* in Israel *Phytoparasitica* **7,** 31–32

Goos M (1975) Influence of aphicides used in sugar-beet plantations on arthropods IV Studies on the flying insects captured in yellow receptacles *Polskie Pismo Entomologiczne* **45,** 423–444

Gotô T (1983a) Observations on the aerial swarming of *Woodiphora* sp (Diptera, Phoridae) *Makunagi, Acta Dipterologica* **10,** 9–11

Gotô T (1983b) A new phorid fly of the genus *Crinophleba* from Japan (Diptera, Phoridae) *Kontyû* **51,** 376–383

Gotô T (1984a) Observation on the mating behaviour of *Beckerina* sp (Phoridae, Metopininae) *Makunagi, Acta Dipterologica* **12,** 20–22

Gotô T (1984b) Systematic study of the genus *Phora* Latreille from Japan (Diptera, Phoridae) I Description of a new species, with discussion of the terminology of the male genitalia *Kontyû* **52,** 159–171

Gotô T (1984c) Two new species of the genus *Trophithauma* Schmitz from Japan (Diptera, Phoridae) *Bulletin of the Kitakyushu Museum of Natural History* **5,** 75–84

Gotô T (1985a) *Megaselia (Aphiochaeta) araneivora* sp nov , an egg predator of the orb-web spider *Argiope aemura* (Walckenaer) in Japan (Diptera, Phoridae) *Esakia* **23,** 77–84

Gotô T (1985b) A review of the Oriental genus *Rhopica,* with description of a new species from Papua New Guinea (Diptera, Phoridae) *Esakia* **23,** 85–92

Gotô T (1985c) Systematic study of the genus *Phora* Latreille from Japan (Diptera, Phoridae) II Descriptions of four new species *Kontyû* **53,** 1–11

Gotô T (1986a) Systematic study of the genus *Phora* Latreille from Japan (Diptera, Phoridae) V *Kontyû* **54,** 128–142

Gotô T (1986b) *Hypogeophora macrothrix* gen et sp nov , a myrmecophilous phorid fly (Diptera, Phoridae) associated with *Pheidole fervida* Fr Smith (Hymenoptera, Formicidae) in Japan *Kontyû* **54,** 303–308

Gotô T, Takeno K (1983b) A new *Triphleba* species with bifurcate antennae from Japan (Diptera, Phoridae) *Esakia* **20,** 149–156

Grandi G (1951) *Introduzione allo studio dell'entomologia* Volume 2 *Endopterigoti* Bologna

Grasse P-P (1986) *Termitologia Anatomie–physiologie–biologie–systematique des termites* Tome III *Comportement–socialite–ecologie–evolution–systematique* Paris Masson

Gratwick M (ed) (1992) *Crop Pests in the UK – collected edition of MAFF leaflets* London Chapman & Hall

Greathead DJ (1963) A review of the insect enemies of Acridoidea (Orthoptera) *Transactions of the Royal Entomological Society of London* **114,** 437–517

Greenberg B (1991) Flies as forensic indicators *Journal of Medical Entomology* **28,** 565–577

Greene CT (1938) A new genus and two new species of the Dipterous family Phoridae *Proceedings of the United States National Museum* **85,** 181–185

Gregor F (1977) Synanthropy and faunistics of some Phoridae (Diptera) from Cuba *Folia Parasitologica* **24,** 73–80

Gregorio R de, Leonide JC (1980) Un nouveau cas de phoride parasite d'Orthopteres adultes (Dipt) *Bulletin de la Societe Entomologique de France* **85,** 103–105

Grensted LW (1945) Predators and other Diptera on cow-parsnip *Entomologist's Monthly Magazine* **81,** 229

Grensted LW (1957) Species of *Spiniphora* (Dipt , Phoridae) in Gloucestershire *Entomologist's Monthly Magazine* **92,** 405

Grimaldi D (1989) The genus *Metopina* (Diptera Phoridae) from Cretaceous and Tertiary ambers *Journal of the New York Entomological Society* **97,** 65–72

Grimaldi D, Beck CW, Boon JJ (1989) Occurrence, chemical characteristics, and palaeontology of the fossil resins from New Jersey *American Museum Novitates* **2948,** 1–27

Grove JF, Blight MM (1983) The oviposition attractant for the mushroom phorid *Megaselia halterata* the identification of the volatiles present in mushroom house air *Journal of the Science of Food and Agriculture* **34,** 181–185

Guthrie M (1989) *Animals of the Surface Film* London Richmond Publishing Co Ltd , Naturalists' Handbooks 12

Hackman W (1963) Studies on the Dipterous fauna in burrows of voles (*Microtus, Clethrionomys*) in Finland *Acta Zoologica Fennica* **102,** 1–64

Hackman W (1964) On reduction and loss of wings in Diptera *Notulae Entomologicae* **44,** 73–93

Hackman W, Meinander M (1979) Diptera feeding as larvae on macrofungi in Finland *Annales Zoologici Fennici* **16,** 50–83

REFERENCES

Hall DW, Brown BV (1993) Pollination of *Aristolochia littoralis* (Aristolochia Aristolochiaciae) by males of *Megaselia* spp (Diptera Phoridae) *Annals of the Entomological Society of America* **86,** 609–13

Hamm AH, Richards OW (1926) The biology of British Crabronidae *Transactions of the Entomological Society of London* **74,** 297–331

Hammond HE, Smith KGV (1955) On some parasitic Diptera and Hymenoptera bred from Lepidopterous hosts Part II Misc records of Phoridae, Larvaevoridae (Dipt), Braconidae, Ichneumonidae, and Eulophidae (Hym) *Entomologist's Gazette* **6,** 168–174

Handschin E (1944) Insekten aus den Phosphoriten des Quercy *Schweizerische palaontologische Abhandlungen* **64,** 1–23

Hanski I (1989) Fungivory, Insects and ecology In Wilding *et al* (eds), Chapter 2 25–68

Hardy AC, Milne PS (1938) Studies in the distribution of insects by aerial currents Experiments in aerial tow-netting from kites *Journal of Animal Ecology* **7,** 199–229

Hardy DE (1951) A case of apparent human myiasis *Proceedings of the Hawaiian Entomological Society* **14,** 212–213

Hardy DE, Beyer EM (1964) Family Phoridae In Zimmerman EC (eds), *Insects of Hawaii* Volume 11 Honolulu, 262–302

Hawksworth DL, Sutton BC, Ainsworth GC (1983) *Ainsworth and Bisby's Dictionary of the Fungi (including the Lichens)* 7th Edition Kew Commonwealth Mycological Institute

Haworth WE (1924) Mosquitoes and coconut palms *Transactions of the Royal Society of Tropical Medicine and Hygiene* **18,** 162–198

Healey IN, Russell-Smith A (1970) The extraction of fly larvae from woodland soils *Soil Biology and Biochemistry* **2,** 119–129

Hennig W (1971) Zur Situation der biologischen Systematik *Erlanger Forschungen (B)* **4,** 7–15

Herbert F, Braun C (1958) Moospolster als Winterquartier europaischer Phoriden – Imagines (Phoridae, Diptera) *Broteria* **27,** 17–29

Hering EM (1951) *Biology of the Leaf Miners* Gravenhage Junk

Hesjedal K (1984) *Megaselia rufipes* Meigen (Dipt , Phoridae) as a prepupae parasite of the winter moth and allied species in deciduous forests in western Norway *Fauna Norvegica series B* **31,** 113–114

Heslop-Harrison Y (1976) Carnivorous plants a century after Darwin *Endeavour* **35,** 114–122

Heynen C (1988) Zur Biologie eines Buchenwaldbodens II Die Dipterenlarven *Carolinea* **46,** 115–130

Heynen C (1990) Verbesserung der Extraktion von Dipterenlarven aus Ackerboden *Verhandlungen der Gesellschaft fur Okologie, Osnabruck* **19,** 282–288

Heynen C, Wubbeler H (1990) Integrieter Pflanzenschutz im Ackerbau Das Lautenbach-Projekt V Schlag-spezifischer Vergleich der Dipteren-Fauna auf integriert und konventionell bewirtschafteten Flachen *Zeitschrift fur Pflanzenkrankheiten und Pflanzenschutz* **97,** 367–380

Hicks EA (1959) *Check-list and Bibliography on the occurrence of Insects in Birds' nests* Ames The Iowa State College Press

Hime NDaC, Costa EDeL (1985) On *Megaselia (M) aristolochiae* n sp (Diptera, Phoridae) with larvae rearing on the flowers of *Aristolochia labiata* Willd (Aristolochiaceae) *Revista Brasiliera de Biologia* **45,** 621–625

Hinton HE (1946) A new classification of insect pupae *Proceedings of the Zoological Society of London* **116,** 282–328

Hobby BM (1931a) The British species of Asilidae (Diptera) and their prey *Transactions of the Entomological Society of the South of England* **6,** 1–42

Hobby BM (1931b) The prey of *Coenosia tigrina* F and other species of the genus *Proceedings of the Entomological Society of London* **6,** 13–15

Hobby BM (1932) A study of the prey of *Dioctria rufipes* Deg (Dipt , Asilidae) in an Oxford community *Journal of Animal Ecology* **1,** 77–82

Hobby BM (1933) Supplementary list of the prey of Asilidae (Dipt) *Journal of the Entomological Society of the South of England* **1,** 69–74

Hodek I (1973) *Biology of Coccinellidae* The Hague Junk

Hodkinson ID, Casson D (1991) A lesser predilection for bugs Hemiptera (Insecta) diversity in tropical rain forests *Biological Journal of the Linnean Society* **43,** 101–109

Hoffmann AA, O'Donnell S (1992) Heritable variation in the attraction of *Drosophila melanogaster* to fruit in the field *Biological Journal of the Linnean Society* **47,** 147–159

Holldobler B, Wilson EO (1990) *The Ants* London Springer-Verlag

Holldobler K (1928) Zur Biologie der diebischen Zwergamiese (*Solenopsis fugax*) und ihrer Gaste *Biologisches Zentralblatt* **48,** 129–142

Holloway JD (1985) The moths of Borneo family Noctuidae subfamilies Euteliinae, Stictopterinae, Plusiinae, Pantheinae *Malayan Nature Journal* **38,** 157–317

Hong Y-C (1981) *Eocene fossil Diptera Insecta in amber of Fushun coalfield* Peking Geological Publishing House

Hovemeyer K (1984) Die Dipterengemeinschaft eines Buchenwaldes auf Kalkgestein Produktion an Imagines, Abundanz und raumliche Verteilung insbesondere der Larven *Pedobiologia* **26,** 1–15

Hovemeyer K (1989) Der Einfluss von Streumenge und Streuqualitat auf die Siedlungsdichte von Dipterenlarven ein Freilandexperiment im Kalkbuchenwald (Zur Funktion der Fauna in einen Mullbuchenwald 4) *Verhandlungen Gesellschaft fur Okologie, Gottingen* **17,** 229–236

Howard NF, Landis BJ (1936) Parasites and predators of the Mexican bean beetle in the United States *United States Department of Agriculture Circular* **418,** 1–12

Howden HF, Vogt GB (1951) Insect communities of standing dead pine (*Pinus virginiana* Mill) *Annals of the Entomological Society of America* **44,** 581–595

Hsin CS (1935) Beitrage zur Naturgeschichte der Blattwespen *Zeitschrift fur angewandte Entomologie* **22,** 253–294

Hussey NW (1960) Biology of mushroom phorids *Mushroom Science* **4,** 260–269 (1959)

Hussey NW (1961a) Identification of Phoridae (Diptera) attacking cultivated mushrooms in Britain *Annals and Magazine of Natural History (13)* **3,** 599–603 (1960)

Hussey NW (1961b) Recent work on the control of the Worthing Phorid *Mushroom Growers' Association Bulletin* **144,** 495–505

Hussey NW (1965) Economics of pest control suggested by a study of phorid flight activity *Mushroom Growers' Association Bulletin* **182,** 3–11

Hussey NW, Gurney B (1963) Mushroom phorids *Report of the Glasshouse Crops Research Institute* **1963,** 76–77

Hussey NW, Gurney B (1964a) Rearing techniques for mushroom fly *Plant Pathology* **13,** 38–39

Hussey NW, Gurney B (1964b) Mushroom phorids *Report of the Glasshouse Crops Research Institute* **1964,** 76–77

Hussey NW, Read WH, Hesling JJ (1969) *The Pests of Protected Cultivation Biology and control of glasshouse and mushroom pests* London Edward Arnold

Hussey NW, Wyatt IJ (1962) The interaction between mushroom mycelium and insect pest populations *Mushroom Science* **5,** 509–517

REFERENCES

Iheagwam EU, Nwankiti OC (1980) Dipterous insect pests of pepper, *Capsicum* spp , in the Eastern States of Nigeria *Revue de Zoologie Africaine* **94,** 936–939

Ingram A (1922) *Aphiochaeta xanthina*, Speiser (*A ferruginae*, Brun) as an intestinal parasite in the Gold Coast *Journal of Tropical Medicine and Hygiene* **25,** 113

Iperti G (1964) Les parasites des coccinelles aphidiphages dans les Alpes-Maritimes et les Basses-Alpes *Entomophaga* **9,** 153–180

Iperti G (1965a) The natural enemies of aphidophagous Coccinellids In Hodek I (ed), *Ecology of Aphidophagous Insects* The Hague Junk

Jack RW (1935) The report of the chief entomologist for the year ending 31st December, 1934 *Rhodesia Agricultural Journal* **32,** 558–581

Jaeger EC (1959) *A Source-book of Biological names and terms* Third Edition Springfield, Illinois Charles C Thomas

Janovy J, Daggett PM, Lee KW (1974a) *Herpetomonas megaseliae* architectural rearrangements during amastigote formation *Journal of Parasitology* **60,** 716–718

Janovy J, Lee KW, Brumbaugh JA (1974b) The differentiation of *Herpetomonas megaseliae* ultrastructure observations *Journal of Protozoology* **21,** 53–59

Jansson A (1919) Die Arthropodenfauna der Eichhornchennester *Entomologisk Tidskrift* **40,** 75–92

Johal K, Disney RHL (in press) Phoridae (Diptera) as pests of cultivated oyster mushrooms (Agaricales Pleurotaceae) in India *Bulletin of Entomological Research*

Johnson DS, Mertl HG, Traut W (1988) Inheritance of cytogenetic and new genetic markers in *Megaselia scalaris*, a fly with an unusual sex determining mechanism *Genetica* **77,** 159–170

Johnson MD (1975) Seasonal and microseral variations in the insect populations on carrion *American Midland Naturalist* **93,** 79–90

Jones FM (1918) *Dohrniphora venusta* Coqu in *Sarracenia flava* *Entomological News* **29,** 299–302

Jones SE (1940) An annotated list of the spiders of an East Central Illinois forest (WM Trelease Woods, Univ of Illinois) *Transactions of the Illinois Academy of Science* **33,** 216–220

Judd WW (1975) Insects associated with flowering silky dogwood (*Cornus obliqua* Raf) at Dunville, Haldimand County, Ontario *Ontario Field Biologist* **29,** 26–35

Kalshoven LGE (1938) Weiteres uber das Benehmen der *Misotermes* abdominal larven und der Myiagenen *Entomologische Mededeelingen van Nederlandisch-Indie* **16,** 395–397

Kamm JA, Buttery RG, Robinson WH (1987) An attractant for mushroom flies (Diptera Phoridae) *Journal of the New York Entomological Society* **95,** 19–22

Kaneko K, Furukawa E (1983) Studies on phorid flies (Phoridae, Diptera) in Japan Part V Morphological notes on *Megaselia trivialis* (Brues, 1911) *Japanese Journal of Sanitary Zoology* **34,** 39–42

Kaneko K, Kano R, Okazaki T (1961) Studies on phorid flies in Japan Key to subfamilies and important genera of Phoridae, and description of six species belonging to five genera *Japanese Journal of Sanitary Zoology* **12,** 238–247

Kanervo V (1946) Tutkimuksia Lepan Lehtikuoriaisen, *Melasoma aenea* L (Col , Chrysomelidae), Luontaisista vihollisista *Annales Zoologici Societatis Zoologicae-Botanicae Fennicae* **12,** 1–174

Kanmiya K (1985) Record of courtship sound in a phorid species, *Puliciphora tokyoensis* Kinoshita, with notes on its acoustic properties *Makunagi, Acta Dipterologica* **13,** 1–7

Kano R, Kaneko K, Kawashima K, So N (1962) On some cases of myiasis *Japanese Journal of Sanitary Zoology* **13,** 96–97

Kaston B, Jenks GE (1937) Dipterous parasites of spider egg sacs *Bulletin of the Brooklyn Entomological Society* **32,** 160–163

Kearns CA (1992) Anthophilous fly distribution across an elevation gradient *American Midland Naturalist* **127,** 172–182

Keilin D (1911a) *Phora maculata* Mg est-il parasite de *Gelechia maculatella* Hb ? *Feuilles de Jeunes Naturalistes* **41,** 150–151

Keilin D (1911b) Recherches sur la morphologie larvaire des dipteres du genre *Phora* *Bulletin Scientifique de la France et de la Belgique* **45,** 27–88

Keilin D (1917) Recherches sur les anthomyides à larves carnivores *Parasitology* **9,** 325–445

Keilin D (1919) On the life history and larval anatomy of *Melinda cognata* Meigen, parasitic on the snail *Helicella (Heliomanes) virgata* Da Costa, with an account of the other Diptera living upon molluscs *Parasitology* **11,** 430–455

Keller S, Zimmermann G (1989) Mycopathogens of soil insects In Wilding *et al* (eds), Chapter 10 237–270

Kemner NA (1922) Die Larve der Termitoxenien entdeckt! Vorlaufiger Bericht *Entomologisk Tidskrift* **43,** 58–61

Kemner NA (1926) Uber die Zucht der Larve einer echten *Termitoxenia* *International Congress of Entomology (1925)* **3,** 389–404

Kemner NA (1932a) Eine termitophile Phoride, die ihre Fluegel abwirft *Echidnophora butteli* Schmitz *Arkiv für Zoologi* **24B,** 1–6

Kemner NA (1932b) Zur Kenntnis der javanischen Termitoxenien, mit Bemerkungen ueber ihre Morphologie und Biologie *Entomologisk Tidskrift* **53,** 17–29

Kern RA (1938) Asthma due to sensitization to the mushroom fly (*Aphiochaeta agarici*) *Journal of Allergy* **9,** 604–606

Khalaf FT (1971) Phoridae from Louisiana and Mississipi *Florida Entomologist* **54,** 273–276

Kidd LN, Brindle A (1959) The Diptera of Lancashire and Cheshire – Part I *Lancashire and Cheshire Fauna Society Publication* **1959,** 1–136

Kirk WDJ (1984) Ecologically selective colour traps *Ecological Entomology* **9,** 35–41

Kistner DH (1969) The biology of termitophiles In Krishna K, Weesner FM (eds), *Biology of Termites* New York Academic Press, Chapter 17 525–557

Kistner DH (1982) The social insects' bestiary In Herman HR (ed), *Social Insects* Volume 3 New York Academic Press, 1–244

Kiyoku M (1958) Studies on the insect fauna of the Japanese pine mushroom and the damage due to the insect pests *Scientific Reports of the University of Okayama University Faculty of Agriculture* **11,** 49–59

Kloter KO, Penner LR, Widmer WJ (1977) Interactions between the larvae of *Psychoda alternata* (Dipt , Psychodidae) and *Dohrniphora cornuta* (Dipt , Phoridae) in a trickling filter sewage bed, with descriptions of the immature stages of the latter *Annals of the Entomological Society of America* **70,** 775–781

Knab F (1913) Some earlier observations on the habits of *Aphiochaeta juli* (Brues) (Diptera, Phoridae) *Insecutor Inscitiae Menstruus* **1,** 24

Kneidel KA (1983) Fugitive species and priority during colonization in carrion-breeding Diptera communities *Ecological Entomology* **8,** 163–169

Kneidel KA (1984a) Influence of carcass taxon and size on species composition of carrion-breeding Diptera *American Midland Naturalist* **111,** 57–63

Kneidel KA (1984b) Competition and disturbance in communities of carrion-breeding Diptera *Journal of Animal Ecology* **53,** 849–865

Kneidel KA (1985) Patchiness, aggegation, and the coexistence of competitors for ephemeral resources *Ecological Entomology* **10,** 441–448

Knutson LV (1970) Biology and immature stages of *Tetanura pallidiventris*, a parasitoid of terrestrial snails (Dipt Sciomyzidae) *Entomologica Scandinavica* **1,** 81–89

Kogan M, Helm CG, Kogan J, Brewer E (1989) Distribution and economic importance of *Heliothis virescens* and *Heliothis zea* in North, Central and South America and of their natural enemies and host plants In King EG, Jackson RD (eds), *Proceedings of the Workshop on Biological Control of Heliothis Increasing the Effectiveness of Natural Enemies* New Delhi Aspect Publishing, 241–297

Komori K, Hara K, Smith KGV, Oda T, Karamine, D (1978) A case of lung myiasis caused by larvae of *Megaselia spiracularis* Schmitz (Diptera Phoridae) *Transactions of the Royal Society of Tropical Medicine and Hygiene* **72,** 467–470 [Karamine is an error for Katamine on title page of this work]

Kovac D, Streit B (in press) Arthropod community in bamboo internodes In Dumont H (ed), Peninsular Malaysia microzonation and trophic structure [Proceedings of tropical rainforest conference in Brunei, April 1993]

Krespi L, Deleporte S, Nenon JP (1985) Étude de l'entomocénose prédatrice de *Bradysia confinis* (Winn , Frey) (Diptère, Sciaridae) *Bulletin de la Société de Bretagne* **57,** 11–23

Krombein KV (1964) The natural history of Plummers Island, Maryland XVIII The Hibiscus wasp, an abundant rarity, and its associates (Hymenoptera Specidae) *Proceedings of the Biological Society of Washington* **77,** 73–112

Krombein KV (1967) *Trap-nesting Wasps and Bees* Washington

Kuhlhorn von F (1987) Uber den Dipterenbeflug von Katzenkot und dessen mogliche hygienische Bedeutung *Angewandte Parasitologie* **28,** 93–101

LaBerge WE (1953) A note on a phorid parasite of *Pheidole dentata* Mayr *Journal of the Kansas Entomological Society* **26,** 69

Lack D, Owen DF (1955) The food of the swift *Journal of Animal Ecology* **24,** 120–136

Latreille P (1796) *Precis des caractères generiques des insectes disposés dans un ordre naturel* Paris

Latreille P (1802) *Histoire Naturelle generale et particulière des Crustacés et des Insectes* Volume 3 Paris

Laurence BR (1951) The prey of some tree trunk frequenting Empididae and Dolichopodidae (Dipt) *Entomologist's Monthly Magazine* **87,** 166–169

Laurence SM (1910) Intestinal myiasis *British Medical Journal* **1910 (II),** 376

Lea AM (1911) On a new Australian genus of Phoridae, associated with termites *Proceedings of the Royal Society of Victoria* **24,** 76–77

Leclercq J (1946b) Influence de l'humidité atmosphérique sur les pupes de *Triphleba lugubris* Meigen (Diptére Phoride) *Compte Rendu des Séances de Biologie, Paris* **140,** 585–586

Leclercq J, Leclercq M (1948) Données bionomique pour *Calliphora erythrocephala* (Meigen) et cas d'application a la medecine legal *Bulletin de la Société Entomologique de France* **53,** 101–103

Leclercq M (1945) Notes sur les Diptéres des environs de Liege – Phoridae de l'avant-plateau de Herve *Bulletin de la Société Royale des Sciences de Liege* **1945,** 160–162

Leclercq M (1978) *Entomologie et Medecine Legale Datation de la mort Collection de Médecine Legale et de Toxicologie Medicale No* **108** Paris Masson

Leclercq M, Watrin P (1973) Entomologie et medecine legale Acariens et Insectes trouvés sur un cadavre humain en décembre 1971 *Bulletin et Annales de la Société Royale Entomologique de Belgique* **109,** 195–201

Lee DJ (1968) Human myiasis in Australia *Medical Journal of Australia* **1968,** 170–173

Leefmans S (1927) Indische Processiesmaden *Tropische Natuur Nederlandsch-Indische Natuurhistorische Vereeniging, Weltevreden* **16,** 21–25

Leleup N (1955) Chasses entomologiques dans des biotopes speciaux au Congo Belge, *Memoires de la Societe Royale Entomologique de Belgique* **27,** 361–377

Le Manse G (1941) *Tubicera lichtwardti* Schmitz (Dipt Phoridae) hôte de *Plagiolepis pygmaea* Latr (Hym Formicidae) *Bulletin de la Societe Entomologique de France* **46,** 110–111

Leonardi G (1927) *Elenco delle specie di insetti dannosi e lore parassiti ricordati in Italia fino all'anno 1911 Parte III Insetti parassiti di altri insetti* Italy Modena

Le Pelley RH (1959) *Agricultural insects of East Africa* Nairobi E A High Commission

Leruth R (1936) Phorides cavernicoles de Belgique *Bulletin du Musee Royal d'Histoire Naturelle de Belgique* **12(36),** 1–23

Leruth R (1939) La biologie du domaine souterrain et la faune cavernicole de la Belgique *Memoires du Musee Royal d'Histoire Naturelle de Belgique* **87,** 1–506

Lesne P (1923) Faune entomologique des fosses d'aisances et des excrements humains *Bulletin du Museum d'Histoire Naturelle, Paris* **29,** 161–167

Lever RJAW (1944) Maggots in imported boot polish *Agricultural Journal, Department of Agriculture, Fiji* **15,** 15

Lewis DJ (1952) *Simulium damnosum* and its relation to Onchocerciasis in the Anglo-Egyptian Sudan *Bulletin of Entomological Research* **43,** 597–644

Lewis DJ (1973) Phlebotomidae and Psychodidae (sand-flies and moth-flies) In Smith KGV (ed), *Insects and Other Arthropods of Medical Importance* London British Museum (Natural History), Chapter 3c 155–179

Lewis T (1967) The horizontal and vertical distribution of insects near artificial windbreaks *Annals of Applied Biology* **60,** 23–31

Lewis T, Taylor LR (1965) Diurnal periodicity of flight by insects *Transactions of the Royal Entomological Society of London* **116,** 393–479

Lichtenstein JL (1920) La parasitisme d'*Aphiochaeta (Phora) fasciata* Fallen *Comptes Rendus Hebdomadaire des Seances de l'Academie des Sciences, Paris* **170,** 531–534

Lincoln RJ, Boxshall GA, Clark PF (1982) *A dictionary of ecology, evolution and systematics* Cambridge Cambridge University Press

Lindner E (1969) Die Dipteren einer an einer Gartenrose durch eine Aleuridide in Ostafrika verursachten Biocoenose *Memorie della Societe Entomologica Italiana* **48,** 222–232

Linsley EG (1945) A phorid fly from the nests of *Anthophora* in California *Bulletin of the Brooklyn Entomological Society* **40,** 67–68

Lipa YY, Sem'yanov VP (1967) The parasites of the ladybirds (Col Coccinell) in the Leningrad region *Entomological Review, Washington* **46,** 43–45

Liston AD (1979) On Phoridae (Diptera) from sawfly cocoons (Hym Symphyta) *Entomologist's Record and Journal of Variation* **91,** 303–305

Liston AD (1982) Further records of Phoridae (Diptera) from cocoons of Cimbicidae (Hymenoptera) *Entomologist's Record and Journal of Variation* **94,** 221

Lloyd JE (1973) Firefly (Col , Lampyridae) parasites and predators *Coleopterists' Bulletin* **27,** 91–106

Lundbeck W (1922) *Diptera Danica Part VI Pipunculidae, Phoridae* London Wesley

Lundt H (1964) Ecological observations about the invasion of insects into carcasses buried in soil *Pedobiologia* **4,** 158–180

Mabberley DJ (1989) *The Plant-Book a portable dictionary of the higher plants* Cambridge Cambridge University Press

McAlpine JF (1965) Observations on anthophilous Diptera at Lake Hazen, Ellesmere Island *Canadian Field Naturalist* **79,** 247–252

McAlpine JF (1989) Phylogeny and classification of the Muscomorpha In McAlpine JF, Wood DM (eds), *Manual of Nearctic Diptera* Volume 3 Ottawa Research Branch Agriculture Canada Monograph 32, Chapter 116 1397–1518

McAlpine JF, Martin JEH (1966) Systematics of Sciadoceridae and relatives with descriptions of two new genera and species from Canadian amber and erection of family Ironomyiidae (Diptera: Phoroidea). *Canadian Entomologist* **98,** 527–544.
McClure HE, Lim B-L, Winn SE (1967) Fauna of the Dark Cave, Batu Caves, Kuala Lumpur, Malaysia. *Pacific Insects* **9**, 399–428.
McCrae AWR (1967) Infestation of emulsion paint by the fly *Megaselia scalaris* (Loew) (Dipt., Phoridae). *Entomologist's Monthly Magazine* **102,** 241–243.
MacDonald JF, Akre RD, Hill WB (1975) Nest associates of *Vespula atropilosa* and *V. pensylvanica* in southeastern Washington State (Hymenoptera: Vespidae). *Journal of the Kansas Entomological Society* **48,** 55–63.
MacDougall RS (1920) Insect and arachnid pests of 1919. *Transactions of the Highland and Agricultural Society of Scotland* **32,** 152–192.
Mackenzie A (1992) The evolutionary significance of host-mediated conditioning. *Antenna* **16,** 141–150.
Macquart J (1835) Histoire naturelle des insectes. Diptères. In Roret NE (ed.), *Collections des Suites à Buffon*, Paris **2,** iv, 1–703.
Maeta Y (1969) Some biological studies on the natural enemies of some coccinellid beetles. (II). *Phalacrotophora* sp. *Tohoku Konchu Kenkyu* **4,** 1–6.
Mainx F (1964) The genetics of *Megaselia scalaris* Loew (Phoridae): a new type of sex determination in Diptera. *American Naturalist* **98,** 415–430.
Majerus M, Kearns P (1989) *Ladybirds*. Slough: Richmond Publishing Co. Ltd., Naturalists' Handbooks **10.**
Malin RG, Kistner DH (1970) Behavioral and anatomical studies on adults of *Termitophilomyia* and *Termitosagma* (Diptera: Phoridae). *Annals of the Entomological Society of America* **63,** 1376–1382.
Malloch JR (1908a) Phoridae in Dumbartonshire; with description of a new species. *Entomologist's Monthly Magazine* **44,** 11–13.
Malloch JR (1908b) Notes on Phoridae in Dumbartonshire, with description of a new species. *Entomologist's Monthly Magazine* **44,** 203–205.
Malloch JR (1909b) Some Phoridae in Dumbartonshire in 1908, with description of a new species. *Entomologists Monthly Magazine* **45,** 34–36.
Malloch JR (1910) Scottish Phoridae, with tables of all the British species and notes of localities. *Annals of Scottish Natural History* **1910,** 15–21, 87–92.
Malloch JR (1911) Some observations on the dipterous family Phoridae. *Transactions of the Natural History Society of Glasgow* **8,** 153–156.
Malloch JR (1912b) The insects of the Dipterous family Phoridae in the United States National Museum. *Proceedings of the United States National Museum* **43,** 411–529.
Malloch JR (1914a) Costa Rican Diptera. Paper 1. A partial report on Borboridae, Phoridae and Agromyzidae. *Transactions of the American Entomologiocal Society* **40,** 8–36.
Malloch JR (1914c) Notes on Illinois Phoridae, with descriptions of three new species. *Bulletin of the Brooklyn Entomological Society* **9,** 56–59.
Malloch JR (1915a) Two new Oriental Phoridae. *Psyche* **22,** 27–29.
Malloch JR (1935b) Phoridae, Agromyzidae, Micropezidae, Tachinidae and Sarcophagidae (supplement). *Insects of Samoa. Part VI Diptera*; fasc. **9:** 329–366. London: British Museum (Natural History).
Malo F, Willis ER (1961) Life history and biological control of *Caligo eurilochus*, a pest of banana. *Journal of Economic Entomology* **54,** 530–536.
Mani MS (1968) *Ecology and Biogeography of High Altitude Insects*. The Hague.
Manuel RL (1984) The egg sacs of *Pitohyphantes costatus* (Hents) (Araneae, Linyphiidae) and its phorid parasite. *Journal of Arachnology* **12,** 371–372.

Martelli G (1913) La *Thea 22–punctata* L è solamente micofaga *Giornale di Agricoltura Meridionale, Messina* **6,** 189–185

Martelli G (1914) Notizie su due Coccinellidi micofagi *Bollettino del Laboratorio de Zoologia Generale e Agraria, Portici* **9,** 151–160

Maruszewska U (1976) The effect of spraying against Colorado beetle – *Leptinotarsa decemlineata* Say (Col , Chrysomelidae) on some Diptera *Polskie Pismo Entomolgiczne* **46,** 607–612

Maschwitz U, Steghaus-Kovac S, Gaube R, Hanel H (1989) A South East Asian ponerine ant of the genus *Leptogenys* (Hym , Form) with army ant life habits *Behavioral Ecology and Sociobiology* **24,** 305–316

Matile L (1970) Les Diptéres cavernicoles *Annales de Spéléologie* **25,** 179–222

Matsuda R (1987) *Animal Evolution in Changing Environments with Special Reference to Abnormal Metamorphosis* Chichester Wiley

May R (1989) How many species? In Friday L, Laskey R (eds), *The Fragile Environment* Cambridge Cambridge University Press

Mayr E (1942) *Systematics and the Origin of Species from the viewpoint of a zoologist* New York Columbia University Press

Mayr E (1970) *Populations, Species and Evolution* Cambridge, Massachusetts Harvard University Press

Mayr E (1991) *One Long Argument* Cambridge, Massachusetts Harvard University Press

McClure HE, Lim B-L, Winn SE (1967) Fauna of the Dark Cave, Batu Caves, Kuala Lumpur, Malaysia *Pacific Insects* **9,** 399–428

Medler JT (1965) Biology of *Isodontia (Murrayella) mexicana* in trap-nests in Wisconsin (Hymenoptera Sphecidae) *Annals of the Entomological Society of America* **58,** 137–142

Meigen JW (1804) *Klassifikation und Beschreibung der europaeischen zweifliegelingen Insecten* Braunschweig

Meigen JW (1830) *Systematische Beschreibung bekannten europaeischen zweifliegelingen Insekten* Hamm **6,** iv, 1–401

Meijere JCH de (1910) *Nepenthes*-Tiere I Systematik *Annales du Jardin Botanique de Buitenzorg (Supplement)* **3,** 917–940

Meijere JCH de (1912) Ueber die Metamorphose von *Puliciphora* und ueber neue Arten der Gattungen *Puliciphora* Dahl und *Chonocephalus* Wandolleck *Zoologische Jahrbucher (Supplement)* **15,** 141–154

Meijere JCH de (1938) Uber die Prothrakalhorner der Puppe von *Diploneura cornuta* Big (Dipt Phoridae) *Tijdschrift voor Entomologie* **81,** 230–233

Meijere JCH de (1947) Over eenige Dipterenlarven, waaronder een galmug, die mijngangen maakt, en twee Dipteren die gallen op paddenstoelen veroorzaken *Tijdschrift voor Entomologie* **88,** 49–62

Meinhardt W, Disney RHL (1989) Urogenital myiasis caused by scuttle fly larvae (Diptera Phoridae) *British Journal of Urology* **64,** 547–548

Menhinick EF (1962) Comparison of invertebrate populations of soil and litter of mowed grasslands in areas treated and untreated with pesticides *Ecology* **43,** 556–561

Menozzi C (1921) Una nuova specie del genere *Apocephalus* Coquil (Diptera – Phoridae) parassita del *Crematogaster scutellaris* Oliv *Bollettino della Societé Entomologica Italiana* **51,** 70–77 [Note The author's name is misspelt Minozzi on the title page]

Menozzi C (1927) Contributo all biologia della *Phalacrotophora fasciata* Fall (Diptera – Phoridae) parassita di Coccinellidi *Bollettino della Società Entomologica Italiana* **59,** 72–78

Mergelsberg O (1935) Uber die postimaginale Entwicklung (Physogastrie) und den Hermaphroditismus bei afrikanischen Termitoxenien (Dipt) *Zoologische Jahrbuchern (Anatomie)* **60,** 345–398

Merrit-Hawkes GA (1927) A preliminary account of the life-history of *Coccinella 11–punctata* (L) *Transactions of the Royal Entomological Society of London* **75,** 47–52

Meyers JG (1934) Aggressive parasitism of a millipede by a phorid *Proceedings of the Royal Entomological Society of London* **9,** 62–63

Michailovskaya MV (1986) New data on the fauna of Phoridae (Diptera) of Primorskij Territory *Entomologicheskoe Obozrenie* **65,** 427–432

Michailovskaya MV (1988) A new species of the genus *Abaristophora* (Diptera Phoridae) from Is Sakhalin *Zoologicheskii Zhurnal* **67,** 470–471

Michl E (1911) Ueber termitophile Dipteren *Mitteilungen des naturwissenschaftlichen Vereines an der Universitat Wien* **9,** 53–60, 84–92

Miller PL (1979a) Intraspecific phorid phoresy *Entomologist's Monthly Magazine* **114,** 211–214 (1978)

Miller PL (1979b) A possible sensory function for stop–go patterns of running in phorid flies *Physiological Entomolgy* **4,** 361–370

Miller PL (1984) Alternate reproductive routines in a small fly, *Puliciphora borinquenensis* (Diptera Phoridae) *Ecological Entomology* **9,** 293–302

Moericke V (1951) Eine Farbfalle zur Kontrolle des Fluges von Blattlausen, insbesondere der Pfirsichblattlaus *M persicae* (Sulz) *Nachrichtenblatt des Deutschen Pflanzenschutzdenstes, Braunschweig* **3,** 23–24

Mollon A (1982) Étude d'un écosystème forestier VII Composition et phénologie du peuplement des Dipteres à larves édaphiques *Revue d'Ecologie et de Biologie du Sol* **19,** 289–306

Moore BP (1969) Biochemical studies in termites In Krishna K, Weesner FM (eds), *Biology of Termites,* London Academic Press Chapter 13 407–432

Moreton BD, John ME (1955) The mushroom phorid fly epidemic of 1953 *Plant Pathology* **4,** 9–11

Morley C (1934) Life-history of *Phora* flies *Transactions of the Suffolk Naturalists' Society* **2,** 298

Morley C, Atmore EA (1915) The Diptera of Norfolk and Suffolk *Transactions of the Norfolk and Norwich Naturalists' Society, Supplement* **1915,** 1–180

Morris HM (1922) On the larva and pupa of a parasitic phorid fly, *Hypocera incrassata* Mg *Parasitology* **14,** 70–74

Morris HM (1927) The insect fauna of arable land at Rothamsted Part II *Annals of Applied Biology* **14,** 442–465

Muesebeck CFW (1965) A new diapriid (Hymenoptera Diapriidae) from termite nests from South Africa *Journal of the Entomological Society of Southern Africa* **27,** 188–190

Muller GW (1919) Ueberparasitische Scatopsiden *Zeitschrift fur wissenschaftliche Insektbiologie* **15,** 120

Muma MH (1954) Predators and parasites of the citrus tree snail *Citrus Magazine* **16(10),** 8–9

Muma MH (1955) Observations on the biology of the citrus tree snail *Citrus Industry* **36,** 6–9, 21

Muma MH, Stone KJ (1971) Predation of *Gasteracantha cancriformis* (Arachnida Araneidae) eggs in Florida citrus groves by *Phalacrotophora epeirae* (Insecta Phoridae) and *Arachnophaga ferruginea* (Insecta Eupelmidae) *Florida Entomologist* **54,** 307–310

Naumov AN (1979) New species of Phoridae (Diptera) from the European part of the USSR and Central Asia (in Russian) *Entomologicheskoe Obozrenie* **58,** 417–420

Nelson JM (1965) A seasonal study of aerial insects close to a moorland stream *Journal of Animal Ecology* **34,** 573–579

Nelson JM (1968) Parasites and symbionts of nests of *Polistes* wasps *Annals of the Entomological Society of America* **61,** 1528–1539

Neun S, Weber G (1985) Dipterenbesiedlung einer abgedeckten Bauschuttdeponie-Vergleich einer Rekultivierungsflache mit verschieden alten Sukzessionsflachen *Drosera* **1985,** 77–90

Neveu-Lemaire M (1938) *Traite d'entomologie medicale et veterinaire* Paris Vigot Frères

New TR (1975) An Australian host record for *Phalacrotophora* Enderlein (Diptera Phoridae) *Australian Entomological Magazine* **2,** 84

Nielsen BO (1968) Studies on the Dipterous fauna of the Stinkhorn (*Phallus impudicus* Pers) *Natura Jutlandica* **14,** 89–98

Nielsen E (1934) *Menozziola camponoti*, a phorid from the Riesengebirge *Broteria* **30,** 149–150

Nishida T (1955) Natural enemies of the melon fly, *Dacus cucurbitae* Coq in Hawaii *Annals of the Entomological Society of America* **48,** 171–178

Nogueira-Neto P (1970) *A criação de abelhas indigenas sem ferrão (Meliponinae)* São Paulo Chacaras e Quintais

Noirot C (1955) Recherches sur le polymorphisme des termites superieurs (Termitidae) *Annales des Sciences Naturelles, Zoologie* **17,** 399–595

Norton RA, Ide GS (1974) *Scutacarus baculitarsus agaricus* N Subsp (Acarina Scutacaridae) from commercial mushroom houses, with notes on phoretic behaviour *Journal of the Kansas Entomological Society* **47,** 527–534

Notton DG (1991a) Some Diptera host-records for species of *Basalys* and *Trichopria* (Hym , Diapriidae) *Entomologist's Monthly Magazine* **127,** 123–126

Notton DG (1991b) *Aspilota intermedia* Fischer a fungicolous Braconid (Hym) new to Britain *Entomologist's Monthly Magazine* **127,** 159–160

Oberstein O (1916) Über ein Massenauftreten von *Phora rufipes* Meig , Larven bei Keimversuchen mit Woll-Luzerne *Zetschrift für Pflanzenkrankheiten, Stuttgart* **26,** 104–105

Ogawa S, Sato N, Fujiwara M (1959) Three cases of the intestinal myiasis found in the case of the mass-treatment of hookworm *Japanese Journal of Sanitary Zoology* **10,** 80–82

Oldroyd H (1964) *The Natural History of Flies* London Weidenfeld and Nicholson

Olechowicz E (1971) Productivity investigation of two types of meadows in the Vistula valley VIII The number of emerged Diptera and their elimination *Ekologia Polska* **19(14),** 183–195

Ondraschek H (1953) Vererbungstudien an *Aphiochaeta zanthia* Speis (Phoridae) *Zeitschrift für induktive Abstammungs- und Vererbungslehre* **85,** 347–353

Orosi-Pal Z (1938) Humpbacked flies and the honey bee *Bee World* **19,** 64–68

Orr MR (1992) Parasitic flies (Diptera Phoridae) influence foraging rhythms and caste division of labor in the leaf-cutter ant, *Atta cephalotes* (Hymenoptera Formicidae) *Behavioral Ecology and Sociobiology* **30,** 395–402

Osawa N (1992) Effect of pupation site on pupal cannibalism and parasitism in the ladybird beetle *Harmonia axyridis* Pallas (Coleoptera, Coccinellidae) *Japanese Journal of Entomology* **60,** 131–135

Pacioni G (1985) *The Macdonald Encyclopedia of Mushrooms and Toadstools* London Macdonald

Painter HA (1980) A survey of filter fly nuisances and their remedies *Water Research Centre Technical Report* **155,** 1–40

Parmenter L (1965) Notes on the distribution of Phoridae (Diptera) in Britain *Proceedings of the South London Entomological and Natural History Society* **1965,** 35–42

Paterson HE (1953) A new species of the genus *Puliciphora* Dahl *Ecuatoriana de Entomologia y Parasitologia* **1,** 61–67

Patton WS (1922) Notes on some Indian *Aphiochaetae* *Indian Journal of Medical Research* **9,** 683–691

Patton WS, Evans AM (1929) *Insects, Ticks, Mites and Venomous Animals of Medical and Veterinary Importance Part I – Medical* Croydon

Paulian R (1949) Recherches sur les insectes d'importance biologique de Madagascar III *Memoires de l'Institut Scientifique de Madagascar (A)* **3,** 347–391

Paulian R (1958) Recherches sur les insectes d'importance biologique à Madagascar XXV à XXVIII *Mémoires de l'Institut Scientifique de Madagascar (E)* **9,** 1–19

Payne JA, Crossley DA (1966) Animal species associated with pig carrion *Oak Ridge National Laboratory – Technical Monograpgh* **1432,** 1–65

Payne JA, King EW, Beinhart G (1968) Arthropod succession and decomposition of buried pigs *Nature, London* **219,** 1180–1181

Pereira C, Castro MP de (1945) Contribuiçao para o conhecimento da espécie tipo de '*Macrocheles* Latr ' ('Acarina') '*M muscaedomesticae* (Scopoli, 1772)' emend *Archivos do Instituto Biologico, São Paulo* **16,** 153–186

Pergande T (1901) The ant-decapitating fly *Proceedings of the Entomological Society of Washington* **4,** 497–501

Persson T, Lohn U (1977) Energetical significance of soil- and litter-inhabiting animals and arthropods in a swedish grassland ecosystem *Ecological Bulletins* **23,** 1–211

Pesquero S, Campiolo S, Fowler HG (1993) Phorids (Diptera Phoridae) associated with mating swarms of *Solenopsis saevissima* (Hymenoptera Formicidae) *Florida Entomologist* **76**, 179–181

Peterson BV (1982) A new species of *Coniceromyia* (Diptera Phoridae) from Panama *Memoirs of the Entomological Society of Washington* **10,** 136–138

Peterson BV (1987) Phoridae In McAlpine JF, Peterson BV, Shewell GE, Teskey HJ, Vockeroth JR, Wood DM (eds), *Manual of Nearctic Diptera*, Volume 2 Ottawa Research Branch Agriculture Canada Monograph 28, Chapter 51 689–712

Peterson BV, Arntfield PW (1971) A new species of *Coniceromyia* from Chiapas, Mexico (Dipt , Phoridae) *Studia Entomologica, Petropolis* **14,** 395–398

Peterson BV, Robinson WH (1976) A new North American genus and species of the family Phoridae (Diptera) *Canadian Entomologist* **108,** 119–121

Phillips R (1981) *Mushrooms and Other Fungi of Great Britain and Europe* London Pan Books

Picard F (1930) Sur le parasitisme d'un Phoride (*Megaselia cuspidata* Schmitz) aux depens d'un myriapode *Bulletin de la Société Zoologique de France* **55,** 180–183

Pickel B (1928) Contribuição para a biologia dos Phorideos *Boletim do Museu Nacional do Rio de Janeiro* **4,** 67–68

Piffer CR, Lello E, Toledo LA, Foresti F (1972) Consideraçóes sobre o parasitismo de *Apocephalus attophilus* (Diptera Phoridae) em saúva (Hymenoptera Formicidae) *Ciência e Cultura, São Paulo* **24** (Supplement) 348–349

Poinar GO (1992) *Life in Amber* California Stanford University Press

Poisson R (1937) A propos de *Thaumatoxena Wasmanni* Breddin and Borner 1904 Insecte Diptère commensal des Termites *Bulletin de la Société Entomologique de France* **42,** 201–208

Poisson R (1939) Sur une nouvelle espèce du genre *Thaumatoxena* (Diptère Thaumatoxenidae) commensal des termites *Thaumatoxena schmitzi*, n sp , *Bulletin de la Société Zoologique de France* **64,** 243–247

Popov GB (1959) Ecological studies on oviposition by *Locusta migratoria migratorioides* (R and F) and its outbreak area in French Sudan *Locusta* **6,** 5–63
Portugal AV de (1977) Contribuição para o conhecimento da biologia de *Pseudohypocera kerteszi* (Enderlein, 1912), sen acasalamento e captura (Diptera, Phoridae) *Acta Amazonica* **7,** 153–155
Postner M (1952) Biologische-okologische Untersuchungen an Hummeln und ihren Nestern *Veroffentlichungen aus dem Ubersee-Museum Bremen (A)* **2,** 45–86
Potts GR, Vickerman GP (1974) Studies in the cereal ecosystem *Advances in Ecological Research* **8,** 107–197
Prado AP do (1976) Records and descriptions of phorid flies, mainly of the Neotropical Region (Diptera, Phoridae) *Studia Entomologica, Petropolis* **19,** 561–609
Prawirodisastro M, Benjamin DM (1979) Laboratory study of the biology and ecology of *Megaselia scalaris* (Diptera, Phoridae) *Journal of Medical Entomology* **16,** 317–320
Prescher S (1992) *Okologie und Biologie der Diptera, insbesondere der Brachycera, eines klarschlamm gedungten Ackerbodens* Germany der Technischen Universitat Carolo-Wilhelmina zu Braunschweig (Ph D thesis)
Priyanond P, Sangsook P, Panpairoj S (1973) A case report of acute urethral obstruction caused by Phoridae larva *Navy Medical Journal (Bangkok)* **13,** 115–121
Procter M, Yeo P (1973) *The Pollination of Flowers* London Collins, New Naturalist **54**
Prŷs-Jones OE, Corbet SA (1991) *Bumblebees* London Richmond Publishing Co Ltd, Naturalists' Handbooks **6**
Pugh CHW (1941) *Phalacrotophora fasciata* Fall bred from *Coccinella septempunctata* L *Entomologist's Monthly Magazine* **77,** 244
Ramirez W (1982) *Bombus mexicanus* Cresson, un hospedante de *Melaloncha*, moscas parasitas de la abeja de miel en el nuevo mundo *Revista de Biologia Tropicale, San Jose, Costa Rica* **30,** 177
Rao VP, Ghani MA (1972) Studies on predators of *Adelges* spp in the Himalayas *Miscellaneous Publications of the Commonwealth Institute of Biological Control* **3,** 1–116
Rao YR (1961) Notes on the life-history and habits of *Chonocephalus depressus* De Meijere (Diptera Phoridae) in South India *Proceedings of the Royal Entomological Society of London (A)* **36,** 163–167
Reichensperger A (1931) Beitrag zur Kenntnis der Termitoxeniidae (Dipt) *Zoologischer Anzeiger* **93,** 289–306
Reinhard H (1882) Beitrage zur Gruber-Fauna *Verhandlungen der kaiserlich-koniglichen zoologisch-botanischen Gesellschaft in Wien* **31,** 207–210 (1881)
Remy PA (1950) On the enemies of myriapods *Naturalist, Hull* **834,** 103–108
Rettenmeyer CW, Akre RD (1968) Ectosymbiosis between phorid flies and army ants *Annals of the Entomological Society of America* **61,** 1317–1326
Reyes OF (1983) A new record of *Pseudohypocera kerteszi*, a pest of honey bees in Mexico *American Bee Journal* **123,** 119–120
Rhoads AS (1924) Report of an examination of hickory trees killed by hickory bark borers *Technical Publication No 17 of the New York State College of Forestry at Syracuse University* **24(22),** 36–269
Rhodes-Jones R (1957) Intestinal myiasis in Tanganyika *Journal of Tropical Medicine and Hygiene* **60,** 169
Richards OW (1926) A note on a dipterous parasite of ladybirds *Entomologist's Monthly Magazine* **62,** 99
Richards OW (1932) Some breeding and habitat records of British Diptera *Journal of the Entomological Society of the South of England* **1,** 11–14

Richards OW (1947) Seasonal variations in the numbers of warehouse insects *Proceedings of the Royal Entomological Society of London (A)* **22,** 30–33

Richardson PN (1987) Susceptibility of mushroom pests to the insect-parasitic nematodes *Steinenema feltiae* and *Heterrhabditis heliothis* *Annals of Applied Biology* **111,** 433–438

Richardson PN, Chanter DO (1979) Phorid fly (Phoridae *Megaselia halterata*) longevity and the dissemination of nematodes (Allantonematidae *Howardula husseyi*) by parasitised females *Annals of Applied Biology* **93,** 1–11

Richardson PN, Hesling JJ (1977a) Laboratory rearing of the mushroom phorid *Megaselia halterata* Wood (Diptera Phoridae) *Annals of Applied Biology* **88,** 211–217

Richardson PN, Hesling JJ (1977b) Studies on an overwintering population of the mushroom phorid *Megaselia halterata* (Diptera Phoridae) parasitized by *Howardula husseyi* (Nematoda Allantonematidae) *Annals of Applied Biology* **86,** 321–327

Richardson PN, Hesling JJ, Riding IL (1977) Life cycle and description of *Howardula husseyi* n sp (Tylenchida Allantonematidae), a nematode parasite of the mushroom phorid *Megaselia halterata* (Diptera Phoridae) *Nematologica* **23,** 217–231

Richerson JV (1970) A world list of parasites of Coccinellidae *Journal of the Entomological Society of British Columbia* **67,** 33–48

Ricou G (1978) The Dipteran community of a newly sown grassland *Scientific Proceedings of the Royal Dublin Society Series A* **6,** 259–271

Riding IL, Hague NGM (1974) Some observations on a tylenchid nematode *Howardula* sp parasitizing the mushroom phorid *Megaselia halterata* (Phoridae, Diptera) *Annals of Applied Biology* **78,** 205–211

Rinker DL, Snetsinger RJ (1984) Damage threshold to a commercial mushroom by a mushroom-infesting phorid (Diptera Phoridae) *Journal of Economic Entomology* **77,** 449–453

Roberg DN (1915) I The role played by the insects of the Dipterous family Phoridae in relation to the spread of bacterial infections II Experiments on *Aphiochaeta ferruginea* Brunetti, with cholera vibrio *Philippine Journal of Science (B) Tropical Medicine* **10,** 309–336

Roberts RA (1935) Some North American parasites of blowflies *Journal of Agricultural Research, Washington* **50,** 479–494

Robinson GE (1981) *Pseudohypocera kerteszi* (Enderlein) (Diptera Phoridae), a pest of the honey bee *Florida Entomologist* **64,** 456–457

Robinson WH (1967) A phytophagous phorid *Proceedings of the North Central Branch of the Entomological Society of America* **22,** 158

Robinson WH (1970) *A revision of the Nearctic species of groups I and II of* Megaselia *(Diptera Phoridae)* Iowa Iowa State University (Ph D thesis)

Robinson WH (1971) Old and new biologies of *Megaselia* species (Diptera, Phoridae) *Studia Entomologica, Petropolis* **14,** 321–348

Robinson WH (1975) *Megaselia (M) scalaris* (Diptera Phoridae) associated with laboratory cockroach colonies *Proceedings of the Entomological Society of Washington* **77,** 384–390

Robinson WH (1977) Phoridae (Diptera) associated with cultivated mushrooms in Eastern North America *Proceedings of the Entomological Society of Washington* **79,** 452–462

Robinson WH (1978a) Identification of *Megaselia agarici* (Lintner) and *Megaselia halterata* (Wood) (Diptera Phoridae) *United States Department of Agriculture Cooperative Plant Pest Report* **3,** 14–16

Robinson WH (1978b) Terminalia of some North American species of *Megaselia*

(Aphiochaeta) and descriptions of two new species (Diptera Phoridae) *Proceedings of the Entomological Society of Washington* **80,** 216–227

Robinson WH (1981) Terminalia of North American species of Group II *Megaselia (Aphiochaeta)* and descriptions of four new species (Diptera Phoridae) *Proceedings of the Entomological Society of Washington* **83,** 489–504

Robinson WH, Brown BV (1993) Life history and immature stages of two species of *Megaselia* (Diptera Phoridae) predatory on gall-inhabiting insects *Proceedings of the Entomological Society of Washington* **95,** 404–411

Robinson WH, Foote BA (1968) Biology and immature stages of *Megaselia aequalis*, a phorid predator of slug eggs *Annals of the Entomological Society of America* **61,** 1587–1594

Robinson WH, Wisseman RW (1983) A new species of *Megaselia* in Group VII (Diptera Phoridae) *Proceedings of the Entomological Society of Washington* **85,** 282–285

Rocha UF, Belo M, Moraes JRE, Sogorb A, Baruch A (1984) Tick ecology VI The influence of environmental moisture on the invasion of engorged females of *Boophilus microplus* (Canestrini) (Acari, Ixodidae) by larvae of *Megaselia scalaris* Loew (Diptera, Phoridae) and upon the prolificness of such arthropods *Naturalia, São Paulo* **9,** 93–100

Rohdendorf B (1974) *The Historical Development of Diptera* Canada The University of Alberta Press

Rohlfien K, Ewald B (1975) Katalog der in den Sammlungen des ehemaligen Deutschen Entomologischen Institutes aufbewahrten Typen – XIII *Beitrage zur Entomologie, Berlin* **25,** 151–161

Ronna A (1936) Observaçôes biologicas sobre dois dipteros parasitas de *Apis mellifica* L (Dipt Phoridae, Sarcophagidae) *Revista de Entomologia, Rio de Janeiro* **6,** 1–9

Ronna A (1937a) Novos dados sobre os habitos de *Melaloncha ronnai* Borgmeier (Dipt Phoridae), endoparasita de *Apis mellifica* L *Revista de Entomologia, Rio de Janeiro* **7,** 409–413

Ronna A (1937b) *Melaloncha ronnai* Brgm 1935 (Phoridae) endoparasita de *Apis mellifica* L (abelha domestica) *Revista do Departamento Nacional da Produção Animal* **4,** 113–126

Rotheray GE (1989) E B Basden's collection of Diptera from bird and mammal nests, and mammal runs, burrows and droppings *Entomologist's Monthly Magazine* **125,** 5–8

Roubik DW (1989) *Ecology and Natural History of Tropical Bees* Cambridge Cambridge University Press

Ruelle JE (1970) A revision of the genus *Macrotermes* from the Ethiopean region (Isoptera Termitidae), *Bulletin of the British Museum (Natural History) Entomology* **24,** 365–444

Ruelle JE, Coaton WGH, Sheasby JL (1975) National survey of the Isoptera of Southern Africa 8 The genus *Macrotermes* Holmgren (Termitidae Macrotermitinae) *Cimbebasia (A)* **3(8),** 73–94

Ryckman RE (1953) Notes in the ecology of *Bombus sonorus* in Orange County, California, and new parasite records *Pan-Pacific Entomologist* **29,** 144–146

Saba F (1970) Parasites, predators and diseases in a rearing culture of *Diabrotica balteata* (Col , Chrysomelidae) *Journal of Economic Entomolgy* **63,** 1674

Salt G (1929) A contribution to the ethology of the Meliponinae *Transactions of the Entomological Society of London* **77,** 431–470

Salt G, Hollick FSJ, Raw F, Brian MV (1948) The arthropod population of pasture soil *Journal of Animal Ecology* **17,** 139–150

Samson RA, Evans HC, Latge J-P (1988) *Atlas of Entomopathogenic Fungi* Heidelberg Springer-Verlag

Sandhu GS, Arora PK (1990) Studies on mechanical and chemical control of mushroom flies – important pests of white button mushroom in Punjab *Journal of Insect Science* **3,** 92–96

Sandhu GS, Bhattal DS (1989) Chemical control of phorid fly *Megaselia sandhui* Disney and its effect on the yield of white button mushroom *Mushroom Journal for the Tropics* **9,** 115–120

Sasaki C (1935) On a new phorid fly infesting our edible mushroom *Proceedings of the Imperial Academy of Japan* **11,** 112–114

Saunders LG (1928) Some marine insects of the Pacific coast of Canada *Annals of the Entomological Society of America* **21,** 521–545

Schiner JR (1864) *Fauna Austriaca Die Fliegen (Diptera) II Theil* Wien

Schlee D (1977) Florale und extraflorale Nektarien sowie Insektenkot als Nahrungsquelle für Chironomidae-Imagines (und andere Diptera) *Stuttgarter Beitrage zur Naturkunde aus dem Staatlichen Museum für Naturkunde in Stuttgart (A)* **300,** 1–16

Schmitz H (1913a) Eine neue termitophile Phoriden-Gattung und -Art, *Bolsiusia termitophila* n g n sp , aus Ostindien *Zoologischer Anzeiger* **42,** 268–273

Schmitz H (1913b) Beschreibung von *Termitophora velocipes* (Wasmann in litt), einer termitophilen Phoride aus Vorderindien *Entomologiske Meddelelser* **10,** 9–16

Schmitz H (1914d) Die myrmecophilen Phoriden der Wasmann'schen Sammlung *Zoologischen Jahrbuchen Abteilung für Systematik, Geographie und Biologie der Tiere* **37,** 509–566

Schmitz H (1915b) Neue Gattungen und Arten termitophiler Phoriden und Termitoxeniiden (Neue Beitraege etc No 1) *Societas entomologica, Stuttgart* **30(7),** 35–36

Schmitz H (1915d) Neue Beitraege zur Kenntnis der myrmecophilen und termitophilen Phoriden (No 2–15) *Deutsche entomologische Zeitschrift* **1915,** 465–507

Schmitz H (1915e) Neue Beitrage zur Kenntnis der myrmecophilen und termitophilen Phoriden (No 16–22) Neue Phoriden aus Vorderindien, gesammelt von J Assmuth S J , Bombay *Wiener entomologische Zeitung* **34,** 311–330

Schmitz H (1916a) Neue termitophile Dipteren aus den Familien der Termitoxeniiden und Phoriden *Zoologischen Jahrbuchern Abteilung für Systematik und Biologie der Tiere* **39,** 211–266

Schmitz H (1916b) Neue Phoriden aus Belgisch Kongo, gesammelt von Josef Bequaert *Zoologische Mededeelingen, Leiden* **2,** 1–10

Schmitz H (1916c) Bemerkungen zu einigen termitophilen und myrmecophilen Phoriden *Zoologische Mededeelingen, Leiden* **2,** 27–32

Schmitz H (1916f) Zur Kenntnis einiger Phoridenarten und ihrer Synonyme *Wiener entomologische Zeitung* **35,** 227–234

Schmitz H (1917a) Biologische Beziehungen zwischen Dipteren und Schnecken *Biologischen Zentralblatt* **37,** 24–43

Schmitz H (1920c) Die Phoriden von Hollandisch Limburg mit Bestimmungstabellen aller bisher Kenntlich beschreiben europaischen Phoriden IV Teil *Jaarboek Natuurhistorisch Genootschap in Limburg* **1919,** 91–153

Schmitz H (1921b) *Stethopathusa* (n g) *phoridarum* (Dipt) *Boletin de la Sociedad Entomologica de España* **4,** 95–98

Schmitz H (1921f) *Aptinandria*, eine neue, in beiden Geschlechtern flugel- und schwingerlose Phoridengattung aus Afrika (Phoridae, Diptera) *Entomolgiske Meddelelser* **13,** 318–322

Schmitz H (1923b) Zwei neue Phoriden aus Australien und Brasilien *Entomologische Berichten, Amsterdam* **6,** 188–192

Schmitz H (1923c) *Pseudacteon borgmeieri* n sp (Diptera – Phoridae) *Vozes de Petropolis* **17,** 714–715

Schmitz H (1923e) Zwei neue myrmecophile Phoriden *Deutscher Verein für Wissenschaft und Kunst in São Paulo* **3,** 149–153

Schmitz H (1924b) Een nieuwe Phoride, *Cremersia zikani* n g n sp *Natuurhistorisch Maandblad* **13,** 32–34

Schmitz H (1924d) Drei neue Platyphorinen (Phoridae, Diptera) und eine neue Wirtstermite *Zoologischer Anzeiger* **59,** 289–304

Schmitz H (1924h) Mitteilungen uber allerlei Phoriden No 6–8 *Natuurhistorisch Maandblad* **13,** 138–143

Schmitz H (1924i) Phorideos Ecitophilos de Minas Geraes (Dipt) *Museu Nacional do Rio de Janeiro Publicação* **4,** 1–35

Schmitz H (1925a) Sumatranische Insekten (Beitrag III) Phoriden I *Entomologische Mitteilungen* **14,** 58–62

Schmitz H (1926b) A new termitophilous phorid from India *Entomologist's Monthly Magazine* **62,** 15–16

Schmitz H (1926c) Neue Gattungen und Arten europaischer Phoriden *Encyclopedie Entomologique B,II Dipteres* **2,** 73–85 (1925)

Schmitz H (1927e) Eine neue myrmecophile *Metopina*, nebst anderen neuen Arten dieser Gattung *Zoologischer Anzeiger* **74,** 241–248

Schmitz H (1928a) Revision der Phoridengattungen, mit Beschreibung neuer Gattungen und Arten *Natuurhistorisch Maandblad* **17,** 12, 20–22, 38–41, 49–54, 66–70, 87–92, 101–105

Schmitz H (1928d) Verslag der Maandelijksche vergadering 5 September L L – *Megaselia oxybelorum* n sp *Natuurhistorisch Maandblad* **17,** 121–122, 131–132

Schmitz H (1928f) Phoriden in doodkisten *Natuurhistorisch Maandblad* **17,** 150–153

Schmitz H (1929b) Eine neue ecitophile *Xanionotum*-Art aus Brasilien (Phorid , Dipt) *Zoologischen Anzeiger* **82,** 228–232

Schmitz H (1929e) Zur Kenntnis einiger von Dr Jos Bequaert gesammelter afrikanischer Phoriden *Revue de Zoologie et de Botanique Africaines* **18,** 37–43

Schmitz H (1929g) *Revision der Phoriden* Berlin Ferd Dummlers

Schmitz H (1929i) *Diptera of Patagonia and South Chile, based mainly on material in the British Museum (Nat Hist) Part VI fasc 1 Sciadoceridae and Phoridae* London British Museum (Natural History)

Schmitz H (1930b) Phoriden aus Eipaketen von *Locusta migratoria* in Daghestan *Natuurhistorisch Maandblad* **19,** 67–69

Schmitz H (1931b) *Diptera of Patagonia and South Chile, based mainly on material in the British Museum (Nat Hist) Part VI fasc 2 Phoridae (supplement), Platypezidae, Pipunculidae, Sphaeroceridae (Borboridae), Ephydridae* London British Museum (Natural History)

Schmitz H (1931c) Neue termitophile Dipteren von Buitenzorg, Java *Natuurhistorisch Maandblad* **20,** 176

Schmitz H (1931d) *Nepenthes*-Phoriden In Thienemann A, Tropische Binnengewasser *Archiv für Hydrobiologie, Supplement-Bande* **9,** 449–471

Schmitz H (1932a) Zur naeheren Kenntnis der von Dr C Franssen in Buitenzorg (Java) gesammelten termitophilen Dipteren *Natuurhistorisch Maandblad* **21,** 27–32, 38–41

Schmitz H (1932d) Beschreibung einiger neuen asiatischen Phoriden *Natuurhistorisch Maandblad* **21,** 127–131

Schmitz H (1932e) Neue *Stichillus*- und *Phalacrotophora*-Arten, mit einer Aufteilung von *Phalacrotophora* in drei Untergattungen *Tijdschrift voor Entomologie* **75** *(supplement)* 115–127

Schmitz H (1933b) Zwei neue exotische Phoriden (Diptera) *Natuurhistorisch Maandblad* **22,** 34–36, 43–44

Schmitz H (1933e) *Javanoxenia franssenì* n sp , eine neue Termitoxeniide aus der Umgebung von Batavia *Natuurhistorisch Maandblad* **22,** 102–105

Schmitz H (1934e) [Aanwinsten voor de Phoridenfauna van Nederland] *Tijdschrift voor Entomologie* **77,** LX, LXV-LX1

Schmitz H (1934g) Zwei neue *Puliciphora*-Arten (Diptera, Phoridae) *Natuurhistorisch Maandblad* **23,** 150–152

Schmitz H (1935d) Zwei neue Phoriden aus Sudbrasilien (Diptera) *Revista de Entomologia, Rio de Janeiro* **5,** 19–23

Schmitz H (1935j) Ein neuer Beitrag zur Kenntnis der *Paraspiniphora*-Arten *Broteria* **31,** 155–173

Schmitz H (1938a) Zwei neue europaische *Megasilien* (Phoridae, Diptera) *Natuurhistorisch Maandblad* **27,** 8–9

Schmitz H (1938b) Beitrage zu einer Monographie der Termitoxeniidae (Diptera) I *Broteria* **34,** 22–40

Schmitz H (1938c) On the Irish species of the Dipterous family Phoridae *Proceedings of the Royal Irish Academy* **44B,** 173–204

Schmitz H (1938g) Drei neue aus toten Schnecken gezuechtete japanische Phoriden *Natuurhistorisch Maandblad* **27,** 80–83

Schmitz H (1938i) *Misotermes exenterans* n g n sp , eine parasitische fliege aus der familie der Phoriden, welche die Entstehung myiagener Soldaten bei *Macrotermes gilvus* Hag in Java verursacht *Treubia* **16,** 369–397

Schmitz H (1938m) Die ersten hundert Phoriden von Portugal I *Broteria* 34 163–179

Schmitz H (1938n) Phoridae In Lindner E (ed), *Die Fliegen der palaearktischen Region* **4(33)** (Lieferung 123) 1–64

Schmitz H (1939a) Neuseelandische Phoriden *Natuurhistorisch Maandblad* **28,** 34–37, 55–56, 67–68, 75–76, 86–89, 98–101, 111–116, 124–129

Schmitz H (1939c) A new species of Phoridae (Diptera) associated with millipedes, from the Yemen *Proceedings of the Royal Entomological Society of London (B)* **8,** 43–45

Schmitz H (1939d) Beitrage zu einer Monographie der Termitoxeniidae (Diptera) V *Broteria* **35,** 53–63

Schmitz H (1940a) Uber einige afrikanische Phoriden (Diptera) *Natuurhistorisch Maandblad* **29,** 9–10, 22–24, 32, 48–50, 57

Schmitz H (1940e) Eine neue ostasiatische *Spiniphora* (Phoridae, Diptera) *Natuurhistorisch Maandblad* **29,** 78–79

Schmitz H (1941c) Phoridae In Lindner E (ed), *Die Fliegen der palaearktischen Region* **4(33)** (Lieferung 141) 65–128

Schmitz H (1941d) Uber die Larve von *Chaetopleurophora pygidialis* m Erste Mitteilung *Natuurhistorisch Maandblad* **30,** 63–66

Schmitz H (1941e) Eine neue Aenigmatiinen-Gattung aus Westafrika *Natuurhistorisch Maandblad* **30,** 77–80

Schmitz H (1941f) Zwei neue *Aenigmatistes* aus Abessinien, nebst einem Verzeich aller bisher beschriebenen Aenigmatiinae *Natuurhistorich Maandblad* **30,** 104–108

Schmitz H (1943) Phoridae In Lindner E (ed), *Die Fliegen der palaearktischen Region* **4(33)** (Lieferung 147, 149) 129–192

Schmitz H (1948b) Zur Kenntnis der fungicolen Buckelfliegen *Natuurhistorisch Maandblad* **37,** 37–44

Schmitz H (1949c) Phoridae In Lindner E (ed), *Die Fliegen der palaearktischen Region* **4(33)** (Lieferung 160) 193–240

Schmitz H (1950a) *Chonocephalus elongatus* n sp (Diptera, Phoridae) *Natuurhistorisch Maandblad* **39,** 21–24

Schmitz H (1950b) Ueber die Gattung *Thaumatoxena*, nebst Beschreibung einer neuen

Art aus dem Uelle-Gebiet (Phoridae, Diptera) *Revue de Zoologie et de Botanique Africaines* **43,** 39–58

Schmitz H (1951a) Beitrage zu einer Monographie der Termitoxeniidae (Diptera) VI *Broteria* **47,** 65–86

Schmitz H (1951b) Nochmals die Gattung *Thaumatoxena* mit Beschreibung neuer Arten (Phoridae, Diptera) *Revue de Zoologie et de Botanique Africaines* **44,** 254–266

Schmitz H (1951c) Phoridae In Lindner E (ed), *Die Fliegen der palaearktischen Region* **4(33)** (Lieferung 165) 241–272

Schmitz H (1951d) Myrmekophile und termitophile Phoriden (Diptera), von S Patrizi und F Meneghetti in Africa gesammelt *Bollettino dell'Istituto di Entomologia della Universita di Bologna* **18,** 128–166

Schmitz H (1951e) Japanische Phoriden 1 *Broteria* **47,** 145–158

Schmitz H (1952c) Über W Lundbecks Sammlung und Beschreibung Danischer Phoriden *Entomologiske Meddelelser* **26,** 350–379

Schmitz H (1953a) Phoridae In Lindner E (ed), *Die Fliegen der palaearktischen Region* **4(33)** (Lieferung 171) 273–320 (1952)

Schmitz H (1953d) Ueber die Gattung *Plethysmochaeta* Schmitz, mit Beschreibung neuer Arten (Diptera, Phoridae) *Revue de Zoologie et de Botanique Africaines* **47,** 297–310

Schmitz H (1954b) *Coridophora convexa* n sp vom Belgischen Kongo, zugleich ein Beitrag zur naheren Kenntnis von *C semicimex* Schmitz 1951 (Diptera, Phoridae) *Revue de Zoologie et de Botanique Africaines* **49,** 13–16

Schmitz H (1954c) Uber afrikanische Phoriden der Gattungen *Aenigmatistes* und *Termitostroma* (Diptera) *Annales du Musee Royal du Congo Belge, Tervuren, nouvelle serie in 4° Sciences Zoologiques* **1,** 514–519

Schmitz H (1955a) Sechs neue termitophile Phoriden aus Angola und S W Afrika (Phoridae, Diptera) *Publicacoes Culturais da Companhia Diamantes de Angola (Museu do Dundo Estudos Diversos X)* **24,** 33–60

Schmitz H (1955b) Phoridae In Lindner E (ed), *Die Fliegen der palaearktischen Region* **4(33)** (Lieferung 180) 321–368

Schmitz H (1955c) Øber Phoriden in *Nepenthes*-Kannen (Phoridae, Diptera) *Broteria* **51,** 57–87

Schmitz H (1955d) Eine neue *Thaumatoxena*-Art (Phoridae, Diptera) *Revue de Zoologie et de Botanique Africaines* **51,** 229–239

Schmitz H (1956b) Phoridae In Lindner E (ed), *Die Fliegen der palaearktischen Region* **4(33)** (Lieferung 187) 369–416

Schmitz H (1957a) Phoridae In Lindner E (ed), *Die Fliegen der palaearktischen Region* **4(33)** (Lieferung 196) 417–464

Schmitz H (1958b) Phoridae In Lindner E (ed), *Die Fliegen der palaearktischen Region* **4(33)** (Lieferung 202) 465–512

Schmitz H (1958e) Acht neue und einige bekannte Phoriden aus Angola und dem Belgischen Kongo (Phoridae, Diptera) *Publicacoes Culturais da Companhia de Diamantes de Angola (Museu do Dundo Estudos Diversos XV)* **40,** 13–62

Schmitz H, Beyer E (1965) Phoridae In Lindner E (ed), *Die Fliegen der palaearktischen Region* **4(33)** (Lieferung 258, 260) 513–608

Schmitz H, Beyer E (1974) Phoridae In Lindner E (ed), *Die Fliegen der palaearktischen Region* **4(33)** (Lieferung 301) 609–637

Schmitz H, Delage A (1974) Phoridae In Lindner E (ed), *Die Fliegen der palaearktischen Region* **4(33)** (Lieferung 301) 638–664

Schmitz H, Delage A (1981) Phoridae In Lindner E (ed), *Die Fliegen der palaearktischen Region* **4(33)** (Lieferung 325) 665–712

Schmitz H, Mjoberg E (1924) Results of Dr E Mjoberg's Swedish Scientific Expedition to Australia 1910–1913 35 Sciaridae und Phoridae *Arkiv för Zoologi* **16(9),** 1–8

Schmitz H, Mjoberg E (1925) Zwei neue myrmekophile Phoriden von Borneo (Diptera) *Zoologischer Anzeiger* **65,** 41–48

Schmutterer H (1952) *Plastophora rufa* (Wood) (Dipt , Phoridae) als Eirauber und Parasit von *Eulecanium corni* (Bché) (Homopt , Coccoidea) *Anzeiger für Schadlingskunde* **25,** 145–148

Schmutterer H (1972) Eine *Megaselia*-Art als Parasit von *Chilocorus quadrimaculatus* Weise (Col Coccin) in Kenia *Zeitschrift für angewandte Entomologie* **72,** 110–112

Scopoli JA (1763) *Entomologia carniolica exhibens insecta carniolae indigena et distributa in ordines, genera, species, varietates Methodo Linneana,* Vindobonae

Scott H (1920) Notes on the biology of some inquilines and parasites in a nest of *Bombus derhamellus* Kirby with a description of the larva and pupa of *Epuraea depressa* Illig (= *aestiva* Auctt Coleoptera, Nitidulidae) *Transactions of the Entomological Society of London* **68,** 99–127

Scott H (1951) Swarming of *Megaselia meconicera* Speiser (Dipt , Phoridae) in houses *Entomologist's Monthly Magazine* **87,** 269–270

Seevers CH (1941) New termitophilous Diptera from the Neotropics *Field Museum of Natural History Zoological Series* **24(18),** 175–193

Séguy E (1927) Diptères (Brachycères) (Asilidae) *Faune de France* **17,** 1–190

Seifert B (1992) A taxonomic revision of the Palaearctic members of the ant subgenus *Lasius* s str (Hymenoptera Formicidae) *Abhandlungen und Berichte des Naturkundemuseums Gorlitz* **66(5),** 1–67

Semenza L (1953) *Aphiochaeta xanthina* Speiser Contributo alla conoscenza morfologia e del ciclo biologica *Rendiconti dell'Istituto Lombardo di Scienze e Lettere* **86,** 320–330

Senior-White R (1924) New Ceylon Diptera *Spolia Zeylanica* **12,** 375–406

Shattuck SO (1992a) Review of the Dolichoderine ant genus *Iridomyrmex* Mayr with descriptions of three new genera (Hymenoptera Formicidae) *Journal of the Australian Entomological Society* **31,** 13–18

Shattuck SO (1992b) Generic revision of the ant subfamily Dolichoderinae (Hymenoptera Formicidae) *Sociobilogy* **21,** 1–181

Shiraki T (1925) Some insects living in termites' nests *Transactions of the Natural History Society of Formosa, Taihoku* **15,** 208–210 (In Japanese)

Sholdt LL (1966) Insects associated with the flowers of the coconut palm, *Cocos nucifera* L in Hawaii *Proceedings of the Hawaiian Entomological Society* **19,** 293–296

Shorrocks B (1991) Competition on a divided and ephemeral resource a cage experiment *Biological Journal of the Linnean Society* **43,** 211–220

Silvestri F (1906) Contribuzione alla conoscenza dei Termitidi et Termitofili dell'Eritrea, *Redia, Giornale di Entomologia, Firenze* **3,** 341–359 (1905)

Silvestri F (1911) Contributo alla conoscenza dei mirmecofili del Messico *Bollettino del Laboratorio di Zoologia Generale e Agraria della R Scuola Superiore d'Agricoltura, Portici* **5,** 172–195

Silvestri F (1920) Contribuzione alla conoscenza dei Termitidi e Termitofili dell'Africa occidentale *Bollettino del Laboratorio di Zoololgia Generale e Agraria della R Scuola Superiore d'Agricoltura, Portici* **14,** 265–319

Silvestri F (1926) Descrizione di particolari individui (Myiagenii) di *Termes gilvus* Hag parassitizzati da larva di Dittero *Bollettino del Laboratorio di Zoologia Generale e Agraria della R Scuolo Superiore d'Agricoltura, Portici* **19,** 3–18

Silvestri F (1945) Primo contributo all conoscenza dei termitofili vivente con specie di *Syntermes* *Commentationes Pontificia Academia Scientiarum* **9,** 515–559

Silvestri F (1947a) Contributo alla conoscenza dei Phoridae termitofili del Brasile *Acta Pontificia Academia Scientiarum* **10,** 281–296

Silvestri F (1947b) Seconda nota su alcuni termitofili dell'Indocina con una appendice sul *Macrotermes Barneyi* Light *Bolletino del Laboratorio di Zoologia Generale e Agraria della R Scuola Superiore d'Agricoltura, Portici* **7,** 13–40

Simôes D, Bego LR, Zucchi R, Sakagami SF (1980) *Melaloncha sinistra* Borgmeier, an endoparasitic phorid attacking *Nannotrigona (Scaptotrigona) postica* Latreille (Hymenoptera, Meliponinae) *Revista Brasileira de Entomologia* **24,** 137–142

Sinclair BJ (1992) A phylogenetic interpretation of the Brachycera (Diptera) based on the larval mandible and associated mouthpart structures *Systematic Entomology* **17,** 233–252

Singh TS, Rana D (1989) Urogenital myiasis caused by *Megaselia scalaris* (Diptera Phoridae) a case report *Journal of Medical Entomology* **26,** 228–229

Sivinski J (1982) Prey attraction by luminous larvae of the fungus gnat *Orfelia fultoni* *Ecological Entomology* **7,** 443–446

Sivinski J (1988) Unusual female-aggregated mating systems in phorid flies *Journal of Insect Behavior* **1,** 123–128

Sivinksi J, Stowe M (1980) A kleptoparasitic cecidomyiid and other flies associated with spiders *Psyche* **87,** 337–348

Skidmore P (1991) Insects of the British cow-dung community *Field Studies Council Occasional Publication* **21,** 1–166

Sladen FWL (1912) *The Humble-bee, its life history and how to domesticate it* London

Smith KGV (1956) On the Diptera associated with the stinkhorn (*Phallus impudicus* Pers) *Proceedings of the Royal Entomological Society of London (A)* **31,** 49–55

Smith KGV (1968) The immature stages of *Rhamphomyia sulcata* Meigen (Diptera Empididae) and their occurrence in large numbers in pasture soil *Entomologist's Monthly Magazine* **104,** 65–68

Smith KGV (1977) Notes on some British Phoridae (Diptera) including two species of *Megaselia* Rondani new to science *Entomologist's Record and Journal of Variation* **89,** 161–168

Smith KGV (1984) Review (of Disney 1983a) *Entomologist's Monthly Magazine* **120,** 196

Smith KGV (1986) *A Manual of Forensic Entomology* London, Ithaca, New York British Museum (Natural History), Cornell University Press

Smith MR (1928) *Plastophora crawfordi* Coq and *Plastophora spatulata* Malloch (Diptera Phoridae), parasitic on *Solenopsis geminata* Fabr *Proceedings of the Entomological Society of Washington* **30,** 105–108

Snetsinger R (1972) Laboratory studies of mushroom infesting arthropods *Mushroom Science* **8,** 199–208

Snyder TE (1949) Catalog of the termites (Isoptera) of the world *Smithsonian Miscellaneous Collections* **112,** 1–490

Soderstrom TR, Calderon CE (1971) Insect pollination in tropical rain forest grasses *Biotropica* **3,** 1–16

Somchoudhury AK, Das TK, Sarkar RK, Mukherjee AB (1988) Life history of three Dipteran flies infesting edible mushroom *Pleurotus sajor-caju* *Environment and Ecology* **6,** 463–464

Spahr U (1985) Erganzungen und Berichtigungen zu R Keilbachs Bibliographie und Liste der Bernsteinfossilien – Ordnung Diptera *Stuttgarter Beitrage zur Naturkunde aus dem Staatlichen Museum für Naturkunde in Stuttgart* **111,** 1–146

Spencer S, McDaniel JS, Burbutis PP (1977) A laboratory, insecticide screening technique for *Megaselia halterata*, a mushroom fly (Dipt , Phoridae) *Journal of Economic Entomology* **70,** 101–103

Spradbery JP (1973) *WASPS – An account of the biology and natural history of social and solitary wasps* London Sidgwick and Jackson

Springer R (1967) Konkurrenz versuche mit den drei Geschlechtsbestimmungstypen von *Megaselia scalaris* Loew *Molecular and General Genetics* **99,** 125–132

Stadler H (1948) Les inquilins et parasites des nids de l'hirondelle de rivage, *Riparia riparia* (L) *Alauda* **16,** 40–54

Steghaus-Kovac S, Disney RHL (1990) A further case of a *Rhynchomicropteron* (Diptera, Phoridae) associated with a ponerine ant (Hymenoptera, Formicidae) *Sociobiology* **16,** 285–287

Stork NE (1988) Insect diversity facts, fiction and speculation *Biological Journal of the Linnean Society* **35,** 321–337

Stradling DJ (1978) The influence of size on foraging in the ant, *Atta cephalotes*, and the effect on some plant defence mechanisms *Journal of Animal Ecology* **47,** 173–188

Stricklan C, Roy DN (1941) Myiasis-producing Diptera in man *Indian Journal of Medical Research* **29,** 863–865

Stuckenberg B (1953) An association of phorid flies with the prey of a mantid *Entomologist's Monthly Magazine* **89,** 131

Stuckenberg BR (1974) Class Insecta order Diptera In Coaton WGH, ed Status of the taxonomy of the Hexapoda in Southern Africa *Republic of South Africa Department of Agricultural Technical Services, Entomology Memoir* **38,** 105–113

Svensson BG, Petersson E, Frisk M (1990) Nuptial gift size prolongs copulation duration in the dance fly *Empis borealis* *Ecological Entomology* **15,** 225–229

Swezey OH (1934) Insects from fruits of *Alectryon* *Proceedings of the Hawaiian Entomological Society* **8,** 385–386

Szadziewska MM (1977) Geofilne muchowki (Diptera) rezerwatu 'Las Piwnicki' kolo Torunia *Acta Universitatis Nicolai Copernici Biologia* **19(39),** 107–115

Szwejda J (1980) Diptera occurring on Brussels sprouts *Polskie Pismo Entomologiczne* **50,** 569–597

Taylor E (1944) *Megaselia (Aphiochaeta) rufipes* Meig (Dipt , Phoridae) bred from a dead honeybee *Entomologist's Monthly Magazine* **80,** 120

Thiede U (1977) Untersuchungen uber die Arthropodenfauna in Fichtenforsten (Populationsokologie, Energieumsatz) *Zoologischer Jahrbucher Abteilung fur Systematik* **104,** 137–202

Thienemann A (1932) Die Tierwelt der *Nepenthes*-Kannen *Archiv fur Hydrobiologie Supplement-Bande* **11,** 1–54

Thienemann A (1934) Die Tierwelt der tropischen Pflanzengewasser *Archiv fur Hydrobiologie Supplement-Bande* **13,** 1–91

Thomson AJ, Davies DM (1973) The biology of *Pollenia rudis*, the cluster fly (Diptera Calliphoridae) II Larval feeding behaviour and host specificity *Canadian Entomologist* **105,** 985–990

Timon-David J (1938) Sur un Phoride parasite de la sauterelle verte *Annales de Parasitologie Humaine et Comparee* **16,** 193–195

Tingle CCD, Lauer S, Armstrong G (1992) Dry season, epigeal invertebrate fauna of mopane woodland in northwestern Zimbabwe *Journal of Arid Environments* **23,** 397–414

Trager JC (1991) A revision of the fire ants, *Solenopsis geminata* group (Hymenoptera Formicidae Myrmicinae) *Journal of the New York Entomological Society* **99,** 141–198

Traut W, Khuong NT, Schneider S (1990) Karyotypes of *Megaselia scalaris* (Diptera) wild-type and translocation strains *Genetica* **83,** 77–84

Traut W, Willhoeft U (1990) A jumping sex determining factor in the fly *Megaselia scalaris* *Chromosoma* **99,** 407–412

Truitt GW (1951) The mushroom fly as a cause of bronchial asthma. *Annals of Allergy* **9,** 513–516.
Trumble JT, Pienkowski RL (1979) Development and survival of *Megaselia scalaris* (Diptera: Phoridae) at selected temperatures and photoperiods. *Proceedings of the Entomological Society of Washington* **81,** 207–210.
Tuck WH (1896) Inquiline and other inhabitants in nests of aculeate Hymenoptera. *Entomologist's Monthly Magazine* **32,** 153–155.
Tumrasvin WS, Sucharit S, Vutikes S (1977) Studies on the life history of *Megaselia scalaris* (Loew) in Thailand. *Southeast Asian Journal of Tropical Medicine and Public Health* **8,** 74–76.
Tuomikoski R (1957) Beobachtungen über einige Sciariden (Dipt.), deren Larven in faulen Holz oder unter der Rinde abgestorbener Bäume leben. *Annales Entomologici Fennici* **23,** 3–35
Ulrich H (1991) The present state of comparative morphology of the thoracic musculature in Diptera. In: Weismann L, Országh I, Pont AC (eds.), *Proceedings of the Second International Congress of Dipterology (Bratislava, 1990).* The Hague: SPB Academic Publishing, 315–326.
Upton MS (1993) Aqueous gum-chloral slide mounting media: an historical review. *Bulletin of Entomological Research* **83,** 267–274
Urbani CB, Bolton B, Ward PS (1992) The internal phylogeny of ants (Hymenoptera: Formicidae). *Systematic Entomology* **17,** 301–329.
van Achterberg C (1973) A study about the Arthropoda caught by *Drosera* species. *Entomologische Berichten. Nederlandische Entomologische, Amsterdam* **33,** 137–140.
van Achterberg C (1988) The genera of the *Aspilota*-group and some descriptions of fungicolous Alysiini from the Netherlands (Hymenoptera: Braconidae: Alysiinae) *Zoologische Verhandelingen, Leiden* **247,** 1–88.
van Achterberg C, Bin F (1981) Notes on two species of *Dinotrema* Foerster (Hym., Braconidae, Alysiinae) with observations on the hymenopterous parasite-complex of *Spiniphora dorsalis* Becker (Dipt., Phoridae) in dead *Helix* spp. (Mollusca). *Entomologische Berichten* **41,** 104–112.
van de Geijn J (1983) Prevention and control of pests on Dutch mushroom farms. *Mushroom Journal* **131,** 390–393.
van Emden FI (1950) Dipterous parasites of Coleoptera. *Entomologist's Monthly Magazine* **86,** 182–206.
Vane-Wright RI (1992) Systematics and the global biodiversity strategy. *Antenna* **16(2),** 49–56.
van Slype W (1932) Trois cas de myiase intestinale à *Aphiochaeta scalaris.* **12,** 581–582.
van Zaayen A (1983) Diseases and pests of mushrooms in Holland. IV. Literature, research and observations in Dutch mushroom farms. *Mushroom Journal* **131,** 1123–1133.
Villa J (1980) 'Frogflies' from Central and South America with notes on other organisms of the amphibian egg microhabitat. *Brenesia* **17,** 49–67.
Villa J, Townsend DS (1983) Viable frog eggs eaten by phorid fly larvae. *Journal of Herpetology* **17,** 278–281.
Vincent J-J (1975) Étude de la faune entomologique d'un verger de pommiers du bassin parisien a l'aide de pièges attractifs colorés. *Annales de la Société Entomologique de France* **11,** 303–334.
Vitzhum HG (1932) Terrestrische Acarinen (unter Ausschluss der Oribatiden und Ixodiden) der Deutschen Limnologischen Sunda-Expedition. *Archiv für Hydrobiologie Supplement-Bande* **9(35),** 57–134.

Wada S (1991) Morphologische Indizien für das unmittelbare Schwestergruppenverhaltnis der Schizophora mit den Syrphoidea (Aschiza) in der phylogenetischen Systematik der Cyclorrhapha (Diptera Brachycera) *Journal of Natural History* **25,** 1531–1570

Wakeford D, Disney RHL (1994) A Cuban *Megaselia* species (Diptera Phoridae) with wing-spots *Entomologica Scandinavia*

Waller DA, Moser JC (1990) Invertebrate enemies and nest associates of the leaf-cutting ant *Atta texana* (Buckley) (Formicidae, Attini) In Vander Meer RK, Jaffe K, Cedeno A (eds), *Applied Myrmecology* Boulder, Colarado Westview Press, Chapter 25 255–273

Walter EV, Wene GP (1951) Tests of insecticides to control larvae of *Euxesta stigmatias* and *Megaselia scalaris Journal of Economic Entomology* **44,** 998–999

Wandolleck B (1898) Die Stenopathidae, eine neue fluegel- und schwingerlose Familie der Diptera *Zoologische Jahrbucher Abteilung für Systematik* **11,** 412–441

Warburton C (1919) Note on the solitary wasp, *Crabo cephalotes Proceedings of the Cambridge Philosophical Society* **19,** 296–299

Wasmann E (1900a) *Termitoxenia*, ein neues flugelloses, physogastres Dipterengenus aus Termitennestern *Zeitschrift für wissenschaftliche Zoologie* **67,** 599–617

Wasmann E (1902a) Zur naheren Kenntnis der termitophilen Dipterengattung *Termitoxenia* Wasm *International Congress of Entomology (1901)* **5,** 852–872

Wasmann E (1902b) Termiten, Termitophilen und Myrmekophilen, gesammelt auf Ceylon, mit andern ostindischen Material bearbeitet *Zoologische Jahrbucher Abteilung für Systematik* **17,** 99–164

Wasmann E (1904b) Termitophilen aus dem Sudan In Jagerskiold LA (ed), *Results of the Swedish Zoological Expedition to Egypt 1901* **13,** 1–21, Upsala

Wasmann E (1913) Revision der Termitoxeniinae von Ostindien und Ceylon *Annales de la Société Entomologique de Belgique* **57,** 16–22

Wasmann E (1918b) Zur Lebenweise und Fortpflanzung von *Pseudacteon formicarum* Verr (Diptera, Phoridae) *Biologisches Zentralblatt* **38,** 317–329

Watson JAL, Gay FJ (1980) The identities of *Termes australis* Walker and *Termes fumipennis* Walker (Isoptera) *Journal of the Australian Entomological Society* **19,** 19–25

Weber D (1989) Die Hohlenfauna und -flora des Hohlenkastergebietes RheinlandPfalz/Saarland, 2 Teil *Abhandlungen zur Karst- und Hohlenkunde* **23,** 1–250

Weber G, Prescher S (1990) Studies on the ecology of Phoridae (Diptera) *Pedobiologia* **34,** 183–189

Weber NA (1972) The Gardening Ants, the Attines *Memoirs of the American Philosophical Society* **92,** I–XVI, 1–146

Webster FM (1890) Notes on a species of necrophagous Diptera *Insect Life* **2,** 356–358

Weiss MA (1911) Sur l'existence, à Djerba, du *Phora chlorogastra* Becker et sur ses metamorphoses *Bulletin de la Société d'Histoire Naturelle de l'Afrique du Nord* **1911,** 34

Weissmair W, Hause E (1993) Fauna der Rettenbachhohle (Biospelaologie, Oberosterreich) *Linzer Biologische Beitrage* **25,** 373–385

Wesché W (1908) The systematic affinities of the Phoridae and of several Brachycerous families in Diptera *Transactions of the Entomological Society of London* **1908,** 283–296

Wheeler WM (1901) An extraordinary ant-guest *American Naturalist* **35,** 1007–1016

Wheeler WM (1907) The polymorphism of ants, with an account of some singular abnormalities due to parasitism *Bulletin of the American Museum of Natural History* **23,** 1–93

Wheeler WM (1910) *Ants their structure, development and behavior* New York Columbia University Press

White P (1993) Problems with flies? *Mushroom Journal* **525,** 23

White PF (1979) Pot tests with methoprene and permethrin against the mushroom phorid (*Megaselia halterata*) and the mushroom sciarid (*Lycoriella auripila*) *Entomologia Experimentalis et Applicata* **26,** 332–338

White PF (1981) Spread of the mushroom disease *Verticillium fungicola* by *Megaselia halterata* (Diptera Phoridae) *Protection Ecology* **3,** 17–24

White PF (1983) Automatic spraying for phorid control – commercial trials *Mushroom Journal* **128,** 293–294

White PF (1985) Pests and pesticides In Flegg PB, Spencer DM, Wood DA (eds), *The Biology and Technology of the Cultivated Mushroom* Chichester Wiley

White PF (1987) The mushroom pest complex – its effect on yield and quality In Wuest PJ, Royse DJ, Beelman RB (eds), *Cultivating Edible Fungi* Oxford Elsevier

Whitfield DG (1987) A cheap robust blow pooter *Entomologist's Monthly Magazine* **123,** 45–46

Whitten JM (1955) A comparative morphological study of the tracheal system in larval Diptera Part I *Quarterly Journal of Microscopical Science* **96,** 257–278

Wildermuth VL (1914) The Alfalfa caterpillar *Eurymus eurytheme* Boisd *Bulletin of the United States Department of Agriculture, Washington* **124,** 1–40

Wilding N, Collins NM, Hammond PM, Webber JF (eds) (1989) *Insect–Fungus Interactions* London Academic Press

Willhoeft U, Traut W (1990) Molecular differentiation of the homomorphic sex chromosomes in *Megaselia scalaris* (Diptera) detected by random DNA probes *Chromosoma* **99,** 237–242

Williams DF, Banks WA (1987) *Pseudacteon obtusus* (Diptera Phoridae) attacking *Solenopsis invicta* (Hymenoptera Formicidae) in Brazil *Psyche* **94,** 9–13

Williams EC (1941) An ecological study of the floor fauna of the Panama rain forest *Bulletin of the Chicago Academy of Science* **6,** 63–124

Williams RN, Panaia JR, Gallo D, Whitcomb WH (1973) Fire ants attacked by phorid flies *Florida Entomologist* **56,** 259–262

Williams RN, Whitcomb WH (1974) Parasites of fire ants in South America *Proceedings of the Tall Timbers Conference on the Ecology of Animals and its Contribution to Habitat Management* **5,** 49–59 (1973)

Wilson EO (1971) *The Insect Societies* Cambridge, Massachusetts Belknap Press

Wilton DP (1961) Refuse containers as a source of flies in Honolulu and nearby communities *Proceedings of the Hawaiian Entomological Society* **17,** 477–481

Winder JA (1977) Some organic substrates which serve as insect breeding sites in Bahian cocoa plantations *Revista Brasileira de Biologia* **37,** 351–356

Wojcik DP, Jouvenaz DP, Lofgren CS (1987) First report of a parasitic fly (Diptera Phoridae) from a red imported fire ant (*Solenopsis invicta*) alate female (Hymenoptera Formicidae) *Florida Entomologist* **70,** 181–182

Wolf KW Blackman RL (1992) Delayed disassembly of spindle microtubules in two insect species, *Megaselia scalaris* (Phoridae, Diptera) and *Amphorophora tuberculata* (Aphididae, Hemiptera) *Journal of Submicroscopic Cytology and Pathology* **24,** 483–488

Wood JH (1906) On the British species of *Phora* *Entomologist's Monthly Magazine* **42,** 186–196, 262–266

Wood JH (1907) *Phora gracilis,* a new species belonging to Becker's group I *Entomologist's Monthly Magazine* **43,** 228

Wood JH (1908) On the British species of *Phora* (Part II) *Entomologist's Monthly Magazine* **44,** 164–174, 215–218, 253–254

Wood JH (1909) On the British species of *Phora* Part II contd *Entomologist's Monthly Magazine* **45,** 24–29, 59–63, 113–120, 143–149, 191–195, 240–244

Wood JH (1910) On the British species of *Phora* (contd) *Entomologist's Monthly Magazine* **46,** 149–154, 195–202, 243–249

Wood JH (1911) Is leaf-mining one of the larval habits of *Aphiochaeta*? *Entomologist's Monthly Magazine* **47,** 69

Wood JH (1912) Notes on British *Phora* (corrections and additions) *Entomologist's Monthly Magazine* **48,** 94–99, 166–181

Wood JH (1914) Notes on British *Phora* (further additions) *Entomologist's Monthly Magazine* **50,** 152–154

Wood TG, Thomas RJ (1989) The mutualistic association between Macrotermitinae and *Termitomyces* In Wilding *et al* (1989), Chapter 3 67–92

Wright RE (1927) Myiasis with chronic degeneration of cornea *American Journal of Opthalmology Series 3* **10,** 411–412

Wylie HG (1958) Observations on *Aphidecta obliterata* (L) (Coleoptera Coccinellidae) a predator of conifer-infesting Aphidoidea *Canadian Entomologist* **90,** 518–522

Yagi AI, Sanousi SM, Osman OM, Adlan AM (1977) Some observations on the biology of *Megaselia curtineura* (Brues), (Diptera, Phoridae) in the Sudan *Revue Zoologique Africaine* **91,** 147–151

Yakovlev EB (1980) Successions of trophic groups of larvae of Diptera in fruit bodies of mushrooms *Doklady Akademii Nauk SSSR* **253,** 506–509

Yakovlev EB (1986) Nasekomye-mitsebionty yuzhoi Karelii (ekologofaunisticheskii spisok) In Yakovlev EB, Uzenbaev SD (eds), *Fauna i ekologiya Chlenistonogikh Karelii* Petrozavodsk Karelbskii Filial AN SSSR, 83–123

Yakovlev EB (1988b) Plodonoshenie gribov i sezonnaya aktivnost' dvukrylykh nasekomykh v sosnovykh i osinovykh molodnyakakakh *Nauchnye Doklady Akademii Nauk SSSR Karel'skii Filial* **1988,** 1–68

Yarkulov F (1972) Mukha-gorbatka *Phora holosericea* (Diptera, Phoridae) – khishchnik kornevykh tley *Zoologicheskii Zhurnal* **51,** 1415–1418

Young AM (1984) Mechanism of pollination by Phoridae (Diptera) in some *Herrania* species (Sterculiaceae) in Costa Rica *Proceedings of the Entomological Society of Washington* **86,** 503–518

Young AM (1986) Distribution and abundance of Diptera in flypaper traps at *Theobroma cacao* L ((Sterculiaceae) flowers in Costa Rican cacao plantations *Journal of the Kansas Entomological Society* **59,** 580–587

Young AM, Erickson BJ, Erickson EH (1987a) Steam-distilled floral oils of *Theobroma* species (Sterculiaceae) as attractants to flying insects during dry and wet seasons in a Costa Rican cocoa plantation *Proceedings of the 10th International Cocoa Research Conference (Santo Domingo, RD)* **1987,** 289–296

Young AM, Erickson E, Erickson B (1987b) A trap survey of flying insects in 'Finca Experimental La Lola' in Costa Rica *Turrialba* **37,** 337–356

Young AM, Erickson BJ, Erickson EH (1989) Pollination biology of *Theobroma* and *Herrania* (Sterculiaceae) – III Steam-distilled floral oils of *Theobroma* species as attractants to flying insects in a Costa Rican cocoa plantation *Insect Science and its Application* **10,** 93–98

Zaitzev VF (1977) Some phorid flies (Diptera, Phoridae) from Mongolia *Insects of Mongolia* **5,** 694–697

Zumpt F (1965) *Myiasis in Animals and Man in the Old World* London Butterworths

Author index

Phoridae index

This index also serves as a checklist for the authorities for the specific epithets Genera are primarily listed in section 8 3 (pages 265–94), in the keys to genera in section 8 2, or the key to fossil genera in section 8 4

Index of other organisms

Individual species and genera are only listed when they appear in the text other than in the alphabetically arranged reviews for the higher taxon in question Thus, most fungi are dealt with in section 3 3 Only where fungus species or genera are mentioned elsewhere in the text are they listed below

Subject index

MIX
Papier aus verantwortungsvollen Quellen
Paper from responsible sources
FSC® C105338

FSC
www.fsc.org

If you have any concerns about our products,
you can contact us on
ProductSafety@springernature.com

In case Publisher is established outside the EU,
the EU authorized representative is:
Springer Nature Customer Service Center GmbH
Europaplatz 3, 69115 Heidelberg, Germany

Printed by Libri Plureos GmbH
in Hamburg, Germany